APPLIED MYCOLOGY AND BIOTECHNOLOGY

VOLUME 3

FUNGAL GENOMICS

APPLIED MYCOLOGY AND BIOTECHNOLOGY

VOLUME 3

FUNGAL GENOMICS

Edited by

Dilip K. Arora
Department of Botany
Banaras Hindu University
India

George G. Khachatourians
Department of Applied Microbiology and Food Sciences
College of Agriculture
University of Saskatchewan
Saskatoon, SK, Canada

2003
ELSEVIER
Amsterdam – Boston – Heidelberg – London – New York – Oxford
Paris – San Diego – San Francisco – Singapore – Sydney – Tokyo

ELSEVIER SCIENCE B.V.
Radarweg 29, PO Box 211, 1000 AE Amsterdam, The Netherlands

First edition 2003

Library of Congress Cataloging in Publication Data
A catalog record from the Library of Congress has been applied for.

British Library Cataloguing in Publication Data
A catalogue record from the British Library has been applied for.

ISBN 13: 978-0-444-51442-4

Transferred to Digital Printing 2007

Editors

Dilip K. Arora
Department of Botany
Banaras Hindu University
India
Fax: +91 542 2368141
Tel : + 91 542 2369570
E-mail: aroradilip@yahoo.co.in

George G. Khachatourians
Department of Applied Microbiology
and Food Sciences
College of Agriculture
University of Saskatchewan
Saskatoon, Canada
Tel : +1 306 966 5032
E- mail: khachatouria@sask.usask.ca

Editorial Board

Contents

Contributors

Paul V. Attfield Microbiogen Pty Ltd, c/- Department of Biological Sciences, Macquarie University, Sydney, NSW 2109, Australia (pattfiel@rna.bio.mq.edu.au)

Philip J. L. Bell Microbiogen Pty Ltd, c/- Department of Biological Sciences, Macquarie University, Sydney, NSW 2109, Australia (pattfiel@rna.bio.mq.edu.au)

Deborah Bell-Pedersen Program in Biological Clocks, Department of Biology, Texas A&M University, College Station, TX 77845, USA (dpedersen@mail.bio.tamu.edu)

Deepak Bhatnagar U.S. Department of Agriculture, Agricultural Research Service, Southern Regional Research Center, New Orleans, LA 70124, USA (desjarae@ncaur.usda.gov)

Charles E. Bullerwell Department of Biochemistry and Molecular Biology, Dalhousie University, Halifax (Nova Scotia), B3H 4R2 Canada

Alejandro Correa Program in Biological Clocks, Department of Biology, Texas A&M University, College Station, TX 77845, USA (dpedersen@mail.bio.tamu.edu)

Anne E. Desjardins U.S. Department of Agriculture, Agricultural Research Service, National Center for Agriculture Utilization Research, Peoria, IL 61604, USA (desjarae@ncaur.usda.gov)

Andrew V. Greene Program in Biological Clocks, Department of Biology, Texas A&M University, College Station, TX 77845, USA

Georg Hausner Department of Microbiology, University of Manitoba, Winnipeg, MB, R3T 2N2, Canada (hausnerg@cc.umanitoba.ca)

Weiming Hu Chipscreen Biosciences, Ltd., Research Institute of Tsinghua University, Suite C301, Shenzhen, Guangdong 518057, China (lmshi@chipscreen.com)

Frank Kempken Abteilung für Botanik mit Schwerpunkt Genetik und Molekularbiologie, Botanisches Institut und Botanischer Garten, Christian-Albrechts-Universität zu Kiel, Olshausenstrasse 40, D-24098 Kiel, Germany (fkempken@bot.uni-kiel.de)

Bernard Lamb Department of Biological Sciences, Imperial College of Science, Technology and Medicine, London SW7 2 AZ, UK (b.lamb@ic.ac.uk)

B. Franz Lang Program in Evolutionary Biology, Canadian Institute for Advanced Research; Département de Biochimie, Université de Montréal, 2900 Boulevard Edouard-Montpetit, Montréal, H3T 1J4 Canada (Franz.Lang@Umontreal.ca)

Jessica Leigh Program in Evolutionary Biology, Canadian Institute for Advanced Research, Département de Biochimie, Université de Montréal, 2900 Boulevard Edouard-Montpetit, Montréal, H3T 1J4, Canada (Franz.Lang@Umontreal.ca)

Zachary A. Lewis Program in Biological Clocks, Department of Biology, Texas A&M University, College Station, TX 77845, USA (dpedersen@mail.bio.tamu.edu)

Brendan Loftus The Institute for Genomic Research (TIGR), 9712 Medical Centre Drive, Rockville, MD 20850, USA (bjloftus@tigr.org)

Joyce E. Longcore Department of Biological Sciences, University of Maine, Orono, ME 04469-5722, USA

Xianping Lu Chipscreen Biosciences, Ltd., Research Institute of Tsinghua University, Suite C301, Shenzhen, Guangdong 518057, China (lmshi@chipscreen.com)

Kevin McCluskey Fungal Genetics Stock Center, Department of Microbiology, University of Kansas Medical Center, Kansas City, KS, USA (kmcclusk@kumc.edu)

Jeff Murrell Novozymes Biotech Inc., 1445 Drew Avenue, Davis, CA 95616-4880, USA

Ross N. Nazar Department of Molecular Biology and Genetics, University of Guelph, Guelph, Ontario, Canada N1G 2W1 (rnnazar@uoguelph.ca)

K.M. Helena Nevalainen Department of Biological Sciences, Macquarie University, Sydney, NSW 2109, Australia

Elias Seif Département de Biochimie, Université de Montréal, 2900 Boulevard Edouard-Montpetit, Montréal, H3T 1J4 Canada (Franz.Lang@Umontreal.ca)

Amir Sharon Department of Plant Sciences, Tel Aviv University, Tel Aviv 69978, Israel

Leming Shi Chipscreen Biosciences, Ltd., Research Institute of Tsinghua University, Suite C 301, Shenzen, Guangdong 518057, China (lmshi @ chipscreen.com)

Zhenqiang Su Chipscreen Biosciences, Ltd., Research Institute of Tsinghua University, Suite C301, Shenzhen, Guangdong 518057, China

Valentino S. Jnr. Te'o Department of Biological Sciences, Macquarie University, Sydney, NSW 2109, Australia

Weida Tong National Center for Toxicological Research, Food and Drug Administration, 3900 NCTR Road, Jefferson, AR 72079, USA (wtong@nctr.fda.gov)

Paul Tudzynski Institut für Botanik, Schlossgarten 3, D-48149 Münster, Germany (tudzyns@uni-muenster.de)

Barbara Weber Novozymes Biotech Inc., 1445 Drew Avenue, Davis, CA 95616-4880, USA

Debbie Sue Yaver Novozymes Biotech Inc., 1445 Drew Avenue, Davis, CA 95616-4880, USA

Preface

Fungi have been pivotal to the development of societies. Presently they have contributed much to the development of various industrial materials and processes, agri-food commodities and human health products. Thus, mycology and its pursuit through modern biotechnology have led to the practical application in broad sense to many spheres of human enterprise.

Fungi represent the second largest species in the biological world after the insects. They number over 1.5 million, of which fewer than 10% of species have been described and only about 1% of known species have been deposited in various collections. With these facts in mind fungi are significant contributors to the vitality of the biosphere. As indicated in our previous volumes, fungi and their study have taught us their value in contemporary production and post-production agriculture. It is unusual that utilization of a small percentage of fungi could have such an enormous intellectual and practical drive and value. We would like to inferentially suggest that knowledge of diversity of fungi and their genome sequences could have a dramatic multiplier effect on their value in all spheres of life and economy. While it is unlikely that in the next quarter century we will characterize any more than a fraction of fungi, we believe it to be highly possible that we will have the genomic sequences of many more fungi.

The field of genomics is developing at an unparalleled rate. Recent accomplishments in the sequencing of the human genome, and that of other animals, several plants, microorganisms and the elucidation of the relationship between biological and ecological or environmental interactions have presented massive new information. Genomes of fungi, as compiled in this volume of Applied Mycology and Biotechnology, contain diverse genes and sequences. Each genome with its compliment of genes and sequences encodes for products that determine the types and influence the quality of interactions to bring about an organism's survival, communication and evolution. Further, the establishment of differences in gene number, structure, conservation, homologue and ortholog, regulatory type and network, and the huge differences in structure and function relationships are the most significant scientific accomplishments in fungal biology.

The methodologies of genomic sequence determination have changed in strategies available and the speed with which this is now accomplished. The commercial and governmental incentives have generated an industry of sizable significance. The allied fields of proteomics and computational analysis of sequences for intelligent use have had a synergistic effect in genomics. Comprehensive genomic map development is no longer the technological challenge of the decade earlier. Structural genomics is a field that intersects other interdisciplinary fields; DNA sequencing, cloning, gene expression, NMR, X-ray crystallography, use of high flux synchrotron for x-ray beam lines, computational sciences and computers. Genomics development now requires high throughput and automated experimental devices. They need microarray technology for identification of functions, mass spectrometry and automated sequencing machines. PCR amplification of small DNA pieces can be analyzed very quickly by electrospray ionizing mass spectrometry for the determination of allele specific changes. Matrix assisted laser desorption ionization time-of-flight mass spectrometry (MALDI-TOF-MS) has become invaluable in reliable multiplexing of single nucleotide polymorphic DNA (fragments of smaller than 100 analysis). Genomics data input also needs high performance computers and computational programs. All of the above capabilities need information processing technologies and modeling programs. What the genomics paradigm does in a day now, perhaps was the research of a team of scientists or doctoral students just a quarter a century ago. In spite of the progress, there are still potential bottlenecks in the ingredient fields that feed into genomics, the only down side of this enterprise.

In contrast to the complexity of the technology, the knowledge gained offers several paths of pursuit for the biotechnological and process-oriented use of fungi. The genomic analysis of the whole genome of *Saccharomyces cerevisiae*, the first single-celled fungus to be sequenced, helped to understand gene duplication events from an evolutionary perspective. This work also lead to the identification of paralogous genes. One hundred and thirteen sequences, which accounted for the 2% of the total genome of *S. cerevisiae*, were identified and shown to be involved in signal transduction and to have their homologous proteins in other eukaryotes. Exploitation of signal transduction has a vast number of target applications, whether in disruption of fungal pathogen functions, or in fungal growth and product synthesis. Use of fungal genomics for comparative studies can help in selection of the appropriate or best organism for bioprocess functions.

Through the use of comparative genomics and structure functions researchers can define a core set of genes among a set of fungi that have the same molecularly distinctive functionality. Such global approaches can be refined to increase our options in deriving more cost and process effective biotechnological use of fungi. Finally, fungal genomics can help in the systematic discovery and analysis of the taxonomic relationship of 1.4 million fungi remaining undiscovered and unknown.

Fungi that serve the tropical or temperate forests and serve as mycorrhiza, endophytes, phytopathogens, entomopathogens, or simple saprophytes to turn over biological matter are a significant and unknown resource. This could be the source for many bioproducts including secondary metabolites, antibiotics and catabolic enzymes of enormous impact. Compared to terrestrial fungi, those in aquatic habitats are some of the most neglected yet important in applied mycological and biotechnological research. Knowledge of fungal genomics is the area for the new pioneers of mycology, microbiology and allied sciences that are worth exploration and mining. There are much good these fungi had to offer to the biosphere and to our well being such that the most expedited action is warranted.

The frontiers of applied mycology as with many other disciplines is proceeding at an unprecedented rate with an array of new tools, with their own particular variety and complexity, to create a wealth of disciplines and sub-disciplines, fungal genomics, proteomics and bioinformatics. Although much of the agenda has a mission-oriented direction, the information gathered and the knowledge gained should offer new solution to many areas of production and postproduction agricultural, food science, pharmaceuticals, natural products, and animal, plant, and environmental health.

In this volume of Applied Mycology and Biotechnology we have chosen the coverage of fungal genomics. We recognize that there are serious difficulties in developing a comprehensive volume on genomics because of the range and complexity of the emerging knowledge. However, an attempt has been made throughout to bring together pertinent information that will serve the needs of the reader, provide a quick reference to material that might otherwise be difficult to locate, and furnish a starting point for further study. In this volume we have coverage of several major questions related to fungal genomics: (1) organization of genomes, introns, transposons, plasmids, germplasms and databases; (2) molecular genetics of development and chromosomal mechanics: circadian clock, ribosome biogenesis, gene silencing, genetic mutation, repair, recombination and expression; (3) genomics strategies used in gene regulation and metabolism, biosynthesis of mycotoxins, pathogenicity determinants, and enzyme hyperproduction and technology; (4) the employment and impacts of genomics in drug discovery and development, expression system for combinatorial biology, production of biochips and use in microarray technology.

In a field where the turnover of literature is less than 2 years, we hope this compilation is only a beginning as we continue with the preparation of the next volume. Together, these volumes should help us arrive at comprehensive, in depth information on Applied Mycology and Biotechnology. With several thousand citations, we hope this will serve as a useful reference for veterans and beginners as well as for those crossing disciplinary boundaries and getting into the exciting field of fungal biotechnology.

We are indebted to the members of editorial board for their valuable assistance in compiling this volume. We thank Ms. Hetty Verhagen and Ana-Bela Sa Dias of Elsevier Life Sciences for their technical assistance.

Dilip K. Arora
George G. Khachatourians

Applied Mycology & Biotechnology
An International Series. Volume 3. Fungal Genomics

Fungal Genomics: An Overview

Anne E. Desjardins[a*] and Deepak Bhatnagar[b]
[a]U.S. Department of Agriculture, Agricultural Research Service, National Center for Agriculture Utilization Research, Peoria, Illinois 61604, USA (desjarac@ncaur.usda.gov); [b]U.S. Department of Agriculture, Agricultural Research Service, Southern Regional Research Center, New Orleans, Louisiana 70124, USA.

Fungi dominate our world as plant and animal pathogens, as sources of food and other useful products, and as critical components of natural and agricultural ecosystems. Genomics technologies such as high-throughput DNA sequencing, expressed sequence tags and microarrays provide powerful tools to elucidate the structures and functions of fungal genomes. As representative, but relatively simple eukaryotes, fungi will continue to play an essential role in the application of genomics for understanding the fundamental processes of biology, and for the development of novel technologies and products for industry, agriculture, and human health.

1. INTRODUCTION

With the advent of the genomics era during the last decade, we have witnessed a revolution in our understanding of biological processes. Since 1995, the genomes, or genetic make-up, of dozens of bacteria and a few model eukaryotes have been completely sequenced. With the exception of *Saccharomyces cerevisiae,* fungal genomics was not a priority in these early efforts. Despite the slow start, however, fungal genomics has gained significant momentum in recent years. Genomic sequencing efforts now are underway on dozens of fungi, including species that are of fundamental biological interest, species that are important to industry and agriculture, and species that cause opportunistic human infections (Bennett and Arnold 2001). The Kingdom Mycota contains a diverse array of multicellular microorganisms, or fungi. It is estimated that there are over one million species of fungi and that a large proportion of species have yet to be identified (Hawksworth, 1991). The interactions of fungi with other organisms have played a vital role in the evolution of microorganisms, plants, and animals.

Fungi are adapted to acquire nutrient molecules from their environments as decomposers (saprophytes) or parasites, or both. Fungi are also very important to global ecosystems (Price *et al.*, 2001; Souciet *et al.*, 2000). Many fungi are beneficial and are used as sources of food (e.g. mushrooms), chemicals (e.g. gluconic acid, citric acid), and pharmaceuticals (e.g. penicillin). *Aspergillus oryzae* and other food grade fungi are used for large-scale fermentation for industrial enzymes (e.g. amylases, pectinases and proteases), and *A. sojae* for soy sauce fermentation. Common yeast, *S. cerevisiae,* is used in brewing and in baking. Other fungi are harmful because they are pathogens of plants, animals, and humans, or produce metabolites that are toxic to plants (phytotoxins) or animals (mycotoxins) (Souciet *et al.*, 2000; Orke *et al.*, 1994; Bhatnagar *et al.*, 2002; Richard and Payne, 2002). Because of the significant impact of fungi on the world economy and on human health, tremendous efforts have been made to exploit the benefits of fungi and to reduce their potential harmful effects (for reviews see Bennett, 1998; as well as http://www.cbs.know.nl/search_ fdb.html).

The rapid development of high-throughput DNA sequencing technology has provided a powerful tool for genetic research, from single gene cloning to whole genome sequencing. With the application of genomics and, in particular, of expressed sequence tag (EST) and microarray technologies, we are able to study fungi on the molecular genetic level far more rapidly than ever could be achieved with traditional and biochemical genetic approaches. Genomics has accelerated development of effective strategies to control opportunistic fungal infections of humans, and to maximize industrial use of fungi and reduce mycotoxin contamination of food and feed, resulting in a sustainable, nutritious, safe, and economical food supply for the ever-increasing world population. This article is intended to provide the reader with some basic concepts of genomics, followed by an overview of the history of fungal genetics and genomics. Excellent recent reviews provide additional information on these topics (Bennett and Arnold, 2001; Fakhoury and Payne, 2003).

2. GENOMICS

The concept of the genome originated in the field of cell cytology, in reference to a complete set of chromosomes in a single cell of an organism (Sybenga, 1972). The modern concept of the genome, however, refers to all of the DNA sequence information in a single cell, often designated as nucleotides or mega base pairs (Mb). Genomics is the study of the genome of an organism, and includes the sequencing and annotation of the entire genome. The term GENOMICS, coined in 1986 by Thomas Roderick to provide a name for a new journal, included mapping, sequencing, and analysis of genomes (Hieter and Boguski, 1997). In a broader sense, however, sequencing and annotation are only a part of genomics, which can be divided into three major components: structural, comparative, and functional. Structural genomics is the physical sequencing and annotation of all of the genetic material of an organism. Structural genomics has been defined as the "initial phase of genomic analysis: with a clear end point that results in the construction of high resolution genetic, physical and transcript maps of an organism" (Hieter and Boguski, 1997). Comparative or evolutionary genomics is the comparison of DNA sequences of related organisms through advanced computer technologies, or bioinformatics. Functional genomics is the identification of the functions of each coding sequence through analysis of gene expression by using libraries of ESTs and microarray technologies, and by targeted gene knock-out experiments. Functional genomics utilizes information and reagents provided by structural genomics to develop and apply global (genome-wide or system-wide) experimental approaches to assess gene function. The combination of high-throughput technologies such as microarrays with statistical and computational analysis of results expands the scope of biological investigation from studying single genes or proteins to simultaneously studying all genes or proteins in a systematic fashion. Genomics combined with bioinformatics enables the identification of all of the genes in an organism and the study of their functions. In addition, novel terminology, new acronyms, and innovative techniques have been added to the scientific vocabulary (Tables 1 and 2).

2.1 DNA Sequencing

Advances in computer technology coupled with significant innovations in engineering, chemistry, and molecular biology have made possible the sequencing of whole genomes of microorganisms at affordable costs, in desired time frames, and with relatively rapid analysis of the massive amounts of data generated from DNA sequencing. In brief, DNA sequencing is the ability to determine the succession of different bases forming a strand of DNA. The basic procedure requires synthesis of a multitude of DNA fragments complementary to the DNA strand to be sequenced (Maxam and Gilbert, 1977; Sanger *et al.*, 1977; Mullis *et al.*, 1986),

Table1. Primer of acronyms regularly encountered in genomics research*

BAC	Bacterial artificial chromosome
PAC	P1 (phage) artificial chromosome
YAC	Yeast (*S. cerevisiae*) artificial chromosome
EMBL	European Molecular Biology (now EBI)
EBI	European Bioinformatics Institute
BLAST	Basic local alignment search tool
EST	Expressed sequence tags
EUROFAN	European Functional Analysis Network (for *S. cerevisiae*)
FGDB	Functional Genome Data Base
GRAIL	Gene Recognition and Analysis Link
ORF	Open reading frame
PCR	Polymerase chain reaction
HGP	Human Genome Project
MIPS	Martinsried Institute for Protein Sequences (now Munich Information Center for Protein Sequences)
NCBI	National Center for Biotechnology Information (GenBank)
RFLP	Restriction fragment length polymorphism
STS	Sequence tagged sites
TIGR	The Institute for Genome Research (Rockville, Maryland)
UTR	Untranslated regions
XML	eXtensible markup language
YPD	Yeast (*S. cerevisiae*) proteome data base

*(modified from Bennett and Arnold, 2001)

separation of the fragments by electrophoresis (Fitch and Sokhansanj, 2000), and a chemical sequencing reaction. Several advances in chemistry of the sequencing reaction have led to cheaper, more reliable, and more reproducible DNA sequencing methods (Rayner et al, 1998; Meldrum, 2000 a, b; Hunkapiller *et al.*, 1991; Smith et al, 1985, 1986; Fitch and Sokhansanj, 2000). Innovation in electrophoresis has been the most significant factor in the development of automated DNA sequencing machines with extremely high throughput capabilities (Mitnik et al, 2001; Fitch and Sokhansanj, 2000; Esch, 2000; Green, 2001; Meldrum 2000 a, b; Righetti *et al.* 2002). Technical limitations, however, allow single reads of only 500-800 bases in one sequencing reaction, which affects the speed with which genomes can be sequenced (Righetti *et al.* 2002). But advances in sample preparation and in automation of most of these procedures have increased sequencing efficiency. Fakhoury and Payne (2003) state "Generation of large amounts of sequences also required the development of software to call the different bases resulting from the sequencing reactions to assemble the different fragments in adjoining contigs. The codes Phred, Phrap, and Conred were developed by Phil Green's group at the University of Washington (Ewing and Green 1998; Ewing *et al.* 1998;

http://www.phrap.org/; http://www.phrep.com/phred/), and are widely used for base calling contigs assembly, and viewing of sequence assemblies, respectively". For a comprehensive

Table 2. Primer of genomics terminology*

BAC (bacterial artificial chromosome)	Vector used to clone DNA fragments in *E. coli* with inserts ranging from approximately 100-300 kb
Genome	"the totality of a cell's genetic information including both genes and other DNA sequences" (Berg and Singer, 1992)
Gen Bank	Public database operated by the National Center for Biotechnology Information (U.S. National Institutes of Health)
Contig	Group of cloned DNAs representing overlapping segments of a particular chromosome region and providing unbroken coverage of that region: the continuous DNA sequence generated from these DNA clones. A contig contains no gaps
Contig map	Map depicting the relative order of a linked library of overlapping clones
Depth of coverage	Number of times a particular DNA is sequenced (1x means that on average a base pair (bp) has been sampled once; 8x means on an average a particular bp has been sequenced eight times)
Structural genomics	Mapping and sequencing stages of genome analysis; also used to describe projects that aim to solve the structure of all possible proteins
Genomic library	Collection of clones containing the entire genome of an organism cut up into many pieces, e.g., a BAC library
ORF (open reading frame)	Series of triplets coding for amino acids without any stop condons; these sequences have the potential to be translated into polypeptides
Homologous	In evolutionary biology, refers to genes that descend from a common ancestral gene; in genomics homology is used to describe DNA that has the same or nearly the same nucleotide sequence
Orthologous	Homologous sequences that descend from a single ancestral gene
Paralogous	Homologous sequences that arise through gene duplication
Functional genomics	Determining the function of genes through the use of microarrays and other methods that can study the function of many genes simultaneously
Proteome	Complete set of proteins that a living cell can synthesize
Proteomics	Identification and characterization of each protein, its structure and its interactions with other proteins
Synteny	"On the same thread": the presence of sets of genes showing the same order in different species, often used as shorthand for saying that a group of genes shows conservation of linkage
Bioinformatics	"the use of mathematics, statistics, and computer science to model, analyze, store, retrieve, and distribute biological data" (Bennett and Arnold, 2001)
Annotation	The analysis of the sequence of A's, T's, G's and C's (DNA sequence) in a given organism to find all of the predicted genes in an organism
DNA Microarrays (gene chips)	Assemblage of assorted short sequences of DNA or polypeptides embedded onto a solid medium such as glass or plastic slides/silicon wafers or nylon membranes

* (modified from Bennett and Arnold, 2001).

review of various sequencing strategies, we refer you to articles by Anderson (1981), Gardner *et al.* (1981), and Green (2001).

2.2 Gene Expression Measurement

Despite technological advances, whole genome sequencing is still a relatively expensive and time-consuming process. An alternative and less expensive approach is to study the pattern and level of gene expression in an organism. These studies can be conducted in as much detail as required by the investigation, and as limited by availability of funds to

Table 3. Uniform resource locators (URLs) for most commonly used major

Name of data base	URL
Human genome database	http://www.gdb.org/
Genome Sequence database	http://www.ncgr.org/gsdb/
National Center for Biotechnology (NCBI/Gen Bank)	http://www.ncbi.nlm.nih.gov/
European Molecular Biology Laboratory (EMBL)	http://www.embl-heidelberg.de/
DNA Databank of Japan	http://ddbj.nig.ac.jp
Kyoto Encyclopedia of genes and genomes (KEGG) (metabolic pathway database)	http://www.genome.ad.jp/kegg/
Washington University database (Pfam, a protein domain database)	http://pfam.wustl.edu/index.html

*(adapted from Skinner, *et al.* 2001; Fakhoury and Payne, 2002; Bennett and Arnold, 2001). Databases on the World Wide Web*

produce sequence data. Many techniques have been devised for this purpose, the most popular of which are EST sequencing, serial analysis of gene expression (SAGE), differential display and microarray analysis.

The large-scale EST sequencing technique is based on generating complementary DNA from a population of RNA extracted from the tissue of interest under the desired experimental conditions. The cDNAs produced are subsequently cloned and sequenced in whole or in part. For a detailed discussion of this technique refer to Bohnert *et al.* (2001) and Ohlrogge and Benning (2000). In the SAGE technique developed by Velculescu in 1995, sequence tags are generated for specific cDNAs of interest, followed by cloning, sequencing, and a rigorous computational analysis (Donson *et al.* 2002; http://www.sagenet.org). In differential display analysis, the mRNA of interest is collected and submitted to reverse transcription, cDNA is amplified by PCR, and the products are separated on a matrix (Liang and Pardee, 1992). The profiles of cDNAs obtained from mRNAs generated under different conditions are compared to identify the unique genes that are expressed under a particular condition (Matz and Lukyanov, 1998).

Microarray or chip analysis is a DNA hybridization-based approach (Donson *et al.* 2002). This technique allows the ability to monitor the expression of thousands of genes in parallel, making it ideal for gene profiling in a genomics context (Blohm and Guiseppi-Elie 2001; Lockhart and Winzeler 2000; Schena, *et al.* 1995; Hegde *et al.* 2000; Baldi and Long 2001). DNA microarrays are " short sequences of DNA or peptide nucleic acids, embedded onto a solid support such as glass or plastic slides, silicon wafers or nylon membranes. A typical microarray analysis involves the exposure of an 'Immobile Phase' that could be either PCR amplified genomic sequences, cDNAs, or oligonucleotides concentrated within a solid background, to a 'mobile phase' of flourescently labeled DNA probe. The resultant binding of complementary DNA sequences is visualized as a 'signal' that is then counted for as an appraisal of gene expression. A single microarray unit allows for the surveillance of expression among thousands of genes from a single tissue specimen or that of a single gene in several tissues" (Joseph *et al.*, 2002).

Genomics requires considerable computation power for data generation, collection, and analysis. The submission of the sequence data to a public database is essential for the widest use of the volumes of data generated in genome sequencing projects. The availability and accessibility of such databases to researchers at large has proven to be a vital part of scientific discoveries in the last decade. Several of the major databases most commonly used by researchers worldwide are listed in Table 3. These databases have standardized protocols and formats for data deposition, storage, and retrieval.

3. HISTORY OF FUNGAL GENETICS AND GENOMICS

The foundation of genetics as a science was established almost 150 years ago by two English naturalists who had explored and keenly observed the biological diversity of South America and by a Moravian monk who conducted meticulous plant breeding experiments in a garden in what is now Czechoslovakia. During his voyage of exploration in H.M.S Beagle, Darwin observed the facts and patterns of distribution of closely related species of plants and animals in the Galapagos Islands and in other areas separated by geographical barriers. Wallace also observed patterns of biogeography as he collected birds, beetles, and butterflies in the Amazon region, and then in the remote islands of the Malay Archipelago. In 1858, Darwin and Wallace coauthored a paper on their independent discoveries of major aspects of evolutionary genetics, including common descent and variation of species and mechanisms of natural selection. The next year, Darwin published *On the Origin of Species by Means of Natural Selection* (Darwin 1859). Although his classic text is the basis for every modern discussion of genetics, Darwin did not understand the source of genetic variation or the particulate inheritance of genetic traits. Even in later editions of *Origin*, Darwin apparently was not aware of the significance of plant breeding experiments published in 1866 by Mendel. By careful selection of plant model systems and statistical analysis of his data, Mendel demonstrated the particulate inheritance and independent assortment of characters. Inheritance of each member of a pair of alleles via transmission of chromosomes became known as Mendelian genetics. Mendel also understood the distinction between what are now called genotype and phenotype (Henig, 2000). In 1905, Bateson invented the word "genetics" for the new field of study that would combine the insights of Darwin, Wallace, Mendel, and others on the nature of heredity, variation, and natural selection (Mayr and Provine, 1998).

3.1 Fungal Model Systems for Genetics and Genomics

Geneticists of the early 20th century turned to systems other than fungi to investigate heredity and variation. The fruit fly *Drosophila* was used by Morgan as a model system for generating and mapping mutations to particular chromosomes. The science of fungal genetics finally began in 1927 when Dodge, who had trained in maize and *Drosophila* genetics, proposed *Neurospora* as a model system for the study of genetics in a haploid organism (Perkins 1992). In a visit to Cornell University, Dodge communicated his enthusiasm for *Neurospora* to an audience that included graduate students Beadle and McClintock, both of whom later worked with *Neurospora* and both of whom later won the Noble Prize for research in genetics. From the 1920s, geneticists increasingly turned to fungi as a group of organisms especially suited to formal genetic analysis. Fungi have the basic characteristics of all eukaryotic organisms, accompanied by relatively low structural and genetic complexity. Fungi typically have a short life cycle and many are relatively easy to maintain and manipulate using standard microbiological techniques. Fungi are amenable to sexual genetic analysis and are, almost uniquely, amenable to tetrad analysis since the products of a single meiosis are kept together within the ascus. Because the progeny are haploid, complications of dominance are not present. Other experimentally useful features of fungi include DNA-mediated transformation systems that allow both specific gene inactivation by homologous recombination and random mutagenesis by integration at nonhomologous sites throughout the genome. The genome sizes of filamentous fungi are relatively small, ranging from 15 Mb to 45 Mb (Table 4) and contain 5,000 to 15,000 functional genes. The gene density on average is about one gene per every 3000 base pairs.

Fungal genomes appear to contain fewer introns and fewer repetitive sequences than higher eukaryotes such as plants. Due to the small size of most fungal genomes, genes often can be cloned directly and relatively easily by mutant complementation with either plasmid or cosmid vectors.

Table 4. Genomes of selected fungal species

Species	Chromosome	Genome Size (Mb)	References*
Aspergillus flavus	6-8	33-36	Keller *et al.*, 1992
Aspergillus parasiticus	5-7	40	Keller *et al.*, 1992
Aspergillus nidulans	8	28.5	Brody & Carbon, 1989
Aspergillus oryzae	8	35	Kitamoto *et al.*, 1994
Aspergillus niger	8	37.5	Debets *et al.*, 1990; 1993
Aspergillus sojae	6-8	35.5 – 38.5	
Aspergillus fumigatus	8 ?	32	http://www.tigr.org/tdb/mdb/mdbinprogress.html
*Fusarium verticilliodes**	12	46	Xu & Leslie, 1996
Fusarium sporotrichioides	6	27.7	Fekete *et al.*, 1993
Fusarium graminearum	9	35 – 40	Jurgenson *et al.*, 2002
Saccharomyces cerevisiae	16	12	Goffeau *et al.*, 1996
Neurospora crassa	7	42.9	http://www.mips.biochem.mpg.de/proj/*Neurospora*
Magnaporthe grisea	7	40	Talbot *et al.*, 1993; Orbach *et al.*, 1996
Candida albicans	8	16-17	Chu *et al.*, 1993; http://alces med.umn.edu/candida.html

**Fusarium verticillioides* former name: *F. moniliforme*.

During the second half of the 20th century, species of *Neurospora*, *Aspergillus*, and *Saccharomyces* became preferred experimental systems for a variety of genetic studies, including biochemical genetics and metabolic regulation, and mechanisms of non-Mendelian genetics (Ainsworth, 1976; Perkins, 1992). *Neurospora* biochemical genetics was founded by Beadle who worked with Tatum to produce *Neurospora* mutants that were altered at particular steps in metabolic pathways. In 1941, Beadle and Tatum published their *Neurospora crassa* mutant analysis and their hypothesis that individual enzymes are specified by single genes. McClintock used *Neurospora* to show the similarity of chromosome cytology in fungi, plants, and animals. Tetrad analysis in *Neurospora* and *Saccharomyces* provided the first proof of meiotic gene conversion and a mechanism for non-Mendelian inheritance. *Aspergillus* genetics was founded in the 1950s by Pontecorvo, a former *Drosophila* geneticist, who developed methods for selecting diploid *Aspergillus nidulans* strains for parasexual genetic analysis without sexual recombination. Pontecorvo and colleagues used both parasexual and meiotic analysis to create the first complete fungal chromosome map in 1958. The parasexual cycle characterized in *Aspergillus* allows genetic recombination in asexual fungi and has been widely used in breeding new fungal strains for various practical applications. During the past 50 years, *Aspergillus* and *Neurospora* have been model systems for studies of the genetic mechanisms that control cell development and cell differentiation in multicellular eukaryotes. Genetics of *S. cerevisiae* began with studies of the yeast beer-fermentation process at the Carlsberg laboratory in Denmark in the 1930s. Because of its single-cell form and the fixed relationship between cells and nuclei, *Saccharomyces* has been invaluable for study of the genetics of cycles of cell growth and cell division. *Saccharomyces* also has provided a model system for genetics of non-Mendelian extranuclear inheritance, for analysis of the variation and recombination of mitochondrial DNA, and for analysis of heritable, non-DNA elements now called prions (Couzin 2002).

In 1996, a European-based multinational consortium published the *S. cerevisiae* genome sequence, the first publicly available sequence of a fungus or, indeed, of any eukaryote (http://genome_www.stanford.edu/Saccharomyces) (Dujon, 1996; Goffeau *et al.* 1996). Since the initial release and update (Goffeau *et al.*, 1997a,b), the newly discovered yeast genes have been systematically investigated by DNA microarray technologies and by gene-knockout experiments. All of the 6116 unique genes identified were spotted onto microarray slides for functional studies (Gross *et al.* 2000). An international consortium has generated thousands of tagged gene deletion mutations of putative genes (open reading frames) of *S. cerevisiae* (Winzeler, 1999). Multiple approaches to functional analysis of mutant strains include direct tests of fitness under different growth conditions and gene expression profiling using microarrays. The era of fungal genomics that began with *Saccharomyces* has continued with other model organisms. *Aspergillus nidulans* was chosen as the first filamentous fungal genome to be sequenced by an industrial consortium; a partial sequence has been released under restricted conditions (Table 5) (Bennett 1997 a,b). Publication of the complete and annotated genome sequence of *N. crassa* is scheduled for autumn 2002 (Table 5). Databases of ESTs and various genomic libraries of *N. crassa* and *A. nidulans* have been published (Table 5).

3.2 Genomics of Fungal Biodiversity

As representative but relatively simple eukaryotes, fungi have played an essential role in the development of genetics and genomics as experimental sciences. But the most striking feature of fungi is the biological diversity that has resulted from their great evolutionary antiquity. Fungi and land plants first appear in the fossil record from 480 to 460 million

Table 5. Selected Fungal Genome Projects.

Fungus	Status (reviewed August 2002)	Sources
Aspergillus flavus	Public, expressed sequence tags (ESTs)	3,4
Aspergillus fumigatus	Public, in progress	1
Aspergillus nidulans	Public, ESTs and private, partial genome	3
Aspergillus niger	Private	1
Botrytis cinerea	Private	
Candida albicans	Public, draft genome	6
Cochliobolus heterostrophus	Private	
Coccicioides immitis	Public, in progress	1
Cryptococcus neoformans	Public, partial genome	2
Fusarium sporotrichioides	Public, ESTs	3
Gibberella moniliformis	Public, ESTs	4
Gibberella zeae	Public, ESTs and genome in progress	4
Magnaporthe grisea	Public, draft genome	5
Neurospora crassa	Public, complete genome	2
Phanerochaete chrysosporium	Public, draft genome	7
Phytophthora infestans	Public, ESTs	8
Pneumocystis carinii	Public, in progress	1
Saccharomyces cerevisiae	Public, complete genome	1
Schizosaccharomyces pombe	Public, in progress	1
Ustilago maydis	Private	1

1.The Institute for Genome Research, Microbial Database, www.tigr.org; 2. Whitehead Institute Center for Genome Research, www-genome.wi.mit.edu; 3. Fungal Genetics Stock Center, www.fgsc.net; 4. USDA, Agricultural Research Service-funded genomics; 5. Fungal Genomics Laboratory, NC State University, www.fungalgenomics.ncsu.edu; 6. Stanford Genome Technology Center, sequence-www.stanford.edu; 7. DOE-funded Microbial Genomes, www.sc.doe.gov; 8. Phytophthora Genome Consortium, www.ncgr.org/pgc/.

years ago, but phylogenetic analyses indicate that major groups of fungi such as the Basiomycota and the Ascomycota were present more than 900 million years ago (Blackwell, 2000). The fossil record also documents the antiquity and ubiquity of the extraordinary symbiotic interactions of fungi with cyanobacteria or green algae in the form of lichens and as mycorrhiza with plants. The fossil record may yet provide evidence of the antiquity of fungal interactions with insects and other animals.

With the discovery of the microscope in the 17th century the extraordinary diversity of microfungi became apparent. In the late 1600s, pioneer microscopist Leeuwenhoek described his observations of single and budding cells of *Saccharomyces* in fermenting beer, and Hooke published the first drawings of fungal sporangia and teliospores in his *Micrographia.* There followed numerous publications with beautiful and accurate drawings of rusts, smuts, ergot, and other fungi associated with plant materials. Nineteenth century mycology culminated in treatises on classification based on morphology, such as the massive *Sylloge Fungorum omnium hucusque cognitorum* published from 1882 to1925 in which Saccardo provided Latin names and descriptions of most known fungi (Ainsworth 1976).

With the development of analytical chemical methods in the 20th century, the extraordinary chemical diversity of the fungi was discovered. In particular, many filamentous fungi produce a bewildering array of biologically active secondary metabolites. Classic examples of medically useful fungal metabolites are the antibiotic penicillins from *Penicillium* species and cephalosporins from *Cephalosporium* species, and the cholesterol-lowering lovastatin from *A. terreus.* On the other hand, many fungi produce toxins that can cause mycotoxicoses in humans and animals that consume contaminated agricultural commodities (Hudler, 1998; Bhatnagar *et al.*, 2002). The most notorious mycotoxicosis in human history is ergotism, which is caused by consumption of grain contaminated with sclerotia of *Claviceps purpurea.* Ergotism was responsible for medieval epidemics of the disease called St. Anthony's Fire, which included gangrene of the extremities, convulsions, psychoses, and death. Sclerotia can contain a complex mixture of biologically active alkaloids, which are the principal causes of ergot poisoning. The modern era of mycotoxicology began in England in the 1960s with Turkey X disease and the discovery of aflatoxins. Toxicity of animal feeds containing contaminated peanut meal led to the deaths of more than 100,000 turkeys by acute liver necrosis. Scientists in England identified the toxic agents as polyketides produced by *A. flavus.* Subsequent studies have shown that aflatoxins produced by *A. flavus* and related species are potent liver toxins and carcinogens. For more than 100 years, both acute and chronic mycotoxicoses in farm animals and in humans have been associated with consumption of grains contaminated with *Fusarium* species. Between 1970 and 1990, the toxic agents were identified as trichothecenes and fumonisins. Trichothecenes inhibit protein synthesis, causing emesis, hemorrhage, anemia, and immunosuppression, whereas fumonisins alter sphingolipid metabolism, causing equine leucoencephalomalacia, porcine pulmonary edema, and kidney and liver cancer in rodents.

Because of the potential impact of mycotoxins on human health, efforts are underway to sequence genomes of several mycotoxigenic fungi and related species. *Aspergillus* species *flavus*, *fumigatus*, *nidulans*, *niger, oryzae*, and *sojae*, and *Fusarium* species *graminearum* and *verticillioides* (formerly *F. moniliforme*) are among the high priority fungi for genomic sequencing proposed by The American Phytopathological Society (http://www.apsnet.org) and the Whitehead Institute, Center for Genomic Research (http://www-genome.wi.mit.edu/seq/fgi/candidates.html); (http://www-genome. wi.mit.edu/seq/ fgi/FGI_ whitepaper_Feb8.pdf). In 2002, the Microbial Genome Sequencing Program of the U.S. Department of Agriculture and the National Science Foundation provided financial support for complete sequencing and public release of the genome of *F. graminearum.*

During the 19th century, attention began to focus on determining whether various diseases of plants were caused by fungi or by unusual atmospheric conditions or other environmental factors. By 1846, Berkeley strongly believed that "The decay is the consequence of the presence of the mould, and not the mould of the decay" (Berkeley, 1846). Berkeley conclusively showed that the fungus *Phytophthora infestans* was the cause of the devastating epidemics of potato late blight that contributed to the great famines in Ireland from 1845. Although *P. infestans* is an Oomycete and no longer placed in the kingdom Fungi, it shares many characteristics of the true Fungi and remains a serious pathogen of potatoes worldwide. Because of their agricultural relevance, genomes of *P. infestans, Ustilago maydis, Botrytis cinerea*, and other plant pathogenic fungi are being sequenced with support from several government agencies and private companies (Table 5). The highest priority was given to sequencing the genome of *Magnaporthe grisea*, which causes blast, the most serious disease of rice worldwide. A draft genome of *M. grisea* was released in 2002 and is the first publicly available complete sequence of a plant pathogenic fungus (Table 5).

Nineteenth century mycologists also determined the fungal basis of some human diseases (Ainsworth, 1976). Gruby founded the field of medical mycology by demonstrating the fungal nature of human skin diseases caused by *Trichophyton* and of infant oral infections caused by *Candida.* Systemic human mycoses caused by *Aspergillus* species and *Coccidioides immitis* were discovered by 1900 and those caused by *Histoplasma capsulatum* were discovered in 1906. As the 21st century begins, fungal infections are an emerging threat to healthy human populations and to the increasing population of immunocompromised individuals. Among humans who are immunocompromised due to cancer chemotherapy, transplantation surgery, or HIV infection, these systemic mycoses often can be fatal. Because of their relevance to human health, genomes of *A. fumigatus*, *Candida albicans*, *C. immitis*, *Pneumocystis carinii* and other human pathogens are being sequenced (Table 5).

The young science of genetics after 1900 followed a Mendelian approach of laboratory and field experimentation with pure lines of fungi, plants and other organisms. Genetic research was dominated by experimentalists such as Bateson and Morgan, who had a typological species concept in which all individuals of a species are essentially alike. After the 1930s, however, the typological species concept gradually was replaced by the populational species concept of Darwin and Wallace (Mayr and Provine, 1998). The biogeography of South America gave Darwin evidence to write in the 1859 *Origin* "No one supposes that all the individuals of the same species are cast in the very same mould. These individual differences are highly important for us, as they afford materials for natural selection to accumulate…" One-hundred years after Darwin, the discovery of DNA sequence polymorphisms within and between natural populations revolutionized the science of population genetics, especially among fungi with little visible morphological diversity. With the development of new genetic and genomic tools such as ESTs and microarrays, fungi are emerging as a group of organisms particularly well suited for analysis of the genetic and phenotypic diversity of natural populations, agricultural populations, and populations associated with human disease. Many fungi have extremely large natural populations, and widespread strain collections are available in fungal stock centers worldwide. Even closely related fungal species can demonstrate diverse modes of sexual and asexual reproduction, diverse patterns of geographical distribution, and diverse acquisition of virulence to plants and humans. Thus, the fungal tree of life offers unique opportunities to address the complexities of the horizontal evolution of species in space by diversification and the vertical evolution of species in time by adaptation. "There is grandeur in this view of life, with its several powers, having been originally breathed into a few forms or into one; and that, whilst this planet has gone cycling on according to the fixed law of gravity, from so simple a beginning endless forms most beautiful and most wonderful have been, and are being, evolved."

4. CONCLUSIONS

Fungi dominate our world as plant and animal pathogens, as sources of food and other useful products, and as critical components of natural and agricultural ecosystems. During the 20th century, fungi became model systems for genetic and biochemical research that has elucidated the fundamental biology of eukaryotic organisms. As the 21st century begins, fungal genomics is becoming a major focus for research in the biological sciences that are vital for development of new technologies for industry, agriculture, and human health. Because of the high cost of genome sequencing and limited public resources, national programs and professional societies are identifying and prioritizing fungi that are of industrial importance or that present an emerging and significant threat to agriculture or to human health through accidental or deliberate introduction. Knowledge of the genomes of plant and human pathogens is expected to elucidate the genetic basis of fungal-host interactions and to assist the development of novel strategies for disease control. Knowledge of multiple fungal genomes also is expected to provide information critical for understanding, engineering, and exploiting the biological diversity of fungal populations in natural and agricultural environments.

REFERENCES

Ainsworth G.C (1976). Introduction to the History of Mycology, Cambridge Univ Press, Cambridge, UK.

Anderson S (1981). Shotgun DNA sequencing using cloned DNAse I-generated fragments. Nucleic Acids Res 9:3015-3027.

Baldi P, and Long AD (2001). A bayesian framework for the analysis of microarray expression data: Regularized t-test and statistical inferences of gene changes. Bioinformatics 17: 509-519.

Bennett J W (1997a). Open letter to fungal researchers. Fungal Genet Biol 21:2.

Bennett J W (1997b). White paper genomics for filamentous fungi. Fungal Genet Biol 21:3-7.

Bennett J W (1998). Mycotechnology: the role of fungi in biotechnology. J Biotechnol 66:101-107.

Bennett J W, and Arnold J. (2001). Genomics of fungi. In. The Mycota VIII, Biology of fungal cell. Howard/Gan (eds). Springer-Verlag, Berlin 286-297.

Berg P, and Singer M (1992). Sealing with genes. The language of heredity. The language of heredity. University Science Books, Mill Valley, California. p 247.

Berkeley M.J (1846). Observations, botanical and physiological, on the potato murrain, reprinted 1948, Phytopathological Classics, American Phytopathological Society, East Lansing, MI.

Bhatnagar D, Yu J, Ehrlich, K.C (2002). Toxins of filamentous fungi. In Fungal Allergy and Pathogenicity (M. Breitenbach, R. Crameri and S.B. Lehrer, eds) Chem Immunol Basel:Karger 81:167-206.

Blackwell M, Terrestrial life- fungal from the start? Science (2000) 289:1884-1885.

Blohm D H, and A.Guiseppi-Elie (2001). New developments in microarray technology. Curr Biotechnol 12:41-47.

Bohnert D J, Ayoubi P, Borchert C, Bressan R A, and. Burnap R.L (2001). A genomics approach towards salt stress tolerence. Plant Physiol Biochem 39: 295-311.

Brody H, and Carbon J (1989). Electrophoretic karyotype of *Aspergillus nidulans*. Proc Natl Acad Sci USA. 86: 6260-6263.

Chu W S, Magee B B, and Magee P T (1993). Construction of an Sfil macorestriction map of the *Candida albicans* genome. J Bacteriol 175 (20):6637-6651.

Couzin J (2002). In yeast, prions' killer image doesn't apply. Science 297:758-761.

Darwin C (1859). On the Origin of Species by Means of Natural Selection, facsimile of the first edition, 1964, Harvard University Press, Cambridge, MA.

Debets A J, Holub E F, Seart K, van den Broek H W, and Bos C J (1990). An electrophoretic karyotype of *Aspergillus niger*. Mol. Gen. Genet. 224:246-268.

Debets F, Swart K, Hockstra R F, and. Bos C J (1993). Genetic maps of eight linkage groups of *Aspergillus niger* based on mitotic mapping. Curr Genet 23:47-53.

Donson J, Fang Y W, Espiritu-Santo G. Xing W M, A. Salazar, et al. (2002). Comprehensive gene expression analysis by transcript profiling. Plant Mol Biol 48: 75-97.

Dujon B (1996). The yeast genome project: what did we learn? Trends Genet 12:263-270 .

Esch J (2000). Genomics engineering: Moving beyond DNA sequence to function. Proc. IEEE 88:1947-1948.

Ewing B, and Green P (1998). Base-calling of automated sequencer traces using phred. II. Error probabilities Genome Res. 8: 186-194.

Ewing B, Hillier L, Wendl M.C, and Green P (1998). Base-calling of automated sequencer traces using phres. I. Accuracy assessment. Genome Res 8: 175-185.

Fakhoury A M, and Payne G A (2003). Genomics of filamentous fungi: a general review. In Handbook of Fungal Biotechnology (D.L. Arora, P.D. Bridge and D.Bhatnagar, eds). Marcel Dekker Inc., New York (in press).

Fekete C, Nagy R, Debets A.J, Hornok L (1993). Electrophoretic karyotypes and gene mapping in eight species of the Fusarium sections Arthrosporiella and Sporotrichiella. Curr. Genet. 24:500-504.

Fitch J P, and Sokhansanj B (2000). Genomic engineering: Moving beyond DNA sequence to function. Proc. IEEE 88:1949-1971.

Gardner R C, Howarth A.J, Han P, Brownluedi M, Shepherd R.J, and Messing J (1981). The complete nucleotide-sequence of an infectious clone of cauliflower mosaic-virus by m13mp7 shotgun sequencing, Mucleic Acids Res 9:2871-2888.

Goffeau A, Barrell B G, Bussey H, and David R W et al. (1996). Life with 6000 genes. Science 274:546-567.

Goffeau A et al. (1997a). The yeast genome directory. Nature 387 (Suppl): 1-105.

Goffeau A, Park J, Paulsen I T, Jonniaux J L, Dinh T, Mordant P, and. Saier M.H (1997b). Jr., Multidrug-resistant transport proteins in yeast: complete inventory and phylogenetic characterization of yeast open reading frames with the major facilitator superfamily. Yeast 13: 43-54.

Green E D (2001). Strategies for the systematic sequencing of complex genomes. Nat. Rev. Genet, 2: 573-583.

Gross C, Delleher M, Iyer V R, Brown P O, and Winge D R (2000). Identification of the copper regulon of *Saccharomyces cerevisiae* by DNA microarrays. J Biol Chem 275:32310-32316.

Hawksworth D L (1991). The fungal dimension of biodiversity magnitude, significance and conservation. Mycol Res 95:641-655.

Hegde P, Qi R, Abernathy K, Gay C, Dharap S, et al. (2000). A concise guide to cDNA microarray analysis. BioTechniques 29:548-562.

Hieter P, and Boguski M (1997). Functional genomics: its all how you read it. Science 278:601-602.

Henig R.M (2000). The Monk in the Garden, Houghten Miflin, NY.

Hudler G W (1998). Magical Mushrooms, Mischievous Molds. Princeton Univ. Press, Princeton, NJ.

Hunkapiller T, Baiser R J, Koop B F, and Hood L (1991). Large-scale and automated DNA-sequence determination. Science 254:59-67.

Joseph B, Shrinivasan A, and Kumaramanickavel G (2002). Microarrays – "chipping" in genomics. Indian J Biotechnol 1:245-254.

Jurgenson J E, Bowden R L, Zeller K A, Leslie J F, Alexander N J, and Plattner R D (2002). A genetic map of Gibberella zeae (*Fusarium graminearum*). Genetics. 2002; 160(4):1451-1460.

Keller N P, Cleveland T E, and Bhatnagar D (1992). Variable electrophoretic karyotypes of members of *Aspergillus* section *Flavi*. Curr Genet 21:371-375.

Kitamoto K S, Kimura K, Gomi K, and Kumagai C (1994). Electrophoretic karyotype and gene assignment to chromosomes of *Aspergillus oryzae*. Biosci. Biotechnol Biochem 58:1467-1470.

Liang P, and Pardee A B (1992). Differential display of eukaryotic messenger-RNA by means of the polymerase chain-reaction. Science 257:967-971.

Lockhart D J, and Winzeler E A (2000). Genomics, gene expression and DNA arrays. Nature 405:827-836.

Matz M V, and Lukyanov S A (1998). Different strategies of differential display: areas of application. Nucleic Acid Res. 26:5537-5543.

Maxam A M, and Gilbert W (1977). A new method for sequencing DNA. Proc Natl Acad Sci USA 74: 560-564.

Mayr E, and Provine W. B (1998). The Evolutionary Synthesis, second edition, Harvard University Press, Cambridge, MA.

Meldrum D (2000a). Automation for genomics, part one: Preparation for sequencing. Genome Res 10: 1081-1092.

Meldrum D (2000b). Automation for genomics, part two: Sequencers, microarrays, and future trends. Genome Res 10: 1288-1303.

Mitnik L, Novotny M, Felten C, Buonocore S, Koutny L, and Schmalzing D (2001). Recent advances in DNA sequencing by capillary and microdevice electrophoresis. Electrophoresis 22: 4104-4117.

Mullis K, Faloona F, Scharf S, Saiki R, Horn G, and H. Erlich (1986). Specific enzymatic amplification of DNA *in vitro* – the polymerase chain-reaction. Cold Spring Harbor Symp Quant Biol 51: 263-273.

Ohlrogge J, and Benning C (2000). Unraveling plant metabolism by EST analysis. Current Opinion in Plant Biology 3: 224-228.

Orbach M J, Chumley F G, and Valent B (1996). Electrophoretic karyotypes of *Magnaporthe grisea* pathogens of diverse grasses. Molec. Plant-Microbe Interactions 9:261-271.

Orke E C, Dehne H W, Schonbeck F, and Eeber A (1994). Crop production and crop protection: Estimated losses in major food and cash crops, Elsevier, Amsterdam.

Perkins D D (1992). Neurospora: the organism behind the molecular revolution. Genetics 130: 687-701.

Price M S, Classen J J, and Payne G A (2001). Aspergillus niger absorbs copper and zinc from swine wastewater. Bioresosur Technol 77: 41-49.

Rayner S, Brignac S, Bumeister R, Belosludtsev Y, and Ward T, et al. (1998). Mermade: An oligodeoxyribonucleotide synthesizer for high throughput oligonucleotide production in dual 96-well plates. Genome Res 8: 741-747.

Richard J L, and Payne G A (2003). Mycotoxins: Risks in plant and animal systems, Council for Agricultural Science and Technology (in press).

Righetti P G, Gelfi, and D'Acunto M R (2002). Recent progress in DNA analysis by capillary electrophoresis. Electrophoresis 23:1361-1374.

Sanger F, Nicklen S, and Coulson A R (1977). DNA sequencing with chain-terminating inhibitors. Proc Natl Acad Sci USA 74: 5463-5467.

Schena M, Shalon D, Davis R W, and Brown P O (1995), Quantative monitoring of gene-expression patterns with a complementary-DNA micoarray. Science 270: 467-470.

Skinner W, Keon J, and J Hargreaves (2001). Gene information for fungal plant pathogens from expressed sequences. Curr Opin Microbiol 4: 381-386.

Smith L M, Fung S, Hunkapiller M W, Hunkapiller T J and Hood L E (1985). The synthesis of oligonucleotides containing an aliphatic amino group at the 5' terminus- synthesis of fluorescent DNA primers for use in DNA-sequence analysis. Nucleic Acids Res 13: 2399-2412.

Smith L M, Sanders J Z, Kaiser R J, Hughes P, Dodd C, et al. (1986). Fluorescence detection in automated DNA-sequence analysis. Nature 321: 674-679.

Souciet J L, Aigle M, Artiguenave F, Blandin G, and M. Bolotin-Fukuhara, et al. (2000). Genomic exploration of the hemiascomycetous yeasts: 1, A set of yeast species for molecular evolution studies. FEBS Lett 487: 3-12 (2000).

Sybenga J (1972). General cytogenetics. Am Elsevier Publ., New York.

Stebbins G L (1966). Chromosome variation and evolution. Science 152:1463-1469.

Talbot N J, Salch Y P, Ma and Hamer J E (1993). Karyotypic variation with clonal lineages of the rice blast fungus, *Magnaporthe grisea.* Appl Environ Microbiol 59:585-593.

Winzeler E.A et al. (1999). Functional characterization of the *S. cerevisiae* genome by gene deletion and parralel analysis. Science 285:901-906.

Xu J-R, and Leslie JF (1996). A genetic map of *Gibberella fujikuroi* mating population A (*Fusarium moniliforme*). Genetics 143:175-189.

Applied Mycology & Biotechnology
An International Series. Volume 3. Fungal Genomics

2

Meiotic Recombination in Fungi: Mechanisms and Controls of Crossing-over and Gene Conversion

Bernard Lamb
Department of Biological Sciences, Imperial College of Science, Technology and Medicine, London SW7 2 AZ, England (b.lamb@ic.ac.uk).

The mechanisms of recombination by crossing-over and gene conversion are fairly well understood in *Saccharomyces cerevisiae*, with initiation by double-strand breaks, the formation of Holliday junctions and heteroduplex DNA, followed by junction resolution and by the chance of mismatch repair at points of heterozygosity. A number of controlling genes and proteins have been identified, including some homologues of better characterised bacterial enzymes. In filamentous fungi, the details are less clear, and one cannot assume that what applies in yeast will be general in all fungi. Other aspects considered here include chromosome pairing, the synaptonemal complex, recombination nodules, recombination models, mismatch repair, recombination controls (including hotspots and coldspots), ectopic recombination, and polarity gradients in gene conversion.

1. INTRODUCTION

Recombination is the production of new combinations of existing parental genes, and is therefore quite distinct from mutation. In most fungi it occurs predominantly during meiosis, but also happens at much lower frequencies in mitosis. For non-syntenic genes (those on non-homologous chromosomes), recombination normally occurs by independent assortment in a diploid nucleus at meiosis. For syntenic loci (those on homologous chromosomes), meiotic recombination occurs in a diploid nucleus by reciprocal crossing-over, involving breakage and reunion of non-sister chromatids, or by the non-reciprocal process of gene conversion. While polyploidy is very common in higher plants, it has occasionally been found in most fungal groups, including in *Saccharomyces*. In fungi it is usually an aberration, giving sterility, so only meiotic recombination in diploids is considered here. Gene conversion has sometimes been defined as a non-reciprocal transfer of genetic information from one member of a homologous pair of chromosomes to another member, and that transfer is usually between non-sister homologous chromatids at meiosis.

There are also rarer processes, such as recombination events between different parts of a single chromosome (intrachromosomal recombination) and recombination events between non-homologous chromosomes (ectopic recombination; translocation). Both those types of event are distinctive in that they can occur in a haploid fungal nucleus, as well as in a diploid nucleus at meiosis.

Some fungi have a parasexual cycle, not involving meiosis, which produces recombination between non-syntenic loci by haploidisation or non-disjunction (usually a double non-disjunction), or recombination between syntenic loci by mitotic crossing-over. For a basic account, see Lamb (2000). Mitotic recombination and the parasexual cycle, and the initiation of meiosis in fungi, are beyond the scope of this article. Most of the work cited here is on Ascomycete fungi as they have been best studied.

In meiosis, the major steps are as follows. There is premeiotic DNA replication, leaving each chromosome composed of two sister chromatids which remain associated until anaphase II. Homologous chromosomes pair in zygotene, with two sets of sister chromatids coming together to form a bivalent, composed of two chromosomes (four chromatids per bivalent). Crossing-over is mainly between non-sister chromatids, but sister chromatids may also cross over (evidence summarised by Lamb 1996a). Meiosis I (MI) is a reduction division, with homologous chromosomes segregating to opposite poles at anaphase I. Those chromosomes are each made up of two sister chromatids joined at their centromeres, but with some DNA differences if there is heterozygosity and if crossing-over has taken place between them and the non-sister homologous chromatids within that bivalent. Meiosis II (MII) is called an equational division, with segregation of sister chromatids to opposite poles at anaphase II. This gives four haploid nuclei after telophase and cell cleavage. In some fungi, the haploid products act as gametes, but in others, such as yeast and *Neurospora*, they are incorporated into resting and dispersive spores such as ascospores.

Independent assortment of non-homologous chromosomes at meiosis results in 50% recombination between genes on the different chromosomes. It arises because of the independent alignment of the non-homologous chromosomes on the meiotic spindle, and their regular segregations at anaphase. In nature, research and biotechnology, it provides a simple and reliable way of generating new combinations of non-syntenic parental genes in the appropriate fungal crosses. For any two loci taken at random, they are more likely to be non-syntenic than syntenic, with this difference increasing as the number of chromosomes and hence the number of linkage groups increases. The majority of meiotic recombination in fungi is thus likely to be by independent assortment. It is not subject to manipulation or genetic or environmental controls. Its mechanisms in fungi are probably very similar to those in other Eukaryotes and will not be discussed further here. In the rest of this article, "recombination" will normally be taken to mean recombination between syntenic loci or alleles.

The major part of this article will be on the controls and mechanisms of recombination between syntenic genes, by crossing-over or gene conversion. The molecular details are best understood, but still not completely in yeast, *Saccharomyces cerevisiae*, with some good details also from fission yeast, *S. pombe*. It is far from clear whether recombination in these two unicellular fungi is typical of others, such as the filamentous Ascomycete fungi *Ascobolus immersus*, *Neurospora crassa*, *Sordaria brevicollis* or *Sordaria fimicola*, or Basidiomycete fungi. For future research on recombination, it is most important to have more information from non-yeast fungi. As this review series is on applied mycology and biotechnology, an emphasis on yeasts is appropriate, given their enormous industrial importance.

The recombination aspects which will be considered are dealt with roughly in the order in which they occur in meiosis: chromosome pairing, the synaptonemal complex, recombination nodules, recombination models and the molecular initiation of recombination, the formation of hybrid DNA (hDNA) and Holliday junctions, correction of mispairs and non-pairs, and resolution of Holliday junctions. Other topics dealt with are recombination controls, including hotspots and coldspots, ectopic recombination, and polarity gradients in gene conversion. Some of these topics have been covered in other reviews in great detail, often with huge numbers of references. For general aspects of recombination in Ascomycete fungi, see Lamb (1996 a). For the synaptonemal complex and recombination nodules, see Zickler and Kleckner (1999). For recombination mechanisms, mutations and involved proteins in yeast, see Pâques and Haber (1999). For mathematical and statistical analysis of ordered tetrads and half-tetrads, see Zhao and Speed (1998 a, b). For recombination hotspots and coldspots, see Petes (2001). For gene conversion disparity, see Lamb (1998).

Some of the findings on recombination mechanisms and controls come from electron and light microscopy, others from observation of phenotypes from crosses, and some from biochemical studies, including the sequencing of critical regions. Much of the work has been done using mutations to disrupt normal mechanisms, and many papers involve combinations of two or more of these techniques. Meiotic mutants in fungi have been extremely useful in research but have had few or no practical applications in biotechnology. Meiotic mutants have been used directly in potato breeding (Peloquin *et al.* 1999).

Some of the basic facts about crossing-over were summarised by Anderson *et al.* (1999), with references. Each bivalent normally has at least one crossover, which helps to ensure proper disjunction of chromosomes at anaphase I. The number of additional crossovers beyond the "obligate" crossover is roughly proportional to a chromosome's length, with longer chromosomes typically having more crossovers than short ones. A crossover in one region generally reduces the chance of a second crossover nearby, which is localised positive chromosome interference. Crossovers are much rarer per unit length of DNA in heterochromatin than in euchromatin, and often differ in frequency between the sexes: an extreme case is the fruit fly, *Drosophila melanogaster*, with normal crossing-over in the female meiosis but no crossing-over in male meiosis, just independent assortment for non-syntenic loci. In humans, according to Broman *et al.* (1998), there is more meiotic recombination in females (total map length 44 Morgans) than in males (total map length 28 Morgans).

A number of recombination-related topics have been largely excluded from this review because of space restrictions. For accounts of interference between crossovers, see Lamb (1996 a) and Zhao and Speed (1998 a, b) for chromatid interference and chromatid interference, Teuscher *et al.* (2000) for chromatid interference; for conversion-generated negative chromosome interference, with a clustering crossovers in *Neurospora crassa*, see Bowring and Catcheside (1999). For strong positive chromosome interference, see Broman *et al.* (2002) and references therein. For a "counting model" of interference, see Foss and Stahl (1995). For corresponding-site interference, see Lamb and Wickramaratne (1973) and Lamb and Shabbir (2002). For map functions in relation to interference, see Barratt *et al.* (1954). For a quantitative analysis of gene conversion in terms of nine parameters relating to hybrid-DNA formation and correction of mispairs, see Lamb (1996 a,b). For the fidelity of gene conversion and the occurrence of mutations in meiosis, see Lamb (1996 a, page 1044). For the way in which long duplications are detected in nature in *Neurospora crassa* and are modified early in meiosis by repeat induced point mutation (RIP), heavily mutating both the original copy and its duplicate, see Watters *et al.* (1999) and references therein. Some quantitative aspects of recombination can be changed by artificial selection for naturally occurring variants (for conversion frequencies, see Zwolinski and Lamb (1995) and for postmeiotic segregation frequencies, see Lamb and Saleem (2002).

The evolution of recombination and its role in natural fungal populations are not considered here, in spite of being very interesting topics. For the effects on natural genetic variation in diploids of different frequencies of recombination, and of local restrictions of recombination, see the tomato studies of Baudry *et al.* (2001). For whether fungi such as *Sordaria fimicola* can evolve an optimum level of recombination to suit their environment, see Saleem *et al.* (2001). For modelling of natural selection in relation to recombination controls, see Hey (1998). For recombination variability and recombination, see Korol and Preygel (1994).

Mehta and Cerda-Olmedo (2001) stated that "No evidence for meiosis has ever been found in Phycomyces" and that in *Phycomyes* "there is no meiosis at all, but that the diploid survivor(s) suffer repeated mitotic divisions in which frequent mitotic recombination and haploidization would occur, leading to haploid progeny nuclei. This is, in essence, the 'parasexual' cycle of *Aspergillus* and other fungi."

Fungi pathogenic on humans include yeast-like *Candida albicans*, which is diploid and asexual, and *Candida glabrata*, which is haploid and can be used in parasexual analysis after spheroplast fusion. Cormack and Falkow (1999) showed that transformation with plasmids worked well in *C. glabrata*, with integration by homologous and non-homologous recombination, but that was not meiotic recombination.

2. CHROMOSOME PAIRING, THE SYNAPTONEMAL COMPLEX, RECOMBINATION NODULES AND EARLY EVENTS IN MEIOSIS

In most eukaryotes, pairs of homologous chromosomes come together during zygotene, with a synaptonemal complex forming between homologues. This facilitates intimate pairing and crossing-over, and stops premature separation of the homologues. Skipper (2002) suggested that there may be some fundamental differences between meiosis in flies and in yeast, as shown by work with meiotic mutants. She stated that in yeast, crossing-over can take place in the absence of the synaptonemal complex (SC), but that in *Drosophila* the SC is essential for the initiation of recombination. She suggests that yeast might use the initiation of recombination to align homologous chromosomes, with the SC stabilising their pairing, but that *Drosophila* needs the SC for the initial alignment of homologues, without which crossing-over does not occur.

For an extremely thorough review of meiotic chromosome structure, pairing and the synaptonemal complex, with detailed references for the next few paragraphs, see Zickler and Kleckner (1999). In fungi, there is an exclusively meiotic synaptonemal complex which joins homologous chromosomes in early prophase. First to appear is the axial element, one per pair of sister chromatids, a rod about 50 nm in diameter. These axial elements form at leptotene, continuously or in short segments which later join. The proteinaceous parts of the central element then assemble in zygotene between two homologous axial elements (which are then termed lateral elements), extending the whole length of the chromosomes and completing their pairing. The central element is about 100 nm wide, complex in structure, and including transverse filaments. All homologues have continuous tripartite SCs along their lengths by pachytene, then the SCs are lost progressively from the bivalents through diplotene, when chiasmata, derived from crossovers, hold the bivalents together.

In *Sordaria macrospora*, *S. humana* and *Neurospora crassa*, full-length axial elements are formed before the SC forms. In other fungi such as yeast and *Coprinus cinereus*, elongation of the axial elements immediately precedes the appearance of the SC. Fungi tend to have multiple interstitial initiations of the SC.

Cytological examination of synaptonemal complexes has been used to study the number and form of chromosomes in species with small chromosomes, such as yeast. The fact that SCs usually form between homologous chromosomes has been used to study chromosome abnormalities and rearrangements, including the identification of breakpoints in translocation heterozygotes. That has been used together with genetic mapping to allocate linkage groups to particular chromosomes.

The first stages of pairing involve some poorly understood long-range recognition of homology between chromosomes, while the chromosomes are more than 300 nm apart, before SCs start forming. In *Sordaria* and *Neurospora*, axial elements are completed during leptotene, and there is complete presynaptic alignment of homologues before SC completion. Long-range early alignment also occurs in *Saccharomyces cerevisiae*. In *Sordaria*, pairs of homologues are aligned at early leptotene at a wide distance before SC formation. SC formation is not essential for pairing of homologues in all organisms. Yeast *hop1* and *spo11-Y135F* mutants lack an SC but have normal meiotic pairing of chromosomes with full alignment. Other mutants in yeast and *Sordaria* lacking SC formation have reduced pairing of chromosomes. Haploid meioses in plants and fungi can result in extensive non-homologous SC formation, and so can some meiotic mutants in yeast.

Trapping of another chromosome or bivalent (zygotene interlocking) within a bivalent during zygotene pairing is rare in fungi with small chromosomes such as *Saccharomyces*, *Sordaria* and *Coprinus*, but is more common in translocation heterozygotes and in some SC mutants. There are mechanisms to resolve such interlocks, as they are less frequent later in meiosis. In fission yeast, *Schizosaccharomyces pombe*, at the diploid stage there are only three pairs of chromosomes. These chromosomes pair but do not form typical tripartite synaptonemal complexes of axial elements in each chromosome and a central component. Instead, there are discontinuous patches of filamentous structures known as linear elements, similar to axial elements. Krawchuk *et al.* (1999) studied three recombination loci, *rec8*, *rec10* and *rec11* in *S. pombe*. The wild type allele of *rec8* was required for sister chromatid cohesion and homologue pairing, mainly affecting meiosis I segregation, while *rec10* and *rec11*, in addition to affecting recombination in meiosis I, mainly affected meiosis II segregation. All three genes affected meiotic sister chromatid cohesion and were required for a normal frequency and distribution of crossovers, especially towards the centres of each of the three chromosomes. Mutations at any of these genes gave aneuploid nuclei and spores. Those authors proposed a model of a “meiotic chromatid cohesion pathway”, linking together sister chromatid cohesion, pairing of the internal regions of homologous chromosomes, centromere proximal recombination, and the proper segregation of chromosomes at each of the two meiotic divisions. It is surprising that three unrelated fungi with regular sexual cycles and recombination do not have SCs. They are *Aspergillus nidulans*, *Ustilago maydis* and *Schizosaccharomyces pombe*.

Krawchuk *et al.* (1999) reported the action of the spindle pole body in *S. pombe* in the initial alignment of homologous chromosomes in meiosis. In prophase, the telomeres become clustered into a bouquet structure before other regions pair. Migration of the spindle pole body then drags the clustered telomeres forwards and backwards, with the unpaired central regions of the chromosomes trailing behind in a “horsetail” structure. This causes the internal parts of chromosomes of similar length to align roughly, which may help local searches for homology. Components of the meiotic chromosome cohesion pathway are proposed to stabilise the interstitial contacts and to promote further pairing. Discontinuous patches of linear elements complete the synapsis, there being no synaptonemal complex.

Jiao *et al.* (1999) showed in *Saccharomyces cerevisiae* that early exchange genes were required for the initiation of meiotic recombination. Mutations in several such genes caused a relatively earlier meiosis I, suggesting that the initiation of meiotic recombination is involved in the proper timing of the division. Mutations in *RAD50* or *REC102* gave a very early meiosis I; mutations in *REC104* or *REC114* had a lesser effect; mutations in *ME14* did not alter the timing of MI but the wild-type allele was required for the formation of meiotic double strand breaks (DSBs) which initiate homologous recombination events in yeast. It was not the double-strand breaks which were the signal for the normal delay in MI. Jiao *et al.* (1999) provided a model for the interaction between the initiation of recombination and the timing of meiosis I. Genes in the first group of early genes generate signal “S” which has partial activity in delaying MI. The genes in the second group modify this signal to a more active “S*” which is responsible for the normal 2-hr transient delay of MI. The genes in the third group are not involved with this signal. It was suggested that the delay in meiosis I allowed time for recombination to occur before homologues separated from each other at anaphase I. In yeast, failure to initiate recombination does not block the progress of cells through meiosis, but very greatly reduces the proportion of viable spores, which shows the importance of recombination in the proper segregation of chromosomes and chromatids.

In zygotene, homologous chromosomes pair, with crossing-over in pachytene, then at metaphase I the chromosomes attach to the spindle before segregating at anaphase I. The role of centromere alignment in meiotic chromosome pairing was investigated in yeast by Guerra and Kaback (1999), using diploids containing one normal copy of chromosome I and one copy bisected into two functional centromere-containing fragments. The centromere on one fragment was aligned with the centromere on the intact chromosome, while the centromere on the other fragment was misaligned by 50 or 100 kbp. Not surprisingly, the aligned centromeres segregated efficiently from each other, while misaligned ones segregated much less efficiently, with chromosomes having the centromere misaligned by 100 kbp segregating randomly. Random segregation of all misaligned centromeres was correlated with crossovers between the intact chromosome and the other fragment in the region separating the centromeres. When there were no crossovers in such regions, or where the DNA in that region had been deleted in one homologue to prevent recombination, segregation was good. The authors suggested that the inability of chromosomes with misaligned centromeres to segregate properly could serve as a control which prevents ectopically recombined chromosomes segregating, favouring the production of balanced products of meiosis and increasing reproductive fitness.

Truncated chromosome fragments in yeast can pair with and recombine with the intact chromosomes from which they were derived (see references in Arbel *et al.* 1999). That interferes with their meiotic segregation, resulting in increased non-disjunction (failure to segregate correctly). The increased non-disjunction was correlated with the length of shared homology, but was not affected by the position of the centromere on the fragment. Arbel *et al.* (1999) found that a single truncated fragment underwent frequent ectopic recombination in meiosis between markers located near the ends of the fragment, often resulting in the loss of markers from the fragment. The authors found intensive meiotic recombination between the two termini of the truncated fragment, and suggested that this was initiated by the telomeric sequences or telomeric-associated sequences.

Genes encoding parts of the transverse elements of the SC have been cloned from yeast, with corresponding proteins Zip1p and SCP1/Syn1p. The lateral elements of the SC sometimes exhibit banding, as in *Neotiella*, *Sordaria* and *Ascobolus* (references in Zickler and Kleckner 1999). In *Sordaria humana* the lateral elements are tubular from leptotene to diplotene with numerous bulges, but as this fungus is self-fertile, the bulges are unlikely to reflect non-homologous pairing regions. In yeast, regions with reduced DNA homology have reduced crossing-over, with even 1% difference in DNA sequence reducing crossing-over several fold (references in Zickler and Kleckner 1999). Those authors quote unpublished yeast data of A. Adjiri, E. Coic and F. Fabre on the effects of sequence variation over 2.1 kb in the *ARG4* gene on recombination, with the rest of the chromosomes being homologous. Recombination-initiating double-strand breaks were not affected but intragenic recombination (usually arising through hybrid-DNA) was reduced 40-fold by the reduced homology. Mismatch repair mutations reduced this reduction to 5-fold, suggesting that mismatch repair systems were involved in the reduced recombination between chromosomes with reduced DNA homology.

In pachytene, fungal chromosomes normally have a complete SC and stiff chromatin. At the end of pachytene, the chromosomes become more diffuse and the SCs start to disassemble. By diplotene, the homologous chromosomes are separating, held together at the chiasmata. Various SC proteins are lost as the homologues separate. For details and references, see Zickler and Kleckner (1999). The sister chromatids remain closely adpressed as the homologues repel each other in the rest of prophase I.

Recombination nodules (RNs) have been studied in a range of organisms including plants, animals and fungi (see references in Anderson *et al.* 2001). These nodules are complex proteinaceous ellipsoids ranging from 50 to 200 nm in length. Early nodules associate with axial elements and SCs from leptotene to early pachytene, and late nodules are found at sites of crossing-over in mid- to late-pachytene. The two types of nodule differ in shape, size, distribution, number and time of existence. Late nodules are generally believed to be organelles containing packages of enzymes involved in crossing-over.

Early nodules are more frequent than late ones, are common in euchromatin and rare in heterochromatin, and are usually shed in early pachytene to leave one or a few late nodules per bivalent. Early nodules are probably involved in DNA homology searching prior to intimate pairing and crossing-over, because at least some of them have RecA-related proteins (references in Anderson *et al.* 2001). One theory is that early nodules assemble at sites of double-strand breaks, and those early nodules which turn into late nodules are those at the sites of crossing-over. For details of chromosome pairing, recombination nodules and chiasma formation in *Coprinus cinereus*, see Holm *et al.* (1981).

In yeast, the *ZIP1* gene encodes protein Zip1p, a component of the central region of the synaptonemal complex. A series of *zip1* in-frame deletions were studied by Tung and Roeder (1998). The results showed that the extent of chromosome synapsis correlated closely with the effects on sporulation, spore viability, crossing-over and crossover interference. Higher levels of synapsis gave higher levels of crossing-over, possibly through favouring the resolution of recombination intermediates in the direction of crossovers. The effects on crossing-over were not uniform in different intervals. In *zip1* null mutations, the chromosomes failed to pair and crossover interference was eliminated. In the *zip1* deletions, all the mutants which made full-length synaptonemal complex had crossover interference. This is one a number of lines of evidence connecting a functional synaptonemal complex with chromosome interference.

According to Pâques and Haber (1999), double-strand breaks (DSBs) "are the sole instigators of recombination in meiotic cells and are a major factor in recombination in mitotic cells, although the origin of spontaneous mitotic recombination remains unknown." In yeast, there are several mechanisms of repairing DSBs by homologous recombination, with less efficient methods taking over from more common ones if the latter are disabled by mutation. In crosses of two heteroallelic auxotrophic mutations in repulsion within a gene, prototrophic recombinants usually arise largely (typically 90% or more) by gene conversion to wild-type at one site, rather than by reciprocal crossover, as demonstrated by examining the segregation of linked outside markers. As a broad generalisation, the closer two markers are on a chromosome, the more likely they are to recombine by gene conversion at one site, rather than by reciprocal crossing-over, especially for markers within a single locus. In meiosis, gene conversion tracts are on average 1 to 2 kb (references in Pâques and Haber 1999), while in mitosis they vary from very short to hundreds of kilobases. The same authors summarise genetic and molecular evidence that crossing-over between sister chromatids is usually suppressed. It does, however, happen at low frequencies (see discussion in Lamb 1996 a, p. 1037).

Key observations that crossovers and gene conversions were associated came from Mortimer and Fogel (e.g., Fogel *et al.* 1981) with yeast, and Kitani *et al.* (1962) with *Sordaria fimicola*, where they showed that crossovers often accompanied conversions and often involved the same chromatids. In the latter work, in samples of asci with normal 4+:4*g* (*gray*) segregation, the recombination frequencies for two pairs of outside (flanking markers) were 4.4% and 3.8%. Out of 23 asci with 5:3 or 3:5 segregations at the heterozygous *g* marker, 44% had a crossover between *g* and *mi*, 22 times the expected number, and 10 had a crossover between *g* and *cor*, six times the expected number. The increase in crossing-over was very local to the converting site, with the same chromatids being usually involved in conversions and crossovers. High associations between crossing-over and conversion were found in *A. immersus* (Rizet and Rossignol 1966; Stadler *et al.* 1970), and in *Sordaria brevicollis* (Sang and Whitehouse 1979).

In a whole range of Ascomycete fungi, roughly 25% to 75% of gene conversions had associated crossovers (references in Lamb 1996 a). It is now widely accepted that crossovers and gene conversions generally arise from common mechanisms, although one can happen without the other being detected, e.g., from repair of hDNA giving correction 4+:4*m* segregations not distinguishable from no recombination initiation, or from resolution of Holliday junctions to non-crossover forms. In mitotic recombination, only about 0% to 20% of gene conversions have associated crossovers (Pâques and Haber 1999).

3. RECOMBINATION MODELS: INITIATION, RECOMBINATION INTERMEDIATES, HOLLIDAY JUNCTION RESOLUTION AND MISMATCH REPAIR

These early data mentioned above gave rise to a number of recombination models, including the pioneering symmetric hybrid DNA model of Holliday (1964), which included a single Holliday junction, a half-crossover intermediate between non-sister chromatids: see Lamb (1996 a). Two models are given here as Figs. 1 and 2 as background to the discussion of later models. Fig. 1 shows the model of Meselson and Radding (1975), in which recombination is initiated by a single-strand nick, leading to strand displacement as DNA polymerase extends the broken strand. The displaced strand invades the non-sister homologous chromatid, forming a D-loop. The unpaired strand in the D-loop is degraded, and the invading strand is integrated into the other chromatid, forming asymmetric (in one chromatid only) hDNA, with mispairs (heterozygous base substitution/wild-type) or non-pairs (heterozygous frame shift/wild-type) at any points of heterozygosity. Branch migration can give symmetric hDNA, formed in two chromatids. An isomerisation step may be involved. The final result is asymmetric hDNA, possibly with symmetric hDNA, and a Holliday junction giving a half-crossover. That can be resolved by cutting the "crossed" strands to give no crossover, or by cutting the "outside" strands, giving a complete crossover. Conversions from combinations of repair and non-repair of mispairs or non-pairs can therefore be associated with a crossover or with no crossover, depending how the Holliday junction was resolved. There are many figures and descriptions of other recombination models and their variants in the other papers quoted.

The double-strand break-repair (DSBR) model of Szostak *et al.* (1983) is shown and described in Fig. 2, with a double-strand break leading to a double-strand gap, with one 3' tail invading the non-sister duplex, initially in the manner of the Meselson-Radding model (iii in Fig. 2). The gap is repaired in two stages, with DNA replication driving one strand off towards the gap, where it could eventually base-pair with the other 3' tail, but only if that tail has not been degraded. As on the previous model, there is no clear suggestion as to how or why the end of the newly synthesised DNA should accurately switch up to and accurately join the other chromatid's 5' end (Fig. 2, iv to v). The gap is then filled by DNA synthesis off the displaced single strand. The two Holliday junctions could be resolved by cutting crossed or uncrossed strands, or one pair of crossed and one pair of uncrossed strands (see later for details).

The DSBR model can explain asymmetric and symmetric hDNA, polarity gradients in gene conversion (see later), and that the initiating chromatid is usually the recipient of genetic information. Orr-Weaver and Szostak's (1985) analysis of this model is partly based on statements that in yeast there is parity in direction of conversion and that postmeiotic segregation is rare, which does not fit the extensive evidence (Lamb 1987, 1996a and 1998). Where the basic version of a recombination model fails to explain certain data, it is possible to have modified versions, such as the modified version of the Meselson-Radding model (Radding 1982; Radding *et al.* 1982; Nicolas and Petes 1994) to explain the initiating chromatid being the recipient of the genetic information. The modified model accommodates this by having a single-strand gap on the initiating chromatid, repaired from the intact donor chromatid, but it is not obvious why the donor strand should invade the other chromatid's gap.

To explain most of the fungal gene conversion data, the DSBR model would need to have very short lengths of gap relative to the lengths of flanking hybrid DNA, in yeast as well as filamentous fungi. There is much evidence for double-strand breaks being involved in recombination, especially in yeast and Prokaryotes, but if the break is extended to a gap, then recombination will only avoid causing chromosome breakage and deletions if each of the following events takes place. One 3' end is not degraded and finds a region of homology on a different chromatid (two non-sister candidates), and a *RecA*-type enzyme. A D-loop is successfully formed. DNA synthesis in the D-loop continues to beyond the other 3' end. That 3' end is not degraded. The displaced strand finds the gap in the gapped chromatid (one candidate) and anneals with the single strand. The 3' end of the newly synthesised DNA (lower chromatid, iv in Fig. 2) gets displaced and also manages to reach the other chromatid and to get ligated to the broken 5' end (or, not part of the model, the left-hand crossed strand in (iv) must break down, with repair of any gaps in both chromatids, leaving only one Holliday junction). The remaining single-strand gap in the upper chromatid (iv and v) must be repaired by synthesis off the transferred strand. While all of these steps are possible, it is not easy to see why they should all occur with nearly 100% probability. Other difficulties with the original DSBR model were pointed out by Lamb (1987).

A modified and much-improved double-strand break-repair model was proposed by Sun *et al.* (1991), who stated that most conversion is now viewed as the result of mismatch repair of heteroduplex DNA, instead of being a direct result of double-strand gap repair. They suggested that the initial DSB has 5' to 3' exonuclease degradation in both directions to leave extensive 3' overhanging single-stranded tails, up to 800 nucleotides long but of variable length, which invade a non-sister chromatid to form extensive hybrid DNA. There may or may not be a gap between the two 3' ends. The revised model is based on their work on the *ARG4* recombination initiation site in yeast and represents a big change in emphasis from a large gap and short flanking hDNA in yeast in the original DSBR model to a short or no gap flanked by long hDNA in the modified model, which is much more in accord with the genetic data on postmeiotic segregation, yeast having frequent disparity in conversion, and other phenomena described in Lamb (1987, 1996a).

Another kind of model has been termed "synthesis-dependent strand annealing" (SDSA) and was originally proposed because most mitotic gene conversions are not associated with crossovers. In these models, the newly synthesised DNA strands are displaced from the templates and return to the broken molecules, permitting the two newly synthesised strands to anneal with each other, perhaps with topoisomerase or helicase enzymes pulling apart the replication structures. The initial D-loop does not break down but its displaced strand returns to its original partner. In the present Fig. 2(iv), that would mean the newly synthesised part copied off the bottom DNA strand unwinding and pairing with the strand about to be synthesised off the top part of the D-loop.

Fig. 1. A slightly simplified meselson and radding 9195) recombination model.

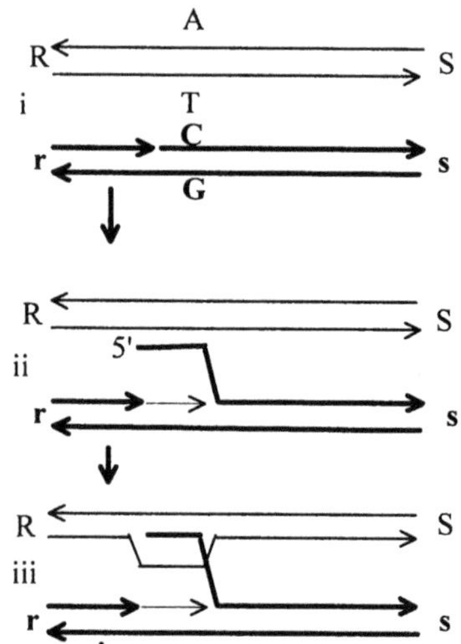

Two non-sister chromatids, each of one double helix, with about one gene length, are shown. The arrows indicate polarity. The wild-type allele has base-pair AT and the mutant allele has CG at the point of heterozygosity for this base-substitution mutation. In the lower chromatid in (i), a single-strand nick is shown, to the 3' end of which a DNA-polymerase enzyme attaches. DNA synthesis (dotted line) in (ii) and (iii) then displaces a single strand which invades the non-sister chromatid's helix, forming a D-loop (iii). DNA synthesis continues to promote strand transfer to the other chromatid (iii and iv). The D-loop is degraded and so is the right-hand end of the resulting gap (iii and iv).

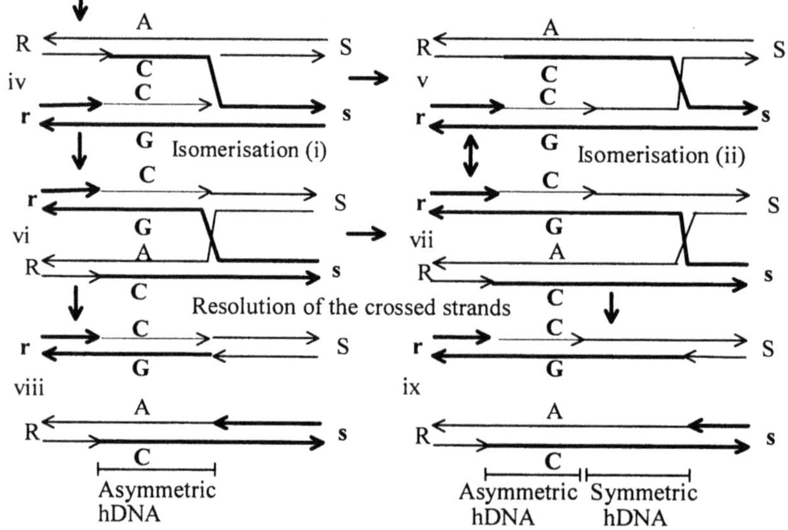

Branch-migration of the cross-point, after a joining of free ends by ligase when the polymerase dissociates, can give symmetric hDNA to the right of the initial asymmetric hDNA. Rotary diffusion can produce (v) from (iv) and (vii) from (vi). Stage (iv), with parental flanking markers, RS and rs, can isomerise to (vi), with recombinant flanking markers, Rs and rS. Stage (v), with parental flanking markers, can undergo a more complicated isomerisation to (vii), with recombinant flanking markers. It was proposed that half-crossovers were resolved by the inner strands breaking and rejoining.

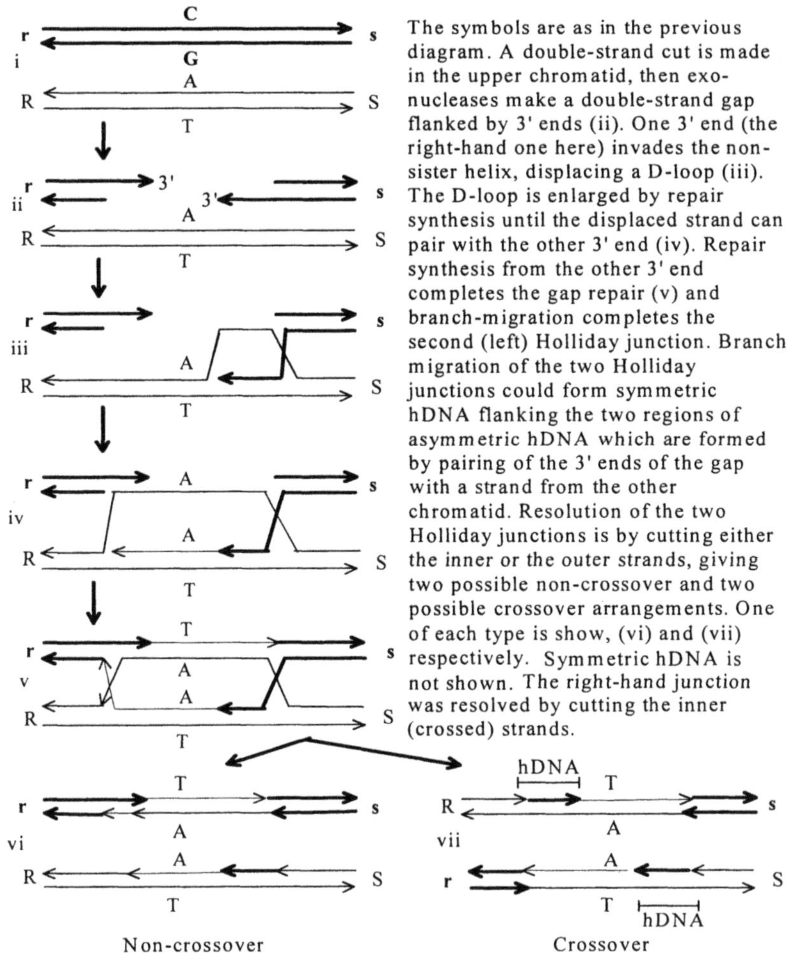

The symbols are as in the previous diagram. A double-strand cut is made in the upper chromatid, then exonucleases make a double-strand gap flanked by 3' ends (ii). One 3' end (the right-hand one here) invades the non-sister helix, displacing a D-loop (iii). The D-loop is enlarged by repair synthesis until the displaced strand can pair with the other 3' end (iv). Repair synthesis from the other 3' end completes the gap repair (v) and branch-migration completes the second (left) Holliday junction. Branch migration of the two Holliday junctions could form symmetric hDNA flanking the two regions of asymmetric hDNA which are formed by pairing of the 3' ends of the gap with a strand from the other chromatid. Resolution of the two Holliday junctions is by cutting either the inner or the outer strands, giving two possible non-crossover and two possible crossover arrangements. One of each type is show, (vi) and (vii) respectively. Symmetric hDNA is not shown. The right-hand junction was resolved by cutting the inner (crossed) strands.

Fig. 2. A Slightly simplified Szostak *et al.* (1983) double-strand break-repair model.

Some of the evidence for SDSA models was as follows. The DSBR models predict that two regions of heteroduplex will be on different chromatids, as in Figure 2 (vii). Studies such as that of Gilbertson and Stahl (1996) often found both heteroduplex regions on a single chromatid, which could be explained by SDSA models (see Figs. 8 and 9, Pâques and Haber 1999). The SDSA models are better at explaining conversions without crossovers than conversions with crossovers, although most models are flexible enough to be adaptable to explaining many things with some additional postulates.

As mentioned earlier, the best data on models come from yeast, which has about 100 crossovers per meiosis, distributed over 16 pairs of chromosomes. There is normally at least one crossover per bivalent, giving a chiasma which joins the homologous chromosomes and helps to ensure proper segregation at anaphase I. In a series of papers, Szostak and colleagues (e.g., Sun *et al.* 1991) reported on the initiation of recombination at the *ARG4* locus in yeast.

Double-strand breaks occurred in the promoter region in early prophase of meiosis I. The ends of the double-strand breaks were resected to produce long single-stranded 3' ends, up to 800 nucleotides long, not double-strand gaps as on the original DSBR model. Later studies (references in Pâques and Haber 1999) showed that DSBs were site-specific but not sequence-specific. Hotspots for DSBs were usually in promoters containing DNase I- or micrococcal nuclease-sensitive sites, with transcription factor remodelling of chromatin, but not active transcription, involved in hotspot activity.

For details of the yeast genes and proteins involved in starting meiotic recombination in yeast, see Pâques and Haber (1999) and references therein. Spo11p seems to be the initial endonuclease in yeast meiotic recombination, and multiprotein complexes carry out most of the recombination steps, rather than single enzymes.

Recombination models often predict the existence of double Holliday junctions as recombination intermediates, as shown in Fig. 2 (v). They have been identified in yeast, forming soon after double-strand breaks (e.g., Schwacha and Kleckner 1995), and disappear when the synaptonemal complex dissociates, by when crossovers have been established. If the double Holliday junctions are artificially denatured, only parental arrangements of markers are found, but if they are cleaved with the bacterial RuvC resolvase, both parental and recombinant strands are found.

Double-strand breaks occur in leptotene, during the axial element formation stage of the synaptonemal complex. The double Holliday junctions are found at the start of pachytene, with heteroduplexes and recombinant chromatids by the end of pachytene (references in Pâques and Haber 1999), confirming the classical picture of recombination by crossing-over occurring during pachytene. In yeast, chromosome synapsis and synaptonemal complex formation depend on recombination, with no synapsis in mutants which do not make double-strand breaks in early meiosis (e.g., Rockmill *et al.* 1995). Some yeast mutants, *zip1* and *zip2*, can carry out recombination without forming a synaptonemal complex (Sym *et al.* 1993). It seems that some major functions of the synaptonemal complex in yeast involve the regulation of recombination, especially the timing, frequency and distribution of crossovers.

As shown by Sym and Roeder (1994) in yeast, mutations which decrease the frequency of crossing-over or which eliminate crossing-over greatly increase the proportion of inviable ascospores, because of large increases in chromosome non-disjunction in meiosis. In fission yeast, *Schizosaccharomyces pombe*, there are only three pairs of chromosomes as opposed to 16 pairs in *Saccharomyces cerevisiae*, and there is no synaptonemal complex, although there are linear elements, and no interference between crossovers. Molnar *et al.* (2001) studied the effects of a *rec7* mutation in *S. pombe* which strongly reduces double-strand break formation. It severely reduced intragenic and intergenic meiotic recombination in all regions tested. This caused frequent non-disjunction of homologous chromosomes at anaphase I, and some diploid colonies from omission of the second division. On spreads of prophase nuclei, about 50 foci of Rec7-GFP (green fluorescent protein) were found, which is similar to the total number of crossovers across the whole genome. These findings show again the importance of crossovers for correct chromosome segregation.

DNA replication is essential for meiosis. If replication is blocked in yeast by hydroxyurea, recombination and meiosis do not occur: see references in Lamb and Mitchell (2001). Those authors list a series of meiosis-specific genes acting at different stages. Gerecke and Zolan (2000) pointed out the similarities of meiotic chromosome behaviour and recombination with DNA double-strand break repair, as they both involve identification of homologous sequences and repair of breaks, often with an exchange of genetic material. In yeast, double-strand breaks are initiated in meiosis by Spo11, a type II topoisomerase-like protein, together with the Mre11/Rad50/Xrs2 protein complex, which is also needed for processing the breaks. Mutations in the relevant genes result in defects in meiotic recombination and in viable spore formation, and give increased sensitivity to ionising radiation. Gerecke and Zolan (2000) used the filamentous fungus *Coprinus cinereus*. They found that *rad11* was a homologue of *MRE11*, which is required for meiosis and DNA repair in many organisms, including yeast. The gene is induced during prophase of meiosis I and after gamma irradiation. The *rad11* mutants had defects in chromatin condensation, homologue pairing and in synaptonemal complex formation. Neither axial elements nor mature complexes were normal or complete, with delays to meiosis.

Studies on non-fungal organisms often show the universality of processes which have been mainly worked out from technically convenient fungi. Li and Baker (2000) investigated the repair of double-strand breaks in mammalian cells. Repair did not usually involve a long double-strand gap. In 43% of recombinants, the results were consistent with a crossover at or near the double-strand break, and in the remaining recombinants, there was a hybrid-DNA intermediate. Individual hDNA tracts were either long or short and asymmetric or symmetric on the one side of the double-strand break examined.

Some fungi have developed special features of recombination which suit some aspect of their biology. In *Neurospora tetrasperma*, the asci produce four binucleate dual mating-type (*A* + *a*) large ascospores per ascus, instead of eight uninucleate ascospores of single mating-type as in *N. crassa*. This secondary homothallism in *N. tetrasperma* results from first division segregation of the mating-type locus and overlapping nuclear spindles at subsequent meiotic and mitotic divisions (see Fig. 1, Gallegos *et al.* 2000). Merino *et al.* (1996) showed in *N. tetrasperma* that crossing-over is suppressed in much of the mating-type chromosome, preventing second division segregation of mating-type. Interestingly, autosomal regions were largely homoallelic as a result of repeated selfing combined with crossing-over, while sequences on much of the mating-type chromosome were heteroallelic as a result of long-maintained suppressed crossing-over. Gallegos *et al.* (2000) confirmed that crossing-over is suppressed in a large segment (exceeding 100 map units) of the *N. tetrasperma* mating-type chromosome, from *nit-2* in the left arm to *al-1* in the right arm, including the centromere-to-mating-type interval. They also found a region in the far end of the left arm where one crossover always occurred, between *cyt-21* and *nit-2*. It was always one crossover, as if there was complete chromosome interference in that interval. Always having a crossover there would compensate for no crossovers in most of the chromosome, ensuring proper segregation. Suppressed recombination was correlated with an extensive unpaired region at pachytene, up to half the length of the chromosome.

The mismatch-repair system (MMR) is an important component of the recombination mechanisms, because heterozygosity within a region of hybrid DNA gives mismatched base pairs. Sequence divergence has been found to decrease recombination in bacteria, yeast and mammalian cells. A single mismatch within a region of otherwise perfect base sequence homology can inhibit transformation in *Bacillus* or mitotic recombination in yeast. In bacteria and yeast, there is a log-linear relation between the frequency of recombination and the level of sequence divergence (references in Chen and Jinks-Robertson 1999). It is not clear how the MMR machinery inhibits recombination when it finds mismatches, but it has been suggested that it might trigger helicase-unwinding of hDNA or immediate resolution of the recombination intermediates. Inactivation of part of the MMR can increase the frequency of recombination (references in Chen and Jinks-Robertson 1999).

The latter authors used yeast to study the rates of mitotic and meiotic recombination between pairs of 350-bp substrates varying from 82% to 100% in sequence identity. Single mismatches reduced recombination about 5-fold in mitosis and about 2-fold in meiosis. Mitotic recombination was affected more than meiotic recombination by single or three mismatches, although both were affected, but having four or more mismatches affected mitotic and meiotic recombination about equally, with reductions in recombination of about

21.5-fold for four mismatches, then with increasingly larger reductions for more mismatches, e.g., more than a thousand-fold for 82% sequence identity. The extent of meiotic hDNA formation in a MMR-defective strain was 65% longer than in wild-type. That is consistent with the MMR machinery interfering with the formation or extension of heteroduplex intermediates during recombination. Higher levels of sequence divergence impeded recombination by action of the MMR system and also by an additional MMR-independent process, perhaps by action on the initiation of recombination through the requirements of homology for initiating pairing.

Somewhat similar results were obtained in a very different system by Lukacsovich and Waldman (1999) with correction of herpes simplex virus genes in mouse cells. Interruption of a region of 232 bp of homology by two single nucleotide heterologies 19 bp apart reduced recombination nearly 20-fold, while on their own they only reduced recombination 2.5-fold, so that there were synergistic effects of multiple heterologies. Different pairs of non-adjacent single nucleotide heterologies acting together reduced recombination from 7- to 175-fold. Substrates leading to G-G or C-C mispairs in hDNA gave particularly low rates of recombination. Increased sequence divergence gave shorter gene conversion tracts. The authors suggested that the suppression of recombination between diverged sequences was mediated via processing of a mispaired hDNA intermediate. They also explained the concept of the "minimum efficient processing segment", the minimum length of perfect homology needed for recombination, quoting examples of about 30 base pairs in *E. coli* for the RecBCD pathway, and between 134 and 232 bp for mammalian cells.

Colaiacovo *et al.* (1999) pointed out that repair of a double-strand break by recombination depends on the invasion of a 3'-ended strand into an intact template to initiate DNA synthesis, and that when the invading end is not homologous, the non-homologous sequences must be removed before new DNA synthesis can begin. In yeast, removal of those ends depends on the nucleotide excision repair endonuclease Rad1p/Rad10p, and on the mismatch repair proteins Msh2p/Msh3p. They found that in *rad1* or *msh2* mutants, when both ends of the break have non-homologous ends, repair is reduced about 90-fold compared with a plasmid with perfect ends. If only one end was non-homologous, the reduction was only about five-fold. They deduced that yeast has a less efficient alternative way of removing a non-homologous tail from the second end taking part in gene conversion. They invoked a synthesis-dependent strand annealing mechanism.

By using mutants of mismatch repair genes *MSH2* and *PMS1* in yeast, Vedel and Nicolas (1999) found an involvement of mismatch-repair in meiotic recombination at the *CYS3* locus, as the mutations relieved the gradient of polarity in conversion frequencies (see later) within this locus, which is a hotspot for gene conversion. The frequency of double-strand breaks is about 8% in the *CYS3* promoter. Neither mutations in *CYS3* nor the absence of the mismatch repair functions affected the distribution or frequency of nearby recombination-initiating double-strand breaks. Those breaks were processed in similar ways in wild-type and mismatch-repair mutants. The authors concluded that mismatch repair functions did not control the distribution of gene conversion events at the initiating steps.

One aspect of gene conversion with a bearing on recombination models is whether a particular heterozygous site shows parity or disparity in the direction of gene conversion. For example, disparity in favour of conversion to wild-type (+) over conversion to *mutant* (*m*) would be shown in a + x *m* cross of an eight-ascospored fungus if the number of 6+:2*m* and 5+:3*m* asci significantly exceeded the number of 2+:6*m* and 3+:5*m* asci. It has been claimed many times by yeast workers (e.g., Szostak *et al.* 1983; Nicolas and Petes 1994; Kearney *et al.* 2001) that yeast does not show disparity in conversion direction, and this has been used as evidence for double-strand-break-repair models of recombination, such as that of Szostak *et al.* (1983), and for some more recent models. The evidence summarised by Lamb (1998) demonstrated very clearly that yeast frequently shows significant and extensive conversion disparity. All types of mutations in yeast - base-substitutions, frame-shifts and longer additions and deletions - can show significant 6:2/2:6 and/or 5:3/3:5 (+:*m*) disparity. Surprisingly, there was little correlation between a mutation's molecular nature and its disparity properties, which seem unpredictable. The issue of conversion disparity and its causes, implications and effects were discussed by Lamb (1998) with equations and with details of how gene conversion disparity could change allele ratios in populations, with evolutionary effects. Kearney *et al.* (2001 and references therein) gave information on very high gene conversion frequencies in yeast, on the repair of large unpaired DNA loops, and on conversion disparity. Heterozygous markers near the 5' end of the *HIS4* gene have very high rates of meiotic gene conversion, about 50%, because of a very high frequency of meiosis-specific double-strand breaks forming about 200 bp upstream of the initiating codon for *HIS4*. Heteroduplexes of hybrid DNA initiated there are regularly extended through the gene's coding region, about 2.4 kb. Repair of mismatches in that region gives gene conversion or restoration of parental sequences, while mispair-correction failure gives postmeiotic segregation. The main findings were that: heteroduplexes formed during meiotic recombination could include large (e.g., 5.6 kb) insertions; heteroduplexes could form between alleles that included two different large insertions; the efficient repair of the heterozygous loops required proteins Rad1p, Rad10p, Msh2p and Msh3p, but not several other nucleotide-excision repair enzymes (Rad2p, Rad14p) or mismatch repair proteins (Msh4p, Msh46p, Mlh1p, Pms1p, Mlh2p, Mlh3p); gene conversions involving large insertions usually duplicate rather than delete the insertions, and double-strand breaks within insertions did not stimulate recombination between homologues. The group found that small (26 bp) non-palindromic inserts at position +469 in the coding region of *HIS4* had 26% gene conversion and 4% postmeiotic segregation, with mutations in *RAD1* or *MSH2* increasing the frequency of postmeiotic segregation and decreasing gene conversion. A 1.5 kb insertion in *HIS4* gave 12% conversion, with strong disparity to mutant (one 6:2 tetrad to 11 2:6 tetrads) and no postmeiotic segregation. A 5.6 kb insertion gave similar results. In both cases, the efficiency of heteroduplex formation was reduced by the large heterozygous insertions, since the conversion frequency was reduced. Disparity could be caused by differences in double-strand-break frequencies in the wild-type chromatid relative to breaks in the mutant chromatid, or to differences in the frequencies of excision of the looped or unlooped strands in mismatch repair. The excess of conversions to mutant (the insert) over conversions to wild-type came at least partly from the shorter, unlooped strand being preferentially cut in repair, rather than the looped inserted strand, so the large insertions are duplicated, not deleted (see Kearney *et al.* 2001, Fig. 1). Clikeman *et al.* (2001) gave a good summary of various groups' work on mismatch repair in yeast and in bacteria, and on the well-conserved proteins in common to bacteria and higher organisms. In yeast, strand exchange in meiotic recombination is probably carried out by Rad51p, homologous with *E. coli*'s RecA, both binding to 3' single-stranded tails. Both have DNA-dependent ATPase activity and can pair or transfer complementary DNA strands *in vitro*. In *E. coli*, *mutHLS* is responsible for most mismatch repair involving excision and new synthesis of long DNA tracts which can exceed 1 kb. MutS (yeast homologues include *MSH2*, *MSH3*, *MSH6*) acts in mismatch recognition and MutL (yeast homologues *PMS1*, *MLH1*) joins the MutS protein to proteins involved in later stages of repair. Msh2p in conjunction with Msh6p or Msh3p binds to single-base or loop mismatches, respectively. In *E. coli*, most single-base mismatches and small loops (less than four bases) are efficiently repaired, but C-C and larger loops are not repaired unless repair is triggered by another mismatch. In yeast also, most single-base mismatches except C-C are repaired well, but palindromic loop mismatches forming stable stem-loop structures are not easily repaired unless repair is triggered by another nearby mismatch. Clikeman *et al.*

(2001) found in yeast that mismatch repair was normally very efficient in meiosis and mitosis for small heterologies (single-base differences or insertions of less than 15 bases). The repair of larger loop mismatches in plasmid substrates or coming from replication slippage was inefficient or did not involve Pms1p/Msh2p mismatch repair. During meiotic recombination, heterozygous large insertions converted readily, without postmeiotic segregation. Their results suggested that Rad51p easily incorporated large heterologies into hDNA. They proposed that there was a Msh2-independent large loop-specific mismatch repair system biased towards loop loss. Large heterologies did not influence recombination frequencies, gene conversion tract spectra or rates of chromosome loss in mitosis. They even converted more efficiently than equidistant (from an initiating double-strand break) small heterologies.

For mispairs in hybrid DNA, two types of correction are possible, and no correction leads to postmeiotic segregation. Repair to the genotype of the donor strand (conversion-type repair) gives gene conversion, whereas repair to the genotype of the recipient strand (called restoration-type repair) results in normal Mendelian 4:4 segregation, and the fact that there has been hDNA formed and corrected at that point of heterozygosity may well be missed. Conversion-type repair is easily detected by the non-Mendelian segregation ratios produced in tetrads and octads, showing gene conversion. The existence of restoration-type repair was demonstrated in multiply-marked crosses of *Ascobolus immersus* by Hastings *et al.* (1980). Its existence in yeast was shown in a similar way by Kirkpatrick *et al.* (1998). They found that a mismatch located near the beginning of the *HIS4* gene had less restoration-type repair than one near the middle of the gene.

At various stages of meiosis, there are genes with checkpoint functions, arresting meiosis under certain conditions. Gruschcow *et al.* (1999) studied this in yeast. They found that checkpoint genes *MEC1*, *RAD17* and *RAD24* were required for normal meiotic recombination partner choice. When recombination was blocked in meiosis by mutations in the *recA* homologue *DMC1*, these checkpoint genes prevented the progression of meiosis. Strains with mutations in these three checkpoint genes had increased levels of ectopic recombination.

Using yeast, Thompson and Stahl (1999) found that meiotic recombination checkpoint mutations had elevated levels of unequal sister-chromatid recombination, as if the wild-type checkpoint genes direct recombination events in meiosis to homologues, not to sister chromatids. They stated that mitotic recombination occurs preferentially between sister chromatids, while meiotic events are mainly between homologous non-sister chromatids. Their findings suggested that *DMC1* functions to bias the repair of meiosis-specific double-strand breaks to homologues, not to sister chromatids. *DCM1* codes a meiosis-specific *recA* homologue, while *RAD51* species a ubiquitous *recA* homologue.

The mismatch repair system is thought to scan hybrid DNA and to abort recombination when too many mismatches are found (hDNA rejection; references in Nickoloff *et al.* (1999). The presence of heterozygous markers affects meiotic features such as crossover frequencies, conversion frequencies and conversion tract lengths (see Borts and Haber 1989). In mitotic conversion in yeast, Nickoloff *et al.* (1999) found that nearly all double-strand break repair was by gene conversion, usually involving mismatch repair of heteroduplex DNA. Extra markers increased gene conversion tract lengths. Also with mitotic gene conversion in yeast, Weng and Nickoloff (1998) suggested that mismatch repair on opposite sides of a double-strand break involved distinct repair tracts.

On some models, the resolution of Holliday junctions is affected by mismatch repair. For example, Alani *et al.* (1994) proposed a version of a heteroduplex-rejection model in which well-repairable mismatches cause Holliday junctions to be resolved double-strand-break proximal to the mismatch. On the other hand, Hillers and Stahl (1999) proposed a restoration conversion variant where mismatch repair has no influence on junction resolution, with mismatches further from the double-strand break preferentially undergoing restoration-type repair, unlike those near the break.

Double-strand breaks can be repaired by several mechanism in Eukaryotes. In lower Eukaryotes such as yeast, recombinational repair is the major method, while in higher Eukaryotes such as mammals, non-homologous end joining is the main pathway (Tsutsui *et al.* 2000). Those authors give an account of homologies between repair genes of *S. cerevisiae*, *S. pombe* and humans. According to Yeadon and Catcheside (1998), it has not been established in *Neurospora* whether recombination is initiated by double-strand breaks, although there is no evidence against it. Those authors used the multiple polymorphic differences between the Emerson and Lindegren strains of *N. crassa* to look at the parental origins of DNA sequences in a 6.9 kb region in and around the *his-3* gene, in prototrophic progeny from crosses heterozygous for auxotrophic mutations. Forty-one percent of the conversion tracts were interrupted, not continuous. When the recombination hotspot *cog* was active, conversion appeared to originate at *cog*, and conversion tracts were up to 5.9 kb long. The chromosome bearing cog^L, the dominant allele which gives a high recombination frequency, is nearly always the recipient of information, i.e., is the invaded chromatid. The presence of different alleles at conversion control loci *rec-2* and *cog* affected conversion tract length, whether or not conversion tracts were initiated at *cog*, and which chromosome was more likely to be converted.

Conversion tracts usually extend in both directions from an initiation site (see references in Yeadon *et al.* 2002). Grimm *et al.* (1994) found in *S. pombe* that the frequency of co-conversion of a silent marker with a selected mutant marker decreased exponentially with increasing distance from the mutation, with a minimum average tract length of about 1 kb. Similar co-conversion data at the *rosy* locus of *Drosophila melanogaster* showed an exponential relation between the distance between two sites and the chance that both sites would convert, with an average length of 352 bp for unselected tracts and 706 bp for selected tracts (Hilliker *et al.* 1994).

Yeadon *et al.* (2002), using *Neurospora crassa*, made a deletion of 1.8 kb in the region between cog^L and *his-3*, with replacements of different lengths to get strains varying in length between cog^L and the selected recombination site from 1.7 kb to nearly 6 kb. The frequency of His^+ prototrophs in the progeny of repulsion phase heteroallelic crosses was inversely proportional to the distance between cog^L and *his-3*. As that distance decreased, the frequency of interallelic recombination increased exponentially, as in *Drosophila* and *S. pombe*, indicating that the extension of recombination events might be a stochastic process. Recombination was estimated to be initiated at cog^L in more than 17% of meioses, with most conversion tracts being very short, with few extending to more than 14 kb.

For a detailed consideration and many diagrams of the consequences of different ways of resolving the pair of Holliday junctions (as shown in the present Fig. 2 (v)) and the relations between resolution methods and mismatch repair, see Hillers and Stahl (1999), Stahl and Hillers (2000) and Foss *et al.* (1999). Crossovers with adjacent hDNA are produced by cutting the two junctions in the opposite sense, i.e., the left junction by cutting the outside (non-crossed) strands and the right junction by cutting the inner (crossed) strands, or by cutting the inner strands at the left junction and the outer ones at the right junction. Non-crossovers with regions of hybrid DNA are formed by cutting both junctions in the same sense, either cutting the inner strands at both junctions, or the outer strands at both junctions. Foss *et al.* (1999) also illustrate a non-crossover outcome from cutting one junction followed by sliding the other junction to that position before religation, or by action of topoisomerase.

Foss *et al.* (1999) studies of recombination at the *ARG4* locus in yeast produced the following conclusions, giving a variation of the classic double-strand break repair model. Holliday junction cutting is biased in favour of the strands on which DNA synthesis occurred during Holliday junction formation, with this bias ensuring that resolution usually leads to crossing-over. Cutting only one of the two junctions gives non-crossovers. Repair of mismatches which are poorly repaired and/or are far from the double-strand break site is mainly directed by junction resolution. The bias in resolution of Holliday junctions favours restoration of 4:4 segregation when those mismatches and the directing junction are on the same side of the DSB site. Their studies on *HIS4* in yeast confirmed the predicted influence of this bias in Holliday junction resolution on the conversion gradient (see later under polarity), on the type of mismatch repair and the frequency of aberrant 5:3 segregation, as well as on the relations between mismatch repair and crossing-over.

Malagon and Aguilera (2001) summarised evidence that in mitosis the main mechanism leading to gene conversion is synthesis-dependent strand annealing, at least in *Drosophila*, *Ustilago* and yeast, and that mitotic gene conversion is generally not associated with crossing-over. They discussed the possibility of meiotic and mitotic recombination being by different mechanisms, with mitotic recombination perhaps not requiring the formation and resolution of Holliday junctions. They showed that in yeast mitotic recombination, certain mutations affecting chromatin structure and transcription stimulated recombination between inverted repeats. The effects of various *RAD* loci were examined.

4. RECOMBINATION CONTROLS, INCLUDING HOTSPOTS AND COLDSPOTS

Recombination events at meiosis are not distributed evenly along chromosomes. We have already seen that there is usually at least one crossover per bivalent, with longer chromosomes tending to have more crossovers. Understanding the factors which influence crossover distribution is important in realising how genetic distances may not accurately reflect physical distances, with practical implications for map-based technologies, including gene identification. For individual and sex-specific differences in recombination in humans, see Broman *et al.* (1998). The number of crossovers per chromosome can also affect fertility through its effects on regular chromosome segregation versus non-disjunction in meiosis. The role of crossovers in producing recombinants between loci or within genes has already been mentioned, and many techniques in applied mycology and biotechnology rely on obtaining recombinant progeny, for example, to increase yields of useful metabolites or to get a desired genotype for research experiments.

The elegant pioneering work on crossover frequencies per unit physical length was done by Bridges (1935) and others. They used the cytologically visible bands for known genes on the giant polytene salivary gland chromosomes in *Drosophila melanogaster* as markers for physical distances, and compared these with the corresponding distances on genetic maps based on meiotic crossover frequencies in female flies. The loci were in the same order on the physical and genetic maps, but the relative distances were often quite different on the two types of map. For example, *y* and *pn* are fairly close on the genetic map but far apart on the physical map, indicating few crossovers per physical unit of distance, while *fa* and *ec* are far apart on the genetic map but fairly close on the physical map, indicating a high frequency of crossovers per unit of physical distance. This is illustrated in Rédei (1982).

Crossover frequencies have long been known to be influenced by chromosome aberrations, since the early work on *Drosophila* by Sturtevant and others (see Srb *et al.* 1965). Crossover-suppressors were discovered which only suppressed crossing-over when they were heterozygous, and many were found to be chromosome inversions, or to be associated with such inversions. Crossing-over within a heterozygous inversion often leads to about 50% of the meiotic products being inviable through duplications, deletions, dicentrics or acentrics, and there would be natural selection for alleles which suppressed crossing-over within the region of the inversion. The centromere or its associated heterochromatin may also reduce crossing-over and/or gene conversion locally in various organisms, not just in fungi (e.g., the yeast chromosome III centromere locally represses crossing-over and gene conversion, Lambie and Roeder 1988). Recombination may also be restricted near telomeres.

There are also genes which suppress recombination by crossing-over and/or gene conversion nearby when they are heterozygous, and which are not associated with chromosome aberrations nor with reduced viability. In fungi, these recombination control genes include *conversion control factor 5* in *Ascobolus immersus*, where alleles *A* and *B* give about 3% conversion at the very closely linked target locus, *w9*, when they are heterozygous, compared with about 10.7% when they are homozygous (Lamb and Shabbir 2002). Other heterozygous suppressors include *ss* in *Neurospora crassa* (Catcheside 1981), and in *A. immersus*, *cv* (Girard and Rossignol 1974), *ccf-1* acting on *w62* (Emerson and Yu-Sun 1967) and *ccf-6* acting on *wBHj* (Howell and Lamb 1984). Their mode of action is not fully understood.

The main data on meiotic recombination hotspots and coldspots have been well summarised by Petes (2001), with a heavy emphasis on yeast. In *Saccharomyces cerevisiae* meiotic recombination is initiated by a double-strand break in DNA, catalysed by Spo11p, a topoisomerase-II-related protein and at least 11 other proteins. The exposed ends are digested 5' to 3', leaving 3' tails, which invade a chromatid of the homologous chromosome, forming a heteroduplex as described earlier. Heterozygosity within the heteroduplex results in a mismatch, which may or may not be corrected. The frequency of gene conversion in yeast is typically about 4% to 5%, but ranges from less than 0.5% to an astonishing 72% (Lichten and Goldman 1995).

In yeast, preferred double-strand break sites have been identified associated with particular loci, and tend to be between genes rather than within them (Wu and Lichten 1994). Deletions removing DSB sites usually reduce gene conversion frequencies, as expected. A number of studies in yeast, *S. pombe*, *A. immersus* and *N. crassa* have shown that the hotspot high recombination frequency alleles usually act as recipients (i.e., are on the invaded chromatid) during heteroduplex formation in heterozygotes for the hotspot, as predicted by DSBR models (references in Lamb 1996 a; Petes 2001). Conversion events therefore give a loss of the hotspot allele more often than a gain of it.

There are no specific consensus sequences for DSBs, and in all hotspots studied in detail, DSBs occur in a range of many positions over a distance of 100-500 bp (Petes 2001), seeming to be more position-specific than sequence-specific. Factors increasing recombination include being transcriptionally active (the binding of transcription factors is more important than actually having transcription), and the chromatin region being more sensitive to nucleases and perhaps being restructured in meiosis (references in Petes 2001). In *S. pombe*, *ade6-M26* is a much-studied recombination hotspot, differing from wild-type by only a single base substitution which alters the binding of a particular heteromeric transcription factor required for its hotspot activity, although the increased recombination does not arise from increased transcription (Kon *et al.* 1997). In yeast, hotspots are often in regions of high G - C content but are not associated with repetitive DNA such as in replication origins, tRNA genes or transposable elements. One type of recombination hotspot and transcription insert consists of 12 tandem repeats of (CCGNN), which are poor substrates for nucleosome formation and give hypersensitivity to DNase I. In contrast, 48 tandem repeats of (CCGNN) stimulate gene expression but suppress hotspot activity (references in Petes 2001). Recombination coldspots include centromere regions, telomere regions, and include a 15 kb mating-type region between *mat2* and *mat3* in *S. pombe* (Egel 1984).

There are large differences between organisms in recombination frequencies per unit physical distance. For example, yeast has an average of one centiMorgan per 3 kb, while humans have 1 cM per 1,000 kb. This may in some way be related to the fact that human DNA is compacted 20-fold in meiosis compared to yeast DNA (references in Petes 2001). In the present author's opinion, the discrepancy may just reflect the fact that yeast has 12.1 Mb DNA per haploid genome of 16 chromosomes and humans have 3,200 Mb per genome over 23 chromosomes, with greater compaction being needed to avoid over-long chromosomes for mechanical reasons during meiosis, and only one or a few crossovers being needed per bivalent to ensure regular segregation. Some of the hotspots studied in mammals resemble yeast ones in being associated with transcription-binding sites, regions of DNase I hypersensitivity and possibly with G + C rich repetitive DNA sequences (references in Petes 2001).

The packaging of DNA into nucleosomes reduces its accessibility to sequence-specific DNA-binding proteins, including transcription factors. The post-translational modification of histones such as H3 and H4 affects the openness of chromatin structure and hence access to DNA by proteins such as recombination enzymes and transcription factors. Histone acetylation is frequently associated with transcription activation, and deacetylation with repression of transcription (references in Petes 2001). Petes suggested that hotspots have a chromosomal region with highly modified histones which promote the initial interactions between chromosomes and recombination enzyme complexes, and also that intergenic sequences in the region are unbound by nucleosomes or transcription factors, with the naked DNA being susceptible to Spo11 protein. The efficiency of binding of DSB-initiating mechanisms might be related to the number of chromatin-loosening histone modifications or to a pattern of modification There might be other mechanisms of chromatin remodelling (Petes 2001). He suggests that coldspots might lack the histone modification, or have silencing modifications such as methylation of histones, or just be poor substrates for Spo11p, like poly (A) sequences. Kirkpatrick *et al.* (1999) studied recombination at the yeast *HIS4* locus. DNA sequences upstream of the gene formed a very strong meiotic recombination hotspot which required the transcription activator Rap1p, but the levels of transcription and of recombination were not directly related. Maximal stimulation of recombination by the transcription factor required the transcription activation domain and a DNA-binding domain. The authors suggested that yeast has two types of recombination hotspots, transcription-factor dependent and transcription factor-independent.

The relations between hotspot activity and DNA sequences have been extensively studied by Fox *et al.* (2000) and others, mainly with the *M26* hotspot in *Schizosaccharomyces pombe*. This hotspot comes from a single base substitution with a G to T transversion in the coding region of the *ade6* gene, raising intragenic recombination by up to 15 times, and raising gene conversion about 10 times, with *M26* being preferentially converted to wild-type. The heptamer sequence including that thymine, ATGACGT, is needed for the meiotic hotspot and is associated with binding of a heterodimeric transcription factor Atf1.Pcr1 to *M26*. Sequences (C/T/G) TGACGT also bound that factor and acted as meiotic hotspots if followed by A or C, unlike *M26* which does not depend on surrounding sequences. *M26* and CTGACGTA were sites of micrococcal nuclease hypersensitivity in meiotic chromatin, so perhaps they create an open chromatin structure during meiosis at their sites, facilitating access of recombination enzyme complexes. The *M26* hotspot also stimulates ectopic recombination: see the work of Virgin and Bailey (1998) in the section on ectopic recombination.

5. ECTOPIC RECOMBINATION

All fungi have repetitive DNA sequences, including ribosomal and transfer RNA genes, multigene families, transposable elements and repeats in centromeric and telomeric DNA. Ectopic recombination between DNA in non-homologous positions can occur by crossing-over, when it generates chromosome rearrangements, interferes with meiotic chromosome and chromatid segregation, and can cause inviable or grossly abnormal products. Human translocations, as between chromosomes 21 and 14, giving translocation Down syndrome in 14, 14, 21, 21 individuals (see Lamb 2000), often arise through ectopic recombination which can therefore cause reproductive and hereditary disorders. Ectopic gene conversion has a role in the spread or elimination of mutations in gene families (e.g., see Murti *et al.* 1994).

Ectopic recombination can be of three kinds: (i) intrachromosomal, between two sites on the same chromosome; (ii) interhomologue, between two different sites on homologous chromosomes; (iii) interheterologue, between different sites on non-homologous chromosomes. All three types have been found in mitotic and meiotic cells in yeast (e.g., Goldman and Lichten 1996), where meiotic ectopic recombination frequencies may be only 2-to-17-fold lower than that of allelic recombination (e.g., Jinks-Robertson and Petes 1985). In yeast, interhomologue ectopic recombination is three-to-six-fold more common than inter-sister chromatid recombination (Haber *et al.* 1984). Davis *et al.* (2000) found that meiotic ectopic recombination occurred at roughly equal frequencies among many sites in the yeast genome, suggesting that most loci were equally accessible to homology searching. As an exception, they found that *his3* sequences put into the rDNA locus *RDN1* were poor at recombining with other *his3* sites, because *RDN1::his3* made a poor donor in meiotic ectopic recombination. They suggested that *RDN1* is largely inaccessible to meiotic homology search mechanisms, so there is some variation between loci in participation in ectopic recombination.

Like normal homologous allelic recombination at meiosis, ectopic recombination requires regions of sequence homology and is much more frequent at meiosis than at mitosis. The mechanisms of the two types of recombination have many similarities (see references in Virgin and Bailey 1998). For naturally occurring repeats, ectopic recombination is much less frequent than allelic recombination. For example, meiotic gene conversion between non-allelic tRNA genes in *S. pombe* is 50-200-fold less than for allelic tRNA genes. Similarly, in yeast, meiotic gene conversion between dispersed Ty retrotransposons is 100-fold less frequent than allelic Ty gene conversion. Reciprocal ectopic recombination in yeast and in *S. pombe* between natural repeats, giving translocations or other gross chromosome aberrations, is very rare (references in Virgin and Bailey 1998). With artificial repeats, however, ectopic recombination in yeast varies from equal to allelic recombination to a 20-fold reduction and there may be frequent crossing-over between artificial repeats (e.g., see Goldman and Lichten 1996). Virgin and Bailey (1998) used artificially dispersed copies of *ade6* in *S. pombe* to study hotspot activity in meiotic ectopic recombination. Ectopic recombination was reduced 10-1000-fold relative to allelic recombination and was similar to the low frequencies of ectopic recombination between natural repeats in that organism. The *M26* hotspot increased ectopic recombination in some but not all integration sites, with similar actions in ectopic and allelic recombination. Crossing-over in ectopic recombination was associated with 35-60% of recombination events and was stimulated 12-fold by *M26*, giving chromosome rearrangements. Their results showed a lot of similarity of ectopic and allelic recombination, and showed that hotspots could cause chromosome rearrangements through stimulating ectopic recombination.

Human gene therapy and many biotechnological processes involving transforming cells with DNA could create duplicated sequences in the same or different chromosomes. Those duplicated regions could lead to chromosomal aberrations by ectopic recombination, perhaps at mitosis.

6. POLARITY GRADIENTS IN GENE CONVERSION

Three kinds of polarity have been described: (i) which of two linked allelic sites shows conversion in two-point crosses; (ii) a gradient of conversion frequencies at sites across of locus; (iii) a gradient of relative frequencies of asymmetric and symmetric hDNA across a locus. For an account of each of these, with references, see Lamb (1996 a); much of the work was from *Ascobolus immersus*, *Neurospora crassa*, *Sordaria fimicola* and yeast. Some genes showed polarity from one end only (unipolar), as if hDNA entered the gene from one end only, and some were bipolar, as if hDNA could come from either end. Alani *et al.* (1994) suggested that polarity reflects the frequency of hDNA formation and/or the processing of hDNA by mismatch-repair processes.

Using *A. immersus*, Paquette and Rossignol (1978) used 15 conversion spectrum type C mutants, probably base substitutions, in the *b2* locus in + x *m* crosses. Mutants mapping towards the left end of the locus gave 30% total conversion, while those towards the middle and right gave about 15% conversion. The aberrant 4:4s (from symmetric hDNA) were generally least frequent for mutations on the left, more frequent in the middle, and most frequent on the extreme right of the locus. The authors estimated that hDNA was about 90% asymmetric and 10% symmetric for the left-most group of mutants with a steady decline in asymmetric hDNA for more rightwards groups of mutants, reaching 30% asymmetric and 70% symmetric at the right-hand end, as if hDNA was initiated asymmetrically at the left end of the locus, with an increasing chance of becoming symmetric as it spread to the right.

Most of the early results on polarity could be explained in terms of a set or preferred region of recombination initiation for a given gene, together with a variable length of hDNA formed from it. Sites nearest the initiation point therefore tended to have higher conversion frequencies than sites further away from it as the hDNA had less chance of reaching the further sites. The high meiotic conversion frequency end of the gene was the 5' end in *niaD* and *brlA* in *Aspergillus nidulans*. In yeast it was the 5' (promoter) end for *ARG4* and *HIS4* (conversion frequencies from 17 to 50%, Petes *et al.* 1991), but was the 3' end for *HIS2* (conversion frequencies 5 to 14%). In yeast *ARG4*, there is a DSB site near the high end of the polarity gradient, while *DED81* (conversion frequencies from 4 to 15%) has a U-shaped polarity gradient and a DSB site near each end. References for all these are given in Lamb (1996 a) or Vedel and Nicolas (1999). Polarity in yeast *ARG4* in conversion frequency was 5' (promoter region) to 3', with CFs (conversion frequencies) for four mutations of 9.1, 7.4, 2.8 and 0.4%, in sequence order, with 68% single-strandedness at the 7.4% conversion site and 35% single-strandedness at the 2.8% conversion site, as if the conversion frequencies reflected the chance of the site getting into a single-stranded region which then can form heteroduplex, with sites nearest the double-strand break getting single-stranded and into hDNA most often. If the double-strand breaks are not processed to single-stranded tails, as in the *rad50S* mutant, recombination is blocked, as is a complete synaptonemal complex (Sun *et al.* 1991). Porter *et al.* (1993) concluded that their yeast data on Rap1-stimulated recombination at *BIK1* and *HIS4* fitted the modified Meselson-Radding model better than a double-strand break model, being consistent with single-strand gaps or asymmetrically processed double-strand breaks. The work of Rossignol and Haedens (1980) with the *b2* locus in *A. immersus* showed that asymmetric hDNA and symmetric hDNA were often present in the same gene at the same time, not arising by two independent events nor by different mechanisms. The work of Nicolas *et al.* (1989) with polarity in the *ARG4* locus in yeast showed that the conversion frequency depended largely on a site's position within the gene, rather than on its own properties.

Vedel and Nicolas (1999) looked at meiotic conversion at the *CYS3* locus in yeast. This locus is a hotspot for conversion, with a 5' to 3' gradient of conversion frequencies. Because the conversion gradient was relieved by *msh2* and *pms1* mutations (as it was for yeast *ARG4* and *HIS4*), the authors deduced that mismatch repair was involved in recombination. The frequency and distribution of DSBs, and the processing of DSBs, were unaffected by the absence of mismatch repair. Vedel and Nicolas therefore concluded that mispair repair functions do not control the distribution of meiotic conversion events at the initiating steps.

The Msh2 protein can bind to artificial Holliday junctions (Alani *et al.* 1997), so that is another stage at which mismatch repair proteins could affect recombination.

Detloff *et al.* (1992) found at the *HIS4* locus in yeast that conversion frequencies at the 5' end of the gene were roughly equal for well-repaired and poorly-repaired mismatches, while in the middle or end of the polarity gradient, conversion frequencies were higher for well-repaired mismatches than for poorly-repaired ones. The authors suggested that the level of heteroduplex was similar from one end of *HIS4* to the other end, with the polarity gradient for poorly-repaired mismatches reflecting a change in the ratio of conversion-type to restoration-type repair, relating to the distance of the mismatch to the initiating DSB. Some evidence for that was provided by Kirkpatrick *et al.* (1998). Conversion-type repair was higher for a marker at the 5' end of the *HIS4* gene than for a marker in the middle of the gene, as if the ratio of conversion-type to restoration-type repair was important in generating polarity gradients in gene conversion.

Hillers and Stahl (1999) examined causes of the polarity gradient at the *HIS4* locus in yeast, attempting to distinguish between the heteroduplex rejection model, in which the recognition of mismatches by mismatch repair enzymes limits hDNA flanking a DSB, and the gradient of restoration repair model. Data for one set of well-repaired mismatches failed to show restoration repair but did show a reduction in the length of hDNA, supporting the heteroduplex rejection theory. A different subset of data showed restoration repair, with a relation between Holliday junction repair and mismatch repair.

Foss *et al.* (1999) tried to account for an excess of opposite-sense sense resolution of pairs of double Holliday junctions over same-sense resolution by proposing that each junction has a structural asymmetry biasing which strands are cut, and that the presence of strand ends from cutting junctions stimulates mismatch repair and directs it to occur on the discontinuous strand. They stated that yeast *ARG4* data suggest that Holliday junction cutting is biased towards strands on which DNA synthesis occurred in the formation of the joint molecule, so that junction resolution usually leads to crossovers, and that junction resolution mainly directs repair of mismatches which are poorly repaired and/or far from the DSB site. They also stated that studies at *HIS4* in yeast confirmed the predicted influence of biased junction resolution on conversion gradients and type of mismatch repair, as well as predicted relations between mismatch repair and crossing-over.

7. CONCLUSIONS

Recombination is a very important process in the lives of most fungi, producing new genotypes upon which natural selection can act. Recombination may be achieved at meiosis through independent assortment for non-syntenic loci, and by crossing-over or gene conversion for syntenic loci. It can also be achieved through mitotic recombination, and in some fungi through the parasexual cycle. Crossing-over and gene conversion are fairly well understood in yeast, although much of the evidence comes from artificial constructs and the use of mutants impaired in recombination. Work is progressing on identifying many genes and proteins involved in the mechanisms and controls of recombination in yeast. Molecular studies are less well developed in the filamentous fungi, but those have provided much of the key evidence about recombination, especially from tetrad or octad analysis, where work with ascospore colour markers was particularly productive. For future work, it is important to study recombination in all groups of fungi, not just those that are technically most convenient, as yeast models of recombination may not apply to all groups.

Acknowledgements: I am grateful to Dr Lewis Frost for giving me a love of genetics, especially the intellectual joys of octad analysis in filamentous fungi. I express my appreciation of *Saccharomyces cerevisiae* and related species for their contributions to my enjoyment of life, from my own wines and beers, and commercial ones. At the risk of offending all those not listed, I particularly admire the many contributions made to research in fungal meiotic recombination by Fogel, Holliday, Lissouba, Nicolas, Rossignol, Perkins, Petes, Smith (G R), Stadler, Stahl and Whitehouse.

REFERENCES

Alani E, Lee S, Kane MF, Griffith J, and Kolodner RD (1997). *Saccharomyces cerevisiae* MSH2, a mispaired base recognition protein, also recognises Holliday junctions in DNA. J Mol Biol 265:289-301.

Alani E, Reenan RA, and Kolodner RD (1994). Interaction between mismatch repair and genetic recombination in *Saccharomyces cerevisiae*. Genetics 137:19-39.

Anderson LK, Hooker, KD, and Stack, SM (2001). The distribution of early recombination nodules on zygotene bivalents from plants. Genetics 159:1259-1269.

Anderson, LK, Reeves, A, Webb, LM, and Ashley, T (1999). Distribution of crossing-over on mouse synaptonemal complexes using immunofluorescent localization of MLH1 protein. Genetics 151:1569-1579.

Arbel, T, Shemesh, R and Simchen, G (1999). Frequent meiotic recombination between the ends of truncated chromosome fragments of *Saccharomyces cerevisiae*. Genetics 153:1583-1590.

Barratt, RW, Newmeyer, D, Perkins, DD, and Garnjobst, L (1954). Map construction in *Neurospora crassa*. Adv Genet 6:1-93.

Baudry, E, Kerdelhué, C, Innan, H, and Stephan, W (2001). Species and recombination effects on DNA variability in the tomato genus. Genetics 158:1725-1735.

Borts, RH and Haber, JE (1989). Length and distribution of meiotic gene conversion tracts and crossovers in *Saccharomyces cerevisiae*. Genetics 123:69-80.

Bowring, FJ and Catcheside, DEA (1999). Evidence for negative interference: clustering of crossovers close to the *am* locus in *Neurospora crassa* among *am* recombinants. Genetics 152:965-969.

Bridges, CB (1935). Salivary chromosome maps with a key to the banding of the chromosomes of *Drosophila melanogaster*. J Hered 26:60-64.

Broman, KW, Murray, JC, Sheffield, VC, White, RL, and Weber, J (1998). Comprehensive human genetic maps: individual and sex-specific variation in recombination. Am J Hum Genet 63:861-869.

Broman, KW, Rowe, LB, Churchill, GA, and Paigen, K (2002). Crossover interference in the mouse. Genetics 160:1123-1131.

Catcheside, DEA (1981). Genes in *Neurospora* that suppress recombination when they are heterozygous. Genetics 98:55-76.

Chen, W and Jinks-Robertson, S (1999). The role of mismatch repair machinery in regulating mitotic and meiotic recombination between diverged sequences in yeast. Genetics 151:1299-1313.

Clikeman, JA, Wheeler, SL, and Nickoloff, JA (2001). Efficient incorporation of large (2 kb) heterologies into heteroduplex DNA: *Pms1/Msh2*-dependent and -independent large loop mismatch repair in *Saccharomyces cerevisiae*. Genetics 157:1481-1491.

Colaiacovo, MP, Pâques, F, and Haber, JE (1999). Removal of one nonhomologous DNA end during gene conversion by a *RAD1*- and *MSH2*-independent pathway. Genetics 151:1409-1423.

Cormak, BP and Falkow, S (1999). Efficient homologous and illegitimate recombination in the opportunistic yeast pathogen *Candida glabrata*. Genetics 151:979-987.

Davis, ES, Shafer, BK, and Strathern, JN (2000). The *Saccharomyces cerevisiae RDN1* locus is sequestered from interchromosomal meiotic ectopic recombination in a *SIR2*-dependent manner. Genetics 155:1019-1032.

Detloff, P, White, MA, and Petes, TD (1992). Analysis of a gene conversion gradient at the *HIS4* locus in *Saccharomyces cerevisiae*. Genetics 132:113-123.

Egel, R (1984). Two tightly-linked silent cassettes in the mating-type region of *Schizosaccharomyces pombe*. Curr Genet 8:199-203.

Emerson, S and Yu-Sun, CCC (1967). Gene conversion in the Pasadena strain of *Ascobolus immersus*. Genetics 55:39-47.

Fogel, S, Mortimer, R, and Lusnak, K (1981). Mechanisms of meiotic gene conversion, or "wanderings on a foreign strand." In: JN Strathern, EW Jones and JR Broach, eds. The Molecular Biology of the Yeast Saccharomyces cerevisiae. Cold Spring Harbor, NY: Cold Spring Laboratory Press, pp 289-339.

Foss, EJ and Stahl, FW (1995). A test of a counting model for chiasma interference. Genetics 139:1201-1209.

Foss, HM, Hillers, KJ, and Stahl, FW (1999). The conversion gradient at *HIS4* of *Saccharomyces cerevisiae*. II. A role for mismatch repair directed by biased resolution of the recombinational intermediate. Genetics 153:573-583.

Fox, ME, Yamada, T, Ohta, K, and Smith, GR (2000), A family of cAMP-response-element-related DNA sequences with meiotic recombination hotspot activity in *Schizosaccharomyces pombe*. Genetics 156:59-68.
Gallegos, A, Jacobson, DJ, Raju, NB, Skupski, MP, and Natvig, DO (2000). Suppressed recombination and a pairing anomaly on the mating-type chromosome of *Neurospora tetrasperma*. Genetics 154:623-633.
Gerecke, EE and Zolan, ME (2000). An *mre11* mutant of *Coprinus cinereus* has defects in meiotic chromosome paring, condensation and synapsis. Genetics 154:1125-1139.
Gilbertson, LA and Stahl, FW (1996). A test of the double-strand break model for meiotic recombination in *Saccharomyces cerevisiae*. Genetics 144:27-41.
Girard, J and Rossignol, J-L (1974). The suppression of gene conversion and intragenic crossing over in Ascobolus immersus: evidence for modifiers acting in the heterozygous state. Genetics 76:221-243.
Goldman, AS and Lichten, M (1996). The efficiency of meiotic recombination between dispersed sequences in *Saccharomyces cerevisiae* depends on their chromosomal location. Genetics 144:43-55.
Grimm, C, Bahler, J, and Kohli, J (1994). *M26* recombinational hotspot and physical conversion tract analysis in the *ade6* gene of *Schizosaccharomyces pombe*. Genetics 135:41-51.
Gruschcow, JM, Holzen, TM, Park, KJ, Weinert, T, Lichten, M, and Bishop, DK (1999). *Saccharomyces cerevisiae* checkpoint genes *MEC1*, *RAD17* and *RAD24* are required for normal meiotic recombination partner choice. Genetics 153:607-620.
Guerra, CE and Kaback, DB (1999). The role of centromere alignment in meiosis I segregation of homologous chromosomes in *Saccharomyces cerevisiae*. Genetics 153: 1547-1560.
Haber, JE, Thornburn, PC, and Rogers, D (1984). Meiotic and mitotic behavior of dicentric chromosomes in *Saccharomyces cerevisiae*. Genetics 106:185-205.
Hastings, PJ, Kalogeropoulos, A, and Rossignol, J-L (1980). Restoration to the parental genotype of mismatches formed in recombinant DNA heteroduplex. Curr Genet 2:169-174.
Hey, J (1998). Selfish genes, pleiotropy and the origin of recombination. Genetics 149:2089-2097.
Hillers, KJ and Stahl, FW (1999). The conversion gradient at *HIS4* of *Saccharomyces cerevisiae*. I. Heteroduplex rejection and restoration of Mendelian segregation. Genetics 153:555-572.
Hilliker, AJ, Harauz, G, Reaume, AG, Gray, M, Clark, SH (1994). Meiotic conversion tract length distribution within the *rosy* locus of *Drosophila melanogaster*. Genetics 137:1019-1026.
Holliday, R (1964). A mechanism for gene conversion. Genet Res 5:282-304.
Holm, PB, Rasmussen, SW, Zickler, D, Lu, BC, and Sage, J (1981). Chromosome pairing, recombination nodules and chiasma formation in the basidiomycete *Coprinus cinereus*. Carlsberg Res Commun 46:305-346.
Howell, WM and Lamb, BC (1984). Two locally acting genetic controls of gene conversion, *ccf-5* and *ccf-6*, in *Ascobolus immersus*. Genet Res 43:107-121.
Jiao, K, Bullard, SA, Salem, L, and Malone, RE (1999). Coordination of the initiation of recombination and the reductional division in meiosis in *Saccharomyces cerevisiae*. Genetics 152:117-128.
Jinks-Robertson, S and Petes, TD (1985). High-frequency meiotic gene conversion between repeated genes on non-homologous chromosomes in yeast. Proc Natl Acad Sci USA 82:3340-3344.
Kearney, HM, Kirkpatrick, DT, Gerton, JL, and Petes, TD (2001). Meiotic recombination involving heterozygous large insertions in *Saccharomyces cerevisiae*. Genetics 158:1457-1476.
Kirkpatrick, DT, Dominska, M, and Petes, TD (1998). Conversion-type and restoration-type repair of DNA mismatches formed during meiotic recombination in *Saccharomyces cerevisiae*. Genetics 149:1693-1705.
Kirkpatrick, DT, Fan, Q, and Petes, TD (1999). Maximal stimulation of meiotic recombination by a yeast transcription factor requires the transcription activation domain and a DNA-binding domain. Genetics 152:101-115.
Kitani, Y, Olive, LS, and El-Ani, AS (1962). Genetics of *Sordaria fimicola*. V. Aberrant segregation at the *g* locus. Amer J Bot 49:697-706.
Kon, N, Krawchuk, MD, Warren, BG, Smith, GR, and Wahls, WP (1997). Transcription factor Mts1/Mts2 (Atf1/Pcr1) activates the M26 meiotic recombination hotspot in *Schizosaccharomyces pombe*. Proc Natl Acad Sci USA 94:13765-13770.
Korol, AB and Preygel, IA (1994). Recombination Variation and Evolution. Andover, UK: Chapman & Hall.
Krawchuk, MD, DeVeaux, LC, and Wahls, WP (1999). Meiotic chromosomes dynamics dependent upon the *rec8*$^+$, *rec1D*$^+$ *and rec11*$^+$ genes of the fission yeast *Schizosaccharomyces pombe*. Genetics 153:57-68.
Lamb, BC (1987). Tests of double-strand gap repair as a major source of meiotic gene conversion in fungi. Heredity 59:63-71.
Lamb, BC (1996 a). Ascomycete genetics: the part played by ascus segregation phenomena in our understanding of the mechanisms of recombination. Mycol Res 100:1025-1059.
Lamb, BC (1996 b). The nine-parameter gene conversion model: simpler equations, validity tests, and multiple fits. Genetica 98:65-73.

Lamb, BC (1998). Gene conversion disparity in yeast: its extent, multiple origins, and effects on allele frequencies. Heredity 80:538-552.

Lamb, BC (2000). The Applied Genetics of Plants, Animals, Humans and Fungi. London: Imperial College Press.

Lamb, BC and Saleem, M (2002). Responses to selection for postmeiotic segregation frequencies in *Ascobolus immersus*. Genet Res: in press.

Lamb, BC and Shabbir, G (2002). The control of gene conversion properties and corresponding-site interference: the effects of *conversion control factor 5* on conversion at locus *w9* in *Ascobolus immersus*. Hereditas 137: in press.

Lamb, BC and Wickramaratne, MRT (1973). Corresponding-site interference, synaptinemal complex structure, and 8+:0*m* and 7+:1*m* octads from *wild-type* x *mutant* crosses of *Ascobolus immersus*. Genet Res 22:113-124.

Lamb, TM and Mitchell, AP (2001). Coupling of *Saccharomyces cerevisiae* early meiotic gene expression to DNA replication depends upon *RPD3* and *SIN3*. Genetics 157:545-556.

Lambie, EJ and Roeder, GS (1988). A yeast centromere acts in *cis* to inhibit meiotic gene conversion of adjacent sequences. Cell 52:863-873.

Li, J and Baker, MD (2000). Use of a small palindrome genetic marker to investigate mechanisms of double-strand-break repair in mammalian cells. Genetics 154:1281-1289.

Lichten, M and Goldman, ASH (1995). Meiotic recombination hotspots. Ann Rev Genet 29:423-444.

Lukacsovich, T and Waldman, AS (1999). Suppression of intrachromosomal gene conversion in mammalian cells by small degrees of sequence divergence. Genetics 151:1559-1568.

Malagon, F and Aguilera, A (2001). Yeast *spt6-140* mutation, affecting chromatin and transcription, preferentially increases recombination in which Rad51p-mediated strand exchange is dispensable. Genetics 158:597-611.

Mehta, BJ and Cerda-Olmedo, E (2001). Intersexual partial diploids of phycomyces. Genetics 158:635-641.

Merino, ST, Nelson, MA, Jacobson, DJ, and Natvig, DO (1996). Pseudohomothallism and evolution of the mating-type chromosome in *Neurospora tetrasperma*. Genetics 143:789-799.

Meselson, M and Radding, CM (1975). A general model for genetic recombination. Proc Nat Acad Sci USA 72:358-361.

Molnar, M, Parisi, S, Kakihara, Y, Nojima, H, Yamamoto, A, Hiraoka, Y, Bozsik, A, Sipiczki, M, and Kohli, J (2001). Characterization of *rec7*, an early meiotic recombination gene in *Schizosaccharomyces pombe*. Genetics 157:519-532.

Murti, JR, Bumbulis, M, and Schimenti, JC (1994). Gene conversions between unlinked sequences in the germline of mice. Genetics 137:837-843.

Nickoloff, JA, Sweetser, DB, Clikeman, JA, Khalsa, GJ, and Wheeler, SL (1999). Multiple heterologies increase mitotic double-strand break-induced allelic gene conversion tract lengths in yeast. Genetics 153:665-679.

Nicolas, A and Petes, TD (1994). Polarity of meiotic gene conversion in fungi: contrasting views. Experientia 50:242-252.

Nicolas, A, Treco, D, Schultes, NP, and Szostak, JW (1989). An initiation site for meiotic gene conversion in the yeast *Saccharomyces cerevisiae*. Nature 338:35-39.

Orr-Weaver, TL and Szostak, JW (1985). Fungal recombination. Microbiol Rev 49:33-58.

Pâques, F and Haber, JE (1999). Multiple pathways of recombination induced by double-strand breaks in *Saccharomyces cerevisiae*. Microbiol and Mol Biol Rev 63:349-404.

Paquette, N and Rossignol, J-L (1978). Gene conversion spectrum of 15 mutants giving post-meiotic segregation in the *b2* locus of *Ascobolus immersus*. Mol and Gen Genet 163:313-326.

Peloquin, LS, Boiteux, LS, and Carputo, D (1999). Meiotic mutants in potato: valuable variants. Genetics 153:1493-1499.

Petes, TD (2001). Meiotic recombination hotspots and coldspots. Nature Rev Genet 2:360-369.

Petes, TD, Malone, RE, and Symington, LS (1991). In: JR Broach, EW Jones, and JR Pringle, eds. The Molecular and Cellular Biology of the Yeast *Saccharomyces*. Vol. 1. New York: Cold Spring Harbor Press. pp 407-521.

Porter, SE, White, M and Petes, TD (1993). Genetic evidence that the meiotic recombination hotspot at the *HIS4* locus of *Saccharomyces cerevisiae* does not represent a site for a symmetrically processed double-strand break. Genetics 134:5-19.

Radding, CM (1982). Homologous pairing and strand exchange in genetic recombination. Ann Rev Genet 16:405-437.

Radding, CM, Flory, J, Wu, A, Kahn, R, DasGupta, C, Gonda, D, Bianchi, M, and Tang, SS (1982). Three phases in homologous pairing: polymerization of recA protein on single-stranded DNA, synapsis, and polar strand exchange. Cold Spring Harbor Symposia Quant Biol 47:821-828.

Rédei, GR (1982). Genetics. London: Collier Macmillan.
Rizet, G and Rossignol, J-L (1966). Sur la dimension probable des échanges réciproques au sein d'un locus complex d'*Ascobolus immersus*. Comp Rend Heb des Séances. Acad des Sci 262:1250-1253.
Rockmill, B, Sym, M, Scherthan, H, and Roeder, GS (1995). Roles for two RecA homologs in promoting meiotic chromosome synapsis. Genes Dev 12:2574-2586.
Rossignol, J-L and Haedens, V (1980). Relationship between asymmetrical and symmetrical hybrid DNA formation during meiotic recombination. Curr Genet 1:185-191.
Saleem, M, Lamb, B C, and Nevo, E (2001). Inherited differences in crossing over and gene conversion frequencies between wild strains of *Sordaria fimicola* from 'Evolution Canyon'. Genetics 159:1573-1593.
Sang, H and Whitehouse, HLK (1979). Genetic recombination at the *buff* spore colour locus in *Sordaria brevicollis*, I. Analysis of flanking marker behaviour in crosses between *buff* mutants and wild type. Mol Gen Genet 174:161-178.
Schwacha, A and Kleckner, N (1995). Identification of double Holliday junctions as intermediates in meiotic recombination. Cell 83:783-791.
Skipper, M, (2002). A different exchange-rate mechanism. Nature Rev Genet 3:9.
Srb, A M, Owen, R D, and Edgar, R S (1965) General Genetics. 2nd ed. San Francisco: W. H. Freeman and Company.
Stadler, DR, Towe, AM, and Rossignol, J-L (1970). Intragenic recombination of ascospore color mutants in *Ascobolus* and its relationship to the segregation of outside markers. Genetics 66:429-447.
Stahl, F W and Hillers, K J (2000). Heteroduplex rejection in yeast? Genetics 154:1913-1916.
Sun, H, Treco, D, and Szostak, JW (1991). Extensive 3'-overhanging, single-stranded DNA associated with the meiosis-specific double-strand breaks at the *ARG4* recombination initiation site. Cell 64:1155-1161.
Sym, M and Roeder, GS (1994). Crossover interference is abolished in the absence of a synaptonemal complex protein. Cell 79:283-292.
Szostak, JW, Orr-Weaver, TL, Rothstein, RJ, and Stahl, FW (1983). The double-strand-break repair model. Cell 33:25-35.
Teuscher, F, Brockmann, GA, Rudolph, PE, Swalve, HH, and Guiard, V (2000). Models for chromatid interference with applications to recombination data. Genetics 156:1449-1460.
Thompson, DA and Stahl, FW (1999). Genetic control of recombination partner preference in yeast meiosis: Isolation and characterization of mutants elevated for meiotic unequal sister-chromatid recombination. Genetics 153:621-641.
Tsutsui, Y, Morishita, T, Iwasaki, H, Toh, H, and Shinagawa, H (2000). A recombination repair gene of *Schizosaccharomyces pombe*, *rhp57*, is a functional homolog of the *Saccharomyces cerevisiae RAD57* gene and is phylogenetically related to the human *XRCC3* gene. Genetics 154:1451-1461.
Tung, K-S and Roeder, GS (1998). Meiotic chromosome morphology and behavior in *zip1* mutants of *Saccharomyces cerevisiae*. Genetics 149:817-832.
Vedel, M and Nicolas, A (1999). *CYS3*, a hotspot of meiotic recombination in *Saccharomyces cerevisiae*: Effects of heterozygosity and mismatch repair functions on gene conversion and recombination intermediates. Genetics 151:1245-1259.
Virgin, JB and Bailey, JP (1998). The *M26* hotspot of *Schizosaccharomyces pombe* stimulates meiotic ectopic recombination and chromosomal rearrangements. Genetics 149:1191-1204.
Watters, MK, Randall, TA, Margolin, BS, Selker, EU, Stadler, DR (1999). Action of Repeat-induced point mutations on both strands of a duplex and on tandem duplications of various sizes in *Neurospora*. Genetics 153:705-714.
Weng, Y-S and Nickoloff, JA (1998). Evidence for independent mismatch repair processing on opposite sides of a double-strand break in *Saccharomyces cerevisiae*. Genetics 148:59-70.
Wu, T-C and Lichten, M (1994). Meiosis-induced double-strand break sites determined by yeast chromatin structure. Science 263:515-518.
Yeadon, PJ and Catcheside, DEA (1998). Long, interrupted conversion tracts initiated by *cog* in *Neurospora crassa*. Genetics 148:113-122.
Yeadon, PJ, Koh, LY, Bowring, FJ, Rasmussen, JP and Catcheside, DEA (2002). Recombination at *his-3* in *Neurospora* declines exponentially with distance from the initiator, *cog*. Genetics 162:747-753.
Zhao, H and Speed, TP (1998 a). Statistical analysis of ordered tetrads. Genetics 150:459-472.
Zhao, H and Speed, TP (1998 b). Statistical analysis of half-tetrads. Genetics 150:473-485.
Zickler, D and Kleckner, N (1999). Meiotic chromosomes: integrating structure and function. Ann Rev Genet 33:603-754.
Zwolinski, SA and Lamb, BC (1995). Non-locus-specific polygenes giving responses to selection for gene conversion frequencies in *Ascobolus immersus*. Genetics 140:1277-1287.

Applied Mycology & Biotechnology
An International Series. Volume 3. Fungal Genomics

3

MOLECULAR GENETICS OF CIRCADIAN RHYTHMS IN *NEUROSPORA CRASSA*

Alejandro Correa, Andrew V. Greene, Zachary A. Lewis and Deborah Bell-Pedersen
Program in Biological Clocks, Department of Biology, Texas A&M University, College Station, TX 77845, USA (dpedersen@mail.bio.tamu.edu).

Endogenous circadian clocks provide organisms with the capability to keep in synchrony with the external world. The clock generates a program with a duration of approximately 24 hours, allowing organisms to anticipate cyclic changes in the environment so that they can coordinate biological activities to occur at appropriate times of day. Demonstrations of circadian rhythms are widespread, and in the fungi, the clock has been shown to control daily rhythms in spore development and liberation. Within the fungi, *Neurospora crassa* provides a powerful model organism for investigations into the underlying processes of circadian rhythms. Through genetic and molecular approaches, significant progress has been made in describing the *N. crassa* circadian system. As discussed in this chapter, the analysis of the *N. crassa* clock has provided important details on 1) the autoregulatory transcription-translation feedback loop through which the clock is assembled, 2) how environmental signals are perceived and result in clock resetting, and 3) the identification and function of rhythmically expressed genes regulated by clock output pathways.

1. INTRODUCTION

We are all familiar with biological rhythms that occur with clock-like regularity such as our sleep-wake cycles, the daily leaf movement of some plants, the seasonal formation of flowers, the deep sleep of the bear, and the annual reproductive activities of some animals. Research over the past several decades has also demonstrated that much of the physiology and biochemistry of organisms change rhythmically over the course of a day. Some biological rhythms occur in direct response to daily environmental changes, whereas other rhythms persist in the absence of environmental stimuli. Daily (circadian) rhythms, as well as the annual rhythms, that persist in constant conditions are regulated by an internal rhythm generator, composed of one or more oscillators, called the circadian clock. To date, hundreds of circadian rhythms have been described in eukaryotes and even in some rapidly dividing prokaryotes (Edmunds, 1988; Golden *et al.*, 1998; Lakin-Thomas *et al.*, 1990).

The circadian clock allows organisms to anticipate and cope with rhythmic changes in the environment such as the light-dark cycle (Pittendrigh, 1960; Pittendrigh, 1993). A prime example of anticipatory behavior occurs in plants. The clock provides a way for a plant to anticipate the sun's arrival so that it can initiate production of photosynthetic enzymes just before dawn and then shut them off at sunset (Harmer *et al.*, 2000). Furthermore, experiments with cyanobacteria have shown that a circadian clock with an intrinsic period that closely matches that of the environmental cycle improves the competitive fitness of the cell (Ouyang *et al.*, 1998). Moreover, it has been recently shown that loss of circadian clock function decreases reproductive fitness of males of *Drosophila melanogaster* (Beaver *et al.*, 2002). Thus, while circadian clocks are not essential for survival, these findings demonstrate that the circadian clock provides a clear adaptive advantage to organisms.

Circadian rhythms, by virtue of their pervasiveness and significance in human mental and physical well being, have been the subject of widespread research. Today, hundreds of laboratories worldwide study the circadian system using a variety of methods and model organisms. Despite this variety, the research is unified by the fact that circadian rhythms in all organisms studied to date share the same defining properties, which in turn likely reflects similarities among clock mechanisms and a common ancestry. These properties include the persistence of endogenous rhythms in constant conditions with a period length close to a day, and the ability of the rhythm to be reset or entrained by environmental stimuli (e.g. light and temperature). A rhythm that persists under constant conditions is called a "free running rhythm"; its period is called the "free-running period" (FRP). Entrainment results from perception of external time cues or "*zeitgebers*" by one or more clock components, resulting in shifting the circadian clock to a new and stable phase. The intensity of the *zeitgeber* and the time of day that a *zeitgeber* is applied determines the magnitude and direction of the phase change, respectively. A third defining characteristic of circadian rhythms is compensation of the FRP for changes in an organisms natural environment (Pittendrigh, 1993). For example, when an organism is placed in varying temperatures within its physiological range the FRP stays essentially the same. The period is said to be 'temperature compensated'. To prevent the clock from responding inappropriately when temperatures vary, it makes sense that an accurate clock requires a mechanism to maintain its rate at different ambient temperatures. Thus, even in microbes and poikilotherns, the FRP varies little when the organism is placed in different temperatures. Together, these fundamental properties are key for a biological timing mechanism that responds rapidly to multiple environmental cues to maintain an appropriate phase relationship with environmental cycles. These circadian properties may be intrinsic to a single oscillator, but more likely may be generated by interactions between multiple oscillators.

Organisms that are both genetically facile and have circadian rhythms that can be easily assayed in the laboratory provide key experimental organisms for chronobiology. The best studied model organisms span 'the tree of life' and, those that are amenable to genetic analysis include the cyanobacteria *Synechcococcus*, the filamentous fungus *N. crassa*, the fruit fly *D. melanogaster*, the hamster and mouse, and the higher plant *Arabidopsis thaliana*. The circadian system of higher eukaryotes is complex and may be correlated with anatomical complexity. In mammals, the intact circadian system is the product of cross talk between many integrated oscillatory pathways (Harmer *et al.*, 2000). Despite this level of complexity, the basis of the oscillations in all organisms lie within the cell (Dunlap, 1999; Herzog *et al.*, 1998; Welsh *et al.*, 1995). Thus, microorganisms such as *N. crassa*, provide powerful models for investigating the molecular mechanisms of the circadian clock.

Physiological, genetic, and molecular data generated from studies with model organisms have led to the basic description of a circadian clock consisting of a minimum of three parts: 1) a central oscillator that generates a program with a duration of about 24 hours, 2) input pathways that receive and relay environmental cues to the oscillator, and 3) output pathways from the oscillator which establish the overt rhythms, an important aspect of which is clock control of gene expression (Fig. 1). However, this is certainly an oversimplified view of the clock. For example, the plant photoreceptor phytochrome B is a component of the input pathway and is also an output from the clock (Bognar *et al.*, 1999). Furthermore, the mammalian *dbp* (D-site binding protein) gene is expressed with a circadian rhythm; although considered an output of the clock, its product can feed back on the oscillator and affect the FRP (Lopez-Molina *et al.*, 1997). Therefore, while simplified models have provided the basic framework for genetic and molecular studies of clocks, it is clear that the circadian system involves multiple levels of feedback control that likely contributes to the robustness and accuracy of the system. This high level of complexity is even evident in the fungus *N. crassa.*

Over the last several years, experiments in *N. crassa* have been at the forefront of studies aimed at addressing several fundamental questions regarding the circadian clock, including what are the components of the oscillators and how do they function to keep accurate time, what are the signaling pathways though which the cellular clock is synchonized to the external world, and what genes are regulated by the clock and how is control achieved? We are now well on our way to understanding the molecular bases for period length, mechanisms for light and temperature resetting of the clock, and the regulation of rhythmic gene expression by the clock. More recently, the role of multiple oscillators in the circadian clock system has been investigated. Significantly, studies of circadian rhythmicity in *N. crassa* has provided many of the insights into circadian clock mechanisms in mammalian cells, such as the involvement of PAS domain containing proteins in circadian oscillators, light resetting of the clock, and interlocked feedback loops involving the dual roles of clock components as activators and repressors.

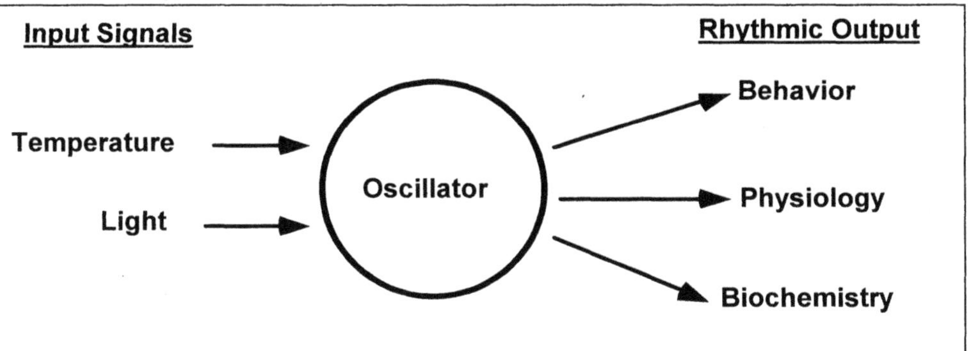

Fig. 1. A simplified view of a circadian clock system.

2. *NEUROSPORA CRASSA* CLOCK

2.1 *N. crassa*, A Model Organsim for Chronobiology

Studies of the filamentous fungus *N. crassa* pioneered the use of microorganisms in genetic analysis and provided the foundations for biochemical genetics (Beadle and Tatum 1941; Davis, 2000; Davis and Perkins, 2002). About 40 years ago, investigations of circadian rhythms in *N. crassa* were initiated. In 1959, Pittendrigh and co-workers demonstrated that *N. crassa* has a rhythm in asexual spore (conidia) development that persists in constant darkness with a period of about 22 h at 25°C. Subsequent to these initial observations, Sargent and coworkers established media conditions and strains for analyzing the circadian rhythm in development that are still widely used today (Sargent *et al.*, 1966). Laboratory strains used for circadian rhythm analysis contain the *band* (*bd*) mutation, which clarifies the developmental rhythm in closed culture tubes. The conidiation rhythm is easily assayed on agar medium contained in long (30 to 40 cm) glass tubes that are bent upwards at a 45° angle at both ends called "race tubes" (Fig. 2). After inoculation, the cultures are incubated for a day in constant light. The growth front is then marked and the race tubes are transferred to constant dark, which synchronizes the cells and sets the clock to dusk or circadian time 12 (CT12)[1]. The mycelial growth front is marked every 24 h under a red safety light, which has no entraining effect on the clock (Sargent et al, 1966). During vegetative growth on the agar surface, some time in the late evening the clock initiates macroconidiation, beginning with the production of aerial hyphae that eventually bud to give rise to the conidiospores. The clock signal for development is turned off sometime later in the day, and the cells that are not determined to differentiate continue to grow down the tube as undifferentiated vegetative hyphae and the cycle renews (see http://www.mrs.umn.edu/~goochv/Circadian/neur.mov for a video of the *Neurospora* circadian rhythm). At the conclusion of an experiment, the center of each conidiation zone (called a band) is marked. The pattern of the conidiation bands can be analyzed later at leisure because they act as a "fossil record" of the state of the clock at the time the conidia were produced. Growth down the tube occurs at a fairly constant rate (~ 3.5 cm/day) at 25°C. Therefore, the period of the rhythm can be calculated from the distance between consecutive bands, and the phase of the rhythm determined from the position of the bands relative to the growth fronts. The center of the band is typically used as the phase reference point. The conidiation rhythm adheres to all of the fundamental properties of a circadian rhythm. The rhythm persists in constant conditions (Pittendrigh *et al.*, 1959), it can be entrained/reset by environmental signals (Francis and Sargent, 1979; Gooch *et al.*, 1994; Nakashima and Feldman, 1980), and the period of the rhythm is temperature compensated (Gardner and Feldman, 1981). It is worthwhile to point out, however, that a circadian oscillator does not generate many of the rhythmic growth patterns that are frequently observed in fungi in the laboratory. For example, many of the rhythms do not persist in constant conditions, or they have periods outside of the circadian range (Loros and Dunlap, 2001).

While the race tube assay of rhythmic development is the most commonly used method to assay the function of the *N. crassa* circadian clock, other methods to monitor rhythmic mRNA and protein accumulation are now widely used. These methods are based on growing mycelia in shaking liquid cultures in which mycelia that is approximately the same developmental age can be harvested at different circadian times (Loros *et al.*, 1989). More recently, the firefly luciferase gene was modified for expression in *N. crassa* and fused to a promoter of a well-characterized circadian clock-controlled gene (*ccg-2*). Transformants containing the chimeric gene display robust, high amplitude rhythms in luciferase activity that can be monitored using automated equipment (Morgan *et al.*, 2003).

The ease of molecular, genetic, and biochemical analyses of *N. crassa*, the readily visible conidiation rhythm, and the recent release of the genome sequence (www-genome.wi.mit.edu/annotation/fungi/*Neurospora*) provide an unparalleled system for revealing the mechanism of the circadian clock at the molecular and biochemical levels. Furthermore, the cell-autonomous clock of *N. crassa* is providing important insights into basic clock mechanisms, which in turn have proven to be applicable to more complex multi-cellular eukaryotes.

---

[1] Circadian time (CT) is used to allow comparison of circadian rhythms in organisms or strains that have different endogenous periods. The period is divided into 24 equal parts, with each part defined as one circadian hour. By convention, CT0 represents subjective dawn, and CT12 represents subjective dusk.

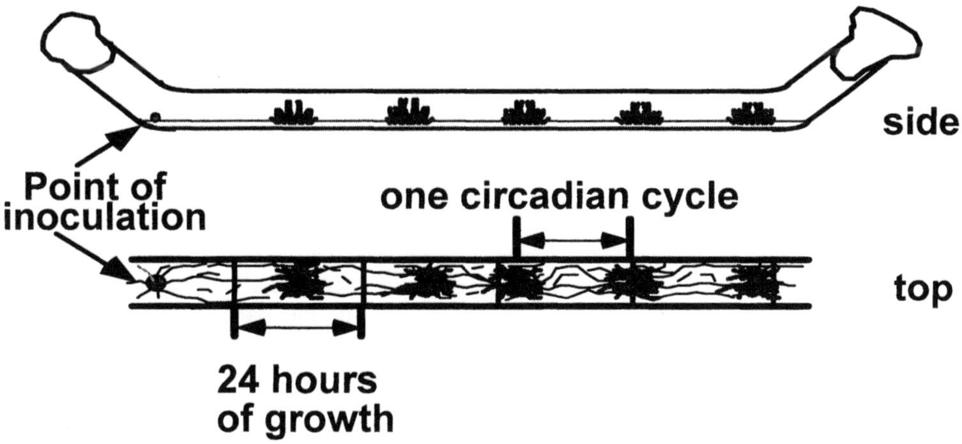

Fig. 2. Diagram of the race tube assay. The race tube assay is used to monitor the phenotypic expression of the *N. crassa* clock. See the text for details of the assay. After a day of growth in constant light, the position of the growth front is marked (solid black line) and the culture is tranferred to constant dark. Following transfer, the growth front is marked every 24 h. The positions of the readily visualized conidial bands (separated by undifferentiated surface mycelia) relative to the marked growth fronts allow determination of period and phase of the rhythm. Figure adapted from Bell-Pedersen (2000).

2.2 The *N. crassa* FRQ-Based Circadian Oscillator

The current model of the *N. crassa* oscillator has the basic signature features of other model systems, including Drosophila and mouse (Dunlap *et al.*, 1999). It consists of a transcription/translation-based feedback loop containing positive and negative elements. The positive elements of the loop activate transcription of the negative elements, while the negative elements feed back to block their own activation through interaction with the positive elements. Moreover, the negative elements regulate the protein levels of the positive elements, forming a positive feedback loop interlocked with the negative loop (Bell-Pedersen, 2000; Young and Kay, 2001).

To identify components involved in *N. crassa* circadian rhythmicity, genetic screens were carried out on race tubes using mutant strains obtained by UV radiation and chemical mutagenesis (Feldman and Atkinson, 1978; Feldman and Hoyle, 1973). Mutants with altered clock parameters, such as period and temperature compensation, were isolated. More than 20 mutant loci were identified, suggesting that many genes and genes products are capable of affecting the normal functioning of the clock (Loros and Dunlap, 2001) (Table 1). One of these loci, the *frequency* (*frq*) locus, was represented by several alleles with periods ranging from 16 to 29 h. Furthermore, some of the mutations in *frq* alter temperature compensation. Moreover, none of the *frq* alleles appear to affect other cellular functions. Together, these data indicated that the *frq* gene encodes a central circadian clock component and significant effort went into describing the role of *frq* in the circadian clock system (Dunlap, 1996; Feldman, 1982; Feldman and Hoyle, 1973; Loros *et al.*, 1986). Cloning of the *frq* gene, the construction of null-alleles (which are viable and typically arrhythmic) and molecular studies established that FRQ is a negative component of a molecular autoregulatory feed back loop required for normal circadian rhythmicity (Fig. 3). It was shown that both *frq* mRNA and FRQ protein levels cycle with a 22 h period in wild type strains grown in constant darkness, and the period of the oscillation is appropriately changed in both short- and long-period mutant strains (Aronson *et al.*, 1994b; Garceau *et al.*, 1997). Negative feedback was demonstrated using a strain containing the *frq* gene under an inducible promoter in which it was found that overexpression of *frq* at an ectopic locus reduced the amount of *frq* transcript from the native promoter. Furthermore, constant high levels of *frq* transcripts are observed at all times of day in strains that lack a functional FRQ protein. (Aronson *et al.*, 1994b). The central role of *frq* in the oscillator was confirmed by showing that rhythmic *frq* mRNA accumulation is essential for overt rhythmicity, and step reduction in the amount of *frq* mRNA sets the clock to a specific and predictable phase (Aronson *et al.*, 1994b).

Activation of *frq* transcription requires the products of the *white collar-1* (*wc-1*) and *wc-2* genes (Crosthwaite *et al.*, 1995). These genes are involved in all known blue light responses in *N. crassa* and are required for *frq* photoinduction and overt circadian rhythms. Mutations in either gene prevent accumulation of *frq* transcripts in the dark, thereby preventing sustained *frq* mRNA and protein cycling. Together, these data show that the WHITE COLLAR (WC) proteins encode positive elements of the feedback loop that activate *frq* transcription (Crosthwaite *et al.*, 1997) (Fig. 3). Consistent with a role for the WC proteins in regulating *frq* transcription in the dark, both proteins are found in the nucleus in dark-grown cultures (Schwerdtfeger and Linden, 2000) and in response to light, bind to the *frq* promoter (Froehlich *et al.*, 2002).

The biochemical function of FRQ is unknown. However, several sequence motifs suggest a role in transcriptional regulation. These include a nuclear localization signal, a coiled-coil motif, and conserved acidic and basic regions. A sequence comparison among different FRQ homologs has shown that a 30 amino acid region (aa 145–174) near the N-terminus of the protein, with a potential to form a coiled-coil structure, is the most conserved region (Lewis *et al.*, 1997). FRQ interacts with itself and with the WC proteins *in vivo*, and deletion of the coiled-coil region abolishes these interactions and results in the loss of the overt rhythm (Cheng *et al.*, 2001a). these data suggest that the formation of the FRQ-FRQ and FRQ-WC complexes is essential for the function of the *N. crassa* clock. Consistent with a role for FRQ in transcriptional regulation, the protein was shown to enter the nucleus; this requires a nuclear localization signal (Luo *et al.*, 1998). Additional motifs in FRQ include a TG/SG repeated amino acid sequence that is also found in *Drosophila* PER (period) protein (McClung *et al.*, 1989), a central component of the fly clock; however, the importance of this motif in FRQ function has not yet been examined. Aternative use of translational initiation sites gives rise to two forms of FRQ at 989 ad 889 amino acids (Garceau *et al.*, 1997). However, no distinct activities have been assigned to the different forms and, no motifs are evident in the first 100 amino acids that might suggest functional differences.

The biochemical function of the positive elements, the WC proteins, is at least partially known. Both genes were cloned by Macino and colleagues (Ballario *et al.*, 1996; Linden and Macino, 1997) and shown to contain functional Zn-finger DNA-binding domains, transcriptional activation domains, PAS domains (two in WC-1 and one in WC-2), and WC-1 contains a LOV domain (Ballario *et al.*, 1998; Talora *et al.*, 1999). Both WC-1 and WC-2 PAS domains are required for the proteins to homodimerize and heterodimerize *in vitro* and *in vivo* (Ballario *et al.*, 1996; Ballario *et al.*, 1998; Cheng *et al.*, 2002). Interestingly, All the identified positive elements of the central oscillator in *N. crassa*, *Drosophila* and mammals are PAS domain-containing transcription factors. The PAS domain was first identified as a common motif among the *Drosophila* clock protein PER, mammalian ARNT (the dimerization partner of the *Drosophila*

Table 1: Rhythm mutants in *Neurospora crassa.*

Allele	Gene	Period (h) at 25°C	Temp-eature	References
wild type		21.5	+	
frequency	*frq[1]*	16.5	+	(Feldman and Hoyle, 1973)
	frq[2]	19.3	+	(Feldman and Hoyle, 1973; Gardner and Feldman, 1980)
	frq[3]	24.0	-	(Feldman and Hoyle, 1973; Gardner and Feldman, 1980)
	frq[7]	29.0	-	(Feldman and Hoyle, 1973; Gardner and Feldman, 1980; Gardner and Feldman, 1981)
	frq[9]	variable	-	(Loros and Feldman, 1986; Loros *et al.*, 1986)
	frq[10]	variable	-	(Aronson *et al.*, 1994a)
	frq[11]	arhythmic at 30°C	-	(Nakashima and Onai, 1996)
period-1	*prd-1*	25.8	-	(Feldman and Atkinson, 1978; Feldman *et al.*, 1979)
period-2	*prd-2*	25.5	+	(Feldman *et al.*, 1979; Gardner and Feldman, 1981)
period-3	*prd-3*	25.1	-	(Feldman *et al.*, 1979; Gardner and Feldman, 1981)
period-4	*prd-4*	18.0	-	(Feldman *et al.*, 1979; Gardner and Feldman, 1981)
period-6	*prd-6*	18.0 at 22°C	+	Morgan and Feldman, 1997
chrono	*chr*	23.5	+	(Feldman *et al.*, 1979; Gardner and Feldman, 1981)
white collar-1	*wc-1 ER53*	arhythmic		(Crosthwaite *et al.*, 1997; Harding and Jr, 1980)
white collar-2	*wc-2 ER33*	arhythmic		(Crosthwaite *et al.*, 1997; Harding and Jr, 1980)
	wc-2 ER24	29.7 at 25°C	-	(Crosthwaite *et al.*, 1997; Harding and Jr, 1980) (Collett *et al.*, 2002)
arginine-13	*arg-13*	19[a]	+	(Lakin-Thomas *et al.*, 1990)
chain elongation	*cel*	variable[b]	-	(Mattern and Brody, 1979)
choline-1	*chol-1*	variable[b]	-	(Lakin-Thomas, 1998)
cytochrome a-5	*cya-5*	19	+	(Brody *et al.*, 1987)
cytochrome b-2	*cyb-2*	18	+	(Brody *et al.*, 1987)
cytochrome b-3	*cyb-3*	20	+	(Lakin-Thomas *et al.*, 1990)
cytochrome-4	*cyt-4*	20	+	(Lakin-Thomas *et al.*, 1990)
cysteine-4	*cys-4*	19[a]	+	(Feldman *et al.*, 1979) (Feldman and Widelitz, 1977)
cysteine-9	*cys-9*	variable [a]	-	(Onai and Nakashima, 1997)
cysteine-12	*cys-12*	19 [a]	+	(Feldman *et al.*, 1979) (Feldman and Widelitz, 1977)
female fertility-1	*ff-1 (glp-3)*	19		(Brody *et al.*, 1987)
maternally inherited	*mi-2, mi-3, mi-5*	18-19	+	(Brody *et al.*, 1987)
oligomycin resistant	*oli[R]*	18-19	+	(Diekmann and Brody, 1980)
phenyl-alanine-1	*phe-1*	19 [a]	+	(Lakin-Thomas *et al.*, 1990)
	rhy-1	arhythmic at 30°C	+	(Chang and Nakashima, 1998)
unknown-18	*un-18*	24.5 at 22°C		(Onai and Nakashima, 1997)

[a] Period length is reduced by increasing starvation for the required supplement; The period length of these strains can be altered by changing the supplementation of the medium; The growth rate was measured at 25°C on standard race tube media containing1X Vogel's salts, 0.3% glucose, 0.5% arginine.

dioxin receptor), and SIM (product of the single- minded gene) (Millar, 1997). The LOV domain is related to the PAS domain and it is associated with light, oxygen and voltage sensing (Christie *et al.*, 1999). The WC-1 LOV motif is similar to the LOV domain of the *Arabidopsis thaliana* blue-light photoreceptor NPH1 that has been shown to bind flavin (Christie *et al.*, 1999). It has recently been shown that WC-1 uses a flavin adenine dinuclotide molecule (FAD) as a cofactor, and binds the *frq* promoter after exposure to light (Froehlich *et al.*, 2002; He *et al.*, 2002). These data suggest that WC-1/FAD is a blue-light photoreceptor in *N. crassa* that mediates light input to the circadian clock.

A model of the *N. crassa* clock has been proposed based on these three central clock components (Fig. 3). At dawn, both *frq* mRNA and protein levels are low, but the amount of *frq* transcript is on the rise (Aronson *et al.*, 1994b). WC-1 and WC-2 dimerize through their PAS domains and activate *frq* transcription. About 4-5 hours later, *frq* mRNA levels reach their peak (CT4) and the long and short forms of FRQ protein accumulate. FRQ protein levels peak around CT8, indicating a post-transcriptional mechanism exists to delay FRQ protein accumulation (Garceau *et al.*, 1997). At this time, *frq* transcript levels begin to decrease. As soon as FRQ protein is synthesized it enters the nucleus (Luo *et al.*, 1998). The two forms of FRQ protein form homodimeric complexes that negatively regulate *frq* mRNA levels (Cheng *et al.*, 2001a). Negative feedback occurs by the interaction of FRQ with WC-1 and WC-2 complexes, interfering with the ability of WC-1/WC-2 complexes to activate *frq* transcription in the dark (Denault *et al.*, 2001; Merrow *et al.*, 2001). For the rest of the day, and into the early evening, FRQ remains at sufficient levels in the nucleus to keep *frq* turned off. FRQ also positively regulates levels of both WC-1 and WC-2 proteins by an unknown post-transcriptional mechanism. Therefore, FRQ serves two roles in the feed back mechanism, interlocking the repression of its own transcription with the up-regulation of the levels of the WC proteins (Cheng *et al.*, 2001b; Cheng *et al.*, 2002; Lee *et al.*, 2000; Merrow *et al.*, 2001). *wc-1* and *wc-2* mRNA levels and WC-2 protein do not show significant cycling. However, the levels of WC-1 protein cycle with a low amplitude, peaking 180° out of phase with FRQ at CT 18. This cycling requires FRQ protein; in FRQ-null strains the level of WC-1 protein is very low and in a long period mutant (*frq*^{7}), WC-1 cycles with a long period (Lee *et al.*, 2000). On the other hand, WC-2 is abundantly expressed and is always in excess of WC-1 and FRQ. To complete the feedback loop, FRQ must be removed from the nucleus so that the positive elements can reestablish the cycle by activating *frq* transcription.

The turnover of FRQ is facilitated by phosphorylation. FRQ protein is progressively phosphorylated over time, and when it is fully phosphorylated it is degraded (Garceau *et al.*, 1997; Liu *et al.*, 2000). This process takes approximately 14 hours; thus, phosphorylation seems to be a major player in the delay of the feed back cycle that takes 22 h in a wild type strain to be completed. To date, three kinases have been shown to phosphorylate FRQ protein both in vitro and in vivo, calcium/calmodulin-dependent kinase (CAMK-1), casein kinase I (CKI), and casein kinase II (CKII). However, these kinases do not fully account for the extensive phosphorylation of FRQ that results in its degradation (Gorl *et al.*, 2001; Yang *et al.*, 2002; Yang *et al.*, 2001). By the middle of the night, most of the FRQ protein has been degraded and the heterodimeric complexes of the WC proteins are now able to bind to the promoter region of frq to activate its transcription (Froehlich *et al.*, 2002; Loros and Dunlap, 2001). frq mRNA levels start to rise and will peak 10 hours later at about CT 4, completing the cycle.

2.3 Input Pahways to the *N. crassa* Circadian Ocillator

The natural light and temperature cycles entrain the endogenous circadian clock of *N. crassa* to local time. Synchronization with the environment allows the fungus to predict and prepare for environmental changes and to coordinate and partition activities to the appropriate times of day. It has been shown that a light pulse given in the late night to early morning advances the conidiation cycle to midday, a light pulse given in the late day to early night delays the cycle, whereas a light pulse given in the middle of the day does not cause an appreciable change in the phase of the conidiation rhythm (Crosthwaite *et al.*, 1995; Pittendrigh, 1993). Temperature changes have also been shown to reset the *N. crassa* clock. Typically, ambient temperature increases at sunrise and decreases at dusk. As might be expected, a temperature step up resets the clock to dawn and a temperature step down resets the clock to dusk (Francis and Sargent, 1979; Gooch *et al.*, 1994).

Blue light regulates several developmental and morphological processes in *N. crassa*, including the induction of the synthesis of carotenoids in mycelia, the formation of asexual spores, and the resetting of the circadian clock. The regulation of these processes occurs primarily at the level of gene expression, and several blue light-regulated genes have been identified (Lewis *et al.*, 2002; Linden *et al.*, 1997; Linden *et al.*, 1999). It has been shown that the blue light photoreceptor WC-1, and the WC-2 protein have primary roles in the blue light signal transduction pathway. *wc-1* and *wc-2* mutants are blind to all of the blue light regulated processes in *N. crassa* (Ballario and Macino, 1997; Degli-Innocenti and Russo, 1984; Harding and Shropshire Jr, 1980). WC complexes bind to consensus GATA elements within the promoters of blue light-regulated genes in *N. crassa* (Carattoli *et al.*, 1994), and the *wc-1* and *wc-2* genes are themselves induced by light (Ballario *et al.*, 1996; Linden and Macino, 1997). This induction results in a transient increase in WC-1 protein levels, but little or no change in the levels of WC-2 (Schwerdtfeger and Linden, 2000; Talora *et al.*, 1999).

In *N. crassa*, like in mammals, light resetting of the clock occurs by rapidly inducing the transcript levels of central oscillator components (Albrecht *et al.*, 1997; Crosthwaite *et al.*, 1995; Shearman *et al.*, 1997; Shigeyoshi *et al.*, 1997). A direct correlation was found between the light-induced levels of *frq* transcript and the magnitude and phase of the shifts in the conidiation rhythm. Light acts to rapidly induce the levels of *frq* transcripts (within 5 min) setting the clock to subjective day, the time when *frq* mRNA levels normally peak in constant darkness. For example, a light pulse given in the late night (when *frq* levels are low) rapidly causes mRNA levels to reach their typical midday levels, resulting in a phase advance of the conidiation cycle (Crosthwaite *et al.*, 1995). The light response of *frq* requires the products of the *wc-1* and *wc-2* genes (Crosthwaite *et al.*, 1997; Collett *et al.*, 2002).
It has recently been suggested that the WC proteins function differently to regulate light responses of clock-associated genes and other photoinducible genes (Collett *et al.*, 2002; Lewis *et al.*, 2002; Merrow *et al.*, 2001). Examination of light induction in several *wc-2* alleles, including those resulting in amino acid substitutions within the Zn-finger domain, revealed that *frq* is photoinducible in these strains, while a gene involved in carotenoid biosynthesis (*al-3*) is not (Collett *et al.*, 2002). These distinctions may be reflected in the interaction of the WC proteins with other clock or light signaling factors, or in the state of modification of the proteins. In addition to increasing levels of WC-1, light also results in phosphorylation of both WC proteins (Schwerdtfeger and Linden, 2000). The light-dependent phosphorylation of WC-1 is transient, whereas phosphorylation of WC-2 is stable in constant light. Transient phosphorylation of WC-1 correlates with the transient induction of some light-regulated genes; however, some light-induced genes are expressed for a long time, corresponding to the length of time that WC-2 levels are high after light induction (Linden *et al.*, 1997; Schwerdtfeger and Linden, 2000; Schwerdtfeger and Linden, 2001). Moreover, microarray analysis was used to show that increasing the levels of WC-1 protein in dark grown cultures is not sufficient to activate all light-responsive genes, and many of the genes induced by overexpression of WC-1 are rhythmically expressed. These data support the notion that WC-1 can mediate both light and circadian responses, with an increase in WC-1 levels affecting circadian clock-responsive gene regulation and other features of WC-1, possibly its phosphorylation, affecting light-responsive gene regulation (Lewis *et al.*, 2002).

Recently, VIVID (VVD) which encodes a novel member of the PAS/LOV protein superfamily was found to be involved in regulating light adaptation responses in *N. crassa* (Heintzen *et al.*, 2001; Shrode *et al.*, 2001). The *vvd* gene itself is rapidly light induced, and is

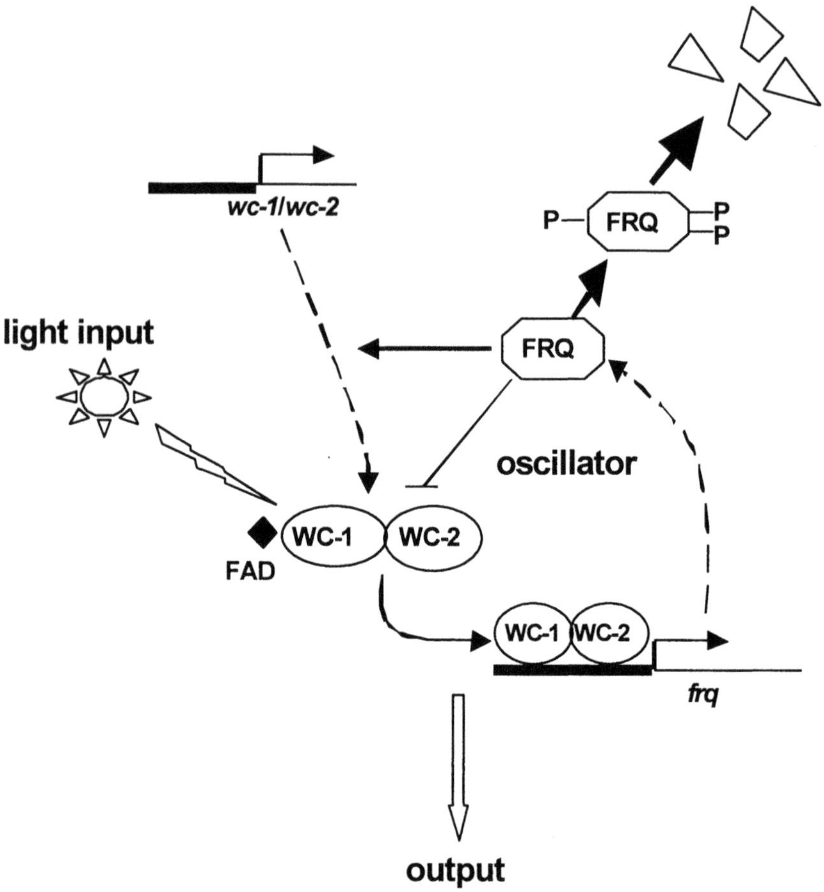

Fig. 3. Model of the transcription/translation-based feedback loop of the *N. crassa* FRQ-based oscillator. WC-1 and WC-2 form a heterodimer to activate the transcription of *frq*. FRQ proteins interact with the WC-1/WC-2 complex to inhibit their transcriptional activation (negative feedback loop). FRQ also positively regulates the levels of both WC-1 and WC-2 (positive feedback loop). The phosphorylation of FRQ promotes its degradation. WC-1 functions as a blue-photoreceptor that signals light information to the oscillator. It is unknown how this oscillator regulates output (see the text for a detailed description of the model).

clock-controlled (Heintzen *et al.*, 2001). Mutation of *vvd* severely dampens the ability of the clock to modulate the light response, a process termed gating. It has been suggested that VVD interacts with the WCC transiently, affecting both input to and output from the clock, but is itself not required for circadian rhythmicity.

Unlike the light input pathway to the clock, the temperature limits of rhythmicity and temperature resetting of the *Neurospora* clock appear to be controlled at the post-transcriptional level. The ratio of the long and short forms of FRQ is dependent on the growth temperature (Liu *et al.*, 1998), and temperature resetting depends on the levels of FRQ in the cell. Approaching 30°C, the total level of FRQ is high and translational initiation at the first ATG (ATG#1) is favored resulting in a higher level of long FRQ. Approaching 18°C, the overall levels of FRQ decrease and initiation at the third ATG (ATG#3) is favored resulting in more short FRQ. At 28°C the amount of total FRQ at the lowest point in the cycle is higher than the highest point in the cycle at 21°C, although the mRNA oscillates with similar levels at both temperatures (Fig. 4). Therefore, a given FRQ level at different temperatures reflects different circadian phases. When cells are shifted from 21°C to 28°C, the overall levels at which FRQ cycles is raised, resulting in the lowest level of FRQ at 28°C being higher than any level at 21°C and phase shifts the clock to dawn. Alternatively, when the temperature is changed from high to low, the clock is reset to the time corresponding to the high point in the new cycle, near dusk (Liu *et al.*, 1998).

2.4 Output Pathways in *N. crassa*

Circadian clocks time many of the daily functions of an organism. Thus, the most biologically relevant property of a circadian oscillator is its ability to direct cellular and organismal activities to occur at the appropriate times of day. The diversity of biological processes regulated by the clock in organisms is vast, ranging from rhythms in the levels of proteins involved in intermediary metabolism to cognitive behavior (Edmunds, 1988). The study of the flow of information from an oscillator to target output genes or proteins serves to 1) identify components of the cell that are regulated by the clock in order to understand the role of rhythms in the life of the organism, and 2) to provide a means to study clock signaling mechanisms by tracing the regulatory pathway(s) from a clock-controlled gene to an oscillator component. A complete understanding of the circadian system therefore requires a detailed description of how circadian oscillators signal time information to regulate diverse output pathways, and work in *N. crassa* is at the forefront of such analyses.

In *N. crassa*, circadian rhythms in the production of CO_2, lipid metabolism, a number of enzymatic activities and heat shock responses have been described (Lakin-Thomas, 1998; Lakin-Thomas and Brody, 2000; Ramsdale, 1999). However, the best-characterized circadian output is the rhythmic formation of conidiospores. Conidial development begins with the vegetative hyphae growing away from the growth medium (Springer, 1993). After several hours of apical aerial growth, the aerial hyphae switch to a budding form of growth that is defined by a series morphological stages distinguished by the diameter of the constrictions between the incipient conidia. At 4 h after induction of conidiation the constrictions are subtle; these proconidial chains are called minor constriction chains. As budding continues, the interconidial constrictions become more pronounced and around 8 h after conidiation is induced major constrictions are observed. The formation of major constriction chains signals the commitment to the formation of conidia. Around 12 h after induction, crosswalls are evident between proconidia of the major constriction chains. Conidial separation takes place about 16 h after the initial developmental switch. Conidiation in *N. crassa* can be induced by several environmental signals, including desiccation, blue light, carbon starvation and

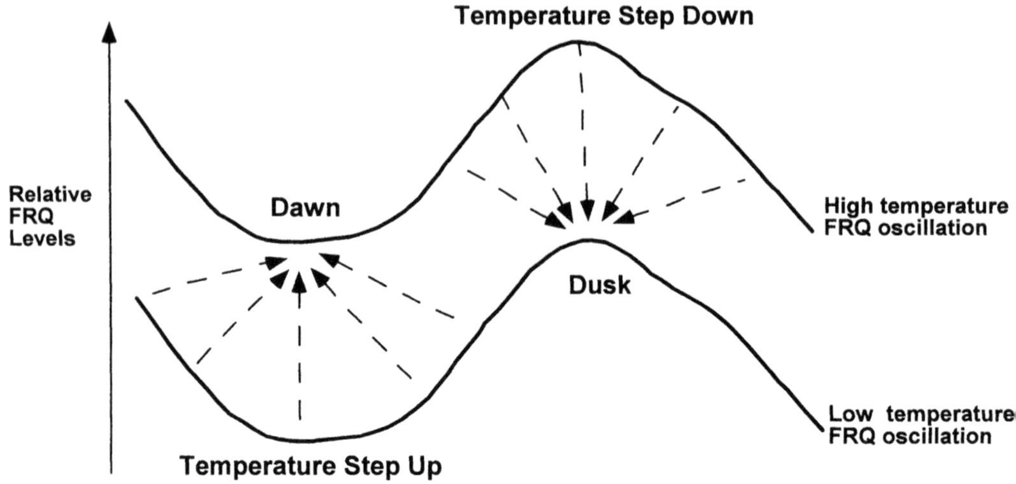

Fig. 4. Temperature resetting of the FRQ-based oscillator. Diagram representing how temperature resets the FRQ-based oscillator. FRQ protein cycles at lower levels at low temperature (bottom curve) and at higher levels at high temperature (upper curve). When the cultures are raised from low to high temperature, the clock is reset to the time corresponding to the low point in the new cycle (arrows pointing up), near dawn. When the temperature is changed from high to low, the clock is reset to the time corresponding to the high point in the new cycle (arrows pointing down), near dusk. This figure was adapted from Dunlap (1999).

nitrogen starvation. However, the only endogenous signal known to induce conidiation is provided by the circadian clock.

To begin to characterize circadian output pathways at the molecular level in *N. crassa,* genes that are rhythmically expressed (i.e. controlled by the clock) were isolated. The term clock-controlled genes (*ccg*s) was used to describe them (Loros *et al.*, 1989). Several *ccg*s were identified by directed approaches (Bell-Pedersen *et al.*, 1996b; Loros *et al.*, 1989; Zhu *et al.*, 2001), and expression of several additional genes have been shown to be rhythmic with circadian periods (Arpaia *et al.*, 1995; Lauter and Yanofsky, 1993) (Table 2). Verification of clock regulation for most of the genes was achieved by demonstrating that the period of the *ccg* mRNA abundance rhythm equals the period of the strain examined. Specifically, in the long period *frq*^{7} background, which has an endogenous period of 29 h, the period of the peak in levels of *ccg* mRNAs approaches 29 h and eventually cycles 180° out of phase with the wild type strain (Bell-Pedersen *et al.*, 1992; Bell-Pedersen *et al.*, 1996b; Loros *et al.*, 1989). In all cases examined, the clock functions normally in strains containing inactivated copies of the *ccg*s, demonstrating that they are part of an output pathway and are not involved in oscillator function (Bell-Pedersen *et al.*, 1992; Lindgren, 1994; Shinohara *et al.*, 1998; Shinohara *et al.*, 2002).

The most highly characterized *N. crassa* ccg is the *eas(ccg-2)* gene. The *eas(ccg-2)* locus was originally identified through mutation, which resulted in easily-wettable (eas) conidiospores (Bell-Pedersen, 2000). The *eas(ccg-2)* gene was independently cloned on the basis of daily rhythms in abundance of the transcript as *ccg-2* (Loros *et al.*, 1989), and as a blue-light-inducible gene, *bli-7* (Sommer *et al.*, 1989). The abundantly expressed *eas(ccg-2)* gene encodes a member of a class of low molecular weight, cysteine rich hydrophobic secreted proteins called hydrophobins (Bell-Pedersen *et al.*, 1992; Lauter *et al.*, 1992). The hydrophobins coat the outer cell wall of fungi and maintain the cell-surface hydrophobicity essential for air dispersal of the mature conidiospores. *eas(ccg-2)* is not only regulated by the circadian clock, it is also induced by the same environmental signals that trigger conidiospore development (Arpaia *et al.*, 1993; Bell-Pedersen *et al.*, 1996a; Lauter *et al.*, 1992). Developmental induction of *eas(ccg-2)* occurs about 1 h after the initiation of conidiation. Similarly, the morning-specific *ccg-1* gene is regulated by the circadian clock and is induced by developmental cues (Lindgren, 1994). Developmental induction of *ccg-1* occurs 1 - 2 h after the initiation of conidiation. Inactivation of *ccg-1* has no obvious affect on conidiation, and *ccg-1*-null strains do not display any discernable phenotypes. In addition, the CCG-1 protein does not share homology to other known proteins. Thus, the function of CCG-1 remains a mystery.

The complexity of the output pathways is suggested by the finding that some clock-controlled genes that are involved in the conidiation pathway are regulated independently of the developmental cascade, whereas others require their upstream developmental regulator for normal rhythmicity. Specifically, both *ccg-1* and *eas(ccg-2)* transcripts peak at the same time of the day, yet are regulated differently in the developmental pathway (Correa and Bell-Pedersen, 2002). High-level developmental induction of the clock-controlled genes *eas (ccg-2)* and *ccg-1* requires the developmental regulatory proteins FLUFFY (FL) and ACON-2, respectively, and normal developmental induction of *fl* mRNA expression requires ACON-2. The circadian clock was shown to regulate rhythmic *fl* gene expression and *fl* rhythmicity requires ACON-2. However, clock regulation of *eas (ccg-2)* is normal in a *fl* mutant strain and *ccg-1* expression is rhythmic in an *acon-2* mutant strain. Together, these data point to the endogenous clock and the environment following separate pathways to regulate conidiation-specific gene expression.

In the initial screens for rhythmically expressed genes in *N. crassa*, only a few times of day were compared and the screens were not saturating. Thus, the *ccg*s represent only a small sampling of clock-regulated genes in *N. crassa*. Current experiments using DNA microarrays representing about 1/7 of *N. crassa* genes revealed that about 20% of the genes represented on the array are rhythmically expressed (AC, ZAL, and DBP, unpublished data), reflecting the importance of the clock in the life of the fungus. The ccgs were found to peak in expression at all phases of the circadian cycle and the functions of the associated proteins involve a wide range of cellular processes, including cell signaling, development, metabolism, and stress responses. Together, these experiments are underscoring the complexity of the outputs and have paved the way to a better description of the role of the clock in the biology of *N. crassa*.

The least understood aspect of the circadian timing system in any organism is how an oscillator signals time information to control the ccgs. One mechanism by which some *ccg*s are predicted to be rhythmically controlled is directly though transcription factors that are known to be components of the oscillator. These immediate clock outputs may in turn regulate downstream outputs in a complex web of events. Direct control of outputs was recently demonstrated in mice. The positive PAS-containing CLOCK/BMAL heterodimers were found to activate transcription of the rhythmically expressed arginine vasopressin gene (Jin *et al.*, 1999). In addition, CLOCK was shown to directly regulate circadian expression of the transcription factor DBP (Ripperger *et al.*, 2000). It is not yet known if the positive elements (WC-1 and WC-2), and/or the negative element (FRQ) of the *Neurospora* FRQ-based oscillator directly regulate rhythmicity of any of the output genes. Furthermore, several transcription factors and signalling components were found to be clock regulated

Table 2: Summary of *Neurospora* clock-controlled genes.

Gene	Average Peak[1]	Identity[2]	Level[3]	Light	References
ccg-1	CT3	unknown	+	+	(Loros *et al.*, 1989; McNally and Free, 1988)
eas (ccg-2)	CT22	hydrophobin	+	+	(Loros *et al.*, 1989; Bell-Pedersen, 1992; Lauter *et al.*, 1992)
ccg-4	CT5	pheromone	+	+	(Bell-Pedersen *et al.*, 1996b; Bobrowicz *et al.*, 2002)
ccg-6	CT19	unknown	+	+	(Bell-Pedersen *et al.*, 1996b)
ccg-7	CT21	GAPDH	-	-	(Bell-Pedersen *et al.*, 1996b; Shinohara *et al.*, 1998)
ccg-8	CT20	unknown	-	-	(Bell-Pedersen *et al.*, 1996b)
ccg-9	CT19	trehelose synthase	+	+	(Shinohara *et al.*, 2002;)
cmt (ccg-12)	CT18	CuMT	-	-	(Bell-Pedersen *et al.*, 1996b; Munger *et al.*, 1987)
ccg-13	CT0	unknown	ND	-	(Zhu *et al.*, 2001)
ccg-14	CT0	unknown	ND	-	(Zhu *et al.*, 2001)
ccg-15	CT4	unknown	ND	-	(Zhu *et al.*, 2001)
lyz	CT2	lysozyme	ND	-	(Zhu *et al.*, 2001)
al-3 [4]	CT10	GGPPS	+	+	(Arpaia *et al.*, 1995)
con-6	ZT20	unknown	+	+	(Lauter and Yanofsky, 1993)
con-10	ZT20	unknown	+	+	(Lauter and Yanofsky, 1993)
vvd	CT3	light repressor	ND	+	(Shrode *et al.*,, 2001; Heintzen *et al.*, 2001)
bli-3	CT3	unknonwn	ND	+	(Eberle and Russo, 1994)
fl	CT3	developmental regulator	+	ND	(Correa and Bell-Pedersen, 2002)
mfa	CT1	pheromone	ND	ND	(Bobrowicz *et al.*, 2002)
frq	CT3	clock component	ND	+	(Aronson *et al.*, 1994b; Crosthwaite *et al.*, 1995)

[1]The peak in message accumulation can vary between experiments and culture conditions. The *con-6* and *con-10* mRNAs peak about 20 hours after a light pulse, representing zeitgeber time (ZT) 20 (Lauter and Yanofsky, 1993); [2]Abbreviations are as follows: GAPDH, glyceraldehyde 3-phosphate dehydrogenase; CuMT, copper metallothionein; GGPPS, geranylgeranyl pyrophosphate synthase; [3]Developmental and light regulation of the ccg's. A + indicates increased transcription following developmental induction and light treatment, a - indicates no effect, and ND means no data; [4]Only the *al-3*c transcript has been shown to be rhythmic (Arpaia *et al.*, 1995).

using microarrays (AC and DBP, unpublished data), and these factors provide good test candidates for critical output components in future experiments.

In any event, the identification of ccgs has provided the tools needed to begin to trace the signaling pathway from a rhythmically expressed gene back to the oscillator. This is accomplished through the identification of clock control regulatory elements (CCREs) in the promoters of the ccgs and determining the trans-acting factors that bind and control them. Most of the progress in this area has been with the *eas(ccg-2)* gene. Nuclear run-on experiments demonstrated that *eas(ccg-2)* is transcriptionally regulated by the circadian clock (Loros and Dunlap, 1991), implicating the involvement of *cis*-acting regulatory elements mediating temporal control. Subsequent dissection of the *eas(ccg-2)* promoter localized a positive-activating clock element (ACE) to within a 45 bp fragment, found to be distinct from other light and developmental elements regulating its expression (Bell-Pedersen *et al.*, 1996a). Using an unregulated promoter/reporter system, it was shown that the ACE element is sufficient to confer high amplitude rhythmicity on the reporter gene. Using a labeled 68-bp *ccg-2* probe containing the ACE, factors present in nuclear extracts from light-grown *Neurospora* were found to interact specifically with these sequences. Examination of the binding factors at different times in the circadian day in either frq^{+} (22 h period) or frq^{7} (29 h period) strains revealed that the amount of binding and the mobility of the complexes changes with time. These data suggest that the amount or activity of the factors, modification of the factors, or the addition of accessory factors, is rhythmic and is consistent with these proteins having a role in clock control of the *eas(ccg-2)* gene. Experiments are in progress to determine the identity of the factors (ZAL and DBP, unpublished data).

In several systems it has been demonstrated that output pathways feed back on the central oscillator (Cassone *et al.*, 1993; Gwinner *et al.*, 1997; Herzog and Block, 1999). Mutations in known *Neurospora ccgs*; however, have not been shown to affect the period of the rhythm. Even mutations that abolish conidiation at early stages do not abolish aerial hyphae formation (Correa and Bell-Pedersen, 2002; Martens and Sargent, 1974), although to date there are no mutations in genes that are known to specifically abolish aerial hyphae formation. However, feedback from an output to the FRQ-based oscillator in *Neurospora* has now been suggested (Ramsdale and Lakin-Thomas, 2000). Diacylglycerol (DAG) levels are rhythmic, and DAG levels are high in a *chol-1* mutant strain that has a long, non-circadian period of 60 h on minimal media lacking choline, suggesting that a correlation might exist between DAG levels and period. The addition of membrane-permeable DAG and inhibitors of DAG kinase further lengthened the period in this strain, hinting that DAG may feedback on the time-keeping mechanism to lengthen the period.

3. COMPLEXITY OF THE *NEUROSPORA* CIRCADIAN SYSTEM

Under most growth conditions, sustained conidiation rhythms are lost in the absence of the FRQ protein. However, under certain media and temperature conditions, FRQ-deficient strains display a conidiation rhythm that ranges between 12 and 30 h (Aronson *et al.*, 1994a; Loros and Feldman, 1986). To explain this residual rhythmicity, the presence of additional oscillators in the *Neurospora* cell has been suggested (Dunlap, 1998; Merrow *et al.*, 1999); however, the exact nature of the putative additional oscillator(s) has not been established.

One hypothesis was that if the residual rhythmicity in *frq*-less strains results from low amplitude, uncompensated, or damped oscillations, perhaps an entraining cycle could bestow an amplifying effect on the rhythm. Indeed, null mutants of *frq* were found to entrained by temperature cycles (Merrow *et al.*, 1999). These data suggested that the entrainment has allowed a cryptic, temperature-entrainable oscillator to be uncovered in the absence of the *frq*-based feedback loop (Iwasaki and Dunlap, 2000; Merrow *et al.*, 1999). Further support for multiple oscillators comes from double mutant studies of *chol-1* or *cel* and *frq* or *wc* nulls. The double mutant strains are arhythmic with full supplementation, but display a long period rhythm on media where the period lengthening effects of the *cel* or *chol-1* mutation are observed (Lakin-Thomas, 1998; Lakin-Thomas and Brody, 1985; Lakin-Thomas and Brody, 2000). With appropriate supplementation, the *cel* and *chol-1* mutations can cause a robust long-period conidiation rhythm (albeit outside of the circadian range) in *frq*-null (or *wc*-null) strains with the same period as the *cel* and *chol-1* single mutants. These data provide additional evidence for the existence of a second oscillator, and further suggests a linkage to cellular metabolism (Lakin-Thomas and Brody, 2000). The two oscillators are likely to be coupled since the period of the system is affected by the *frq* allele. For example, the short period frq^{1} allele shortens the long period observed in the *chol-1* or *cel* backgrounds (Lakin-Thomas, 1998). However, when FRQ is absent, the rhythms lose some circadian characteristics, including light entrainment and compensation for changes in temperature and metabolic state.

In summary, these data indicate that similar to circadian clocks of more complex eukaryotes, the circadian system of *N. crassa* is comprised of a population of oscillatory systems. However, the lack of molecular data sill holds the connection between the FRQ oscillator and the rest of the cell a mystery. In particular, all of what we know about the independent role of the other oscillator(s) is derived from the ability, though mutation and genetically engineered strains, to manipulate or to eliminate altogether the FRQ feedback loop. Until the other oscillators can be similarly manipulated, we are constrained to modeling and phenomenology. Thus, one goal now is to identify components of the other oscillator(s) and we may already have some clues. Genetic data indicates a possible role of the *prd-6* gene in coupling of the FRQ-based oscillator to a temperature-dependent metabolic oscillator (Morgan *et al.*, 2000). Mutations in *prd-6* have an increased range of temperature compensation, suppress the temperature compensation defects of other mutations, and are resistant to some media conditions previously shown to affect period. Furthermore, while the period of most of the ccg rhythms identified in microarray experiments were found to be dependent on the well-characterized FRQ-based oscillator, several ccgs had a wild-type 22-h rhythm in the 29-h period *frq*^{7} strain (AC, ZAL, and DBP, unpublished data). These ccgs accumulate mRNA rhythmically in a *frq*-null allele, further supporting the existence of a FRQ-independent circadian oscillator in *N. crassa* cells and providing molecular tools for identifying components of the novel oscillator(s).

4. CONCLUSIONS

Solving the mechanisms of the circadian clock has become an important goal, mainly because of the ubiquity of clocks and their role in many organisms' lives, including humans. The past few years have seen significant advances in our understanding of the mechanisms of circadian rhythmicity, with the molecular genetic analysis of clocks in *Neurospora* continuing to provide major contributions to the story. Genes that are critical to clock function have been characterized with regard to their roles in generating rhythms, and the molecular mechanisms for entrainment are beginning to be understood. Together, these studies have allowed the formulation of plausible models for the circadian clock. However, there are still several aspects of the clock that we do not understand. For instance, we still know very little about the role of the FRQ-less oscillators in circadian timing, and how these oscillators are coupled to the FRQ-based oscillator. It is possible that some of the genes that were identified in the original genetic screens for clock mutants may participate in these other oscillators. Cloning and characterization of these genes is required in order to determine their potential role in the clock. In addition, we have very little understanding in any system of how circadian rhythms are temperature compensated. The completed genome sequence, coupled with the use of microarray technology to identify ccgs has led to a more detailed description of the processes that are regulated by the clock, and has led to the identification of genes that are regulated by FRQ-independent oscillators. In the future, this technology will allow investigators to fully examine the effects of mutations in oscillator components on rhythmically expressed genes in order to develop detailed maps of the output pathways. While it is clear that different organisms use their clocks to regulate different biological processes, an understanding of the entire clock system in *N. crassa* will continue to be a tractable model for providing fundamental insights into the workings of the clocks in more complex systems.

Acknowledgements. We thank the members of our laboratory for sharing their ideas and unpublished data. Studies in the D. B.-P.'s lab are supported by NIH R01 GM58529-01 and P01 NS39546.

REFERENCES

Albrecht U, Sun ZS, Eichele G and Lee CC (1997). A differential response of two putative mammalian circadian regulators, mper1 and mper2, to light. Cell 91: 1055-1064.

Aronson BD, Johnson KA and Dunlap JC (1994a). Circadian clock locus frequency: protein encoded by a single open reading frame defines period length and temperature compensation. Proc Natl Acad Sci U S A 91: 7683-7687.

Aronson BD, Johnson KA, Loros JJ and Dunlap JC (1994b). Negative feedback defining a circadian clock: autoregulation of the clock gene frequency. Science 263: 1578-1584.

Arpaia G, Carattoli A and Macino, G (1995). Light and development regulate the expression of the albino-3 gene in *Neurospora crassa.* Dev Biol 170: 626-635.

Arpaia G, Loros JJ, Dunlap JC, Morelli G and Macino G (1993). The interplay of light and the circadian clock. Independent dual regulation of clock-controlled gene ccg-2(eas). Plant Physiol 102: 1299-1305.

Ballario P and Macino, G (1997) White collar proteins: PASsing the light signal in *Neurospora crassa.* Trends Microbiol 5: 458-462.

Ballario P, Talora C, Galli D, Linden H and Macino G (1998). Roles in dimerization and blue light photoresponse of the PAS and LOV domains of *Neurospora crassa* white collar proteins. Mol Microbiol 29: 719-729.

Ballario P, Vittorioso P, Magrelli A, Talora C, Cabibbo A and Macino G (1996). White collar-1, a central regulator of blue light responses in *Neurospora,* is a zinc finger protein. Embo J 15: 1650-1657.

Beadle G W and Tatum E L (1941). Genetic control of biochemical reactions in *Neurospora.* Proc Natl Acad Sci USA 27: 499-506.

Beaver LM, Gvakharia BO, Vollintine TS, Hege DM, Stanewsky R and Giebultowicz JM (2002). Loss of circadian clock function decreases reproductive fitness in males of Drosophila melanogaster. Proc Natl Acad Sci U S A 99: 2134-2139.

Bell-Pedersen D (2000). Understanding circadian rhythmicity in *Neurospora crassa*: from behavior to genes and back again. Fungal Genet Biol 29: 1-18.

Bell-Pedersen D, Dunlap JC and Loros JJ (1992). The *Neurospora* circadian clock-controlled gene, ccg-2, is allelic to eas and encodes a fungal hydrophobin required for formation of the conidial rodlet layer. Genes Dev 6: 2382-2394.

Bell-Pedersen D, Dunlap JC and Loros JJ (1996a). Distinct cis-acting elements mediate clock, light, and developmental regulation of the *Neurospora crassa* eas (ccg-2) gene. Mol Cell Biol 16: 513-521.

Bell-Pedersen D, Shinohara ML, Loros JJ and Dunlap JC (1996b). Circadian clock-controlled genes isolated from *Neurospora crassa* are late night- to early morning-specific. Proc Natl Acad Sci U S A 93: 13096-13101.

Bobrowicz P, Pawlak R, Correa A, Bell-Pedersen D and Ebbole DJ (2002). The *Neurospora crassa* pheromone precursor genes are regulated by the mating type locus and the circadian clock. Mol Microbiol 45: 795-804.

Bognar LK, Hall A, Adam E, Thain SC, Nagy F and Millar AJ (1999). The circadian clock controls the expression pattern of the circadian input photoreceptor, phytochrome B. Proc Natl Acad Sci U S A 96: 14652-14657.

Brody S, MacKensie L and Chuman L (1987). Circadian rhythms in *Neurospora crassa*: The effects of mitochondrial mutations and inhibitors. Genetics 116: S30.

Carattoli A, Cogoni C, Morelli G and Macino, G (1994). Molecular characterization of upstream regulatory sequences controlling the photoinduced expression of the albino-3 gene of *Neurospora crassa.* Mol Microbiol 13: 787-795.

Cassone VM, Warren WS, Brooks DS and Lu J (1993). Melatonin, the pineal gland, and circadian rhythms. J Biol Rhythms 8: S73-81.

Chang B and Nakashima H (1998). Isolation of temperature sensitive rhythm mutants in *Neurospora crassa.* Genes Genetic Systems 73: 71-73.

Cheng P, Yang Y, Gardner KH and Liu Y (2002). PAS domain-mediated WC-1/WC-2 interaction is essential for maintaining the steady-state level of WC-1 and the function of both proteins in circadian clock and light responses of *Neurospora.* Mol Cell Biol 22: 517-524.

Cheng P, Yang Y, Heintzen C and Liu Y (2001a). Coiled-coil domain-mediated FRQ-FRQ interaction is essential for its circadian clock function in *Neurospora.* Embo J 20: 101-108.

Cheng P, Yang Y and Liu Y (2001b). Interlocked feedback loops contribute to the robustness of the *Neurospora* circadian clock. Proc. Natl. Acad. Sci. USA 98: 7048-7413.

Christie JM, Salomon M, Nozue K, Wada M and Briggs WR (1999) LOV (light, oxygen, or voltage) domains of the blue-light photoreceptor phototropin (nph1): binding sites for the chromophore flavin mononucleotide. Proc Natl Acad Sci U S A 96: 8779-8783.

Collett MA, Garceau N, Dunlap JC and Loros JJ. (2002). Light and clock expression of the *Neurospora* clock gene frequency is differentially driven by but dependent on WHITE COLLAR-2. Genetics 160: 149-158.
Correa A and Bell-Pedersen D (2002). Distinct signaling pathways from the circadian clock participate in regulation of rhythmic conidiospore development in *Neurospora crassa*. Eukaryot Cell 1: 273-280.
Crosthwaite SK, Dunlap JC and Loros JJ (1997). *Neurospora* wc-1 and wc-2: transcription, photoresponses, and the origins of circadian rhythmicity. Science 276: 763-769.
Crosthwaite SK, Loros JJ and Dunlap JC (1995). Light-induced resetting of a circadian clock is mediated by a rapid increase in frequency transcript. Cell 81: 1003-1012.
Davis R H (2000) *Neurospora* : contributions of a model organism (Oxford University Press, New York).
Davis R H and Perkins D D (2002). Timeline: *Neurospora*: a model of model microbes. Nat Rev Genet 3: 397-403.
Degli-Innocenti F and Russo VE (1984) Isolation of new white collar mutants of *Neurospora crassa* and studies on their behavior in the blue light-induced formation of protoperithecia. J Bacteriol 159: 757-761.
Denault DL, Loros JJ and Dunlap JC (2001). WC-2 mediates WC-1-FRQ interaction within the PAS protein-linked circadian feedback loop of *Neurospora*. Embo J 20: 109-117.
Diekmann C and Brody S (1980). Circadian rhythms in *Neurospora crassa*: oligomycin-resistant mutations affect periodicity. Science 207: 896-898.
Dunlap JC (1998). Circadian rhythms. An end in the beginning. Science 280: 1548-1549.
Dunlap JC (1996). Genetics and molecular analysis of circadian rhythms. Annu Rev Genet 30: 579-601.
Dunlap JC (1999). Molecular bases for circadian clocks. Cell 96: 271-290.
Dunlap JC, Loros JJ, Liu Y and Crosthwaite SK (1999). Eukaryotic circadian systems: cycles in common. Genes Cells 4: 1-10.
Eberle J and Russo VE (1994). *Neurospora crassa* blue light-inducible gene bli-3. Biochem Mol Biol Int 34: 737-744.
Edmunds LN (1988). Cellular and molecular bases of biological clocks. Springer-Verlag, New York.
Feldman JF (1982) Genetic apporaches to circadian clocks. Annual Reviews Plant Physiology 33: 583-608.
Feldman JF and Atkinson CA (1978). Genetic and physiological characteristics of a slow-growing circadian clock mutant of *Neurospora crassa*. Genetics 88: 255-265.
Feldman JF, Gardner GF and Dennison RA (1979). Genetic analysis of the circadian clock of *Neurospora*. In Suda, M. (ed.) Biological Rhythms and their Central Mechanism. Elsevier, Amsterdam, pp. 57-66.
Feldman JF and Hoyle MN (1973). Isolation of circadian clock mutants of *Neurospora crassa*. Genetics 75: 605-613.
Feldman JF and Widelitz R (1977). Manipulation of circadian periodicity in cysteine auxotrophs of *Neurospora crassa*. American Society of Microbiology, Abstract, 158.
Francis CD and Sargent ML (1979). Effects of temperature perturbations on circadian conidiation in *Neurospora*. Plant Physiology 64: 1000-1004.
Froehlich AC, Liu Y, Loros JJ and Dunlap JC (2002). White Collar-1, a circadian blue light photoreceptor, binding to the frequency promoter. Science 297: 815-819.
Garceau NY, Liu Y, Loros JJ and Dunlap JC (1997). Alternative initiation of translation and time-specific phosphorylation yield multiple forms of the essential clock protein FREQUENCY. Cell 89: 469-476.
Gardner GF and Feldman JF (1980).The frq locus in *Neurospora crassa*: a key element in circadian clock organization. Genetics 96: 877-886.
Gardner GF and Feldman JF (1981). Temperature compensation of circadian period length mutants of *Neurospora crassa*. Plant Physiology 68: 1244-1248.
Golden SS, Johnson CH and Kondo T. (1998). The cyanobacterial circadian system: a clock apart. Curr Opin Microbiol 1: 669-673.
Gooch VD, Wehseler RA and Gross CG (1994). Temperature effects on the resetting of the phase of the *Neurospora* circadian rhythm. J Biol Rhythms 9: 83-94.
Gorl M, Merrow M, Huttner B, Johnson J, Roenneberg T and Brunner M (2001). A PEST-like element in FREQUENCY determines the length of the circadian period in *Neurospora crassa*. Embo J 20: 7074-7084.
Gwinner E, Hau M and Heigl S (1997). Melatonin: generation and modulation of avian circadian rhythms. Brain Res Bull 44: 439-444.
Harding RW and Shropshire W Jr. (1980). Photocontrol of carotenoid biosynthesis. Annual Reviews Plant Physiology 31: 217-238.
Harmer SL, Hogenesch JB, Straume M, Chang HS, Han B, Zhu T, Wang X, Kreps JA and Kay SA (2000). Orchestrated transcription of key pathways in Arabidopsis by the circadian clock. Science 290: 2110-2113.
He Q, Cheng P, Yang Y, Wang L, Gardner KH and Liu Y (2002). White collar-1, a DNA binding transcription factor and a light sensor. Science 297: 840-843.
Heintzen C, Loros JJ and Dunlap JC (2001). The PAS protein VIVID defines a clock-associated feedback loop that represses light input, modulates gating, and regulates clock resetting. Cell 104: 453-464.

Herzog ED and Block GD (1999) Keeping an eye on retinal clocks. Chronobiol Int 16: 229-247.
Herzog ED, Takahashi JS and Block GD (1998). Clock controls circadian period in isolated suprachiasmatic nucleus neurons. Nat Neurosci 1: 708-713.
Iwasaki H and Dunlap JC (2000). Microbial circadian oscillatory systems in *Neurospora* and Synechococcus: models for cellular clocks. Curr Opin Microbiol 3: 189-196.
Jin X, Shearman LP, Weaver DR, Zylka MJ, de Vries GJ and Reppert SM (1999). A molecular mechanism regulating rhythmic output from the suprachiasmatic circadian clock. Cell 96: 57-68.
Lakin-Thomas PL (1998). Choline depletion, frq mutations, and temperature compensation of the circadian rhythm in *Neurospora crassa.* J Biol Rhythms 13: 268-277.
Lakin-Thomas PL and Brody S (1985). Circadian rhythms in *Neurospora crassa*: interactions between clock mutations. Genetics 109: 49-66.
Lakin-Thomas PL and Brody S (2000). Circadian rhythms in *Neurospora crassa*: lipid deficiencies restore robust rhythmicity to null frequency and white-collar mutants. Proc Natl Acad Sci U S A 97: 256-261.
Lakin-Thomas PL, Cote GG and Brody S (1990). Circadian rhythms in *Neurospora crassa*: biochemistry and genetics. Crit Rev Microbiol 17: 365-416.
Lauter FR, Russo VE and Yanofsky C (1992). Developmental and light regulation of eas, the structural gene for the rodlet protein of *Neurospora.* Genes Dev 6: 2373-2381.
Lauter FR and Yanofsky C (1993). Day/night and circadian rhythm control of con gene expression in *Neurospora.* Proc Natl Acad Sci U S A 90: 8249-8253.
Lee K, Loros JJ and Dunlap JC (2000). Interconnected feedback loops in the *Neurospora* circadian system. Science 289: 107-110.
Lewis MT, Morgan LW and Feldman JF (1997). Analysis of frequency (frq) clock gene homologs: evidence for a helix- turn-helix transcription factor. Mol Gen Genet 253: 401-414.
Lewis ZA, Correa A, Schwerdtfeger C, Link KL, Xie X, Gomer RH, Thomas T, Ebbole DJ and Bell-Pedersen D (2002). Overexpression of White Collar-1 (WC-1) activates circadian clock- associated genes, but is not sufficient to induce most light-regulated gene expression in *Neurospora crassa.* Mol Microbiol 45: 917-931.
Linden H, Ballario P, Arpaia G and Macino G (1999) Seeing the light: news in *Neurospora* blue light signal transduction. Adv Genet 41: 35-54.
Linden H, Ballario P and Macino G (1997). Blue light regulation in *Neurospora crassa.* Fungal Genet Biol 22: 141-150.
Linden H and Macino G (1997). White collar 2, a partner in blue-light signal transduction, controlling expression of light-regulated genes in *Neurospora crassa.* Embo J 16: 98-109.
Lindgren KM (1994). Characterization of ccg-1, a clock-controlled gene of *Neurospora crassa.* Biochemistry. Dartmouth Medical School, Hanover.
Liu Y, Loros J and Dunlap JC (2000). Phosphorylation of the *Neurospora* clock protein FREQUENCY determines its degradation rate and strongly influences the period length of the circadian clock. Proc Natl Acad Sci U S A 97: 234-239.
Liu Y, Merrow M, Loros JJ and Dunlap JC (1998). How temperature changes reset a circadian oscillator. Science 281: 825-829.
Lopez-Molina L, Conquet F, Dubois-Dauphin M and Schibler U (1997). The DBP gene is expressed according to a circadian rhythm in the suprachiasmatic nucleus and influences circadian behavior. Embo J 16: 6762-6771.
Loros JJ, Denome SA and Dunlap JC (1989). Molecular cloning of genes under control of the circadian clock in *Neurospora.* Science 243: 385-388.
Loros JJ and Dunlap JC (1991). *Neurospora crassa* clock-controlled genes are regulated at the level of transcription. Mol Cell Biol 11: 558-563.
Loros JJ and Dunlap JC (2001). Genetic and molecular analysis of circadian rhythms in *Neurospora.* Annu Rev Physiol 63: 757-794.
Loros JJ and Feldman JF (1986). Loss of temperature compensation of circadian period length in the frq- 9 mutant of *Neurospora crassa.* J Biol Rhythms 1: 187-198.
Loros JJ, Richman A and Feldman JF (1986). A recessive circadian clock mutation at the frq locus of *Neurospora crassa.* Genetics 114: 1095-1110.
Luo C, Loros JJ and Dunlap JC (1998). Nuclear localization is required for function of the essential clock protein FRQ. Embo J 17: 1228-1235.
Martens CL and Sargent ML (1974). Circadian rhythms of nucleic acid metabolism in *Neurospora crassa.* J Bacteriol 117: 1210-1215.
Mattern D and Brody S (1979). Circadian rhythms in *Neurospora crassa*: effects of saturated fatty acids. J Bacteriol 139: 977-983.

McClung CR, Fox BA and Dunlap JC.(1989). The *Neurospora* clock gene frequency shares a sequence element with the Drosophila clock gene period. Nature 339: 558-562.

McNally MT and Free SJ (1988). Isolation and characterization of a *Neurospora* glucose-repressible gene. Curr Genet 14: 545-551.

Merrow M, Brunner M and Roenneberg T (1999). Assignment of circadian function for the *Neurospora* clock gene frequency. Nature 399: 584-586.

Merrow M, Franchi L, Dragovic Z, Gorl M, Johnson J, Brunner M, Macino G and Roenneberg T (2001). Circadian regulation of the light input pathway in *Neurospora crassa*. Embo J 20: 307-315.

Millar AJ (1997) Circadian rhythms: PASsing time. Curr Biol 7: R474-476.

Morgan LW, Feldman JF and Bell-Pedersen D (2001). Genetic interactions between clock mutations in *Neurospora crassa*: Can they help us to understand complexity? PhilosTrans R Soc Lond B Biol Sci 356: 1717-1724.

Morgan LW, Greene AV and Bell-Pedersen D (2003). Circadian and light-induced expression of luciferase in *Neurospora crassa*. Fung Genet Biol, in press.

Munger K, Germann UA and Lerch K (1987). The *Neurospora crassa* metallothionein gene. Regulation of expression and chromosomal location. J Biol Chem 262: 7363-7367.

Nakashima H and Feldman JF (1980). Temperature-sensitivity of light-induced phase shifting of the circadian clock of *Neurospora*. Photochemisty and Photobiology 32: 247-251.

Nakashima H and Onai K (1996). The circadian conidiation rhythm in *Neurospora crassa*. Seminars in Cell and Developmental Biology 7: 765-774.

Onai K and Nakashima H (1997). Mutation of the cys-9 gene, which encodes thioredoxin reductase, affects the circadian conidiation rhythm in *Neurospora crassa*. Genetics 146: 101-110.

Ouyang Y, Andersson CR, Kondo T, Golden SS and Johnson CH (1998) Resonating circadian clocks enhance fitness in cyanobacteria. Proc Natl Acad Sci USA 95: 4475-4480.

Pittendrigh CS (1960). Circadian rhythms and the circadian organization of living things. Cold Spring Harbor Symp Quant Biol 25: 159-184.

Pittendrigh CS (1993). Temporal organization: reflections of a Darwinian clock-watcher. Annu Rev Physiol 55: 16-54.

Pittendrigh CS, Bruce BG, Rosensweig NS and Rubin ML (1959). Growth patterns in *Neurospora*. Nature 184: 169-170.

Ramsdale M (1999). Circadian rhythms in filamentous fungi. In Gow NAR. Robson GD and Gadd GM (eds.), The Fungal Colony. Cambridge University Press, Cambridge, pp. 75-107.

Ramsdale M and Lakin-Thomas PL (2000). sn-1,2-diacylglycerol levels in the fungus *Neurospora crassa* display circadian rhythmicity. J Biol Chem 275: 27541-27550.

Ripperger JA, Shearman LP, Reppert SM and Schibler U (2000). CLOCK, an essential pacemaker component, controls expression of the circadian transcription factor DBP. Genes Dev 14: 679-689.

Sargent ML, Briggs WR and Woodward DO (1966). Circadian nature of a rhythm expressed by an invertaseless strain of *Neurospora crassa*. Plant Physiol 41: 1343-1349.

Schwerdtfeger C and Linden H (2000). Localization and light-dependent phosphorylation of white collar 1 and 2, the two central components of blue light signaling in *Neurospora crassa*. Eur J Biochem 267: 414-422.

Schwerdtfeger C and Linden H (2001). Blue light adaptation and desensitization of light signal transduction in *Neurospora crassa*. Mol Microbiol 39: 1080-1087.

Shearman LP, Zylka MJ, Weaver DR, Kolakowski LF, Jr. and Reppert SM (1997). Two period homologs: circadian expression and photic regulation in the suprachiasmatic nuclei. Neuron 19: 1261-1269.

Shigeyoshi Y, Taguchi K, Yamamoto S, Takekida S, Yan L, Tei H, Moriya T, Shibata S, Loros JJ Dunlap JC and Okamura H (1997). Light-induced resetting of a mammalian circadian clock is associated with rapid induction of the mPer1 transcript. Cell 91: 1043-1053.

Shinohara ML, Correa A, Bell-Pedersen D, Dunlap JC and Loros JJ (2002). *Neurospora* clock-controlled gene 9 (ccg-9) encodes trehalose synthase: circadian regulation of stress responses and development. Eukaryot Cell 1: 33-43.

Shinohara ML, Loros JJ and Dunlap JC (1998). Glyceraldehyde-3-phosphate dehydrogenase is regulated on a daily basis by the circadian clock. J Biol Chem 273: 446-452.

Shrode LB, Lewis ZA, White LD, Bell-Pedersen D and Ebbole DJ (2001). vvd is required for light adaptation of conidiation-specific genes of *Neurospora crassa*, but not circadian conidiation. Fungal Genet Biol 32: 169-181.

Sommer T, Chambers JA, Eberle J, Lauter FR and Russo VE (1989). Fast light-regulated genes of *Neurospora crassa*. Nucleic Acids Res 17: 5713-5723.

Springer ML (1993) Genetic control of fungal differentiation: the three sporulation pathways of *Neurospora crassa*. Bioessays 15: 365-374.

Talora C, Franchi L, Linden H, Ballario P and Macino G (1999). Role of a white collar-1-white collar-2 complex in blue-light signal transduction. Embo J 18: 4961-4968.
Welsh DK, Logothetis DE, Meister M and Reppert SM (1995). Individual neurons dissociated from rat suprachiasmatic nucleus express independently phased circadian firing rhythms. Neuron 14: 697-706.
Yang Y, Cheng P and Liu Y (2002) Regulation of the *Neurospora* circadian clock by casein kinase II. Genes Dev 16: 994-1006.
Yang Y, Cheng P, Zhi G and Liu Y (2001). Identification of a calcium/calmodulin-dependent protein kinase that phosphorylates the *Neurospora* circadian clock protein FREQUENCY. J Biol Chem 276: 41064-41072.
Young MW and Kay SA (2001). Time zones: a comparative genetics of circadian clocks. Nat Rev Genet 2: 702-715.
Zhu H, Nowrousian M, Kupfer D, Colot HV, Berrocal-Tito G, Lai H, Bell-Pedersen D, Roe BA, Loros JJ and Dunlap JC (2001). Analysis of expressed sequence tags from two starvation, time-of-day- specific libraries of *Neurospora crassa* reveals novel clock-controlled genes. Genetics 157: 1057-1065.

Applied Mycology & Biotechnology
An International Series. Volume 3. Fungal Genomics

4

Genome Sequencing, Assembly and Gene Prediction in Fungi

Brendan Loftus
The Institute for Genomic Research (TIGR), 9712 Medical Centre Drive, Rockville, MD 20850, USA (bjloftus@tigr.org).

Genome sequencing and the science of genomics is now being applied to the study of fungi. Although resources have been slow in coming, a number of fungi are now being sequenced and an increasingly diverse array of these organisms are being considered as candidates for whole genome sequencing. Currently there are only two complete fungal genome sequences available, those of *Saccharomyces cerevisiae* and *Schizosaccharomyces pombe* and the ensuing post-genomic resources has transformed research in both organisms. Going forward however, the methodologies initially used to generate and computationally analyze data from both of these projects, may not be the most appropriate for future fungal genome projects. Recent advances in Whole Genome Shotgun sequencing methodology, and improvements in whole genome assemblers appear to make them the most efficient and cost-effective strategies for current and future fungal projects. Similarly, the problems associated with computational gene discovery and annotation in the current and proposed fungal genome projects seem now more akin to those being faced by other large eukaryotic genome projects. This chapter attempts to outline the current state of the art in terms of sequencing, whole genome assembly, and computational gene prediction methodologies as applied to eukaryotic genomes. Given the diversity within the fungal kingdom both in terms of genome size and complexity, the advances and lessons learned from other eukaryotic genome projects can reasonably be expected to inform the methods by which future fungal genome projects are carried out.

1. INTRODUCTION

The recent advances in DNA sequencing and other high throughput technologies have meant that the science of genomics can now be applied to a broader range of species and the result has had a transformational effect on those organisms for which complete genomes have been produced. Fungi are only now beginning to be counted amongst the species for which there is significant genome-related information available. Fungi represent an enormous range of both medically and agriculturally important eukaryotic organisms. Fungal species that cause invasive infections of humans, increasingly important in the second half of the twentieth century are the focus of much of the current genome efforts. However, these represent only a tiny fraction of documented fungi. Much of the impact of fungal pathogens is felt throughout the agricultural sector as many of the most destructive and economically damaging commercial crop disorders are caused by fungi (Pennisi 2001). It can only be hoped that with the ongoing cost reductions in sequencing and the increased throughput of sequencing centres, fungal genome projects will mushroom in number, and grow to be more representative of the diversity found within the fungal kingdom. The arrival of fungal genomics was heralded with the publication of the 12Mb genome of *Saccharomyces cerevisiae*, a watershed event involving the collaboration of more than 600 scientists from over 100 laboratories during the period 1989 to 1996 (Goffeau *et al.* 1996). The sequence data from the 16 chromosomes represented the first complete genome from a fungal species, (Mewes *et al.* 1997) and meant that genome wide expression studies were now for the first time possible in eukaryotes (Banerjee *et al.* 2002; Que *et al.* 2002; Robyr *et al.* 2002). The genome identified all of the genes at once and allowed a glimpse into the metabolism and life cycle of the yeast cell. As a model organism, the genetic tools of the yeast system were now available for the first time to analyze the functions and interactions of all yeast homologs of human proteins (Steinmetz *et al.* 2002). This had, (and continues to have), a dramatic effect on research within the field of yeast genetics, and for those outside the field, the presence of a completed genome allowed association of their research to take account of the new resources available (e.g. http://genome-www.stanford.edu/Saccharomyces/) (Grunenfelder *et al.* 2002). Six years later, the publication of the genome of the fission yeast *Schizosaccharomyces pombe* has marked the completion of the second fungal genome (Wood *et al.* 2002). In the intervening period more than 80 microbial genomes have been sequenced to completion and more than 570 are in progress, (http://ergo.integratedgenomics.com/GOLD/). Additionally, genome drafts of *Anopheles gambiae, Drosophila melanogaster, Caenorhabditis elegans* and two separate drafts of the human genome have been published (The C elegans Consortium 1998; Adams *et al.* 2000; Lander *et al.* 2001; Venter *et al.* 2001; Holt *et al.* 2002). This reflects an imbalance in the quantity and quality of genome sequence from the fungal kingdom being produced or published. Funding agencies have until recently directed research dollars into mammalian projects (mouse and human), model organisms with large research communities, (fly, worm and pufferfish), and the major harbingers of human disease (Fleischmann *et al.* 1995; Tomb *et al.* 1997; Cole *et al.* 1998; Alm *et al.* 1999; Tettelin *et al.* 2000; Tettelin *et al.* 2001; Tettelin *et al.* 2002). The recent lack in funding of fungal genome projects may be attributed to a combination of factors, including the perception that fungal genomes would not provide equivalent results to bacterial projects in terms of the qualitative information generated for disease alleviation. Similarly, the application of whole genome shotgun (WGS) methodology, though effective in bacterial sequencing remained controversial and unproven for larger genomes. This thinking would make fungal genome sequencing projects appear as potentially costly exercises with little in the way of a guaranteed outcome. However, with the winding down of some of the mammalian genome projects (human and mouse), there is a freeing up of resources and a greater emphasis on fungi as suitable targets for sequencing. This is a timely shift in focus as there is an increase in the numbers of mycoses in immuno-compromised populations exposed as a direct result of the AIDS epidemic (Powderly 1990). Additionally, the use of organ transplantation treatment regimes and other immunosuppressive therapies is contributing to a rise in the incidence of fungal disease in the general population (Diamond 1991). With a larger number of ongoing genome projects than ever, the problems and solutions associated with sequencing, assembly and gene identification of fungal genomes are likely to benefit from the experiences with other eukaryotic organisms. This chapter will attempt to outline the rationale behind such approaches in other organisms within the context of the unique challenges posed by fungal projects.

2. GENOME SEQUENCING

2.1 Clone-by-Clone Approach

The strategy adopted to obtain both yeast genomes mirrored that used for sequencing of the human genome, (Linton *et al.* 2001). Individual chromosomes were broken into large fragments and representative libraries containing overlapping clones were generated (Thièrry *et al.* 1992). Following sequence assembly, these clones must then be anchored onto a genetic or physical map of the genome and assigned to their specific locations within the genome. This strategy requires the generation of high quality genomic libraries and the presence of high-resolution physical maps. The yeast genomic libraries were constructed by cloning into bacterial artificial chromosomes (BACs), cosmids or lambda phage vectors. The nested chromosome fragmentation method (Thierry and Dujon 1992) was used to construct a fine resolution physical map of each yeast chromosome. Individual chromosomes were then allocated to various laboratories and inserts sequenced using different strategies, including sequencing of cosmid clones, nested deletions, walking primers, and PCR (Dujon *et al.* 1997). Sequence data from each chromosome was then collected and assembled on a chromosome-by-chromosome basis.

2.2 Whole Genome Shotgun Sequencing

The clone-by-clone methodology is in stark contrast to the whole genome shotgun (WGS) approach used for the completion of most bacterial genomes (Adams *et al.* 1995). WGS involves randomly shearing the DNA of an organism into small size restricted fragments representing the genome several times over and cloning these fragments into suitable plasmid vectors for DNA sequencing (Roach 1995). Following sequencing of each end of the cloned fragments to a sufficient depth of coverage, assembly of the sequences takes place to form sequence contigs or assemblies. The depth of sequence coverage of the genome ensures the integrity of each base in the assemblies. Sequence assemblies can be connected using forward and reverse read information from the sequence reads. The resulting contigs can firstly be grouped together into scaffolds by linking different contigs that have forward and reverse sequence reads from the same clone. Following grouping of contigs into scaffolds, the remaining gaps in the genome can be classified into 'physical gaps' between contigs for which no spanning clone information is available, and 'sequence gaps', where there is a clone linking two contigs. Because the linking clones are available to use in closure of the sequencing gaps, closure efforts are directly proportional to the number of sequencing gaps and exponentially proportional to the number of physical gaps. A successful assembly has few or no mis-assembled regions, a relatively small number of sequencing gaps and little or no physical gaps. In an ideal situation, assembly of the genome would be straightforward however difficulties arise in sequence assembly due to the presence of various repeated portions of the genome. A successful genome assembly in large part depends on the ability to identify and, where possible, correctly orient, these repeated regions of the genome.

2.2.1 Use of different sized plasmid libraries in the WGS methodology

The initial bacterial sequencing projects based on WGS relied on end sequence linking information from clones containing between 2-3 Kilobases (Kb) of DNA, and from larger lambda, or cosmid, clones. The end sequence data from the lambda or cosmid clones provides the longer range clone linkage information necessary to span repeated regions of the genome and to link assemblies. In spite of the demonstration of the utility of this approach towards bacterial genomes, its efficacy for larger eukaryotic genomes remains somewhat controversial. In addition to their larger size, eukaryotic genomes are more complex and often contain a significant number of repetitive regions which make correct sequence assembly more problematic. Weber and Myers (Weber *et al.* 1997) presented a theoretical analysis of the WGS strategy in which they outlined the impact of repetitive sequences and suggested that the WGS methodology could be applied to large eukaryotic genomes including human. The publication of the genome sequence of *Drosophila* (genome size ~130 Mb) in 2000 was a clear demonstration of the utility of the WGS approach for more complex eukaryotic genomes (Celniker *et al.* 2000). In addition to the ability of mate pair information to resolve difficult regions of the genome, a key feature of the *Drosophila* genome sequencing effort was the capacity to clone DNA fragments of approximately 10Kb into plasmid vectors. This ensured that the resolving ability of plasmids could span the size range of most of the repetitive regions of the genome, thus reducing the numbers of physical gaps without the increased costs associated with using BACs. The ability to link assemblies and the consequent decrease in the numbers of physical gaps means that the overall architecture of the genome is easier to decipher and makes assembled drafts of larger genomes more useful to the scientific community. With the recently published draft of the genome of *Anopheles gambiae* (genome size ~290 Mb), WGS has been demonstrated to work in even larger genomes containing a high percentage of repetitive sequences (Subramanian *et al.* 2002).

2.3 Fungal WGS projects

Most of the fungi being considered for sequencing have small genome sizes with an apparently low percentage of repetitive elements, and WGS appears to be the most cost effective method for producing completed or high quality draft genome sequences. Indeed, most of the ongoing fungal genome sequencing projects are using a WGS-based approach in combination with mapping data where present. Examples include, the sequencing of the *Neurospora crassa* genome at the Whitehead Institute, which uses a combination of end sequences from 4Kb and 40Kb libraries (http://www-genome.wi.mit.edu/annotation/fungi/*Neurospora*/). The ongoing *Cryptococcus neoformans* and *Aspergillus fumigatus* projects at TIGR (http://www.tigr.org/tdb/fungal/) use plasmid libraries containing 2-3Kb, 3-4Kb, 8-12Kb and 25-40Kb inserts as well as BAC end sequences. In each case there are physical and/or genetic maps available which can be used to cross-reference the accuracy of the final assemblies. The *Candida albicans* genome project at Stanford (http://sequence-www.stanford.edu/group/candida/index.html) also uses WGS with sequence derived from both plasmid and M13 libraries.

3. WHOLE GENOME ASSEMBLY USING WGS DATA

Following the initial random shotgun sequencing phase, the task for all whole genome assemblers using WGS sequence data is to combine information from the individual sequence reads and use it to re-create the original sequence as it appears in the genome. In cases where polymorphisms exist in the genome, an assembly should be able to provide confidence values for the existence of each polymorphism within the sequence. Genome assembly attempts to incorporate information associated with sequence reads including the quality values associated with each base, the sequence of the read and the directionality of the sequence read within the clone. Following initial sequence comparison, assembly software incorporates clone information including the presence and orientation of mate paired reads within an assembly and the approximate insert sizes of the clones as they appear within an assembly. Although much of the assembly process can be automated, the presence of problematic regions within most (if not all) genomes means that this process still requires significant manual intervention to transit from a computational assembly to a complete genome sequence. It can be hoped that future assembly algorithms will be able to incorporate other genome-related data (e.g. mapping data) in order to improve the integrity of an assembly. There is also a trend towards producing cost-effective high-quality draft sequences for larger genomes where the costs of finishing are currently prohibitive. In these cases, determining the overall architecture of the genome and orienting unknown regions correctly within the genome become the priority. Most of the currently available assembly software utilizes the overlap-layout-consensus approach (Pevzner *et al.* 2001). In the generalized assembly methods described below using examples from various assembly algorithms, initial assembly of those regions that are unique to the genome occurs, and sequence reads that represent problem areas are set aside. Once the unique sequences are assembled and oriented, the outcome represents the majority of the genome sequence, and connected assemblies or scaffolds of these regions are generated. Scaffolding of assemblies of the unique regions of the genome is followed by a series of steps that attempt to place problematic or repeat regions within the pre-established structural framework of the genome.

3.1 Whole Genome Assemblers

Most widely used sequence assemblers, including the TIGR Assembler (TA) (Sutton *et al.* 1995) and the Celera Assembler (CA) (Huson *et al.* 2001), use a 'greedy' strategy for preliminary assembly of the individual sequences. Each read is represented as a collection of 'words' or continuous strings of fixed length. The exact number of base words varies between different assemblers (32 for TA and 40 for CA). The assembly algorithm computes all pair-wise alignments between the input sequences by looking for exact sized base words shared by each pair of sequences, and assigning a score to each such alignment (Fig. 1).

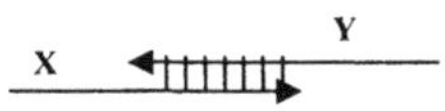

Fig. 1. Anatomy of an overlap between sequence reads X and Y.

Of the sequence overlaps, some are genuine and others represent repeated regions. In true overlaps, the shared sequence involves fragments that come from overlapping sections of the genome and belong together (Fig 2). In repeat-induced overlaps, the shared sequence involves part of a repeat that occurs in several dispersed parts of the genome and do not belong together

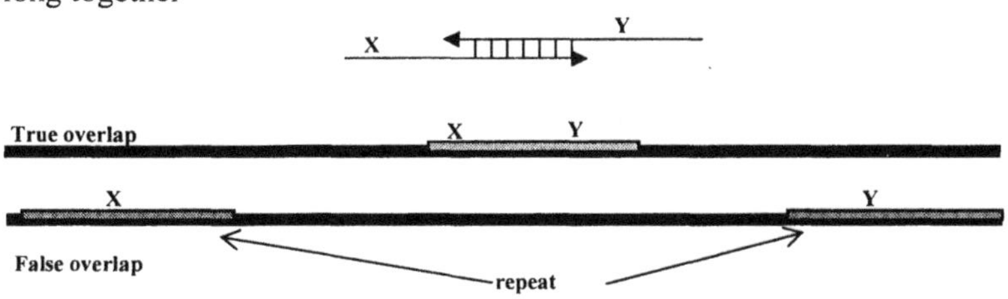

Fig. 2. True versus repeat induced overlaps in the case of a sequence overlap between X and Y.

(Fig 2). The TIGR Assembler scores the alignment taking into account not only the number of 32-mers shared by the two sequences, but also the uniqueness of these 32-mers. Given the fold sequence coverage of the genome the likelihood of a unique 32-mers occurrence can be determined. Intuitively, words that occur too many times in the assembly are indicative of repeat areas, and are therefore given a lower score. This helps ensure that unique regions will be assembled before potential repeats. The pair-wise alignments (matches) are considered in order, the highest scoring first. Each match is checked for feasibility using an implementation of the Smith-Waterman algorithm for sequence alignment (Smith *et al.*

1981). The assembler screens sequence alignments based on length of overlap, maximum length of the overhang, and the Smith-Waterman score of the alignment. If an alignment satisfies all the constraints, the two sequences are merged into a single sequence contig. The contigs corresponding to the matched sequences are merged into a single contig using a technique similar to that of Gribskov (Gribskov *et al.* 1987). The assembler now searches for groups of overlapping fragments that match the contig sequence and don't match other sequence reads that dispute, or contest, the contig. Such uncontested groups of fragments are assembled into what are called unique contigs or "unitigs", (Fig. 3) (Adams *et al.* 2001).

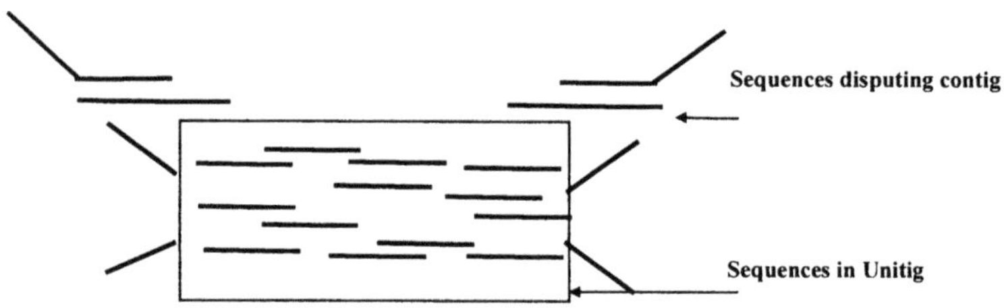

Fig. 3. Anatomy of a Unitig.

The procedure is repeated until the output consists of a set of contigs that cannot be merged any further. At this stage practically all of the Unitigs are correctly assembled, but a small percentage consist entirely of DNA from a number of instances of the same repeat. Identification of incorrectly assembled Unitigs is achieved by looking at the depth of coverage in each of the Unitigs relative to the overall depth of sequencing coverage (Fig 4). Those Unitigs for which the depth of fragment coverage corresponds to approximate genome sequence coverage are called U-Unitigs; the remaining Unitigs are set aside.

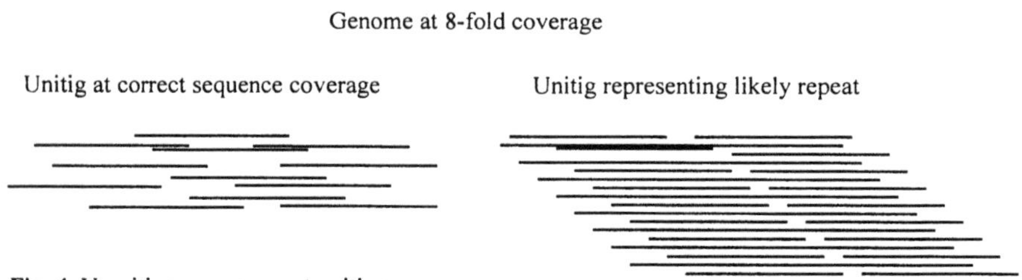

Fig. 4. U-unitigs versus repeat unitigs.

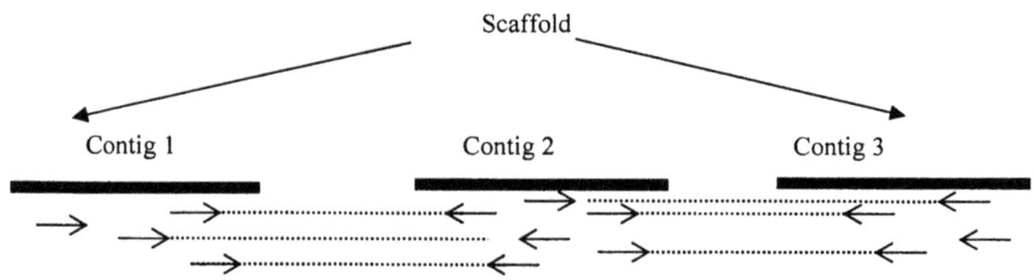

Fig. 5. Scaffolding of assembled contigs using clone mate pair information.

A contiguous sequence of ordered Unitigs is referred to as a contig. During a process termed 'scaffolding' the assembler uses clone mate pair information to orient contigs by ensuring that forward and reverse sequence reads from the same clone face each other (Fig 5). This mate pair information represents a series of internally consistent and reliable landmarks as they constrained by orientation and are generally a distance apart, consistent with the clone size estimate of the library, within the assembly. In addition to providing consistency within a contig, paired end sequences are used to link contigs within a scaffold. If sequence reads from the same clone lie on different contigs, for instance, the contigs are likely to be neighbors about 99% of the time. If two or more mate pairs from different clones enforce each other, that is, they indicate the same orientation of assemblies then the contigs involved are almost certain to be neighbors within the genome.

As the assembler compares more clone mates, the overall architecture of the genome becomes apparent as well as the problem areas within it. At this point, the scaffolding is continuous except for gaps (Fig. 6). Some of these gaps are due to missing sequence reads, and closing them requires further sequencing. The missing sequence may be due to a number of factors including under representation of some sequence data in the plasmid library or sequencing reaction artifacts, due to DNA secondary structure. Other gaps contain repetitive sequences, and can be closed either partially or completely using remaining unitigs that were set aside earlier in the assembly process. The ARACHNE whole genome shotgun assembler developed at the Whitehead/MIT Center for Genome Research has been used to assemble complex genomes including those of fungi (*Magnaporthe grisea* and *Neurospora crassa*) (Batzoglou *et al.* 2002).

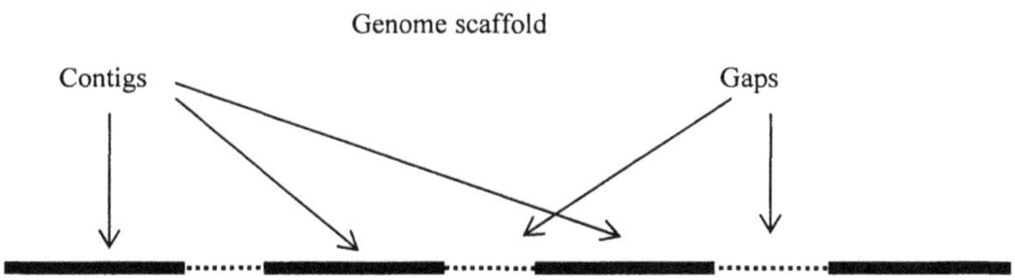

Fig. 6. Genome scaffolding of contigs into supercontigs.

ARACHNE also uses the overlap-layout-consensus approach and shares several similarities with the Celera Assembler, including the merging of contigs into unique contigs which are similar to the Unitigs described above. Similarly, unique contigs are ordered and oriented on the basis of forward-reverse mate pair links to form Supercontigs or Scaffolds which can be merged into larger supercontigs.

3.2 Repeat Incorporation into an Assembly

The Celera assembler classifies repeat sequences by size and reliability, calling the largest and most reliable repeats "rocks", smaller and less reliable repeats "stones" and finally the smallest and least reliable of repeats "pebbles". The assembler initially uses "rocks" for gap closure, placing rocks into the assembly requires linking information from at least two separate clones between the rocks and adjacent contigs. Following placement of the rocks into the assembly the assembler adds the stones which require at least one mate pair matching the stone with the adjacent. Pebbles are placed in a gap based on the quality of the overlaps between each other and the adjoining contigs. ARACHNE also uses the predefined repeat contigs to attempt to fill in the gaps between supercontigs, although the methods by which it identifies these is different from that used by the Celera assembler.

3.3 Error Correction in Genome Assembly

Sequencing errors that cause conflicts within assemblies are computationally expensive and a number of assemblers including ARACHNE and Euler (Pevzner *et al.* 2001; Tang *et al.* 2001) attempt to make genome assembly a simpler problem through the process of error correction (Pevzner *et al.* 2001). ARACHNE and Euler attempt to make consensus generation an element of fragment assembly, whereas other existing assemblers attempt the error correction at the end of the fragment assembly phase. Similar to above, sequence substrings for which there is appropriate representation within the genome are used for overlap analysis, and multiple sequence alignments are generated. These multiple alignments indicate potential sequencing errors in regions where there are alignments of high confidence with the exception of one or two bases. In these cases the base(s) in question is overwhelmingly out-voted by bases that are aligned to it and in these cases the base is modified where there appears to be an error. The assembler uses the short substrings to modify the original reads and to create a new instance of the assembly problem with a greatly reduced number of errors. This reduction in the number of errors sometimes results in the sequence reads being incorrectly modified. From an algorithmic point of view however elimination of the competition between competing bases at these positions greatly reduces the complexity of the assembly problem. The correct nucleotides are later corrected in the final stages of consensus generation using either a majority rule or other approach.

3.4 Genome Assembly of Fungal Genomes

Whole genome assembly draws together unique portions of the genome as an initial step, and characterize sequentially, the remaining difficult-to-assemble regions based on the available evidence. This reduces the overall errors in the individual assemblies to a minimum, while producing the most accurate draft of the overall structure of a genome. Accurate computational assembly of the fungal genomes currently underway should not prove a major technical hurdle, given the demonstrated ability of the assemblers to assemble the human, and other large eukaryotic genomes. Available assembly data from a diverse array of fungal genomes including *Cryptococcus, Neurospora, Magnaporthe, Aspergillus* and *Coccidoides*, indicates that this is indeed the case. In this context, the genome of *Candida albicans* may be considered an exception as its genome is diploid and standard sequence assembly software does not recognize the possibility of diploidy. Therefore, when confronted with sufficiently different alleles, the assembler often assembles them into separate contigs. This problem may become more significant as more diploid or polyploid fungi are sequenced in the future. A number of the above projects have supplemental mapping information which proves to be a great cross referencing method for the veracity of computational assemblies. In general, the presence of a physical map is a useful resource but as sequencing technology is progressing so quickly relative to map availability, the latter is unlikely to be a resource for many future fungal genome projects. Indeed, given the demonstrated benefits of the increased use of clone constraint information from insert libraries of different sizes in computational assembly, the presence of a physical map may not be deemed an essential component for the correct assembly of fungal genomes going forward.

4. GENE STRUCTURE ANNOTATION

Following genome assembly and/or finishing methods, the next step of a genome project is the annotation of biologically significant features onto the sequence. Currently, the most relevant attributes and annotations of genomic sequence are genes and gene structures. From analysis of a number of large eukaryotic genomes (e.g human, mouse, *Anopheles*,

Drosophila, *Fugu*), the identification of genes and gene structures is a huge challenge, and becoming increasingly reliant on purely computational or *ab initio* methods. For bacterial genomes, computational gene structure prediction may in essence be considered a solved problem, as gene structures are simple and a number of algorithms have been developed that work well on prokaryotic sequences (Audic *et al.* 1998; Delcher *et al.* 1999; Besemer *et al.* 2001; Suzek *et al.* 2001). Conversely, the problems in computational eukaryotic gene structure prediction have been well documented and are far from solved (Guigo *et al.* 2000; Korf *et al.* 2001; Rogic *et al.* 2002; Zhang 2002).

4.1 Eukaryotic Gene Structure

The main characteristic of eukaryotic genes is their organization into exons and introns (Fig. 7). The exons can be further subdivided into coding and non-coding exons or 5' exons, internal exons and 3' exons (Zhang 2002).

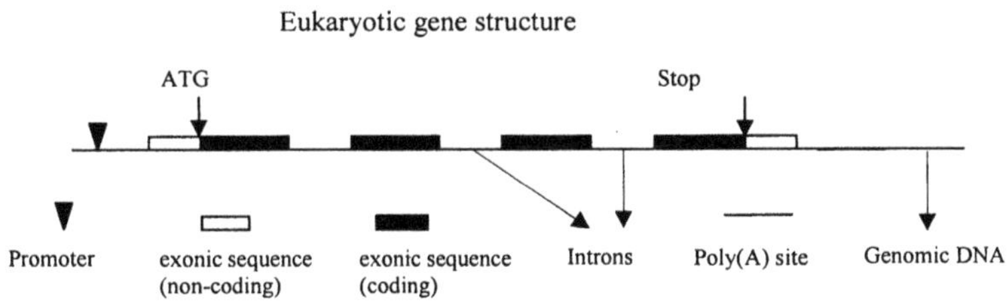

Fig. 7. Model of eukaryotic gene structure within genomic DNA containing introns, exons, promoter, start and stop sites, and a Poly (A) site.

4.2 Challenges in Eukaryotic Gene Structure Prediction

The challenges of eukaryotic gene structure prediction in the genomic context may loosely be defined as identifying all the exons of all the genes and parsing them into the correct structure without overlapping with other genes within the sequence. The current crop of computational gene finders, irrespective of the method used, mainly attempt to identify only the coding portions of genes. This is partly due to the fact that no salient sequence features have been identified that are adequately predictive of a promoter, or of the presence of an alternatively spliced transcript. The bulk of the coding portions of a gene are typically comprised of internal coding exons. To identify these, genefinding has focused on the detection of both intronic and exonic sequences using a variety of methods described below. However, accurate gene structure prediction not only requires the identification of exons and introns but also the determination of the correct exon-intron organization. This makes the task of gene structure analysis considerably more difficult but unfortunately the difficulties do not end there. When the problem of gene structure identification is applied to large tracts of genomic sequence containing many genes, gene boundary prediction is also essential to prevent either the truncation, or the merging, of predicted genes. The gene boundaries may be roughly divided into 3' exons, and 5' exons, both of which may be partially, or completely, non-coding. In the case of the 3' boundary of a gene, exon prediction is helped by the availability of expressed sequence tag (EST) sequence data from the organism. This is a consequence of the way EST sequence data is generated: for most organisms, EST's being more often truncated at the 5' end of the gene. This, in combination with some organism specific characteristics, sometimes allows for the poly-A signal site to be determined for some genes. Identifying and establishing the boundaries of the 5' exon is currently one of the most difficult tasks in computational genefinding (Brent 2002). This is due in part to the difficulty in identifying the promoter and the transcriptional start site. Some methods take advantage of the fact that transcriptional start sites are sometimes found in CpG islands; however this is still a rather low-resolution method. As in the case of the 3' boundary exons, computational genefinding currently only allows for the 5' Coding sequence (CDS) exons to be determined. Integration of the exons into a transcript can be challenging, however, *a priori* accurate first exon prediction can make the detection and prediction of downstream exons within the transcript more accurate. In terms of complexity, genes in fungi are likely to range from the simple as in the case of *Saccharomyces cerevisiae* to the more complex structures observed in the *Cryptococcus neoformans* genome. Therefore, many of the problems, techniques and solutions associated with the more complex eukaryotes and vertebrates in terms of gene structure prediction may be informative for the purposes of fungal genome annotation.

4.3 Measurements of Gene Prediction Accuracy

Measurement of the success of gene structure prediction software depends on the data being analyzed but is often measured in terms of sensitivity and specificity (Burset *et al.* 1996; Guigo *et al.* 2000; Agarwal *et al.* 2000). Sensitivity is the percentage of the coding region of a gene which is captured by the genefinder. Specificity is the measure of how much of the coding region

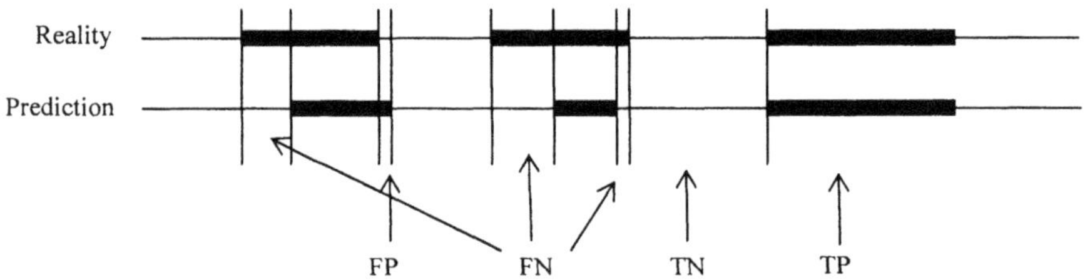

Fig. 8. Measurements of accuracy for gene prediction software. Reality and prediction are compared. FP indicates a false positive prediction. FN indicates a false negative. prediction. TN indicates a true negative prediction and TP indicates a true positive prediction.

predicted by the genefinding software is correct. Sensitivity and specificity can be determined at the nucleotide level, the exonic structure level or at the level of the protein product (Burset and Guigo 1996). In general, as is shown below (Fig. 8), computational genefinding represents a balance between prediction of a sufficient number of correct coding nucleotides and the absence of large numbers of false positive nucleotides. As gene prediction programs predict exons, for practical purposes it is usually the case that much of the exons predicted by computational methods are going to be partially correct or incorrect (Fig. 8).

4.4 Methodologies for Gene Structure Annotation

Most computational genefinding can be categorized into three main methodologies although many of the current genefinders operate by combining features from different methods.

4.4.1 Content based methods

Content based methods rely on analysis using the overall bulk features of the sequence, such as GC content, the location and content of repeated regions, the presence of different isochores within the sequence, and the compositional complexity of the sequence (Fickett *et al.* 1992). These methods work because genes are more often found in compositionally distinct (e.g. GC-rich) regions of a genome. Similarly, different organisms have distinct codon usage biases, which can be used to identify coding regions. Such methods also attempt to capture the reading frame-specific hexamer composition of coding regions as well as the hexamer composition of introns and intergenic regions. Another characteristic of coding regions within a sequence is the length of coding exons, as internal exons are rarely long and are quite restricted in their size distribution (Zhang 2002).

4.4.2 In-site based methods

In-site based methods rely on the identification of certain patterns, such as splice site patterns or branch point signals, which are indicative of the presence of introns or exons. These splice site and/or branch point patterns are conserved to a greater or lesser degree between different species, necessitating in many cases the development of a species specific genefinder. Most of these genefinders have default parameters derived from human, mouse or *Arabidopsis* genomic DNA and may work well or poorly on a given organism depending on the degree of conservation of gene features. Genscan (Burge *et al.*) utilizes a probability model which accounts for many essential features of gene structure such as splicing sites, branch point consensus sequences, gene density, the typical number of exons per gene and the distribution of exon size. Compositional properties of genes such as differences in gene density as well as distinct C+G% compositional regions are also incorporated into the gene model parameters. The program determines the most likely gene structures from a scoring of all the predicted exons and genes within the sequence. Other genefinders originally designed for use in bacterial gene identification have been extended for use in small eukaryotic genomes. GlimmerM is a modification of the Glimmer genefinder (Salzberg *et al.* 1998) used in Bacteria; has been developed specifically for small eukaryotes and has been trained for the genomes of *Plasmodium falciparum*, *Aspergillus fumigatus* and *Cryptococcus neoformans*. GlimmerM uses a dynamic programming algorithm that considers all combinations of possible exons for inclusion in a gene model and chooses the best of these through scoring a combination of the strength of the splice sites and the score of the individual exons. In-site based genefinders used on fungal genomes with varying degrees of success include Pombe (Chen *et al.* 1998), Find Fungal Gene (FFG) (Kraemer *et al.* 2001) and HMMgene (Krogh 1997).

4.4.3 Comparative Methods

Comparative methods identify gene structures by aligning the genomic sequence with a candidate amino acid or DNA sequence. Traditionally, gene structure and exon boundary determination has been modified manually using sequence similarities from related protein or EST matches post computational prediction. Some genefinding software has been developed which attempts to combine the steps in order to achieve better exon boundary prediction. Examples of such programs include Genewise (Birney *et al.* 2000) and Procrustes (Gelfand *et al.* 1996). Genewise combines a hidden Markov Model (HMM) for gene prediction with an HMM for protein profile detection, whereas Procrustes produces a spliced alignment of genomic sequence against a closely related protein homolog. In both cases, accuracy is improved where closely related homologs exist but the degree of conservation between the genomic sequence and the target gene must first be assessed. The genefinder Genomescan (Yeh *et al.* 2001) uses sequence similarity (e.g. BLASTX hits) in combination with the Genscan gene prediction algorithm to identify intron-exon boundaries and gene structures at a genomic scale. By combining the Genscan probability model described above with sequence similarity information, improvements in gene identification and prediction can be achieved where there is reasonable sequence similarity between all the genes and known proteins. The main drawback of these methods is that, for newly sequenced genomes, in most cases protein homologs do not exist in the public databases. This shortcoming is likely to become more pronounced with the ever-increasing ratio of WGS sequence data to curated protein sequences in the public databases. Methods that hold some promise for reducing the dependence on the presence of protein homologs for gene identification make use of comparative genomics to identify genes between the genomes of two related species. These methods are based upon the fact that two recently diverged genomes are likely to contain regions of synteny, where gene content, order and structural components of genes such as numbers and lengths of exons and/or splicing signals have been detectably conserved. These strategies do not rely upon similarity with previously published data, and previously undetected genes can be discovered by directly comparing homologous regions between genomes. Twinscan (Korf *et al.* 2001; Flicek *et al.* 2001) extends the Genscan probability model to incorporate comparative genomic information for the correct assignment of intron-exon and gene-gene boundaries. Gene structure prediction by Twinscan relies on comparison of two genomes that have been separated by sufficient evolutionary time, i.e. where functional gene features have been conserved but where there is a low occurrence of chance conservation of non-coding regions. Originally designed for analysis of the mouse and human genomes, Twinscan has now been trained for use in *Cryptococcus neoformans* using the *Phanerochaete chrysosporium* genome sequence for comparative purposes. Intuitively, comparative genomics methods appear to offer a solution for the problem of novel gene identification in fungi, as increasing numbers of related species are being, or will be sequenced, and comparisons can be made across multiple species. Another program using comparative genomics is Doublescan (Meyer *et al.* 2002) which simultaneously predicts gene structures in two DNA sequences which are homologous to each other and retrieves the subsequences shared between the two. It should be noted that problems occur with comparative methods whenever there is conservation of non-coding regions between the genomes used, as has been reported for the comparison of human and mouse (Zhang 2002).

4.5 Gene Structure Identification in *Cryptococcus neoformans*

The annotation of the *Cryptococcus neoformans* genome serves as a useful example of the issues relating to gene structure identification in fungi. In *Cryptococcus neoformans*, accurate gene structure prediction proves challenging as the organism appears to have an average of 6 coding exons per gene. In spite of having relatively complex structures, the genes are compacted within the relatively small genome (20Mb), and practically all known introns are small, averaging less than 100 bp in length, which makes detection by computational methods difficult. Additionally, as *Cryptococcus* is only the second basidiomycete genome for which there is a significant amount of sequence data publicly available there is little in the way of training data useful for genefinders. Since the presence of an open reading frame in the sequence of a eukaryote is no guarantee of the presence of coding sequence, it was decided to combine the outputs of a number of genefinders, themselves trained on *Cryptococcus*. The training set was generated from known *Cryptococcus* genes and from complete gene structures that could be inferred using the presence of EST and cDNA sequence data. The genefinders used were modified versions of Glimmer-M (Salzberg *et al.* 1999), Phat (Cawley *et al.* 2001) and Twinscan (Korf *et al.* 2001; Flicek *et al.* 2001). After very poor initial measures of specificity and sensitivity, there were significant increases in overall accuracy following training. However, gene structure prediction for this genome remains far from accurate and will require further refinement. Given the extent of the difficulties faced by the genefinders using the *Cryptococcus* data, we would recommend that the generation of a training dataset occurs as a preliminary step in the annotation of future fungal genomes.

4.6 The Use of Expressed Sequence Data in the Gene Finding Process

For instances where the genome of a given organism is not well represented in the public databases, an important feature of the development of a training set is the presence of a correlated EST or cDNA based sequencing project. In cases where a full-length cDNA has been sequenced, the entire gene structure can be easily predicted by aligning the cDNA sequence to the genomic sequence. In cases of limited EST data, where the entire gene structure cannot be deciphered, aligning the sequence to the genomic sequence can assist in determining splice site consensus sequences and/or branch point signal sequence signatures. Sufficient quantities of this kind of data can overcome the barrier of not having a lot of completely sequenced and characterized genes available as a training set. Many EST and/or cDNA sequences are single pass sequencing reads, prone to error and often too short to be informative in terms of identifying significant intron/exon boundaries. Consequently this type of data is considerably more useful when it is assembled and used to construct a high-fidelity non-redundant set of transcript sequences. These transcript sequences are usually longer and of a higher quality than individual EST sequences. This makes them useful for the purposes of genome annotation when aligned to the genomic sequence, as they are more likely to cover intron/exon boundaries and to provide a better representation of the 5' region of genes. There are a number of databases of such assembled EST sequences. These include UniGene, the TIGR gene index and STACK (Schuler *et al.* 1996; Christoffels *et al.* 2001; Quackenbush *et al.* 2001). The TIGR gene indices provide a highly refined rigorous protocol for cleaning, assembling and representing species-specific EST and gene datasets to produce high-fidelity consensus sequences for represented genes while minimizing the numbers of low quality, mis-clustered or chimeric sequences. The resulting tentative consensus (TC) sequences can then be used for the purposes of genome annotation. An additional feature of the TC sequence dataset is that they can be used to incorporate gene based mapping information and identify orthologous genes from related species. An alternative method to using clustered EST and cDNA sequences in the context of a finished genome is to use a Top-down EST clustering method. In this case EST sequences are aligned on an individual basis to the genome and these are then stiched together into a larger alignment (Kent 2002). This has been used to align the almost 4 million human EST sequences against the recently completed draft of the human genome (Kent 2002).

4.7 Construction of a Gene Index for *Cryptococcus neoformans*

The premise behind the gene indices is to treat the sequence data of a transcriptome of an organism as a shotgun sequencing project. EST sequences are downloaded from GenBank (http://www.ncbi.nlm.nih.gov/dbEST), or from other available sources. The sequence data is trimmed to remove vector, Poly A/T tails, adaptor sequences and contaminating bacterial sequences. Gene sequences for a particular organism are downloaded from Entrez. ESTs, complementary DNAs (cDNAs), and gene coding sequences are compared using a rapid sequence similarity program: FLAST (based on dds (Huang *et al.* 1997)). Sequences sharing 95% sequence identity over > 40nt (nucleotides) with < 20 nt mismatched sequence over 20 nt, are grouped into a cluster. For each cluster the component sequences are downloaded and assembled using the assembly algorithm CAP3 (Huang *et al.* 1999). Assembly produces one or more consensus sequences for each cluster and rejects any chimeric, low-quality or overlapping sequences. Each cluster of sequences is assembled separately in a similar fashion until the entire cluster set has been processed and the resulting TCs are loaded into the species-specific TIGR gene index database for annotation. Following assembly, TCs are annotated to provide a provisional functional assignment. TCs representing known genes are assigned the function of that gene. Those TCs without assigned functions are searched against a non-redundant protein database using the search program dps (Huang *et al.* 1997). Those TC's with high scoring database matches are tentatively assigned the function of that gene. The TIGR gene indices currently have representative datasets from seven fungal species including *Aspergillus nidulans, Coccidioides immitis, Cryptococcus neoformans, Magnaporthe grisea, Neurospora crassa, Saccharomyces cerevisiae*, and *Schizosaccharomyces pombe* (http://www.tigr.org/tdb/tgi/). Given the relatively low numbers of ESTs available for *Cryptococcus* and the high degree of similarity between the transcribed sequences of serotypes D and A, a combination of data from both serotypes was used to generate the current gene index.

5. CONCLUSIONS

A Genome project produces vast quantities of information for a given organism. Much of this information is interpreted by comparison with characterized sequences from other related organisms contained within public databases. A consequence however of the compilation of such a large amount of sequence information is that it must be compiled and cataloged properly. Much of the value of a given genome sequence rests on the ability to view it in the light of new information garnered from newly characterized genes or improvements in the methods used to identify genes. As the number and scope of both completed and partially sequenced genomes proliferate and increasing numbers of scientists make use of incomplete data, correct assembly and especially accurate computational gene identification, will be essential to derive meaning from the data. Much of the post-genomic work uses as a start point, identification of the gene set of a particular organism. Comparative genomics between species at the level of coding regions relies almost exclusively on the assumption that a majority of the genes are predicted correctly. Inferences regarding the evolutionary complexity of an organism, prediction of metabolic potential and assumptions about its relative position within the tree of life also are at least dependant on the correct identification of genes. Finally, the generation of DNA-based micro-array chips used in generating a genome-wide expression profile, relies almost totally on the assumption that the most of the genes within a sequenced genome can be identified with a high degree of success. It is therefore essential that, in conjunction with the explosion in the amount of sequence data being produced, significant progress is made in the identification of the genes therein. This is likely to present formidable challenges, at least initially, as the genomes of more organisms underrepresented in the sequence databases are completed. The current and future crop of genomes from the fungal kingdom look set to fall squarely within this group. As more fungal genomes are sequenced, the power of comparative genome analysis to decipher the location, structure and annotation of genes may offer the most expedient and practical first step towards the goal of applying genomics to the study of fungi.

Acknowledgements: I would like to thank the several members of the TIGR staff who read the various drafts and made helpful suggestions.

REFERENCES

Adams MD, Celniker SE, Holt RA, Evans CA, Gocayne JD, Amanatides PG, Scherer SE, Li PW, Hoskins RA (2000). The genome sequence of Drosophila melanogaster. Science 287(5461): 2185-2195.

Alm RA, Ling LS, Moir DT, King BL, Brown ED, Doig PC, Smith DR, Noonan B *et al.* (1999). Genomic-sequence comparison of two unrelated isolates of the human gastric pathogen Helicobacter pylori. Nature 397(6715): 176-180.

Audic S and Claverie JM (1998). Self-identification of protein-coding regions in microbial genomes. Proc Natl Acad Sci U S A 95(17): 10026-10031.

Banerjee N and Zhang MQ (2002). Functional genomics as applied to mapping transcription regulatory networks. Curr Opin Microbiol 5(3): 313-317.

Batzoglou S, Jaffe DB, Stanley K, Butler J, Gnerre S, Mauceli E, Berger B, Mesirov JP and Lander ES (2002). ARACHNE: a whole-genome shotgun assembler. Genome Res 12(1): 177-189.

Besemer J, Lomsadze A and Borodovsky M (2001). GeneMarkS: a self-training method for prediction of gene starts in microbial genomes. Implications for finding sequence motifs in regulatory regions. Nucleic Acids Res 29(12): 2607-2618.

Birney E and Durbin R (2000). Using GeneWise in the Drosophila annotation experiment. Genome Res 10(4): 547-548.

Brent MR (2002). Predicting full-length transcripts. Trends Biotechnol 20(7): 273-275.

Burge C and Karlin S (1997). Prediction of complete gene structures in human genomic DNA. J Mol Biol 268(1): 78-94

Burset M and Guigo R (1996). Evaluation of gene structure prediction programs. Genomics 34(3): 353-367.

Cawley SE, Wirth AI and Speed TP (2001). Phat--a gene finding program for Plasmodium falciparum. Mol Biochem Parasitol 118(2): 167-174.

Chen T and Zhang MQ (1998). Pombe: a gene-finding and exon-intron structure prediction system for fission yeast. Yeast 14(8): 701-710.

Christoffels A, van Gelder A, Greyling G, Miller R, Hide T and Hide W (2001). STACK: Sequence Tag alignment and consensus knowledgebase. Nucleic Acids Res 29 (1): 234-238.

Cole ST and Barrell BG (1998). Analysis of the genome of Mycobacterium tuberculosis H37Rv. Novartis Found Symp 217: 160-172.

Consortium TC (1998). Genome sequence of the nematode C. elegans: a platform for investigating biology. The *C. elegans* Sequencing Consortium. Science 282 (5396): 2012-2018.

Delcher AL, Harmon D Kasif K, White O and Salzberg SL (1999). Improved microbial gene identification with GLIMMER. Nucleic Acids Res 27(23): 4636-4641.

Diamond RD (1991). The growing problem of mycoses in patients infected with the human immunodeficiency virus. Rev Infect Dis 13 (3): 480-486.

Dujon B, Albermann K, Aldea M, Alexandraki D, Ansorge W, Arino J, Benes V, Bohn C, Bolotin-Fukuhara M, Bordonne R, Boyer J *et al.* (1997). The nucleotide sequence of Saccharomyces cerevisiae chromosome XV. Nature 387(6632 Suppl): 98-102.

Fickett JW and Tung CS (1992). Assessment of protein coding measures. Nucleic Acids Res 20(24): 6441-6450.

Fleischmann, RD, Adams MD, White O, Clayton RA, Kirkness EF, Kerlavage AR, Bult CJ, Tomb JF, Dougherty BA and Merrick JM (1995). Whole-genome random sequencing and assembly of Haemophilus influenzae Rd. Science 269 (5223): 49-512.

Fraser CM, Gocayne JD, White O, Adams MD, Clayton RA, Fleischmann RD, Bult CJ, Kerlavage AR, Sutton G, Kelley JM and others. (1995). The minimal gene complement of Mycoplasma genitalium. Science 270 (5235): 397-403.

Gelfand MS, Mironov AA and Pevzner PA (1996). Gene recognition via spliced sequence alignment. Proc Natl Acad Sci U S A 93 (17): 9061-9066.

Goffeau A, Barrell BG, Bussey H, Davis RW, Dujon B, Feldmann H, Galibert H, Hoheisel JD, Jacq C, Johnston M, Louis EL, Mewes HW, Murakami Y, Philippsen P, Tettelin T and Oliver SG (1996). Life with 6000 genes. Science 274 (5287): 546, 563-547.

Gribskov M, McLachlan AD and Eisenberg D (1987). Profile analysis: detection of distantly related proteins. Proc Natl Acad Sci U S A 84 (13): 4355-4358.

Grunenfelder B and Winzeler EA (2002). Treasures and traps in genome-wide data sets: case examples from yeast. Nat Rev Genet 3(9): 653-661.

Guigo R, Agarwal P, Abril JF, Burset M and Fickett JW (2000). An assessment of gene prediction accuracy in large DNA sequences. Genome Res 10 (10): 1631-1642.

Holt RA, Subramanian GM, Halpern A, Sutton GG, Charlab C, Nusskern DR, Wincker P, Clark AG and others (1997). A tool for analyzing and annotating genomic sequences. Genomics 46(1): 37-45.

Huang X and Madan A (1999). CAP3: A DNA sequence assembly program. Genome Res 9(9): 868-877.

Huson DH, Reinert K, Kravitz SA, Remington KA, Delcher AL, Dew IM, Flanigan M, Halpern AL, Lai Z, Mobarry CM, Sutton GG and Myers EW (2001). Design of a compartmentalized shotgun assembler for the human genome. Bioinformatics 17(Suppl 1): S132-139.

Kent WJ (2002). BLAT--the BLAST-like alignment tool. Genome Res 12(4): 656-664.

Korf I, Flicek P, Duan D and Brent MR (2001). Integrating genomic homology into gene structure prediction. Bioinformatics 17(Suppl 1): S140-148.

Kraemer E, Wang J, Guo J, Hopkins J and Arnold J (2001). An analysis of gene-finding programs for *Neurospora crassa*. Bioinformatics 17 (10): 901-912.

Krogh, A (1997). Two methods for improving performance of an HMM and their application for gene

finding. Proc Int Conf Intell Syst Mol Biol 5: 179-186.
Lander ES, Linton LM , Birren B, Nusbaum C, Zody MC, Baldwin J, Devon K, Dewar K, Doyle M, FitzHugh, R Funke, D Gage, K Harris, A Heaford and others (2001). Initial sequencing and analysis of the human genome. Nature 409 (6822): 860-921.
Mewes HW, Albermann K, Bahr M, Frishman D, Gleissner A, Hani J, Heumann J, Kleine K, Maierl A, Oliver SG, Pfeiffer F and Zollner A (1997). Overview of the yeast genome. Nature 387(6632 Suppl): 7-65.
Meyer IM and R Durbin (2002). Comparative ab initio prediction of gene structures using pair HMMs. Bioinformatics 18 (10): 1309-1318.
Oliver SG, Winson MK, Kell DB and Baganz F (1998). Systematic functional analysis of the yeast genome. Trends Biotechnol 16(9): 373-378.
Oliver SG (2002). Functional genomics: lessons from yeast. Philos Trans R Soc Lond B Biol Sci 357 (1417): 17-23.
Pennisi E (2001). The push to pit genomics against fungal pathogens. Science 292(5525): 2273-2274.
Pevzner PA, H Tang and MS Waterman (2001). An Eulerian path approach to DNA fragment assembly. Proc Natl Acad Sci U S A 98(17): 9748-9753.
Pevzner PA and H Tang (2001). Fragment assembly with double-barreled data. Bioinformatics 17(Suppl 1): S225-233.
Powderly WG (1990). Fungal infections in patients infected with HIV. Mo Med 87(6): 348-350.
Quackenbush, J, Cho J, Lee D, Liang F, Holt I, Karamycheva S, Parvizi B, Pertea G, Sultana R and White J (2001). The TIGR Gene Indices: analysis of gene transcript sequences in highly sampled eukaryotic species. Nucleic Acids Res 29(1): 159-164.
Que, QQ and EA Winzeler (2002). Large-scale mutagenesis and functional genomics in yeast. Funct Integr Genomics 2(4-5): 193-198.
Roach, JC (1995). Random subcloning. Genome Res 5(5): 464-473.
Robyr, D, Suka Y, Xenarios I, Kurdistani K, Wang A, Suka N and Grunstein M (2002). Microarray deacetylation maps determine genome-wide functions for yeast histone deacetylases. Cell 109(4): 437-446.
Rogic, S, Ouellette BF and Mackworth AF (2002). Improving gene recognition accuracy by combining predictions from two gene-finding programs. Bioinformatics 18 (8): 1034-1045.
Salzberg, SL, Delcher AL, Kasif S and White O (1998). Microbial gene identification using interpolated Markov models. Nucleic Acids Res 26(2): 544-548.
Salzberg, SL, Pertea M, Delcher ML, Gardner MJ and Tettelin H (1999). Interpolated Markov models for eukaryotic gene finding. Genomics 59(1): 24-31.
Schuler, GD, Boguski MJ, Stewart EA, Stein LD, Gyapay G, Rice K, White RE, Rodriguez-Tome P, Aggarwal A, Bajorek E, Bentolila S, Birren BB, Butler A, Castle AB, Chiannilkulchai N, Chu A, Clee C, Cowles S, Day PJ, Dibling T, Drouot N, Dunham I, Duprat S, East C, Hudson TJ *et al.* (1996). A gene map of the human genome. Science 274(5287): 540-546.
Smith, TF and MS Waterman (1981). Identification of common molecular subsequences. J Mol Biol 147(1): 195-197.
Steinmetz, LM, C Scharfe, AM Deutschbauer, D Mokranjac, ZS Herman, T Jones, AM Chu, G Giaever, H Prokisch, PJ Oefner and RW Davis (2002). Systematic screen for human disease genes in yeast. Nat Genet 31(4): 400-404.
Sutton, G, O White, M Adams and A Kerlavage (1995). TIGR Assembler: A new tool for assembling large shotgun sequencing projects. Genome Sequence Technol 1(1): 9-19.
Suzek, BE, MD Ermolaeva, M Schreiber and SL Salzberg (2001). A probabilistic method for identifying start codons in bacterial genomes. Bioinformatics 17(12): 1123-1130.
Tettelin, H, NJ Saunders, J Heidelberg, AC Jeffries, KE Nelson, JA Eisen, KA Ketchum, DW Hood, JF Peden, RJ Dodson, WC Nelson, ML Gwinn, R DeBoy, JD Peterson, EK Hickey, DH Haft, SL Salzberg, O White, RD Fleischmann, BA Dougherty, T Mason, A Ciecko, DS Parksey, E Blair, H Cittone, EB Clark, MD Cotton, TR Utterback, H Khouri, H Qin, J Vamathevan, J Gill, V Scarlato, V Masignani, M Pizza, G Grandi, L Sun, HO Smith, CM Fraser, ER Moxon, R Rappuoli and JC Venter (2000). Complete genome sequence of Neisseria meningitidis serogroup B strain MC58. Science 287(5459): 1809-1815.
Tettelin, H, KE Nelson, IT Paulsen, JA Eisen, TD Read, S Peterson, J Heidelberg, RT DeBoy, DH Haft, RJ Dodson, AS Durkin, M Gwinn, JF Kolonay, WC Nelson, JD Peterson, LA Umayam, O White, SL Salzberg, MR Lewis, D Radune, E Holtzapple, H Khouri, AM Wolf, TR Utterback, CL Hansen, LA McDonald, TV Feldblyum, S Angiuoli, T Dickinson, EK Hickey, IE Holt, BJ Loftus, F Yang, HO Smith, JC Venter, BA Dougherty, DA Morrison, SK Hollingshead and CM Fraser (2001). Complete genome sequence of a virulent isolate of Streptococcus pneumoniae. Science 293(5529): 498-506.
Tettelin, H, V Masignani, MJ Cieslewicz, JA Eisen, S Peterson, MR Wessels, IT Paulsen, KE Nelson, I Margarit, TD Read, LC Madoff, AM Wolf, MJ Beanan, LM Brinkac, SC Daugherty, RT DeBoy, AS Durkin,

JF Kolonay, R Madupu, MR Lewis, D Radune, NB Fedorova, D Scanlan, H Khouri, S Mulligan and others (2002). Complete genome sequence and comparative genomic analysis of an emerging human pathogen, serotype V Streptococcus agalactiae. Proc Natl Acad Sci U S A 99(19): 12391-12396.

Tettelin, H, V Masignani, MJ Cieslewicz, JA Eisen, S Peterson, MR Wessels, IT Paulsen, KE Nelson, I Margarit, TD Read, LC Madoff, AM Wolf, MJ Beanan, LM Brinkac and others (2002). Complete genome sequence and comparative genomic analysis of an emerging human pathogen, serotype V Streptococcus agalactiae. Proc Natl Acad Sci U S A 99(19): 12391-12396.

Thierry, A and B Dujon (1992). Nested chromosomal fragmentation in yeast using the meganuclease I-Sce I: a new method for physical mapping of eukaryotic genomes. Nucleic Acids Res 20(21): 5625-5631.

Tomb, JF, O White, AR Kerlavage, RA Clayton, GG Sutton, RD Fleischmann, KA Ketchum, HP Klenk, S Gill, BA Dougherty, K Nelson, J Quackenbush, L Zhou, EF Kirkness, S Peterson, B Loftus, D Richardson, R Dodson, HG Khalak, A Glodek, K McKenney, LM Fitzegerald, N Lee, MD Adams and JC Venter (1997). The complete genome sequence of the gastric pathogen Helicobacter pylori. Nature 388(6642): 539-547.

Venter, JC, MD Adams, EW Myers, PW Li, RJ Mural, GG Sutton, HO Smith, M Yandell, CA Evans, RA Holt, JD Gocayne, P Amanatides, RM Ballew, DH Huson, JR Wortman, Q Zhang and others (2001). The sequence of the human genome. Science 291(5507): 1304-1351.

Weber, JL and EW Myers (1997). Human whole-genome shotgun sequencing. Genome Res 7(5): 401-409.

Wood, V, R Gwilliam, MA Rajandream, M Lyne, R Lyne, A Stewart, J Sgouros, N Peat, J Hayles, S Baker, D Basham, S Bowman, K Brooks, D Brown, S Brown, T Chillingworth, C Churcher, M Collins, R Connor, A Cronin, P Davis, T Feltwell, A Fraser (2002). The genome sequence of Schizosaccharomyces pombe. Nature 415(6874): 871-880.

Yeh, RF, LP Lim and CB Burge (2001). Computational inference of homologous gene structures in the human genome. Genome Res 11(5): 803-816.

Zhang, MQ (2002). Computational prediction of eukaryotic protein-coding genes. Nat Rev Genet 3(9): 698-709.

Applied Mycology & Biotechnology
An International Series. Volume 3. Fungal Genomics

5

Fungal Transposable Elements: Inducers of Mutations and Molecular Tools

Frank Kempken
Abteilung für Botanik mit Schwerpunkt Genetik und Molekularbiologie, Botanisches Institut und Botanischer Garten, Christian-Albrechts-Universität zu Kiel, Olshausenstr. 40, D-24098 Kiel, Germany (fkempken@bot.uni-kiel.de).

Transposable elements in fungi so far are mostly restricted to Asco- and Basidiomycota. As in other eukaryotes, class I and II transposons were found. Class I elements transpose via a RNA intermediate, while class II elements excise and reintegrate on the DNA level. Both types of transposons may influence their hosts gene expression and can also trigger chromosomal recombination. Aside of causing mutations, transposons also may have beneficial effects for their hosts, i.e. repair of chromosomal breakage or modifying amino acid sequences thereby promoting protein evolution. Nevertheless, mechanisms have been identified in fungi which are aimed to inactivate transposons. Finally transposons provide important tools for diagnostic and gene tagging purposes.

1. INTRODUCTION

First evidence for the presence of transposable elements in fungi dates back to the 1970s when genetic instability was discovered in two different stocks of the filamentous fungus *Ascobolus immersus* (Decaris *et al.* 1978; Decaris *et al.* 1981; Rossignol *et al.* 1984). Interestingly, the two stocks in question 28 and 50, differ clearly in their characteristics, indicating two different transposable elements involved (Nicolas *et al.* 1987). To date only one element has been characterized in more detail, i.e. the *Ascot-1* element of stock 28 (Colot *et al.* 1995). In addition, numerous transposable elements have been identified in stock 50, albeit no correlation to the previously described genetic instability was established (Goyon *et al.* 1996b; Kempken 2001).

Although evidence for the presence of transposons was present early in filamentous fungi, the first transposons actually were discovered in the yeast ***Saccharomyces cerevisiae***. There, so-called *Ty*-elements (transposon yeast) were identified and, were found to be structurally very similar to retroviral genomes (Fink *et al.* 1981; Clare and Farabaugh 1985; Hauber *et al.* 1985). It was soon demonstrated that indeed *Ty*-elements are true retroelements and transpose via a RNA intermediate. They even possess virus-like capsids (Garfinkel *et al.* 1985).

Molecular data on transposons in filamentous fungi were published in the late 1980s (Kinsey and Helber 1989). Ever since, the scientific community witnessed a large increase of transposable elements from filamentous fungi, both in number and types. These will be described in much detail below. Transposons in fungi have been subject of a number of recent review articles (Daboussi 1996; Kempken and Kück 1998b; Kempken 1999; Pöggeler and Kempken 2003). Some of the many fascinating aspects of transposons, such as their unsolved evolutionary origin, potential horizontal distribution, and their impact on their host genomes will be discussed in detail here. As transposons had already been shown to be quite useful in other taxa, consequently, several attempts were undertaken to employ transposons as molecular tools in fungi as well. This includes the use of transposons as diagnostic markers, which can be useful in biotechnology (Kempken 1999) or with plant and animal pathogens (Fernandez and Langin 2002). Transposons have been shown to be of great value as tagging tools in plants (Haring *et al.* 1991; Ellis *et al.* 1992; Gierl and Saedler 1992; Aarts *et al.* 1993; Fitzmaurice *et al.* 1999), and may be of similar usefulness for fungal applications, in both their natural (Kempken and Kück 2000; Hua-Van *et al.* 2001) and foreign hosts (Migheli *et al.* 1999; Li Destri Nicosia *et al.* 2001; Villalba *et al.* 2001; Windhofer *et al.* 2002). Amazingly, even a transposon from a plant source was successfully employed in a fungal host (Weil and Kunze 2000).

2. CLASSIFICATION AND ORIGIN OF FUNGAL TRANSPOSONS

Transposons, first described by Barbara McClintock in maize in the 1940s (McClintock 1947; McClintock 1951; McClintock 1971), were later on detected in numerous other plants, bacteria, animals, man and fungi. Both, pro- and eukaryotic transposable elements are divided in different classes. There are four classes in bacteria (Brown and Evans 1991), and two classes in eukaryotes (Finnegan 1989). In this review only eukaryotic transposons, particularly fungal ones are covered. For those familiar with bacterial transposons, the most notable differences between transposons from prokaryotic and eukaryotic being (i) eukaryotic transposons do not carry resistance genes, and (ii) the class I eukaryotic transposable elements transpose via RNA intermediates.

2.1. Transposition Employing Reverse Transcriptases

Large amounts of animal and plant genomes consist of various class I transposons. They are characterised by transposition via an RNA intermediate which, upon reverse transcription is reintegrated into the genome, thus creating an additional copy of the element. At the site of integration a target site duplication (TSD) is generated (Kumar and Bennetzen 1999). In the Ascomycota copy numbers of retroelements are much lower than in plant or animal genomes, as repeated elements are less frequent. It would be interesting to analyse retroelements in the Zygomycota, as these have much higher contents of repetitive DNA (Wöstemeyer and Kreibich 2002), however no information about transposons in zygomycetes is available. So far, three different types of class I elements were identified in ascomyctes: Retrotransposons, retroposons and SINE-like elements (Kempken and Kück 1998b). Two of them, retrotransposons and retroposons, are characterized by their sequence similarity to retroviral reverse transcriptases.

Retrotransposons carry two protein-encoding genes. The *pol*-gene encodes a multifunctional polypeptide with a reverse transcriptase, a protease, a RNaseH and an integrase activity. The second gene, called *gag*, encodes a DNA-binding group specific antigen, generating a viral-like capsid. Retrotransposons are flanked by two long terminal repeats, called LTRs (Boeke and Corces 1989) and are similar to retroviruses in both their structure and their retrotransposition mechanism (Whitcomb and Hughes 1992). There are two major retrotransposon subfamilies, the *copia*- and the *gypsy*-families. These two subfamilies differ in the order of their *pol*-gene domains. Some members of the *gypsy*-family also carry incomplete and nonfunctional *env*-genes. In retroviruses this gene encodes an envelope polypeptide, which is responsible for the infectivity of the viruses (Lerat and Capy 1999). In fungi, several LTR retrotransposons have been discovered, with the yeast *Ty*-elements being particularly well known (Fink *et al.* 1981; Clare and Farabaugh 1985; Hauber *et al.* 1985). Most fungal retrotransposons being members of the *gypsy* subfamily (Table 1). Only *Mars2* and *Mars3* of *Ascobolus immersus* and *Tcen* of *N. crassa* are *copia*-like elements (Goyon *et al.* 1996b; Cambareri *et al.* 1998)..

Retroposons or LINE-like elements (long interspersed nuclear elements) lack terminal repeats but usually posses poly-A tails. These elements (also called non-LTR elements) have been characterized as carrying one or two open reading frames. Some, but not all encode an endonuclease or a *gag*-like polypeptide. In mycelial fungi the first member of this group, which was shown to transpose through an RNA intermediate, was the *Tad1* element of *Neurospora crassa* (Kinsey and Helber 1989; Kinsey 1993).

Intact retroposons have also been identified in other ascomycetes ***Ascobolus immersus***, *Magnaporthe grisea* and *Colletotrichum gloeosporioides* (Goyon *et al.* 1996b; He *et al.* 1996; Kachroo *et al.* 1997) and in a basidiomycete, *Tricholoma matsutake* (Murata *et al.* 2001).

Table 1. Selected class I transposons in fungi.

Host	*Transposon*	Family	EMBL Accession	References
Ascobolus immersus	*Mars2*	*copia*	X99082, X99083	(Goyon *et al.* 1996b)
Aspergillus fumigatus	*Afut1*	*gypsy*	L76085, L76086	(Neuveglise *et al.* 1996)
Fusarium oxysporum f. sp. lycopersici	*Foxy*	SINE	AJ250814	(Mes *et al.* 2000)
	Skippy	*gypsy*	L34658	(Anaya and Roncero 1995)
Magnaporthe grisea	*Fosbury*	*gypsy*	U15189, U15190	(Shull and Hamer 1996)
	MAGGY	*gypsy*	L35053	(Farman *et al.* 1996b)
Neurospora crassa	*Tad1-1*	LINE	L25662, L25663	(Kinsey and Helber 1989; Cambareri *et al.* 1994)
Podospora anserina	*Repa*	single LTR	X52957	(Deleu *et al.* 1990)
Saccharomyces cerevisiae	*Ty1*	*copia*	NC_001142	(Clare and Farabaugh 1985;
	Ty3	*gypsy*	M23367	Hauber *et al.* 1985; Hansen *et al.* 1988)

A full compilation of class I transposable elements in fungi was published recently (Pöggeler and Kempken 2003). SINEs (short interspersed nuclear element) are mobile elements which typically possess a RNA polymerase III promoter, and an adenine-rich 3' end of several base pairs (Deininger 1989). SINEs do not encode proteins facilitating their proliferation. Therefore, it is assumed that they use both host-specific and retroposon-specific activities *in trans* to secure their efficient amplification through retroposition and subsequent integration into a new location in the genome (Okada and Hamada 1997; Okada *et al.* 1997). Several SINE elements have been identified in the genomes of filamentous fungi (Pöggeler and Kempken 2003).

2.2. Transposition on the DNA Level

Transposable elements of class II do not possess reverse trancriptases or employ RNA intermediates, but do transpose on the DNA level by excision from one genomic locus and reintegration at a different position. This process is catalysed by a group of enzymes called transposases, which may act as a site-specific endonuclease (Beall and Rio 1997). From studies performed with plant and animal transposons it is known, that the transposase recognizes short direct repeats in the subterminal region of transposons and also the terminal inverted repeats (Ichikawa *et al.* 1987; Kunze and Starlinger 1989; Bravo-Angel *et al.* 1995; Becker and Kunze 1997), which flank all class II elements. In some cases even crystal structures of transposases have been established (van Pouderoyen *et al.* 1997) and common structural motifs have been proposed for transposases (Pietrokovski and Henikoff 1997). Transposition of class II elements may lead to the generation of circular molecules (Ruan and Emmons 1984; Sundaresan and Freeling 1987; Radice and Emmons 1993; Gorbunova and Levy 1997; Kempken and Kück 1998a; Gorbunova and Levy 2000), which are most probably a by-product of the transposition process. The latter is believed to have strong similarity to the V-D-J-recombination (Colot *et al.* 1998). As class II elements transpose via an excision-reintegration mechanism, an increase in copy number is restricted to those transposition events occurring during replication (Kunze 1996). Interestingly, no class II transposons are present in the yeast genome.

In fungi a large number of different class II transposons have been identified and intensively reviewed (Daboussi and Langin 1994; Daboussi 1996; Kempken and Kück 1998b; Kempken 1999; Kempken and Windhofer 2001; Pöggeler and Kempken 2003), some of which are shown in Table 2. The fungal class II transposons resemble those known from other eukaryotes, e.g. the *hAT* family (Kempken and Kück 1996), or the *Tc1/mariner* family (Langin *et al.* 1995). Class II transposons are put together in families based on similarity between (i) their transposases, (ii) their terminal inverted repeats, and (iii) the length of their target site duplication (TSD), generated during transposition. While the *Fot1* transposon from *F. oxysporum* has a two base pair "AT" target site duplication (Daboussi *et al.* 1992), the *hAT* family members are characterized by eight base pair TSDs with varying sequences (Kempken and Windhofer 2001).

The well known *Ac-Ds* transposons from maize (Kunze 1996) resemble autonomous and non-autonomous copies, respectively. Non-autonomous *Ds* copies may be *trans*-activated by autonomous *Ac* elements, if carrying *cis*-sequences necessary for binding of the transposase (Bravo-Angel *et al.* 1995). Similarily, non-autonomous copies of transposons have been observed in fungi. Several copies of the *F. oxysporum Impala* transposons were found to have frame shift mutations in their transposase open reading frame (Langin *et al.* 1995), a truncated *Restless* copy was identified in a Beauveria strain (Kempken *et al.* 1998), and most notably, the generation of *Ds*-like non-autonomous transposons was observed when *Restless* was introduced in *Neurospora crassa* and subsequently also in its natural host *Tolypocladium inflatum* (Windhofer *et al.* 2002).

2.3. Origin of Fungal Transposons

Despite many efforts the evolutionary roots of transposons have not yet been elucidated. It is generally accepted, that retroelements share a common origin possibly dating back to the so-called RNA-world (Xiong and Eickbush 1990; Faßbender and Kück 1995; Flavell 1995). This is most clearly demonstrated when comparing retrotransposons and retroviruses. However, conserved motifs of reverse transcriptases have been used to show relations to group II intron encoded reverse transcriptases as well (Michel and Lang 1985; Pöggeler and Kempken 2003). Some introns are mobile elements themselves, e.g. the pl-DNA of *Podospora anserina* (Stahl *et al.* 1978; Osiewacz and Esser 1984; Kück 1989). There are many mitochondrial plasmids of various kinds which encode reverse transcriptases or appear to be related to retroelements (Kennell *et al.* 1993; Kempken 1994; Chiang and Lambowitz 1997; Walther and Kennell 1999). Consequently, mobile retroelements are ubiquitous genetic elements, which may have evolved from simple intronic sequences to retrotransposons. According to a recent analysis based on ribonuclease H domains, a late chimeric origin of LTR retrotransposons and retroviruses was suggested, well after the origin of non-LTR-retroposons (Malik and Eickbush 2001).

Table 2. Selected class II transposons in fungi.

Host	Transposon	Family	EMBL Accession	References
Ascobolus immersus	*Ascot-1*	*hAT*	AF054897	(Colot *et al.* 1995; Colot *et al.* 1998)
Botrytis cinerea	*Flipper*	*Fot1/Pogo*	U74294	(Levis *et al.* 1996; Levis *et al.* 1997)
Fusarium oxysporum	*Folyt1*	*hAT*	AF057141	(Gómez-Gómez *et al.* 1999)
	Fot1	*Fot1/Pogo*	X64799	(Daboussi *et al.* 1992)
	Hop	*Mu/DR*	---	(Daboussi and Langin 1994; Chalvet *et al.* 2001)
	Impala	*Tc1/mariner*	AF282722	(Langin *et al.* 1995)
Magnaporthe grisea	*MGR586*	*Fot1/Pogo*	U60989	(Farman *et al.* 1996a)
	Pot2	*Fot1/Pogo*	Z33638	(Kachroo *et al.* 1994)
Neurospora crassa	*Guest*	*Tc1/mariner*	---	(Yeadon and Catcheside 1995)
Podospora anserina	*Pat*	*Fot1/Pogo*	AJ270953	(Hamann *et al.* 2000)
Schizophyllum commune	*Scooter*	*hAT*	AF267871, AF267872	(Fowler and Mitton 2000)
Tolypocladium inflatum	*Restless*	*hAT*	Z69893	(Kempken and Kück 1996)

Class II transposons share the possession of transposase enzymes. However, similarities between amino acid sequences of transposase genes are limited to specific families. It is less clear what origin they may have. As similar elements are present in eu- and archaebacteria one may assume, that present day class II elements have evolved from their prokaryotic counterparts, although transposition mechanisms differ considerably. Phylogentic trees have been established for individual families only, e.g. the *hAT*-transposon family (Kempken and Windhofer 2001). Horizontal gene transfer has been suggested in the case of the *Fot1* transposon from *Fusarium oxysporum*, but appears to be limited to genus Fusarium (Daboussi *et al.* 2002). Horizontal transfer was also discussed to explain the distribution of *Restless* in different strains of Tolypocladium and Beauveria (Kempken *et al.* 1998).

3. IMPACT OF TRANSPOSONS ON THEIR HOSTS

3.1. Modified Gene Expression and Chromosomal Rearrangements

A typical characteristic of active transposable elements is inducing mutations due to integration at new insertion sites (see Fig. 1), which may be located in (i) exons, (ii) introns, (iii) regulatory regions, or (iv) transposable activity may even lead to large scale recombinations or deletions. Transposons moving through a cut-and-paste mechanism (eukaryotic class II) are in addition often mutagenic when excising because repair of the empty site seldom restores the original sequence. The characterization of numerous excision events in many eukaryotes indicates that transposon excision from a given site can generate a high degree of DNA sequence and phenotypic variation. Whether such variation is generated randomly remains largely to be determined (Kidwell and Lisch 1997). Owing to the limited studies so far performed, in fungi only few cases of mutations caused by transposons have been investigated: *Scooter*, is an active transposon of the basidomycete ***Schizophyllum commune*** (Fowler and Mitton 2000). One copy of *Scooter* was identified by analysis of a spontaneous mutant in the Bβ2 pheromone receptor gene. The second *Scooter* element generated a mutant, which frequently is observed in *S. commune* and known as the "thin" mutant. The corresponding gene, *thn1*, is believed to encode a regulator of a G protein signalling protein (Fowler and Mitton 2000). As both *Scooter* copies were observed spontaneously, *Scooter* appears to be a rather active transposon, exemplifying the ability of transposable elements to disrupt gene expression. Likewise transposon *Restless* from *Tolypocladium inflatum* inactivated a nitrate regulator gene in a gene tagging approach (Kempken and Kück 2000).

The *Fot1* transposon is active in strains of the plant pathogenic fungus *Fusarium oxysporum*. In a high-copy-number strain, five independent mutants of the nitrate reductase gene (*niaD*) were generated by insertion of *Fot1* into one of the introns (Deschamps *et al.* 1999). The analysis of the effect of *Fot1* insertion in these mutants showed that, depending on the orientation of *Fot1* relative to *niaD*, different truncated chimeric *niaD-Fot1* transcripts were present. Mapping the termini of these transcripts revealed initiation of some transcripts in the 3' part of the *niaD* gene at sites located immediately 3′ of the *Fot1* insertion. Thus, a novel promoter, associated with the end of *Fot1*, directs transcriptional activity outwards from the element into the coding sequence of the *niaD* gene. This demonstrates that *Fot1* insertions provide an additional mechanism controlling fungal gene expression (Deschamps *et al.* 1999).

In the fungus *Ascobolus immersus*, genetic instable strains have long been known (see introduction). One instable locus, the spore color gene *b2* was studied intensively. Instability is due to the insertion of a transposable element named *Ascot-1* (Colot *et al.* 1998). It was shown that this system, which produces many phenotypically and genetically distinct derivatives, results from the excision of the *Ascot-1* element from the spore color gene *b2*. 48 molecularly distinct excision products were identified, which generate at least six phenotypically distinct colored derivatives, due to different footprints caused by the excision of *Ascot-1*. From 72 strains derived from partially colored spores, only four yielded phenotypically wild-type spores (Colot *et al.* 1998). This exemplifies the high degree of mutations which may be generated by transposons even when excising from a genetic locus.

The impact of transposons on chromosomal recombination has been investigated in *Fusarium oxysporum* where many different transposons have been identified (Daboussi 1996). Sequence analysis of stretches of genomic DNA surrounding insertion sites of one transposon family revealed that these are packed with repeated sequences. A number of these repeats are frequently reiterated and several of them are inserted into other elements. Some sections of these regions are also duplicated and appear prone to rearrangement, transposition, and rapid reorganization (Hua-Van *et al.* 2000). Analysis of transposon-mediated chromosome rearrangements using karyotypes of a set of Fusarium strains in which transposition events had occurred led to an exceptional electrophoretic karyotype variability, in both number and size of chromosomal bands. Chromosome length polymorphisms likely resulted from ectopic recombination between transposable elements (Daviere *et al.* 2001).

Transposable elements are also associated with centromere regions of *Neurospora crassa* (Cambareri *et al.* 1998), where a cluster of three new retrotransposon-like elements as well as degenerate fragments from the 3' end of *Tad*, a LINE-like retrotransposon (Kinsey and Helber 1989) were identified. The characteristics and arrangement of these elements are similar to those seen in centromeres of other organism (Cambareri *et al.* 1998).

3.2. Benefits and Evolutionary Impacts of Transposons

The impact of transposons on the evolution of their host have been considered in a number of review articles (e.g. Bennetzen 2000, Federoff 2000, Kidwell and Lisch 1997, Lönning and Saedler 1997). While some regard transposable as selfish DNA (Doolittle and Sapienza 1980) or even as some sort of genetic parasite (Orgel and Crick 1980), other cite evidence for a beneficial role of transposons or an impact on their hosts evolution (Lönning and Saedler 1997; Federoff 2000). As this area of research has not yet found many applications with fungal organisms, research from plant and animal sources is included. It is generally assumed that increased mutation rates may contribute to boost adaptation of the population to changing environments (Kidwell and Lisch 1997), and apparently to chromosome length polymorphism in fungi (Daviere *et al.* 2001).

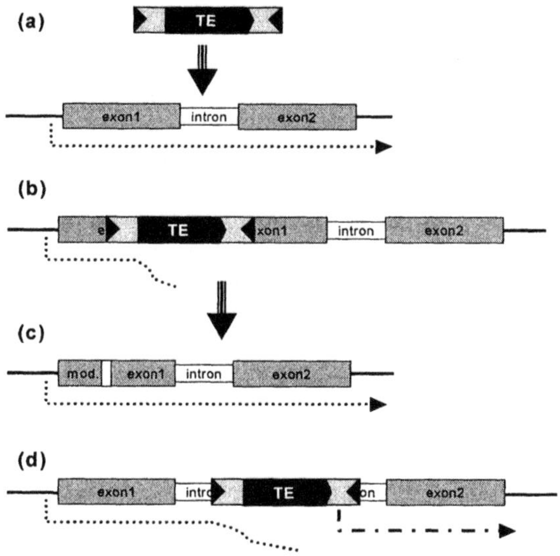

Fig. 1. Consequences upon integration and subsequent excision of a transposon from a host gene. (a) Prior to integration of a transposable element (TE). A protein coding gene with two exons is shown with its main mRNA (dotted line with arrow). (b) Integration of TE interferes with transcription. (c) Upon excision of the TE often footprints occur, which may change the coding sequence (mod. exon1) and hence amino acid sequence of the encoded protein. (d) Integration of the TE into an intron has been shown to led to transcription of downstream exons (arrow) from transposon promoter-like sequences. For more detail see main text.

Transposons may generate modified proteins due to excision footprints (Nordborg and Walbot 1995). An excellent example for modifications to a gene coding sequence from transposon excision was shown for the *Ascot* transposon from *Ascobolus immersus* (Colot *et al.* 1998). Integration of the transposon into the *b2* spore color gene lead to colorless spores. Regarding the type of footprints produced upon excision, different types of revertants were obtained with speckled, banded, spread, blotchy or double-belted spore phenotypes (Colot *et al.* 1998). Over time, modifications to gene sequences due to transposon integration and subsequent excision may thus have a tremendous impact on gene evolution (Lönning and Saedler 1997; Federoff 2000).

Introns may be generated from transposon insertion, which may promote exon shuffling (Giroux *et al.* 1994). And also regulatory elements (Schwarz-Sommer and Saedler 1987) may be generated. This was originally proposed based on the *Activator/Dissociation* system, where transposition of an inactive *Dissociation* element is *trans*-activated by the *Activator* element. This two-element system over time may mutate to a real regulatory system. Recently, *Ds*-like derivatives were also detected for the *Restless* transposon from *Tolypocladium inflatum*, raising the possibility for a similar mechanism in fungi.

Rescue of damaged telomeres, as it has been shown for *Drosophila* (Biessmann *et al.* 1994; Danilevskaya *et al.* 1994), provides an intriguing example of a beneficial role of transposable elements. Similar mechanisms may occur in fungi, as a yeast retrotransposon has been demonstrated to heal a broken chromosome (Garfinkel 1997).

4. HOST RESPONSE TO TRANSPOSONS

Several mechanisms able to inactivate repeated sequences have been described in filamentous fungi. The best known process, called repeat-induced point mutation RIP, has been intensively studied (Selker 1997). It was first detected in *Neurospora crassa*, but is now known also from other filamentous fungi (Hamann *et al.* 2000). In a specific period of the sexual cycle when haploid nuclei of the two mating types are in a common cytoplasm the genome is searched for the presence of repeated sequences. As a consequence of RIP, local mutagenesis leads to G-C→A-T transitions. Remaining cytosines are usually methylated (Selker 1997). Duplications larger then a few hundred basepairs usually suffer RIP when present as a tandem duplication. Unlinked duplications of one to two kilobases are also subject to RIP, but at a lower frequency (Selker 1990). RIP always acts pairwise; duplicated sequences in a nucleus are either both subject to RIP or neither is changed. RIP however does not affect certain sequences such as the rDNA repeats. Consequently, all transposons detected in laboratory strains of *N. crassa* have been inactivated by RIP (Cambareri *et al.* 1998; Margolin *et al.* 1998; Mannhaupt *et al.* 2002). The transposon *Tad* from a field strain *N. crassa* is also subject to RIP (Kinsey *et al.* 1994), indicating that vegetative reproduction may be of greater importance in the wild, as a merely sexual reproduction would have lead to the destruction of this element.

In *A. immersus,* repeated sequences are inactivated by methylation prior to meiosis, a process called MIP (Goyon and Faugeron 1989; Colot *et al.* 1995). The mechanism is clearly different from RIP and involves a cytidine methylase (Malagnac *et al.* 1997; Goyon 1998). Similar to RIP, tandem repeats are more efficiently targeted by MIP than ectopic repeats (Colot *et al.* 1995). Strains of stock 28 of *A. immersus* exhibit genetic instability and harbor a number of different transposons. Among these are several class I transposable elements, including LINE-like elements and LTR-retrotransposons (Goyon *et al.* 1996b). These and other repeated sequences such as rDNA genes have been characterized and monitored for methylation (Goyon *et al.* 1996a; Goyon *et al.* 1996b). The transposons were found to be highly methylated and most probably are inactive. In remarkable contrast, the rDNA cluster which may contain about 100 copies of the rDNA exhibits reduced methylation, i.e. only some of the cytidin residues at a specific position are methylated. Small repeated sequences such as the 5S rDNA and the small non-autonomous transposon *Ascot-1* (Colot *et al.* 1998) are not subject to MIP (Goyon *et al.* 1996a; Goyon *et al.* 1996b). Likewise, the repeated element *Hideaway*, which exhibits structural characteristics of retrotransposons, was shown to be partially methylated only (Kempken 2001). Mechanisms like RIP or MIP are believed to act as host defense mechanism to avoid the accumulation of repeated DNA sequences (Selker 1997; Colot and Rossignol 1999).

Methylation in *N.crassa* may also occur in vegetative mycelia and targets foreign DNA. For example in cells, transformed with plasmids carrying hygromycin phosphotransferase (*hph*), the gene showed reversible inactivation due to cytosine methylation after prolonged growth when present in multiple copies (Pandit and Russo 1992). In contrast to other organisms, methylation does not block transcript initiation, but does block elongation (Rountree and Selker 1997). Methylation of DNA sequences is apparently triggered by A/T content and particularly the TpA content. Sequences with a high TpA/ApT ratio also have a high likelihood of being subject to RIP in *N. .crassa* (Margolin *et al.* 1998). The TpA/ApT ratio of the complete *Restless* transposon sequence is 0.77, but the ratio is much higher in some areas (Windhofer *et al.* 2000). Consequently, integration of *Restless* in *N. crassa* led to strong methylation. Fungal transposon sequences generally are characterized by TpA/ApT ratios of about 0.8 and higher (Windhofer *et al.* 2000). It seems likely these ratios trigger non-RIP methylation in vegetative mycelia of *N.crassa*, and consequently one may assume that this mechanism indeed has evolved in response to invading mobile elements.

The methylation and mutation mechanisms in fungi usually work on sequences which are larger than about 500 bp. Repeated sequences larger than 500 bp are inactivated. This may provide a selection to generate small deleted copies such as the *Guest* mini-transposon of *N. crassa* (Yeadon and Catcheside 1995), which can be *trans*-activated by single full-length elements. Moreover, when transposon *Restless* was introduced into *N. crassa* (see below) the generation of a large number of deleted mini-transposons was observed (Windhofer *et al.* 2002). This also might be a consequence of an adaptation to RIP or MIP mechanisms.

5. FUNGAL TRANSPOSONS AS MOLECULAR TOOLS

5.1. Transposons as Diagnostic Tools

Knowledge of the population genetics of fungi has increased in the past, mostly due to the use of molecular markers (Leung *et al.* 1993; McDonald 1997). A large number of fungi is capable of reproduction both sexually and asexually. The population structure of a species may therefore be influenced by the degree these two modes of reproduction occur for a given species. Many plant pathogens for example display a predominantly asexual reproduction with infrequent sexual reproduction (Brygoo *et al.* 1998). In addition, genetic variability in pathogenic fungal populations is rather important for disease management, epidemiology, identification of individual clones, and to detect dispersal of clones between subpopulations (Rogers 1995; Kempken 2002).

Molecular markers which help to distinguish different fungal populations or even individual clonal lines are obviously of great value in assessing fungal populations or to diagnose fungal pathogens. Transposable elements combine several advantages as molecular markers (Fernandez and Langin 2002), as they are considered to be neutral markers, occur in moderate to high copy numbers, and therefore are useful as fingerprint markers. Here a short overview about the current use of transposons for diagnostic purposes is given.

Repeated sequences have also even been used for epidemiological studies of the human opportunistic pathogen *Aspergillus fumigatus* (Girardin *et al.* 1993; Girardin *et al.* 1994a; Girardin *et al.* 1994b). One of these sequences was identified as a transposable element belonging to the *gypsy* group of retrotransposons (Neuveglise *et al.* 1996). The discovery of transposons in other pathogens will provide additional means for epidemiological studies.

Applications in plant pathology have been covered in a recent comprehensive review (Fernandez and Langin 2002) and for that reason are mentioned only briefly here. *Fot1* elements in *Fusarium oxysporum f. sp. albedians*, the causal agent of Bayoud disease of date palm provide diagnostic PCR targets for the detection of this pathogen. In a study with 286 *Fusarium oxysporum f. sp. albedians* and 25 related non-pathogenic strains, one primer pair gave rise to pathogen-specific amplification (Fernandez *et al.* 1998). MGR586, a *FotI*-like element in *Magnaporthe grisea* (Hamer *et al.* 1989), the causal agent of rice blast disease, occurs in low copy (1-2) in strains which exhibit pathogenicity for wheat or other grasses (Talbot 1998). However, MGR586 elements are preferentially amplified in rice pathogens and are found to have 30 to 50 copies. The differences in distribution of MGR586 are an indication that gene flow between specific host-forms of this pathogen is limited. DNA fingerprints by MGR586 fingerprinting give characteristic multi-locus haplotypes for each strain, which indicate the degree of genetic relatedness between rice blast isolates (Levy *et al.* 1991).

Transposons are also used for population analysis based on PCR with transposon-specific primers, a method termed repetitive element based or repPCR (George *et al.* 1998). Two outward facing primers specific for a transposon, e.g. *Pot2* from *M. grisea* are employed (Kachroo *et al.* 1994). Variable length fragments are generated, which define the sequences lying between two adjacent copies. Depending on the localization of transposons, strain specific fingerprints are generated. The method requires a high copy number of the transposon, as it is the case with the transposon *Pot2* having about 100 copies in the *M. grisea* genome.

Many fungi are used to produce pharmaceuticals. Strain verification and identification is a very important issue for this industry. Transposable elements are quite useful tools here. For example in the cyclosporin producing ATCC34921 strain of *Tolypocladium inflatum*, about 15 copies of the *Restless* transposon were detected (Kempken and Kück 1996) leading to a very specific hybridization pattern, as each integration site results in one specific band. This hybridization pattern was compared to other cyclosporin producing strains. Interestingly, almost identical hybridization patterns were obtained, suggesting that these strains are likely of the same origin (Kempken *et al.* 1998). In addition another DNA element is present in multiple copies in the ATCC34921 producer strain exclusively (Kempken *et al.* 1995). Consequently, repeated DNA elements and most notably transposons can be very valuable tools in strain identification.

5.2. Transposon Aided Gene Tagging

In plants and bacteria, the ability of transposons to cause mutations has long been used to analyse gene function (Luo *et al.* 1991; Bradley *et al.* 1993; MacGinnitie *et al.* 1995) or identify genes (Haring *et al.* 1991; Gierl and Saedler 1992; Aarts *et al.* 1993; Long and Coupland 1998; Fitzmaurice *et al.* 1999). The latter method called gene tagging, is based on the use of a transposon with a known sequence. Using an unmodified, endogenous transposon, a new gene has been successfully tagged using transposon *Restless* from *Tolypocladium inflatum* (Kempken and Kück 2000). Chlorate resistant colonies were physiologically screened to obtain putative nitrate regulator mutants. Inverse PCR was performed using *Restless*-specific oligonucleotides. In one mutant an additional *Restless* integration site was identified and its adjacent sequence analysed. This DNA sequence was used to screen a wild type genomic library. Finally, the nitrogen pathway-specific regulator *tnir1* was identified (see Fig. 2). Complementation of a *Neurospora crassa nit-4* mutant (Yuan and Marzluf 1992) identified *tnir1* as a functional ortholog (Kempken and Kück 2000). While the tagging experiment with *Restless* was the first approach of that kind in fungi, similar approaches are possible and promising in any fungus carrying a well characterized active transposon. Similarly, *Scooter*, an active transposon from *S. commune* spontaneously tagged the *thn1* gene in that fungus, aiding its cloning and sequence analysis. The *thn1* gene may encode a regulator of a G protein signaling protein (Fowler and Mitton 2000). However, in that case the transposon was not intended to be used for that purpose.

5.3. Development of Vector Systems

Not all fungi contain transposons suitable for gene tagging. Therefore vector systems are required which allow the use of transposons in a heterologous host. So far one retrotransposon (Nakayashiki *et al.* 1999) and three class II transposons (Windhofer *et al.* 2000; Hua-Van *et al.* 2001; Li Destri Nicosia *et al.* 2001; Villalba *et al.* 2001; Windhofer *et al.* 2002) have shown activity in foreign hosts. These examples will be discussed in more detail.

The *Maggy* retrotransposon (Farman *et al.* 1996b) is of the *gypsy*-type (see Table 1) and was isolated from *Pyricularia grisea* (teleomorph, *Magnaporthe grisea*). The transposon was introduced into three *P. grisea* isolates previously devoid of the element as well as the heterologous fungi *Colletotrichum lagenarium* and *P. zingiberi* (Nakayashiki *et al.* 1999). Transposition via an RNA intermediate was observed in all fungi, but was comparably rare in *C. lagenarium*, indicating that host-specific factors may influence transposon activity.

Nevertheless, this publication was the first to demonstrate activity of a fungal transposon in a heterologous host.

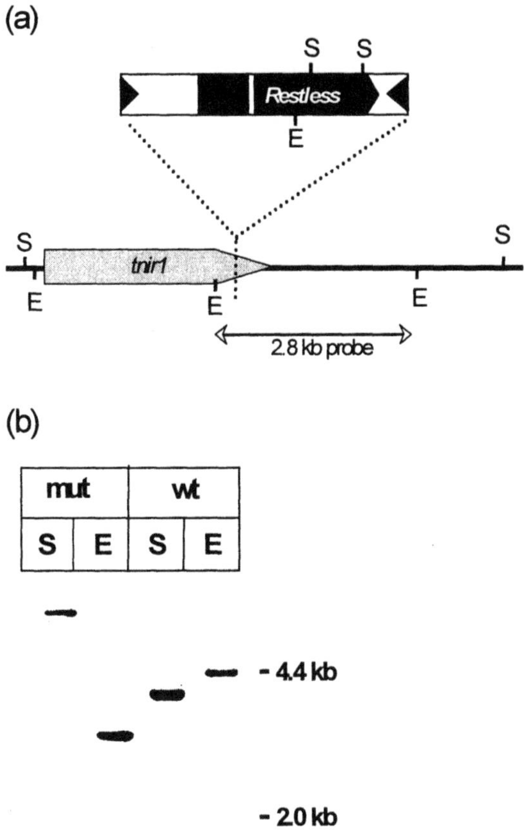

Fig. 2. Tagging of a nitrate regulator gene in *Tolypocladium inflatum*. (**a**) Insertion of the *Restless* transposon in the *tnir1* gene. Restriction sites for *Eco*RI (E) and *Sal*I (S) are given. The probe used for the Southern hybridization shown in "b" is indicated. (**b**) Southern hybridization confirms integration of *Restless* into *tnir1* gene. Depending on the restriction enzyme used, smaller or larger bands appear when DNA from the mutant is used.

The *FotI* element occurs in many copies in certain strains of *Fusarium oxysporum* (Daboussi *et al.* 1992). Therefore identification of an autonomous, active copy was necessary, as usually at least some copies are inactive due to point mutations or deletions (Migheli *et al.* 1999). Similarly an active copy of the *Impala* transposon from *F. oxysporum* was identified (Hua-Van *et al.* 2001). Vectors carrying *FotI* and *Impala* were introduced into *Aspergillus nidulans* (Li Destri Nicosia *et al.* 2001), and *Impala* was also introduced into *Magnaporthe grisea* (Villalba *et al.* 2001). Excision and reintegration of the transposons was observed in both cases. In *A. nidulans* excision frequencies were described between 10^{-4} and 10^{-5}, and reintegration apparently occurred in 90% of all excision events. Tagging of genes was not reported in *A. nidulans* (Li Destri Nicosia *et al.* 2001), but in *M. grisea* a gene believed to be involved in pathogenicity was tagged by the *Impala* transposon (Villalba *et al.* 2001).

Restless based vectors were introduced into *Neurospora crassa* (Windhofer *et al.* 2000; Windhofer *et al.* 2002), *Penicillium chrysogenum* (Windhofer *et al.* 2002), and the phytopathogenic fungus *Botrytis cinerea* (van Kan and Kempken; unpublished data). In *B. cinerea* and *P. chrysogenum* so far excision of the element was observed, often leading to the generation of *Ds*-(*Dissociation*)-like deleted *Restless* elements. In *N. crassa* single copy integration was necessary to avoid methylation and inactivation of *Restless* (Windhofer *et al.* 2000). Southern blot analysis indicated rare reintegration of *Restless* into the *N. crassa* genome (Kempken, unpublished data). No data are currently available regarding the ability of *Restless* to tag genes in a heterologous host.

6. CONCLUSIONS

We have now reached a point where a large number of fungal transposons are characterised, some of them ideally suited to address important questions about the biological role of transposons and their mechanisms of movement. Fungal experimental systems have a variety of advantages, with regard to other eukaryotic systems. This includes a coenocytical organization, short life cycles and small genome sizes. Fungi therefore may permit the solution of problems that are difficult to resolve in other organisms. Future studies of transposons in fungi should and will focus on three main areas: (i) to further elucidate ability of host organisms to inactivate invading transposons and ways to avoid that inactivation, (ii) horizontal transfer between different species which is much easier to approach experimentally in fungi and (iii) the unique environment for transposons in coenocytical organisms as filamentous fungi, which is different from almost all other eukaryotic cells. Finally, transposon tagging in fungi will provide an excellent tool for gene identifications, which is of particular interest with respect to genome sequencing projects, currently being in progress in a number of fungal genomes, e.g. *Aspergillus* or *Neurospora.*

Acknowledgements: Research of the author is funded by the Deutsche Forschungsgemeinschaft. I thank Mrs. Kerstin Stockmeyer for critical reading the manuscript.

REFERENCES

Aarts MGM, Dirkse WG, Stiekema WJ, and Pereira A (1993). Transposon tagging of a male sterility gene in Arabidopsis. Nature 363:715-717.

Anaya N, and Roncero MIG (1995). *Skippy*, a retrotransposon from the fungal plant pathogen *Fusarium oxysporum*. Mol. Gen. Genet. 249:637-647.

Beall EL, and Rio DC (1997). *Drosophila* P-element transposase is a novel site-specific endonuclease. Genes Dev 11:2137-2151.

Becker HA, and Kunze R (1997). Maize *Activator* transposase has a bipartite DNA binding domain that recognizes subterminal sequences and the inverted repeats. Mol Gen Genet 254:219-230.

Biessmann H, Kasravi B, Bui T, Fujiwara G, Champion LE, and Mason JM (1994). Comparison of two active *HeT-A* retroposons of *Drosophila melanogaster*. Chromosoma 103:90-98.

Boeke JD, and Corces VG (1989). Transcription and reverse transcription of retrotransposons. Annu Rev Microbiol 43:403-434.

Bradley D, Carpenter R, Sommer H, Hartley N, and Coen E (1993). Complementary floral homeotic phenotypes result from opposite orientations of a transposon at the plena locus of Antirrhinum. Cell 72:85-95.

Bravo-Angel AM, Becker HA, Kunze R, Hohn B, and Shen WH (1995). The binding motifs for *Ac* transposase are absolutely required for excision of *Ds1* in maize. Mol Gen Genet 248:527-534.

Brown NL, and Evans LR (1991) Transposition in prokaryotes: transposon Tn501. Res Microbiol 142:689-700

Brygoo Y *et al.* (1998). Reproduction and population structure in phytopathogenic fungi. In: Bridge P, Couteaudier Y, Clarkson J (eds) Molecular variability of fungal pathogens. CAB International, Wallingford, pp 133-148.

Cambareri EB, Aisner R, and Carbon J (1998). Structure of the chromosome VII centromere region in *Neurospora crassa*: degenerate transposons and simple repeats. Mol Cell Biol 18:5465-5477.

Cambareri EB, Helber J, and Kinsey JA (1994). *Tad1-1*, an active LINE-like element of *Neurospora crassa*. Mol Gen Genet 242:658-665.
Chalvet F, Kaper F, Langin T, and Daboussi MJ (2001). *Hop*, an active *MuDR*-like element in the filamentous fungus *Fusarium oxysporum*. Fungal Genet Newsl 48(Suppl):86.
Chiang CC, and Lambowitz AM (1997) The Mauriceville retroplasmid reverse transcriptase initiates cDNA synthesis de novo at the 3' end of tRNAs. Mol Cell Biol 17:4526-4535.
Clare J, Farabaugh P (1985) Nucleotide sequence of a yeast Ty element: Evidence for an unusual mechanism of gene expression. Proc Natl Acad Sci USA 82:2829-2833.
Colot V, Goyon C, Faugeron and G, Rossignol JL (1995). Methylation of repeated DNA sequences and genome stability in *Ascobolus immersus*. Can. J. Bot. 73:S221-S225.
Colot V, Haedens V, and Rossignol JL (1998). Extensive, nonrandom diversity of excision footprints generated by *Ds*-like transposon *Ascot-1* suggests new parallels with V(D)J recombination. Mol Cell Biol 18:4337-4346.
Colot V, and Rossignol JL (1999). Eukaryotic DNA methylation as an evolutionary device. Bioessays 21:402-411
Daboussi MJ (1996) Fungal transposable elements: generators of diversity and genetic tools. J. Genet. 75:325-339.
Daboussi MJ, Daviere JM, Graziani S, and Langin T (2002). Evolution of the *Fot1* transposon in the genus Fusarium: discontinuous distribution and epigenetic inactivation. Mol Biol Evol 19:510-520.
Daboussi MJ, and Langin T (1994). Transposable elements in the fungal plant pathogen *Fusarium oxysporum*. Genetica 93:49-59.
Daboussi MJ, Langin T, and Brygoo Y (1992). *Fot1*, a new family of fungal transposable elements. Mol. Gen. Genet. 232:12-16.
Danilevskaya O, Slot F, Pavlova M, and Pardue ML (1994). Structure of the Drosophila *HeT-A* transposon: a retrotransposon-like element forming telomeres. Chromosoma 103:215-224.
Daviere JM, Langin T, and Daboussi MJ (2001). Potential role of transposable elements in the rapid reorganization of the *Fusarium oxysporum* genome. Fungal Genet Biol 34:177-192.
Decaris B, Francou F, Kouassi A, Lefort C, and Rizet G (1981). Genetic instability in *Ascobolus immersus*: modalities of back-mutations, intragenic mapping of unstable sites, and unstable insertion. Cold. Spring. Harb. Symp. Quant. Biol. 45:509-517.
Decaris B, Francou F, Lefort C, and Rizet G (1978). Unstable ascospore color mutants of *Ascobolus immersus*. I. Temporal occcurence and modalities of back-mutations. Mol Gen Genet 162:69-81.
Deininger PL (1989). SINEs short interspesed repeated DNA elemenets in higher eucaryotes. American Society for Microbiology, Washington D.C.
Deleu C, Turcq B, and Begueret J (1990). *Repa*, a repetitive and dispersed DNA sequence of the filamentous fungus *Podospora anserina*. Nucleic Acids Res 18:4901-4903.
Deschamps F, Langin T, Maurer P, Gerlinger C, Felenbok B, and Daboussi MJ (1999) Specific expression of the *Fusarium* transposon *Fot1* and effects on target gene transcription. Mol Microbiol 31:1373-1383.
Doolittle WF, and Sapienza C (1980). Selfish genes, the phenotype paradigm and genome evolution. Nature (London) 284:601-603.
Ellis JG, Finnegan EJ, and Lawrence GJ (1992) Developing a transposon tagging system to isolate rust-resistance genes from flax. Theor Appl Genet 85:46-54.
Farman ML, Taura S, and Leong SA (1996a). The *Magnaporthe grisea* DNA fingerprinting probe MGR586 contains the 3′ end of an inverted repeat transposon. Mol. Gen. Genet. 251:675-681.
Farman ML, Tosa Y, Nitta N (1996b). *Maggy*, a retrotransposon in the genome of the rice blast fungus *Magnaporthe grisea*. Mol. Gen. Genet. 251:665-674.
Faßbender S, and Kück U (1995) Reverse transcriptase activities in mycelial fungi. In: Kück U (ed) The Mycota II, genetics and biotechnology. Springer, Berlin, Heidelberg, New York, Tokyo, pp 247-259
Federoff N (2000). Transposons and genome evolution in plants. Proc Natl Acad Sci USA 97:7002-7007.
Fernandez D, and Langin T (2002). Transposable elements in fungal pathogens: new diagnostic tools. In: Kempken F (ed) The Mycota XI. Agricultural Applications. Springer, Berlin, Heidelberg, New York, pp 171-192.
Fernandez D, Quinten M, Tantaoui A, Geiger JP, Daboussi MJ, and Langin T (1998). *Fot 1* insertions in the Fusarium oxysporum f. sp. albedinis genome provide diagnostic PCR targets for detection of the date palm pathogen. Appl Environ Microbiol 64:633-636
Fink G, Farabaugh P, Roeder G, and Chaleff D (1981). Transposable elements (*Ty*) in yeast. Cold Spring Harb Symp Quant Biol 45 Pt 2:575-580
Finnegan DJ (1989). Eukaryotic transposable elements and genome evolution. Trends Genet 5:103-107
Fitzmaurice WP, Nguyen LV, Wernsman EA, Thompson WF, and Conkling MA (1999). Transposon tagging of the sulfur gene of tobacco using engineered maize *Ac/Ds* elements. Genetics 153:1919-1928.

Flavell AJ (1995) Retroelements, reverse transcriptase and evolution. Comp Biochem Physiol 110:3-15
Fowler TJ, and and Mitton MF (2000). *Scooter*, a new active transposon in *Schizophyllum commune*, has disrupted two genes regulating signal transduction. Genetics 156:1585-1594.
Garfinkel DJ (1997). Genetic loose change: how retroelements and reverse transcriptase heal broken chromosomes. Trends Microbiol 5:173-175.
Garfinkel DJ, Boeke JD, and Fink GR (1985). *Ty* element transposition: reverse transcriptase and virus-like particles. Cell 42:507-517.
George MLC, Nelson RJ, Zeigler RS, and Leung H (1998). Rapid population analysis of *Magnaporthe grisea* by using rep-PCR and endogenous repetitive DNA sequences. Phytopathology 88:223-229.
Gierl A, and Saedler H (1992) Plant-transposable elements and gene-tagging. Plant. Mol. Biol. 19:39-49.
Girardin H, Latgé JP, Srikantha T, Morrow B, Soll DR (1993). Development of DNA probes for fingerprinting *Aspergillus fumigatus*. J Clinc Microbiol 31:1547-1554.
Girardin H, Sarfati J, Kobayashi H, Bouchara JP, and Latgé JP (1994a). Use of DNA moderately repetitive sequence to type *Aspergillus fumigatus* isolates from aspergilloma patients. J Infect Dis 169:683-685.
Girardin H, Sarfati J, Traoré F, Dupouy-Camet J, Derouin F, and Latgé JP (1994b). Molecular epidemiology of nosocomial invasive aspergillosis. J Clin Microbiol 32:684-690.
Giroux MJ, Clancy M, Baier J, Ingham L, McCarty D, and Hannah LC (1994). *De novo* synthesis of an intron by the maize transposable element *Dissociation*. Proc Natl Acad Sci USA 91:12150-12154.
Gómez-Gómez E, Anaya N, Roncero MIG, and Hera C (1999). *Folyt1*, a new member of the *hAT* family, is active in the genome of the plant pathogen *Fusarium oxysporum*. Fungal Genet Biol 27:67-76.
Gorbunova V, and Levy AA (1997). Circularized *Ac/Ds* transposons: formation, structure and fate. Genetics 145:1161-1169.
Gorbunova V, and Levy AA (2000). Analysis of extrachromosomal Ac/Ds transposable elements. Genetics 155:349-359.
Goyon C (1998). Isolation and identification by sequence homology of a second putative C5-DNA-methyltransferase gene from *Ascobulus immersus*. DNA Seq 9:109-112.
Goyon C, Barry CS, Gregoire A, Faugeron G, and Rossignol JL (1996a). Methylation of DNA repeats of decreasing sizes in *Ascobulus immersus*. Mol Cell Biol 16:3054-3065.
Goyon C, Faugeron G (1989). Targeted transformation of *Ascobolus immersus* and *de novo* methylation of the resulting duplicated DNA sequences. Mol Cell Biol 9:2818-2827.
Goyon C, Rossignol JL, and Faugeron G (1996b). Native DNA repeats and methylation in Ascobolus. Nucleic Acids Res. 24:3348-3356.
Hamann A, Feller F, and Osiewacz HD (2000). The degenerate DNA transposon Pat and repeat-induced point mutation (RIP) in *Podospora anserina*. Mol Gen Genet 263:1061-1069.
Hamer JE, Farrall L, Orbach MJ, Valent B, and Chumley FG (1989). Host species-specific conservation of a family of repeated DNA sequences in the genome of a fungal plant pathogen. Proc. Natl. Acad. Sci. USA 86:9981-9985.
Hansen LJ, and Chalker DL, Sandmeyer SB (1988). *Ty3*, a yeast retrotransposon associated with tRNA genes, has homology to animal retroviruses. Mol Cell Biol 8:5245-5256.
Haring MA, and Rommens CMT, Nijkamp HJJ, and Hille J (1991).The use of transgenic plants to understand transposition mechanisms and to develop transposon tagging strategies. Plant Mol Biol 16:449-461
Hauber J, Nelböck-Hochstetter P, and Feldmann H (1985). Nucleotide sequence and characteristics of a Ty element from yeast. Nucl Acids Res 13:2745-2758.
He C, Nourse JP, Kelemu S, Irwin JAG, and Manners JM (1996) *CgT1*: a non-LTR retrotransposon with restricted distribution in the fungal phytopathogen *Colletotrichum gloeosporioides*. Mol. Gen. Genet. 252:320-331
Hua-Van A, Daviere JM, Kaper F, Langin T, and Daboussi MJ (2000). Genome organization in *Fusarium oxysporum*: clusters of class II transposons. Curr Genet 37:339-347.
Hua-Van A, Pamphile JA, Langin T, and Daboussi M-J (2001). Transposition of autonomous and engineered *impala* transposons in *Fusarium oxysporum* and a related species. Mol Gen Genet 264:724-731.
Ichikawa H, Ikeda K, Wishart WL, and Ohtsubo E (1987). Specific binding of transposase to terminal inverted repeats of transposable element Tn3. Proc Natl Acad Sci USA 84:8220-8224.
Kachroo P, Ahuja M, Leong SA, and Chattoo BB (1997). Organisation and molecular analysis of repeated DNA sequences in the rice blast fungus *Magnaporthe grisea*. Curr Genet 31:361-369.
Kachroo P, Leong SA, and Chattoo BB (1994). *Pot2*, an inverted repeat transposon from the rice blast fungus *Magnaporte grisea*. Mol. Gen. Genet. 245:339-348.
Kempken F (1994). Unique features of a linear plasmid of *Ascobolus immersus* and its implications for plasmid evolution in fungi. Curr Topics Mol Genet 2:207-218.
Kempken F (1999). Fungal transposons: from mobile elements towards molecular tools. Appl Microbiol Biotechnol 52:756-760.

Kempken F (2001). *Hideaway*, a repeated element from *Ascobolus immersus* is rDNA associated and may resemble a class I transposon. Curr Genet 40:179-185.
Kempken F (2002) The Mycota XI, Agricultural Applications. Springer, Berlin, Heidelberg
Kempken F, Jacobsen S, and Kück U (1998). Distribution of the fungal transposon *Restless*: full-length and truncated copies in closely related strains. Fungal Genet Biol 25:110-118.
Kempken F, and Kück U (1996). *Restless*, an active *Ac*-like transposon from the fungus *Tolypocladium inflatum*: structure, expression, and alternative RNA splicing. Mol Cell Biol 16:6563-6572.
Kempken F, and Kück U (1998a). Evidence for circular transposition derivatives from the fungal *hAT*-transposon *Restless*. Curr Genet 34:200-203.
Kempken F, and Kück U (1998b). Transposons in filamentous fungi - facts and perspectives. BioEssays 20:652-659
Kempken F, Kück U (2000) Tagging of a nitrogen pathway-specific regulator gene in *Tolypocladium inflatum* by the transposon *Restless*. Mol Gen Genet 263:302-308.
Kempken F, Schreiner C, Schörgendorfer K, and Kück U (1995). A unique repeated DNA sequence in the cyclosporin-producing strain of *Tolypocladium inflatum* (ATCC 34921). Exp Mycol 19:305-313.
Kempken F, Windhofer F (2001). The *hAT* family: a versatile transposon group common to plants, fungi, animals, and man. Chromosoma 110:1-9.
Kennell JC, Moran JV, Perlman PS, Butow RA, and Lambowitz AM (1993). Reverse transcriptase activity associated with maturase-encoding group II introns in yeast mitochondria. Cell 73:133-146.
Kidwell MG, and Lisch D (1997). Transposable elements as sources of variation in animals and plants. Proc. Natl. Acad. Sci. USA 94:7704-7711.
Kinsey JA (1993). Transnuclear retrotransposition of the *Tad* element of *Neurospora*. Proc Natl Acad Sci 90:9384-9387.
Kinsey JA, Garrett-Engele PW, Cambareri EB, and Selker EU (1994). The *Neurospora* transposon *Tad* is sensitive to repeat-induced point mutation (RIP). Genetics 138:657-664.
Kinsey JA, and Helber J (1989) Isolation of a transposable element from *Neurospora crassa*. Proc Natl Acad Sci USA 86:1929-1933.
Kück U (1989). Mitochondrial DNA rearrangements in *Podospora anserina*. Exp Mycol 13:111-120.
Kumar A, and Bennetzen JL (1999). Plant retrotransposons. Annu Rev Genet 33:479-532.
Kunze R (1996) The maize transposable element *Activator* (*Ac*). Curr Top Microbiol Immunol 204:161-194
Kunze R, and Starlinger P (1989). The putative transposase of transposable element *Ac* from *Zea mays* L. interacts with subterminal sequences of *Ac*. EMBO J 8:3177-3185.
Langin T, Capy P, and Daboussi MJ (1995). The transposable element *impala*, a fungal member of the *Tc1-mariner* superfamily. Mol Gen Genet 246:19-28.
Lerat E, Capy P (1999). Retrotransposons and retroviruses: analysis of the envelope gene. Mol Biol Evol 16:1198-1207.
Leung H, Nelson RJ, and Leach JE (1993). Population structure of plant pathogenic fungi and bacteria. Adv Plant Pathol 10:157-205.
Levis C, Fortini D, and Brygoo Y (1996). *Flipper*, a bacterial-like transposable element in *Botrytis cinerea*. Fungal Genet Newsl 43B:46.
Levis C, Fortini D, and Brygoo Y (1997). Flipper, a mobile Fot1-like transposable element in *Botrytis cinerea*. Mol Gen Genet 254:674-680.
Levy M, Romao J, Marchetti MA, and Hamer JE (1991). DNA fingerprinting with a dispersed repeated sequence resolves pathotype diversity in the rice blast fungus. Plant Cell 3:95-102.
Li Destri Nicosia MG, Brocard-Masson C, Demais S, Hua Van A, Daboussi MJ, and Scazzocchio C (2001). Heterologous transposition in *Aspergillus nidulans*. Mol Microbiol 39:1330-1344.
Long D, and Coupland G (1998). Transposon tagging with Ac/Ds in Arabidopsis. Methods Mol Biol 82:315-328
Lönning WE, Saedler H (1997). Plant transposons: contributors to evolution? Gene 205:245-253.
Luo D, Coen ES, Doyle S, and Carpenter R (1991) Pigmentation mutants produced by transposon mutagenesis in *Antirrhinum majus*. Plant J 1:59-69.
MacGinnitie AJ, Anant S, and Davidson NO (1995). Mutagenesis of *apobec-1*, the catalytic subunit of the mammalian apolipoprotein B mRNA editing enzyme, reveals distinct domains that mediate cytosine nucleoside deaminase, RNA binding, and RNA editing activity. J Biol Chem 270:14768-14775.
Malagnac F *et al.* (1997). A gene essential for de novo methylation and development in Ascobulus reveals a novel type of eukaryotic DNA methyltransferase structure. Cell 91:281-290.
Malik HS, and Eickbush TH (2001). Phylogenetic analysis of ribonuclease H domains suggests a late, chimeric origin of LTR retrotransposable elements and retroviruses. Genomic Res 11:1187-1197.
Mannhaupt G *et al.* (2002). What's in the genome of a filamentous fungus? Analysis of the *Neurospora* genome sequence. Nucleic Acids Res, submitted.

Margolin BS *et al.* (1998) A methylated *Neurospora* 5S rRNA pseudogene contains a transposable element inactivated by repeat-induced point mutation. Genetics 149:1787-1797.

McClintock B (1947). Cytogenetic studies of maize and *Neurospora* mutable loci. Carnegie Inst Washington Year Book 46:146-152.

McClintock B (1951). Chromosome organization and genic expression. Cold Spring Harbor Symp Quant Biol 16:13-47.

McClintock B (1971). The contribution of one component of a control system to versatility of gene expression. Carnegie Inst Washington Year Book 70:5-17.

McDonald BA (1997) The population genetics of fungi: tools and techniques. Phytopathology 87:448-453

Mes JJ, Haring MA, and Cornelissen BJC (2000). *Foxy*: an active family of short interspersed nuclear elements from *Fusarium oxysporum*. Mol Gen Genet 263:271-280.

Michel F, Lang BF (1985). Mitochondrial class II introns encode proteins related to the reverse transcriptases of retroviruses. Nature 316:641-642.

Migheli Q *et al.* (1999). Transposition of the autonomous *Fot1* element in the filamentous fungus *Fusarium oxysporum*. Genetics 151:1005-1013.

Murata H, Miyazaki Y, and Yamada A (2001). *MarY2N*, a LINE-like non-long terminal repeat (non-LTR) retroelement from the ectomycorrhizal homobasidiomycete *Tricholoma matsutake*. Biosci Biotechnol Biochem 65:2301-2305.

Nakayashiki H, Kiyotomi K, Tosa Y, and Mayama S (1999). Transposition of the retrotransposon *Maggy* in heterologous species of filamentous fungi. Genetics 153:693-703.

Neuveglise C, Sarfati J, and Latge JP, Paris S (1996). *Afut1*, a retrotransposon-like element from *Aspergillus fumigatus*. Nucleic Acids Res. 24:1428-1434.

Nicolas A, Hamza H, Mekki-Berrada A, Kalogeropoulos A, and Rossignol JL (1987). Premeiotic and meiotic instability generates numerous *b2* mutation derivatives in Ascobolus. Genetics 116:33-43.

Nordborg M, and Walbot V (1995). Estimating allelic diversity generated by excision of different transposon types. Theor Appl Genet 90:771-775.

Okada N, and Hamada M (1997) The 3' ends of tRNA-derived SINEs originated from the 3' ends of LINEs: a new example from the bovine genome. J Mol Evol 44:S52-56

Okada N, Hamada M, Ogiwara I, and Ohshima K (1997). SINEs and LINEs share common 3′sequences: a review. Gene 205:229-243.

Orgel L, and Crick FHC (1980) Selfish DNA - the ultimate parasite. Nature (London) 284:604-607

Osiewacz HD, and Esser K (1984). The mitochondrial plasmid of *Podospora anserina*: a mobile intron of a mitochondrial gene. Curr Genet 8:299-305.

Pandit NN, and Russo VE (1992). Reversible inactivation of a foreign gene, *hph*, during the asexual cycle in . *Neurospora crassa* transformants. Mol Gen Genet 234:412-422.

Pietrokovski S, and Henikoff S (1997). A helix-turn DNA-binding motif predicted for transposases of DNA transposons. Mol Gen Genet 254:689-695.

Pöggeler S, and Kempken F (2003). Mobile genetic elements in mycelial fungi. In: Kück U (ed) THE MYCOTA II, genetics and biotechnology. Springer, Berlin, Heidelberg, New York, Tokyo

Radice AD, and Emmons SW (1993) Extrachromosomal circular copies of the transposon Tc1. Nucl Acids Res 21:2663-2667

Rogers TR (1995). Epidemology and control of nosocomial fungal infections. Curr Opin Infect Dis 8:287-290

Rossignol JL, Nicolas A, Hamza H, and Langin T (1984) Origins of gene conversion and reciprocal exchange in *Ascobolus*. Cold Spring Harb Symp Quant Biol 49:13-21.

Rountree MR, and Selker EU (1997). DNA methylation inhibits elongation but not initiation of transcription in *Neurospora crassa*. Genes Develop 11:2383-2395.

Ruan KS, and Emmons SW (1984). Extrachromosomal copies of the transposon *Tc1* in the nematode *Chaenorhabditis elegans*. Proc Natl Acad Sci USA 81:4018-4022.

Schwarz-Sommer Z, and Saedler H (1987). Can plant transposable elements generate novel regulatory systems? Mol Gen Genet 209:207-209.

Selker EU (1990). DNA methylation and chromatin structure: a view from below. Trends Biochem Sci 15:103-107

Selker EU (1997) Epigenetic phenomena in filamentous fungi: useful paradigms or repeat-induced confusion? Trends Genet 13:296-301.

Shull V, and Hamer JE (1996). Genetic differentiation in the rice blast fungus revealed by the distribution of *Fosbury retrotransposon*. Fungal Genet Biol 20:59-69.

Stahl U, Lemke PA, Tudzynski P, Kück U, and Esser K (1978). Evidence for plasmid-like DNA in a filamentous fungus, the ascomycete *Podospora anserina*. Mol Gen Genet 162:341-343.

Sundaresan V, and Freeling M (1987). An extrachromosomal form of the *Mu* transposon of maize. Proc Natl Acad Sci USA 84:4924-4928.

Talbot NJ (1998) Molecular variability of fungal pathogens: using the rice blast fungus as a case study. In: Bridge P, Couteaudier Y, Clarkson J (eds) Molecular variability of fungal pathogens. CAB International, Oxon, New York, pp 1-18

van Pouderoyen G, Ketting RF, Perrakis A, Plasterk RHA, and Sixma TK (1997). Crystal structure of the specific DNA-binding domain of Tc3 transposase of *C.elegans* in complex with transposon DNA. EMBO J 16:6044-6054.

Villalba F, Lebrun MH, Hua-Van A, Daboussi MJ, and Grosjean-Cournoyer MC (2001) Transposon *impala*, a novel tool for gene tagging in the rice blast fungus *Magnaporthe grisea*. Mol Plant Microbe Interact 14:308-315

Walther TC, Kennell JC (1999). Linear mitochondrial plasmids of *F. oxysporum* are novel, telomere-like retroelements. Mol Cell 4:229-238.

Weil CF, and Kunze R (2000). Transposition of maize *Ac/Ds* transposable elements in the yeast *Saccharomyces cerevisiae*. Nat Genet 26:187-190.

Whitcomb JM, and Hughes SH (1992). Retroviral reverse transcription and integration: progress and problems. Annu Rev Cell Biol 8:275-306.

Windhofer F, Catcheside DEA, and Kempken F (2000). Methylation of the foreign transposon *Restless* in vegetative mycelia of *Neurospora crassa*. Curr Genet 37:194-199.

Windhofer F, Hauck K, Catcheside DEA, Kück U, and Kempken F (2002). *Ds*-like *Restless* deletion derivatives occur in *Tolypocladium inflatum* and two foreign hosts, *Neurospora crassa* and *Penicillium chrysogenum*. Fungal Genet Biol 35:171-182.

Wöstemeyer J, and Kreibich A (2002). Repetitive DNA elements in fungi (Mycota): impact on genomic achitecture and evolution. Curr Genet 41:189-198.

Xiong Y, and Eickbush TH (1990) Origin and evolution of retroelements based upon their reverse transcriptase sequences. EMBO J 9:3353-3362.

Yeadon PJ, and Catcheside DEA (1995). *Guest*: a 98 bp inverted repeat transposable element in *Neurospora crassa*. Mol Gen Genet 247:105-109

Yuan GF, and Marzluf GA (1992). Molecular characterization of mutations of *nit4*, the pathway-specific regulatory gene which controls nitrate assimilation in *Neurospora crassa*. Mol Microbiol 6:67-73.

Applied Mycology & Biotechnology
An International Series. Volume 3. Fungal Genomics

6

Fungal Mitochondrial Genomes, Plasmids and Introns

Georg Hausner
Department of Microbiology, University of Manitoba, Winnipeg, MB, R3T 2N2, Canada (hausnerg@cc.umanitoba.ca).

Within the fungi mitochondrial genomes can exist as either linear or circular molecules, whose size variation is mostly due to the presence or absence of optional introns, and size variation in the intergenic regions. Optional introns can be either group I or group II introns, which are potential ribozymes that, in part, catalyze their own removal from the precursor RNA transcript. Mitochondria can also contain autonomously replicating DNA molecules, that are either derived from the mitochondrial DNA or represent true plasmids that show no homology with the mitochondrial chromosome. True plasmids are mostly cryptic in nature, and may have a different evolutionary origin from that of the mitochondrial host-genome. Amongst true plasmids at least three different categories can be recognized: (1) Circular plasmids encoding a DNA polymerase; (2) linear plasmids with terminal inverted repeats encoding a DNA and an RNA polymerase and; (3) retroplasmids, which are linear or circular plasmids that encode a reverse transcriptase. These different groups of true plasmids probably arose independently of one another, and were either vertically transmitted from the original endosymbiont that gave rise to the mitochondrion, or invaded the mitochondrion at various times during fungal evolution.

1. INTRODUCTION

Recent advances in DNA sequence technology have made possible characterization of entire genomes. As organellar genomes are quite small compared to nuclear genomes, they have been studied more intensively. However, only a few fungal mitochondrial genomes have been characterized to date. From the limited DNA sequence data available, and other studies dealing with additional aspects of fungal mitochondrial genetics, it is clear that mitochondrial genome sizes, structural features, presence and absence of introns and plasmids, are highly variable among the fungi. This review will attempt to provide a broad overview of the structural features that are found within fungal mitochondrial genomes and where data are available, components of the mitochondrial genome will be examined from a functional and evolutionary point of view. Current knowledge of plasmids and plasmid-like elements within fungal mitochondria are examined in this review along with an overview of what is known about introns and their encoded open reading frames and catalytic RNAs. The evolutionary significance of these elements and their potential applications in biotechnology are also addressed.

2. THE BIOLOGY OF FUNGAL MITOCHONDRIA

2.1 Mitochondria as Organelles and their Application to Biotechnology and Basic Research

Mitochondria are semiautonomous organelles dependent for their maintenance and function on genes encoded within both the nuclear and the mitochondrial genomes. Fungi are eukaryotic microorganisms of great economic importance, representatives have served as model systems to study a variety of cellular processes. For example, nucleo-mitochondrial interactions (Grivell 1995), mitochondrial import and export (Neupert 1997), and elucidation of such fundamental cellular processes as respiration and the synthesis of amino acid and other metabolites (Deacon 1997). Fungal mitochondrial genetics appears to have implications for the understanding of components of aging in eukaryotes (Griffiths 1992; Osiewacz 2002). In addition, respiratory defects in some fungal pathogens that are due to mitochondrial DNA (mtDNA) mutations, might be the cause of hypovirulence (attenuated virulence) (Mahanti *et al.* 1993; Monteiro-Vitorello *et al.* 1995). From an applied sense, the latter might have implications in the use of hypovirulent strains as a biocontrol strategy against virulent forms of a fungal pathogen (Baidyaroy *et al.* 2000a,b; Bertrand 2000). Contact between a hypovirulent strain and an aggressive one might permit the defective mitochondria from the hypovirulent strain to enter the aggressive strain and eventually render the recipient strain hypovirulent as the defective mitochondria slowly replace the normal organelles. The infectious nature of this process has been demonstrated in *Cryphonectria parasitica* (Monteiro-Vitorello *et al.* 1995; Baidyaroy 2000a, b). In a recent study of virulent members in the *Heterobasidium annosum* species-complex (causal agent of root and butt rot in conifers), virulence was shown to be controlled by the mitochondrial genome (Olsen and Stenlid 2001). This example, along with mitochondrial mutations have been implicated in some instances of hypovirulence, illustrates the importance and potential applications of mitochondrial genomics in plant pathology.

Features of the mitochondrial genome and its products are also important in biotechnology, such as production of metabolites (citric acid), and food production (Carlile and Watkinson 1994). Mitochondria can also be the target sites for certain fungicides (Deacon 1997). It has also been demonstrated that mtDNA shows high levels of mutations/variations as indicated by length variations (insertions/deletions), DNA sequence variation, and restriction site differences (RFLP). Therefore, detailed characterization of mtDNAs has become a routine strategy in taxonomic or phylogenetic studies that require estimates of genetic/evolutionary distances (Taylor 1986).

2.2 Origin of the Mitochondrial Genome

Sequence analysis of mitochondrial genomes strongly supports the belief that a single, endosymbiotic event, involving an -proteobacterium, gave rise to the mitochondrion (Gray *et al.* 1999; Lang *et al.* 1999). This event was quickly followed by both a reduction in the number of genes originally present within the ancestral mitochondrial genome, and a transfer of some genetic material from the protomitochondrion to the host nuclear genome. Phylogenetic analysis of nuclear-encoded mitochondrial protein sequences of *Saccharomyces cerevisiae* suggest that the mitochondrial proteome has at least two distinct origins: (1) genes relating to bioenergetic and translational processes appear to be related to -proteobacterial genomes; and (2) genes relating to transport and regulatory functions, which allowed the endosymbiont to develop into an ATP exporting organelle appear to have been recruited or co-opted from the original host nuclear genome (Karlberg *et al.* 2000). The fungal mitochondrial genome offers a relatively small chromosomal landscape that includes few genes but harbors many selfish DNA elements such as group I and group II introns. In addition, autonomous DNAs (plasmids) have been noted in some fungal mitochondria and appear to have varied origins (Kempken 1995a). In most instances these plasmids are cryptic (no phenotype), and some could be evolutionary relics that date back to the eubacterial origin of the mitochondrion. A group of plasmids that replicate via an RNA intermediate may be a link to the time of transition from an RNA to a DNA world (Lambowitz and Chiang 1995). Recent advances in comparative mitochondrial genomics offer insights into the evolution and composition of fungal mitochondrial genomes (Paquin *et al.* 1997; Gray *et al.* 1998).

2.3 Members of the Kingdom Fungi

From a modern phylogenetic perspective, fungal species can be assigned to the following groups: Chytridiomycota (zoosporic fungi), Zygomycota, Ascomycota and Basidiomycota. The chytridiomycetes are believed to have evolved first and are viewed as the "lower" fungi, while the ascomycetes and basidomycetes are believed to have arisen later in the course of evolution. But recently, on the basis of phylogeny of various protein_coding genes, the obligately parasitic amitochondriate intracelluar microsporidia have been shown to be highly derived fungi (Keeling and Fast 2002). In addition, molecular data suggest that the true fungi and the Metazoa share a common ancestry (Wainright *et al.* 1993; Paquin *et al.* 1997); both being probably derived independently from choanoflagellate-like protozoan ancestors (Cavalier-Smith 1998). Historically fungus-like organisms, such as members of the Oomycota and Hyphochytriomycota, were included within the "Kingdom Fungi", but it has been shown clearly that these organisms belong to the Kingdom Stramenopila, which includes autotrophic heterokont algae as well as the heterotrophic oomycetes, hyphochytrids, labyrinthulids, thraustochytrids, and bicosoecids (Leipe *et al.* 1994; Hausner *et al.* 2000). This review will focus on the mitochondrial genomes and associated plasmids of true fungi.

2.4 Mitochondrial Dynamics and Inheritance

Mitochondrial inheritance and the factors (nuclear genes) that mediate the movement and segregation of the mitochondrial DNA during mitotic growth or meiotic divisions are still poorly understood. However, some progress has been made in finding the factors that are involved in mitochondrial maintenance and transmission during cell proliferation in *Saccharomyces cerevisiae* (Yaffe 1999; Berger and Yaffe 2000; Contamine and Picard 2000; Boldogh *et al.* (2001); Kang and Hamasaki 2002). Many fungi are obligate aerobes and have filamentous growth patterns, which involve apical extension of potentially coenocytic hyphae; this allows for mixing and fairly free movement of the organelles. Further mixing of organelles can occur within and sometimes between fungal thalli as a result of hyphal fusion (anastomosis), which allows for exchanges of protoplasm. Therefore, the maintenance and proliferation of mitochondrial DNA in filamentous fungi is difficult to follow. Because mitochondria can fuse and then partition by fission, the DNA content from more than one mitochondrion can be mixed in a common environment and then recombination between different mtDNA molecules may occur (Westermann 2002).

In the budding yeast, mtDNA molecules are synthesized throughout the cell cycle and are present as 20-50 copies per mitochondrion. The distribution of the mtDNA molecules is not uniform throughout the mitochondrial matrix. Instead, discrete aggregations of mtDNA (nucleoids) have been observed, associated with the inner membrane (Miyakawa *et al.* 1984). Although the molecular mechanisms that mediate the formation, positioning, and segregation of nucleoids during vegetative growth are still largely unknown, recent studies have shown that a nucleoid structure does affect the inheritance patterns of mitochondrial genes, and that integrity of a nucleoid involves Holliday structures (Lockshon *et al.* 1995; Birky 2001). The latter structures might be functionally relevant because there is continued speculation that in the fungi, mtDNA replication is initiated by recombination. Recent reports on *Saccharomyces cerevisiae* suggest that mtDNA maintenance (segregation and replication) involves recombination (Ling and Shibata 2002), and that mtDNA inheritance from mother cell to bud is similar to that operating in the replication and packaging of phage DNA. However, mitochondrial motility, fusion, and fission are poorly understood in filamentous fungi (Westermann and Prokisch 2002).

Most of the genes required for mitochondrial maintenance reside in the nuclear genome, but for unknown reason mitochondria that are dysfunctional due to so-called "suppressive" mtDNA mutations proliferate rapidly in the fungal hypha and gradually displace organelles that contain wild-type mtDNA. The phenomenon of suppressive mtDNA mutations appears to be relevant to fungal senescence and hypovirulence in fungal plant pathogens (Bertrand 2000), and to degenerative mitochondrial diseases in humans cells. In all cases, the symptoms are elicited by a gradual accumulation of dysfunctional mitochondria and the concomitant progressive deterioration of respiratory capacity in affected tissues.

It is still unknown how rare mutations that occur in a single mtDNA molecule accumulate within an essentially polyploid organelle of filamentous or yeast-like fungi. Nor is it known how dysfunctional mitochondria eventually displace the wild-type mitochondria within the entire fungal thallus or tissue. But it has been demonstrated that mutant mitochondria can be passed on to wild-type strains by hyphal anastomosis (Griffiths 1992; Bertrand 1995) in a manner that resembles an "infectious process". It is postulated that respiratory deficient mitochondria are transmitted by hyphal contact/fusion after which they aggressively replace the normal mitochondria in the heteroplasmic mycelium. A number of mechanisms have been proposed to explain the suppressiveness of some mtDNA mutations in yeasts and filamentous fungi (reviewed in Bertrand 1995): 1. Unidirectional recombination between mutant and normal mitochondrial chromosomes whereby the wild type form is converted to the mutated (usually involves deletions) version. 2. Replicative advantage conferred on mutant mitochondrial chromosomes (for deletion mutations). 3. Biased transmission of mutant mitochondrial chromosomes (non-random segregation due to nucleoids). 4. Dysfunctional mitochondria replicate faster then the wild-type mitochondria due to a nuclear signal(s) that attempts to restore normal oxidative phosphorylation levels. The last of these would explain the suppressiveness of point mutations that can adversely affect ATP production.

In some ascomycetous fungi, crosses can be arranged whereby one strain can be the recipient (maternal) of a nucleus and the other strain can be the donor (paternal). Usually specialized structures (trichogyne = maternal) or cells (microconidia, spermatia etc. = paternal) are involved, so that one parent provides the cytoplasm in addition to a haploid nucleus (i.e., the maternal parent). In the filamentous ascomycetes, such as the heterothallic species *Podospora anserina* and *Neurospora crassa* (Röhr *et al.* 1999), or the homothallic species *Aspergillus nidulans* (Coenen *et al.* 1996), it has been demonstrated that mitochondria are "maternally" derived. In the majority of fungi, mitochondrial genomes are inherited predominantly from only one parent: uniparental inheritance. In members of the homobasidiomycetes or other fungi in which sexual matings rely initially on the fusion of two vegetatively compatible homokaryotic mycelia (plasmogamy) the terms maternal and paternal cannot be readily applied. Here, in theory, the resulting dikaryotic hyphae may contain a mixture of the parental-type mitochondria, potentially allowing for recombination between the parental mtDNAs. Nuclei of opposite mating type (i.e., compatible mating-type loci) will co-migrate and eventually fuse (karyogamy), initiating the process of meiosis resulting in the production of sexual spores. Generally in these situations, uniparental transmission of mitochondria still appears to be the case. So during the development of the dikaryotic mycelium there is a conversion of the initial heteroplasmon into a homoplasmon, due to the elimination of one mitochondrial haplotype. The non-random sorting of mitochondria that results in the selective exclusion, or rentention, of certain parental haplotypes is poorly understood but might be related in some way to either the nature or functioning of either the nuclear genotypes or mitochondrial haplotypes (Griffiths 1996; Barroso and Labarère 1997).

Biparental inheritance of mitochondrial genomes has been observed in the isogamous yeasts *Saccharomyces cerevisiae* and *Schizosaccharomyces pombe*. However, even in *S. cereviseae*, once a zygote starts to generate new buds, there appears to be non-random segregation of mtDNA and, after 20 successive rounds of vegetative growth, homoplasmons will be reestablished (Birky 2001).

3. THE MITOCHONDRIAL GENOME

3.1 Physical Characteristics

Fungal mitochondrial genomes tend to be AT-rich molecules that are highly variable in size, ranging from 19.4 kb in *S. pombe* to 175 kb in *Agaricus bisporus* (Hudspeth 1992). Although it was originally assumed that fungal mitochondrial genomes are circular, linear forms have been described for two yeast species, *Hansenula mrakii* and *Candida rhagii* (Wesolowski and Fukuhara 1981; Kavac *et al.* 1984), and for members of the oomycetes genus *Pythium* (McNabb and Klassen 1988; confirmed by Martin 1995). Recent reports suggest that linear mtDNAs might be more common than previously assumed (Nosek *et al.* 1998). Many mtDNAs yield circular physical maps, but experimental evidence suggests that some of them could be long linear concatemers, that are likely to be products of a rolling-circle mechanism of replication (Maleszka *et al.* 1991; Maleszka and Clark-Walker 1992; Bendich 1993, 1996; Ling and Shibata 2002). In some ascomycetous yeasts, the mtDNA consists of linear monomers that have covalently closed, single-stranded DNA termini (terminal hairpin-like structures) and ends that carry inverted repeats (Dinouel *et al.* 1993). A unique mtDNA architecture has been observed in *Hyaloraphidium curvatum*, a nonphotosynthetic freshwater nanoplankton that has recently been shown to be member of the lower fungi (Chytridiomycota); here the mitochondrial genome is a linear, monomeric molecule with identical inverted repeats at both ends (Forget *et al.* 2002). The mitochondrial genome of the chytrid *Spizellomyces punctatus* is also unusual. In *S. punctatus* the mitochondrial genome is segmented and consists of three circular molecules: a large 58 kpb molecule and two smaller 1.2 kpb molecules (Forget *et al.* 2002).

3.2 Gene Content, Coding Capacity and Composition of mtDNAs

3.2.1 Gene content relating to mitochondrial functions

Few fungal mtDNAs have been characterized by physical mapping and Southern hybridization studies, their gene content has been cataloged by Hudspeth (1992). And, more recently, the gene contents of fungal mtDNAs have been compiled by Paquin and Lang (1996), Gray *et al.* (1998); Lang *et al.* (1999); and Forget *et al.* (2002). Given the species diversity found within the Kingdom Fungi, very few complete mitochondrial genome sequences are actually available, but these can be retrieved from the National Center for Biotechnology Information (NCBI) site http://www.ncbi.nlm.nih.gov/PMGifs/Genomes/futax_short.html, GOBASE (organellar genome database, Shimko *et al.* 2001), and from the Fungal Mitochondrial Genome Project Homepage http://megasun.bch.umontreal.ca/People/lang/FMGP/progress.html (Paquin *et al.* 1997).

Genes encoded by fungal mitochondrial genomes can be categorized as follows: 1. RNA encoding genes involved in translation, e.g. the ribosomal small and large subunit RNAs (*rns* and *rnl*), and the *trnA-W* genes; 2. genes coding for protein involved in the respiratory chain: cytochrome oxidase subunits (*cox1*, *cox2*, and *cox3*), and the apocytochrome b (*cob*), subunits of the NADH dehydrogenase (*nad1* to *nad6*), and components of the ATP synthase (*atp6*, *atp8*, and *atp9*) and; 3. ribosomal proteins required for the assembly of other ribosomal proteins needed for building the small ribosomal subunit (one of the following: *rps3*, *var1*, or *S5*).

Variations among mitochondrial genomes within the fungi are due to the number of tRNA genes present and absence of the *nad* genes in some ascomycetes yeasts (Wolf and Del Giudice 1988; Hudspeth 1992; Wolf 1995). The *atp9* gene is missing in *Podospora anserina* and in *Neuropora crassa* although present, it is not active (Griffiths *et al.* 1995). A mitochondrial gene encoding the RNA subunit of RNase P (a ribozyme that is responsible for the 5' maturation of tRNA precursors) has been noted in members of the zygomycetes and ascomycetes but so far it has not been detected in members of the Chytridiomycota (Forget *et al.* 2002). In general mt rDNA genes are single copy genes (except for oomycetous mtDNAs with inverted repeats) and no 5S mt rRNA gene has been detected. Unusual mtDNA has been reported for *Hyaloraphidium curvatum* which has only 8 tRNA genes and the SSU rRNA is encoded in two segments that are 8 kbp apart (Forget *et al.* 2002). Whereas some eukaryotes, such as the jakobid *Reclinomonas americana* have about 27 ribosomal proteins encoded in the mitochondrion the mtDNAs of higher eukaryotes generally encode fewer mitochondrial ribosomal proteins (Gray *et al.* 1998). Many fungi appear to encode only one such protein (Bullerwell *et al.* 2000). Three mitochondrial ribosomal protein genes have been described within the true fungi. The *rps3* gene (ribosomal small subunit protein 3) has been noted in members of Chytridiomycota and in *Rhizopus stolonifer* (Zygomycota). The *var1* gene has been detected in the ascomycetous yeasts and the *S5* gene is found in the filamentous ascomycetes, encoded within a group I intron that is located in the *rnl* gene (Burke and BajBhandary 1982). While these three ribosomal proteins appear to have few sequence similarities, they have been shown recently to share a novel motif at the C-terminus, this suggests that *var1* and *S5* are homologs of *rps3* (Bullerwell *et al.* 2000).

3.2.2 Optional introns in mitochondrial genomes

Mitochondrial genomes in the true fungi are highly variable both in size and organization. Most of this size variation is due to the presence of introns and intron-encoded open reading frames (ORFs) (Wolf and Giudice 1988; Clark-Walker 1992; Gillham 1994; Belcour *et al.* 1997; Salvo *et al.* 1998). For example, the Oak Ridge laboratory strains of *Neurospora crassa* have a 62-kbp mitochondrial genome that contains ten introns which account for about 20 kb of the DNA (Collins 1993), whereas *Podospora anserina* race A has a mitochondrial genome that is approximately 100 kbp in size, about 60 kbp consists of 36 introns and intron-encoded ORFs (Cummings *et al.* 1990). One extreme example is provided by *Podospora anserina* in which DNA sequence analysis revealed that the *cox1* gene alone extends over 24.5 kilobase pairs and contains sixteen introns (Cummings *et al.* 1989). This is in contrast to the mtDNA of the fission yeast *S. pombe* in which the entire mitochondrial genome is composed of 19 431 nucleotides and contains only three introns (Lang 1984; Lang *et al.* 1985; Lang *et al.* 1999). Furthermore, for both *S. cerevisiae* and *S. pombe* it has been demonstrated that introns are dispensable genetic elements (Séraphin *et al.* 1987; Schäfer *et al.* 1991; Wolf 1994). Comparative studies among the budding yeasts, the fission yeasts, and species of *Aspergillus*, *Neurospora* and allied genera have demonstrated that at least some of these optional introns are mobile and probably even capable of inter species lateral transfer (Dujon 1989; Dujon and Belcour 1989; Clark-Walker 1992; Grivell 1995; Wolf 1994). Intron insertion occurs mainly in highly conserved sites within mitochondral genes. Even phylogenetically unrelated species from different kingdoms can contain introns in identical regions of homologous genes, thus supporting arguments for horizontal intron transmission (Wolff *et al.* 1993).

It has been experimentally demonstrated that in *Aspergillus japonicus* introns can be transmitted among heterokaryon incompatible strains after protoplast fusion (Hamari *et al.*

2001; Hamari *et al.* 2002). In nature, either transient or temporary hyphal anastomosis might allow for heteroplasmons wherein mitochondria from different strains can mix and fuse and introns can mobilize, generating recombinant mtDNAs due to gain or loss of introns (Hamari *et al.* 2002). In *S. cerevisiae* intron mobility can be demonstrated experimentally by crossing compatible strains wherein one parental mtDNA harbors a mobile intron (donor) and the other parent contributes mtDNAs that lack the equivalent intron (recipient) (Butow and Zinn 1986; Gillham 1994; Wolf 1996). A more detailed discussion of fungal mtDNA introns, their biology, RNA components and their encoded ORFs will follow later in this review (see section 4).

3.2.3 Non-coding sequence motifs/elements in mitochondrial genomes

In many fungi mtDNA size variability may also be due to the presence of AT-rich intergenic spacers comprising significant portions of that mtDNAs. More defined repetitive sequence motifs can be found in some mtDNAs, but their biological significance is not understood. Although these sequences might be strictly selfish DNAs, they might have regulatory functions or be DNA elements that are in a symbiotic relationship with their host's genome. One example of a non-coding sequence motif is the GC-rich *Pst*I palindromes that are widely distributed throughout the *Neurospora* mitochondrial chromosome (Yin *et al.* 1981). These palindromes are scattered throughout the AT-rich intergenic spacers and contain two *Pst*I restriction sites in their primary sequences; some are even found within introns of the mtDNA. They can form long highly-stable hairpin structures, and are thought to be preferred sites for recombination. While recombination at GC-rich regions has been implicated in the generation of large mtDNA deletions found in the so-called *stopper* mitochondrial mutants of *Neurospora* (De Vries *et al.* 1986), and also in the generation of plasmid-like elements (Gross *et al.* 1989a; Almasan and Mishra 1990), to date the biological significance of the mitochondrial GC-rich *Pst*I palindromes remains unclear. However, it is tantalizing to suggest that despite their potential for being involved in detrimental mtDNA recombination events, these structures may have been conserved because they are regions that are sites of initiation of either DNA replication, transcription initiation or primary sites for the processing of transcripts. It is worth noting that the positioning of some *Pst*I palindromes (Gross *et al.* 1989a, b) does mimic the arrangement of structural elements usually associated with origins of replication in yeast and animal mtDNAs. Here a GC-rich palindrome is located upstream of an AT-rich area that includes a promoter (De Zamaroczy and Bernardi 1985; Wolf 1995).

A similar structural type of repetitive element is found in *Saccharomyces cerevisiae* and related species, consisting of G + C-rich clusters, many of which can be folded into stem loop structures. These elements have been characterized as preferred recombination sites (Clark-Walker 1989, Weiller *et al.* 1991), and some of them are associated with initiation sites for DNA replication (de Zamaroczy and Bernardi 1985). It has also been shown that these G + C-rich clusters are potentially mobile and this mobility (gain or loss) explains the apparent size variations observed in the yeast *var*1 gene (Butow *et al.* 1985; Butow and Zinn 1986; Wenzlau and Perlman 1990). The mobility of the GC-clusters in yeast is thought to be due to a "cut and paste-like" mechanism analogous to that observed in prokaryotic IS elements (Butow *et al.* 1985; Weiler *et al.* 1989). The *Neuropora Pst*I palindromes are also thought to be mobile elements (Yin *et al.* 1981).

In members of the chytridiomycetous genus *Allomyces,* G + C-rich double-hairpin elements (DHE) have been characterized. Based on their overall distribution pattern within the mtDNA of other members of the chytridiomycetes, and quite possibly within members of the zygomycetes, they appear to be mobile and recombinogenic (Paquin *et al.* 2000). The presence of these enigmatic GC-rich structural elements in representatives of three major groups of fungi (Chytridiomycota, Zygomycota, and Ascomycota) suggests that these motifs might have functional significance.

Another category of dispersed repeated sequences has been noted in the mtDNAs of several *Podospora* species. These are mitochondrial ultra-short elements (MUSEs), e.g. the 11 bp sequence called MUSE1 (GGCGCAAGCTC) of *Podospora anserina*, which are highly recombinogenic and associated with the excision and amplification of short mitochondrial segments during the degenerative phenomenon called senescence (Koll *et al.* 1996). Koll *et al.* (1996) suggested that these MUSEs are highly invasive and contribute to the evolution (sequence polymorphism and mtDNA rearrangements) of the mitochondrial genome in the species of *Podospora.* They also suggest that these elements are mobile and that mobility involves a target, DNA-primed reverse transcription step mediated by mtDNA-intron encoded reverse transcriptases (see section 4.3).

3.2.4 Miscellaneous features: RNA and DNA polymerase segments

Fragmented versions of both RNA and DNA polymerases have also been found in the mtDNAs of a few fungi: *Podospora anserina* (Hermanns and Osiewacz 1994), *Neurospora* sp. (Nargang *et al.* 1992), and *Agaricus* sp. (Robison and Horgen 1996; Robison *et al.* 1991, 1997). These vestigial polymerase genes appear to be related to those found in true mitochondrial plasmids (Nargang *et al.* 1992; Oeser *et al.* 1993; Hermanns and Osiewacz 1994; Kempken 1995a; see section 5). It is possible that these fragmented genes result from plasmid integration via recombination with the chromosomal mtDNA and that integrated plasmid sequences have degenerated over time in the absence of selection pressure.

3.3 Synthesis of mRNA, the Genetic Code and RNA Editing

In general, mitochondrial genomes are transcribed in multigenic segments that are processed to produce mature mRNAs. In *Neurospora crassa* tRNAs and hairpin-like structures appear to be important for the processing of the primary transcripts (Kubelic *et al.* 1990). For detailed treatments of mitochondrial transcription, promoters and mitochondrial RNA processing, the reader is referred to reviews by Wolf (1995), Grivell (1995), and Kennel and Cohen (2003).

The universal genetic code appears to be used in the lower fungi: the Chytridiomycota (*Allomyces macrogynus*, *Hyaloraphidium curvatum*), and in the Zygomycota (*Rhizopus stolonifer*) (Paquin and Lang 1996). *Schizosaccharomyces pombe* an early branching member of the ascomycetes also appears to use the universal genetic code, but the remaining members of the Ascomycotina and Basidiomycotina so far characterized, appear to show some nonstandard uses within the genetic code (Griffiths 1996; Paquin *et al.* 1997). Thus, the universal genetic code appears to be ancestral to the fungi with changes evolving later.

RNA editing of mtDNA encoded tRNA's has been detected in only two members of the Chytridiomycota: *Hyaloraphidium curvatum* and *Spizellomyces punctatus* (Forget *et al.* 2002). In these chytrids RNA editing involves replacement of one to three nucleotides at the 5' end of the tRNA acceptor stem, and is thus analogous to the RNA editing first described in the amoeba *Acanthamoeba castellanii* (Lonergan and Gray 1993).

4. FUNGAL MITOCHONDRIAL INTRONS

4. 1 Overview of Fungal Mitochondrial Introns

In fungal mitochondria genomes, two classes of introns, group I and group II, have been described so far, and these can be distinguished from each other by their sequence, structure, and splicing mechanism (Michel *et al.* 1982; Michel *et al.* 1989; Michel and Ferat 1995). These intervening sequences are located within protein coding genes and ribosomal RNA genes, but unlike chloroplast genomes, the fungal mtDNAs studied to date lack introns within the tRNA genes (Lambowitz *et al.* 1999; Bonen and Vogel 2001; Belfort *et al.* 2002). Mitochondrial group-I and group-II introns have been associated with maternally-inherited senescence in *Podospora anserina* (Griffiths 1992; Gillham 1994), and also with mtDNA rearrangements in yeasts and filamentous ascomycetes (Dujon 1989). They can also be components of plasmid-like elements derived by intramolecular recombination events from the mitochondrial genome (Michel and Cummings 1985). Respiratory defects that arise from intron-splicing deficiencies either due to mitochondrial or nuclear gene mutations, have also been noted in budding yeasts and in *Neurospora crassa* (Dujon and Belcour 1989, Lambowitz and Perlman 1990, Gillham 1994). The introns are removed from the precursor RNA by an autocatalytic RNA splicing event that is mediated by the intron's RNA tertiary structure and proteins; the latter are encoded by either the intron or the nuclear genome.

Group I introns are found in the organelles of fungi, plants, protists, and in bacteria and their bacteriophages, as well as in ribosomal nuclear genes of fungi and protists; they have also been reported recently from the mtDNA's of sea anemones and soft corals (Belfort *et al.* 2002; van Open *et al.* 2000). Although group I introns show minimal primary sequence conservation they do have conserved secondary and tertiary structures. They have been shown to be potentially autocalytic (self-splicing) in vitro, and can therefore be viewed as ribozymes. In group I introns base-pairing interactions between the 5' end of the intron and flanking exon regions define the location of the 5' and 3' splice sites. Splicing of the ribozymic group I intron RNA's is by transesterification with an external guanosine as an initiating nucleophile; this results in a linear, excised intron (see Fig.1; Bonen and Vogel 2001).

Group II introns occur in organelles of fungi, plants, protists, and from bacteria. While group-I introns seem to predominate in the fungal mitochondrial DNAs, several group-II introns have also been noted and characterized (Zimmerly *et al.* 1995a; Shnyreva 1995; Dai *et al.* 2003). The ribozymic group II intron RNAs self-splice by a two-step transesterification involving a bulged adenosine as the initiating nucleophile, the end result being an excised intron in the lariat form (see Fig. 1; Jacquier 1996).

Typically group-II introns contain ORFs that code for reverse-transcriptase-like proteins. In contrast, group-I introns can encode proteins with maturase and/or endonuclease activity. These group I and group II ORFs can be either free standing within the intron, or be fused in frame to an upstream exon. In the latter case, it has been shown in yeast that such chimeric translation products are proteolytically cleaved to liberate the fused peptides, perhaps by a nucleus-encoded ATP dependent protease such as PIM1 (van Dyck *et al.* 1998). A group I intron encoded endonuclease has also been implicated in interspecific transfer of mitochondrial genes between two members in the Chytridiomycota (*Allomyces macrogynus* and *Allomyces arbusculus*) (Paquin *et al.* 1994). Furthermore, group II intron encoded ORFs have attracted considerable attention because they appear to have an evolutionary connection to telomerases and reverse-transcriptase-encoding retroelements such as non-LTR retrotransposons (i.e., LINEs), bacterial retrons, and fungal mitochondrial retroplasmids (Lambowitz 1989; Eickbush 1994, 1997; Eickbush and Malik 2002).

Group I and group II introns in fungal mtDNAs frequently encode ORFs that assist in RNA splicing (maturases) and/or in "intron mobility" (homing endonucleases). There are

Figure 1: Splicing of group I and group II introns

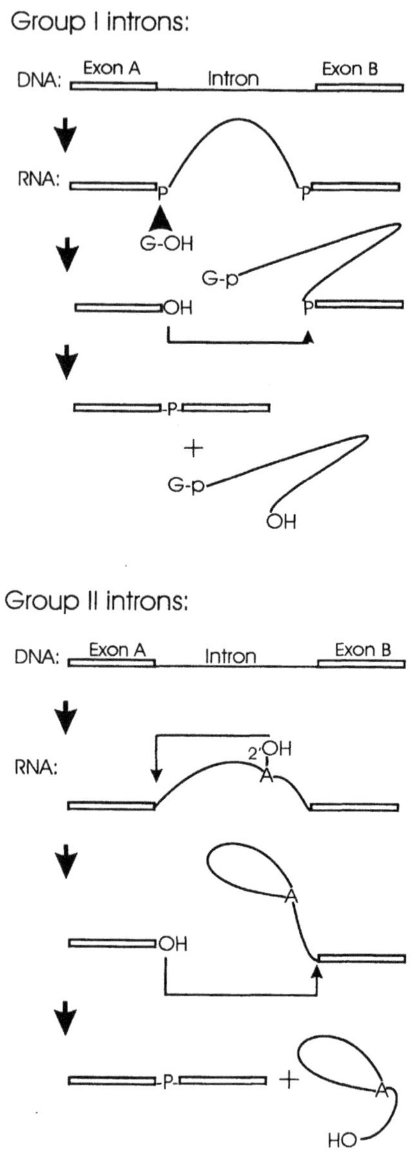

Figure 2: Structural features of fungal mitochondrial plasmids

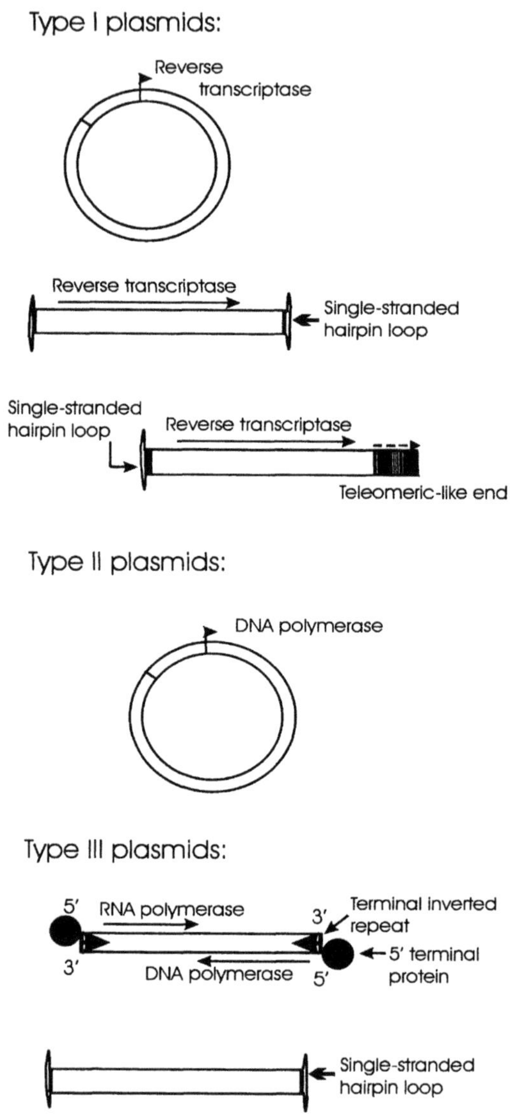

two types of "Intron mobility": (1) Intron homing, where the intron invades a specific site in a cognate intronless allele, implying that a particular intron encodes an endonuclease that has evolved to recognize a specific target site for invasion; and (2) intron transposition, where an intron inserts itself at different sites (alleles), referred to as ectopic integration.

4.2 Group I introns: Structure and mobility

By convention short conserved sequence motifs have been defined within group I intron sequences, including the P-, Q-, R-, and S-sequence motifs, each about 10 to 12 bp in length. These motifs are involved in the formation of paired and unpaired domains that comprise stems and helices within secondary RNA structure (Cech 1988; Michel and Westhof 1990). Based on secondary structure characteristics, group I introns were initially classified into two subdivisions designated I A and I B (Cech 1990; Saldanha *et al.* 1993). However, based on nucleotide sequences within the conserved core regions or on peculiarities within the secondary structure, group I intron classification has been further refined and now at least five classes of group I introns are recognized (IA to IE, which can be further subdivided e.g., IA1, IC3 etc.) (Michel and Westhof 1990; Suh *et al.* 1999). Over 1000 group I introns have been identified in a variety of organisms, and information about group I introns and their secondary structures have been compiled at the Comparative RNA Web Site (R. Gutell; http://www.rna.icmb.utexas.edu/). When ORFs are present within a group I intron, they are usually inserted in any of the several loops that protrude from the core secondary structure.

Group I intron mobility is catalyzed by an endonuclease, usually encoded by an ORF, that is embedded within the mobile intron. Intron-encoded homing endonucleases are usually cis-acting, and have specific target sites, with some allowance for base variation in their homing sites (target cleavage site). This ensures propagation against the forces of evolutionary drift which might modify the homing site within its host genome. Group I intron homing was first recognized for a 1.1 kb group I intron (called omega) found in the mitochondrial large ribosomal RNA (LSU rRNA) gene of *S. cerevisiae* (Dujon 1980). Detailed genetic and biochemical characterization of this intron has revealed that its mobility is mediated by the intron encoded homing endonuclease. The latter generates a double-stranded break at the cognate intronless allele, initiating a repair process involving unidirectional homologous recombination (gene conversion) using the intron-containing gene as a template (reviewed in Dujon 1989). Intron homing is also associated with the co-conversion of the sequences that flank the mobile intron. There is also experimental evidence that, in vitro, group I introns might be able to transpose into new sites in rRNA genes involving RNA intermediates via reverse splicing of RNA (Roman and Woodson 1995). This is a possibility as ribosomal RNAs are present at relatively high levels, offering targets for insertion of an intron RNA by reverse splicing. The resulting recombinant rRNA molecule would then have to be reverse transcribed into DNA and inserted by recombination into the mtDNA. This model of transposition would explain how group I introns that lack ORFs can avoid being lost and can either disperse into new positions or be transferred horizontally between different species. Loss of an intron can be envisioned by a mechanism that involves reverse transcription of a mature RNA (intron removed) followed by a recombination event that replaces the intron-containing DNA sequence with a DNA sequence without an intron.

4.2.1 Group I intron encoded maturases and homing endonucleases

Group-I intron ORFs in fungi have been shown to encode essential cellular proteins such as the rps3 (= S5) ribosomal protein, site-specific endonucleases, or maturases (Lambowitz and Belfort 1993). Most of the maturase-like proteins contain two highly-conserved symmetrically arranged dodecapeptide motifs which include conservative variants of the amino-acid sequence LAGLI-DADG (Michel and Cummings 1985, Pel and Grivell 1993). Maturases are thought to facilitate splicing by promoting proper folding of the intron RNA; actual splicing is catalyzed by the ribozyme itself.

Maturases, and maturase-like proteins, constitute the majority of mitochondrial intron-encoded polypeptides in the fungi. Although group I intron maturases are usually cis-acting there is at least one example of a trans-acting maturase, i.e. the *S. cerevisiae cytb* bI4 intron encoded maturase, which is required for splicing both the *cytb* bI4 intron and the *cox1* aI4 intron. Whereas most such proteins seem to promote RNA splicing, some maturases with LAGLI-DADG motifs actually cleave DNA and might be involved in intron homing (Lambowitz 1989; Lambowitz and Belfort 1993). In *Schizosaccharomyces pombe*, the first intron of *cox1* encodes a dual functional maturase/endonuclease protein, which is involved in RNA splicing and intron homing (Schäfer *et al.* 1994). Similar activities have been proposed for the proteins encoded by the ORFs in the aI4 intron in the *cox1* gene of *S. cerevisiae* (Wenzlau *et al.* 1989; Henke *et al.* 1995) and the intron in the *cytb* gene of *Aspergillus nidulans* (Ho *et al.* 1997). Furthermore, it has been demonstrated that modification of the amino-acid sequence can turn RNA maturases into DNA endonucleases (Goguel *et al.* 1992; Pel and Grivell 1993; Szcezepanek and Lazowska 1996), suggesting that there is an evolutionary link between endonuclease and maturase activity (Lambowitz and Belfort 1993).

The current view is that pre-existing group I introns recruited ORFs that were homing endonucleases, and that this allowed for a more precise dispersal mechanism. However, once an intron is established within a specific host gene, selection might favor the development of maturase activity over endonuclease activity because correct and efficient splicing would lessen the impact on the host gene, thus ensuring survival of the host and, therefore, survival of the intron (Goguel *et al.* 1992; Belfort *et al.* 2002). The reliance of group I introns on host factors for splicing further demonstrates that there is a selection pressure on both the group I intron and on the host genome for the intron to splice both accurately and quickly. Group I introns that lack ORFs may have evolved from ORF containing introns that are now completely dependent on host factors and thus their own maturases became redundant.

4.2.2 Host factors facilitating intron splicing

Genetic screens in *S. cerevisiae* have identified a plethora of nuclear factors that appear to be either directly or indirectly involved in the removal of mitochondrial introns (Pel and Grivell 1993). Such studies demonstrate the interdependence of organellar introns with their host nuclear genomes. However, the best studied system to date is the splicing of the group I intron in the *Neurospora crassa* mitochondrial LSU rDNA gene. Three strains with different nuclear mutations (*cyt-4*, *cyt-18*, *cyt-19*) were isolated that showed defective splicing of this intron (Bertrand *et al.* 1982). Subsequent experiments revealed that this intron would not self-splice in vitro unless mitochondrial lysates were added, indicating that splicing is protein dependent (Garriga and Lambowitz 1986). Cyt-4 appears to be an RNAse II-like protein that might be involved in the turnover of the excised group I intron (Turcq *et al.* 1992). Cyt-18 is a tyrosyl-tRNA synthetase that appears to interact with several group I introns to promote splicing by helping the intron RNA fold into a catalytically active structure (Akins and Lambowitz 1987) by interacting with tRNA-like structural motifs that are found in the catalytic core of group I introns (Mohr *et al.* 2001). Cyt-19 appears to be an ATP-dependent RNA chaperone, that can recognize and destabilize non-native RNA folds that might arise during the Cyt-18 mediated folding of group I intron RNAs (Mohr *et al.* 2002). Overall, from an evolutionary perspective what emerges from these studies is that intron RNAs could have interacted with cellular RNA-binding proteins and fortuitous interactions might have occurred that further promoted RNA folds required for splicing. In some cases such interactions would have been able to compensate for mutations within the intron RNA that prevented self-splicing. Thus the group I introns lost their ability to self-splice and are now dependent on these protein interactions.

4.2.3 Homing endonuclease genes

Homing endonucleases are found in all three domains of life: the Archaea, Eubacteria, and the Eukaryota. These elements have exploited all possible genetic niches available by locating themselves within nuclear, organellar, viral and bacteriophage genomes, as well as within plasmids and transposons (Dalgaard *et al.* 1997; Gimble 2000). This indicates that homing (site specific gene conversion) is a successful strategy to ensure long-term survival of a genetic element within natural populations.

In general, homing endonucleases are a diverse collection of proteins that are found in self-splicing mobile introns (Gimble 2000; Chevalier and Stoddard 2001), but they are also found in self-splicing inteins (protein introns) (Pietrokovski 2001). The target sites recognized by the homing endonucleases are usually quite long (14 to 44 base pairs) (Johansen *et al.* 1997; Belfort and Roberts 1997). Four families of homing endonuclease genes (HEGs) have been identified based on the presence of conserved amino acid sequence motifs, including the H-N-H, the His-Cys box, the GIY-YIG, and the LAGLIDADG type of endonucleases (Jurica and Stoddard 1999). The latter two have been identified within fungal mitochondrial group I introns (Belfort and Roberts 1997; Saguez *et al.* 2000), and the H-N-H domain is present within the ORFs of group II introns in fungi (Belfort and Roberts 1997).

Homing endonuclease genes may be mobile genetic elements in their own right, or they may be embedded within a mobile genetic element. In the latter situation, there might be a degree of mutualism between the two elements as HEGs would confer efficient target specific mobility to the host element in which they are embedded. The host element would provide a neutral position for the HEG that would ensure that the host genome would suffer minimal damage. That group I intron ORFs in fungal mitochondrial genomes are mobile elements in their own right has been illustrated in several studies. Mota and Collins (1988) noted that in two closely related species of *Neurospora*, structurally related group I introns in the *nad1* gene contain different coding sequences located at different positions within the intron. In *Podospora*, and in allied genera, the fourth intron in the *nad1* gene can be organized in one of three ways: intron with one ORF, two ORFs, or lacking ORFs (Sellem *et al.* 1996; Sellem and Belcour 1997). This suggests that the structural group I intron components and the embedded ORFs have evolved independently. A detailed study of 20 phylogenetically-related yeast species, including *S. cerevisiae*, demonstrated that the LSU rDNA gene could be found in four states: intron plus functional HEG (omega), intron with non-functional HEG, intron without an ORF, and intron-free (Goddard and Burt 1999).The four patterns do not match the phylogenetic relationships among the yeast species tested. Several conclusions were reached from these observations: (1) the intron and the HEG are optional sequences, (2) the HEG appears to be gained and lost, (3) gain of a HEG can occur by horizontal transmission, and (4) as both the intron and the HEG are optional, their survival depends upon frequent transposition and horizontal transmission.

4.2.4 Group I intron-encoded rps3 ribosomal protein in the filamentous ascomycetes

Group I introns varying in length from 1.1 to 4.2 kb have been reported at equivalent sites in the mitochondrial LSU rRNA genes (the U11 region near the 3' end) in a diverse set of ascomycetous species such as the yeasts, various *Kluyveromyces* spp. and *Saccharomyces* spp, and *Hansenula wingei* (Dujon 1980; Jacquier and Dujon 1983; Sekito *et al.* 1995; Goddard and Burt 1999), as well as the filamentous fungi *N. crassa, P. anserina, Penicillium chrysogenum, Cryphonectria parasitica* and *Aspergillus nidulans* (Burke and RajBhandary 1982; Cummings *et al.* 1990; Naruse *et al.* 1993; Hausner *et al.* 1999). Indeed, they probably occur in all remaining filamentous ascomycetes that diverged after the evolution of the budding and fission yeasts (Hausner unpublished data). Within the budding and fission yeasts this intron, when present, can encode a potential endonuclease, but in all filamentous ascomycetes so far examined, this intron encodes a putative rps3 ribosomal protein. In *Neurospora crassa*, this polypeptide has been identified as the mitochondrial small subunit ribosomal protein, rps3 (LaPolla and Lambowitz 1981), and it is assumed that the related proteins of the other fungi have similar functions. Encoding an essential host gene within an otherwise optional mobile element ensures the maintenance of the intron, and provides an efficient means of preserving essential host genes. A rather complex ORF is embedded within the LSU rDNA intron of *C. parasitica* where a potential coding region consists of an 851-codon open reading frame that encodes a putative, but complete, rps3 protein of 510 amino acids. This is fused at its carboxyl terminus to a 311 amino-acid polypeptide representing a typical maturase/endonuclease-like protein (Hausner *et al.* 1999). This arrangement suggests a recent insertion of a mobile HEG gene, as other filamentous ascomycetes appear to lack a HEG/maturase-like gene within this intron.

4.3 Group II Introns: Structure, ORFs, Mobility

Group II introns are potentially self-splicing RNAs that are widely believed to be the ancestors of nuclear pre-mRNA introns (Jacquier 1990). In the fungi, group II introns have been detected in protein-coding genes (reviewed in Shnyreva 1995) and, more recently, in the mitochondrial SSU rDNA (Toor and Zimmerly 2002) and LSU rDNA genes (Gonzalez *et al.* 1999). Mobile group II introns consist of approximately 600 nucleotides comprising the ribozymic component which surrounds an ORF-encoding segment of about two kbp. Group II introns, similar to group I introns, have a conserved secondary structure at the RNA level, that can be visualized as six stem-loop domains (domains I to VI) emerging from a central wheel (reviewed in Michel *et al.* 1989; Michel and Ferat 1995; Jacquier 1996). When reverse transcriptase-like ORFs are present they tend to be embedded within domain IV. Primary sequence conservation among group II introns is minimal except at the intron boundaries, with GUGYG and AY (Y = pyrimidines) defining the 5' and 3' ends, respectively. However, the most reliable diagnostic approach for confirming the presence of a group II intron is to search for the domain V consensus structure (Toor and Zimmerly 2002), and by analyzing a putative group II sequence by RNA folding via MFOLD (http://www.bioinfo.rpi.edu/applications/mfold/old/rna/).

Group II intron RNAs found in organellar genomes can be classified into two major subgroups based on specific structural features: subgroups IIA and IIB, which can be further segregated into IIA1, IIA2, IIB1, IIB2 (Michel *et al.* 1989). In general, the fungal-mitochondrial ORF containing group II introns can be assigned to subgroup IIA1. A recent analysis of ORF containing group II intron RNA structures from prokaryotic and eukaryotic sources found a total of six groups of intron structures: three were conventional forms of group IIA1, B1 and B2 secondary structures (Toor *et al.* 2001). There are additional subgroups in the bacteria, possibly associated with the most primitive ORFs (Zimmerly *et al.* 2001), that have unusual structures and hybrid features of group IIA and IIB introns (Toor *et al.* 2001). One current model for the evolution of group II introns, designated the retroelement ancestor hypothesis, predicts that the major structural forms of group II introns developed through co-evolution with the intron-encoded proteins rather than as independent catalytic RNAs, and that most introns lacking ORFs are derivatives of ORF-containing introns (Toor *et al.* 2001).

Mobile group II introns encode a multifunctional protein with three activities. First, nearly all group II-encoded ORFs are related and most of them include a segment homologous to reverse transcriptases (RT) (Michel and Ferat 1995, Lambowitz *et al.* 1999). In addition, a region referred to as domain X has been implicated in RNA splicing (Mohr *et al.* 1993) and therefore can be viewed as the maturase domain. The third activity is provided by the so called Zn domain which contains a potential zinc finger and has been implicated in endonuclease activity. However, the Zn domain is absent in a few of the fungal group II introns that have been characterized so far (see Zimmerly *et al.* 2001, for a compilation of mitochondrial group II intron encoded ORFs).

Both splicing and mobility activity of group II introns require catalytic activity of both the intron RNA, the intron-encoded protein, and possibly host factors (Zimmerly *et al.* 1995a, b; Eickbush 1999). After transcription of the host gene and translation of the encoded ORF, the protein will bind to the intron RNA and induce splicing. The ribo-nucleoprotein (RNP) particle (lariat intron RNA plus protein) recognizes a target homing site, and the first cut is made by the intron RNA. This initiates a reverse splicing reaction whereby the intron RNA is inserted into the sense DNA strand. The Zn domain cleaves the antisense DNA strand, generating a free 3'-OH that serves as a primer for the RT. Eventually, the host DNA repair machinery will remove the RNA and fill in any gaps. This process is termed target DNA-primed reverse transcription and has been reviewed in detail by Lambowitz *et al.* (1999) and Belfort *et al.* (2002). Group II introns can be mobilized by retrohoming (retro = RNA intermediate involved) where the intron moves from an intron containing allele to an intronless allele, or by retrotransposition where the intron is inserted into a new site (ectopic integration). Group II introns have also been shown to retrotranspose by reverse splicing into RNA molecules (see Bonen and Vogel (2001) for the various models of RT-mediated group II intron mobility).

Circular group II DNA introns have been observed as "plasmids" within some fungi (Osiewacz and Esser 1984). Two models are currently available to explain the appearance of such elements. The original model was based on the observation that group II introns can insert into sites that already harbor a homologous intron, creating a tandem duplication (Schmidt *et al.* 1994). Intramolecular recombination could result in the excision of one copy of the intron, releasing it as a circular DNA molecule. However, another model has been proposed recently based on the observation that sometimes group II introns can be removed from the precursor RNAs as stable circular RNA molecules instead of lariats (Murray *et al.* 2001). Thus reverse transcription of the circular RNA offers another mechanism that could generate plasmid-like elements composed of group II introns.

Recently, a new subfamily of group II introns has been detected within fungal mitochondrial ribosomal genes. Based on RNA structural features, these introns can be assigned to the IIB1 class of group II introns but, most strikingly, this subfamily harbors within its IV domain LAGLIDADG ORFs, which typically are associated with group I introns (Toor and Zimmerly 2002). As discussed previously, LADLIDADG type ORFs are homing endonucleases which are quite promiscuous and exhibit their own mobility, independent of the structural RNAs in which they can be embedded. It is not known whether these LADLIDADG ORFs contribute towards the mobility or splicing of the host group II intron.

4.4 Mobile Introns: Applications and Biotechnology

Catalytic RNAs are thought to be remnants of the RNA world. Therefore, group I and group II introns have been studied intensely as potential model systems for understanding ribozyme-catalyzed RNA cleavage reactions, which have implications for understanding early self-replicating systems in the RNA world (Landweber *et al.* 1998; Doudna and Cech 2002). The autocatalytic group II introns are viewed by many as the ancestors of the eukayotic nuclear spliceosomal introns, and their modes of transposition provide working models for intron invasion and dispersal within the eukaryotic nuclear genomes (Pyle 2000). Target DNA-primed reverse transcription is a mobility mechanism used by group II introns and LINE elements. LINE elements are retroelements that can comprise 10 to 40 % of eukaryotic genomes. Thus, detailed biochemical studies of group II introns may shed light on the biology of LINE elements (Eickbush 1997, 1999).

Ribozymes have also been proposed both as gene delivery systems for gene therapy (Guo *et al.* 2001) and as therapeutic agents that target RNA transcripts of mutated genes or viruses (ribozyme-directed chemotherapy, Johansen *et al.* 1997). Guo *et al.* (2001) showed that genetic manipulated group II introns can be retargeted to insert into desired sites. This ground breaking work forms the basis for practical applications of "targeted group II introns" in genetic engineering, functional genomics and gene therapy.

Artificial manipulation of group I introns may also have multiple applications. Group I ribozymes have been modified to reduce their size, had their target specificity for splicing/cleavage altered and their resistance to nucleases has been increased. The goal is to obtain trans-cleaving group I ribozymes that can inactivate specific nuclear or viral gene products by cleaving mRNAs (Johansen *et al.* 1997). Group I intron encoded homing endonucleases are rare cutting endonucleases that may be useful as tools in physical genome mapping (Belfort *et al.* 2002).

5. MITOCHONDRIAL PLASMIDS

5.1 Fungal Mitochondrial Plasmids and Plasmid-like Elements

In fungi mitochondrial plasmids can be defined as autonomously replicating circular or linear double-stranded extrachromosomal DNA molecules. However, unlike bacterial plasmids, most fungal plasmids appear to be cryptic in nature. Nonetheless, some mitochondrial plasmids have been associated with mitochondrial instabilities. Various aspects of the biology of circular and linear fungal plasmids have been reviewed previously (Rubidge 1992; Griffiths 1995; Kempken 1995a; Meinhardt *et al.* 1997; Griffiths 1998). Thus the goal of this review is to provide a general overview that emphases recent advances.

Mitochondrial extrachromosomal DNA molecules can be segregated into two categories: either plasmid-like elements or "true" plasmids. Plasmid-like mitochondrial elements (plMEs) are circular, covalently closed, oligomeric elements that are homologous to regions within the mitochondrial genome. True plasmids are either linear monomeric elements or circular (sometimes oligomeric) molecules that lack any homology with the mitochondrial chromosome; they evolved independently of the mitochondrial genome.

Mitochondrial plasmids are transmitted throughout fungal populations in various ways. They can move vertically, along with their host's mitochondrial genome, through asexual transmisson (mitotic spores) or sexual crosses (meiotic spores) or they can be transmitted horizontally across inter- and intraspecific boundaries (Griffiths *et al.* 1990; Arganoza *et al.* 1994; Kempken 1995b; Van der Gaag *et al.* 1998; Robinson and Horgen 1999; Bok and Griffiths 1999). "Paternal leakage" of plasmids has been noted in crosses where mitochondria are usually maternally inherited (Lee and Taylor 1993; Griffiths 1996), and in some instances plasmid elimination has been noted to occur as a consequence of meiotic transmission of mitochondria through sexual crosses (Debets *et al.* 1995; Chung *et al.* 1996). Independent transfer of mitochondrial chromosomes and plasmids during unstable vegetative fusion has been demonstrated in *Cryphonectria parasitica* (Baidyaroy *et al.* 2000a) and *Neurospora crassa* (Collins and Saville 1990). Factors involved in copy-number control, partitioning of plasmids during mitochondrial fission, or plasmid replication mode (strand displacement replication via rolling circles versus defined origins of replication for bidirectional or unidirectional DNA replication) are still poorly understood.

Plasmid-like mitochondrial elements (plMEs) are usually associated with nuclear or mitochondrial mutations. Although their formation and amplification might be dependent on genetic and physiological factors, very little is actually known about the mechanisms that regulate the maintenance and sexual and asexual transmission of small plasmid-like mtDNA derivatives within fungi. But once present, plMEs accumulate and behave like suppressive mutations (see section 2.4), that can be transmitted asexually via heterokaryosis or during formation of asexual spores (mitosis). However, some plasmid-like elements appear to be eliminated through sexual crosses (Griffiths 1992; Charter *et al.* 1993; Silliker *et al.* 1996).

5.2 Plasmid-like Mitochondrial Elements (plMEs)

Plasmid-like mitochondrial elements that are derived from regions of the mitochondrial DNA have been found in a variety of fungi. For example, in *Neurospora crassa,* mtDNA-derived multimeric plMEs have been discovered in some cultures of several cytoplasmic mutants, e.g. *poky* (Mannella *et al.* 1979) and *stp*-B1 strains (Almasan and Mishra 1990), and in the death-prone *stp*-ruv variant (Gross *et al.* 1989a, b). These nested sets of plMEs are similar to the mtDNAs of the respiration-defective, cytoplasmic *petite* mutants of *Saccharomyces cerevisiae*, which may consist of tandem repeats of as few as 35 bp of the 78-kb wild-type mtDNA (Dujon and Belcour 1989; Fangman *et al.* 1989). In the budding yeast these mtDNAs, can completely displace the wild-type DNA, resulting in the *petite* phenotype. Filamentous fungi are usually obligate aerobic organisms, thus the replacement of the wild-type mtDNA with defective mtDNA would have lethal consequences.

In *Podospora anserina*, the appearance of circular multimeric plMEs derived from mtDNA, the so-called senDNAṣ (Dujon and Belcour 1989; Jamet-Vierny *et al.* 1997a, b; Albert and Sellem 2002) have been correlated with senescence. *Podospora anserina* is viewed by some researchers as a model system for studying "aging" in fungi that is caused by the loss of respiratory competence leading to death (senescence). Analyses of mt DNA revealed that aging in *P. anserina* is associated with progressive mitochondrial dysfunction and with the liberation and amplification of a variety of senDNAs, the most common being derived from the first intron (a group II intron) of the mitochondrial cytochrome-oxidase I gene (*cox1*). This "liberated" group II intron appears as a multimeric, head-to-tail, circular, double-stranded DNA molecule, monomer size 2.54 kb, that has been named senDNẠ. A second frequently encountered plasmid-like element in senescing strains of *Podospora anserina* is called senDNẠ. The senDNẠ monomer size can vary from 4 to 20 kb, however all versions share a common 1.1 kb-long segment derived from the intergenic region downstream of the mtDNA *cox1* gene. It has been argued that the various senDNAs may have different replicative properties, and that this may, in some way, determine their lethality (Jamet-Vierny *et al.* 1997a). Various models for senescence in *Podospora* have been proposed. Some of these argue for nuclear genes being the cause, while others suggest that the accumulation of the senDNẠ ultimately leads to the death of a culture, either by inducing further mtDNA rearrangements or by the out-replication of the full-length mtDNA and, eventually, the displacement of the functional mtDNA. However, in *Podospora anserina,* the appearance of plMEs is not always associated with senescence. The persistence of such elements has been noted in longevity mutants (Turker *et al.* 1987, Silar *et al.* 1997) and their absence has been noted in aging mycelia of *Podospora curvicolla* (Böckelmann and Esser 1986). Detailed studies of long-lived strains of *P. anserina*, where thẹ sequence and the first exon of the *cox1* gene have been deleted from the mtDNA, showed that it was the deletion of segments of the *cox1* gene that increased life span and not the absence of senDNẠ. However, these studies did show that senDNẠ, although not the initiator or cause of senescence, can amplify the senescence process (Begel *et al.* 1999).

In the "ragged" cytoplasmic mutants of *Aspergillus amstelodami*, sets of mtDNA-derived sub-genomic, circular, multimeric molecules have been shown to co-exist with the intact wild-type mitochondrial genome (Lazarus *et al.* 1980). More recently, small circular plMEs that exist in multimeric forms have been found associated with a degenerative disease in the plant pathogenic fungus *Ophiostoma novo-ulmi* (Charter *et al.* 1993, Abu-Amero *et al.* 1995) and with a cytoplasmically-transmissible hypovirulence phenotype in a second plant pathogen, *Cryphonectria parasitica* (Monteiro-Vitorello *et al.* 1995).

In the filamentous fungi, the mechanisms involved in the initial formation of plMEs, their amplification, and the resulting physiological effects are still poorly understood (Griffiths 1992, Griffiths *et al.* 1995). Excision of circular segments of DNA by illegitimate, intra-molecular recombination events (Abu-Amero *et al.* 1995; Jamet-Vierny *et al.* 1997b; Röhr *et al.* 1999), and the synthesis of DNA by reverse transcription of spliced intronic RNAs (e.g., group II intron circles) have been implicated in the generation of plMEs in *Podospora* and *Neurospora* species (Murray *et al.* 2001). Amplification by reverse transcription of RNAs, or the presence of an origin of replication have been invoked to explain the autonomous replication of these elements (Steinhilber and Cummings 1986; Gross *et al.* 1989b; Almasan and Mishra 1990; Jamet-Vierny *et al.* 1999). Overall, several diverse mechanisms may lead to the production and maintenance of these elements.

5.3 Naturally Occurring Fungal Mitochondrial Plasmids

Mitochondrial plasmids are found relatively frequently in the filamentous fungi. They have been reported in members of at least 30 genera, and by far the majority are linear genetic elements (Griffiths 1995; Kempken 1995a; Meinhardt *et al.* 1997). Relatively few circular mitochondrial plasmids have been detected so far, not necessarily because they are scarce, but more likely because they are less easily detected than linear plasmids (Griffiths 1995). *Neurospora* plasmids are the best studied to date, with the nucleotide sequences of two linear plasmids, named *kalilo* and *maranhar*, and four circular plasmids, named *mauriceville*, *varkud*, *fiji* and *laBelle*, having been analyzed so far.

According to Monteiro-Vitorello *et al.* (2000) fungal mitochondrial plasmids can be classified into three groups (see Fig. 2): Type I plasmids that encode a reverse transcriptase, prototypes are the *mauriceville* and *varkud* plasmids; Type II circular plasmids that encode DNA polymerases, prototypes are the *fiji* and *laBelle* plasmids; Type III linear plasmids that encode a DNA and an RNA polymerase, prototypes are the *kalilo* and *maranhar* plasmids.

Recent work suggests that additional subgroups should be recognized within the type I and type III fungal plasmid groups. Most appropriately, Type I plasmids should be referred to as mitochondrial retroplasmids (mRPs, Akins *et al.* 1988) as they code for reverse transcriptases and appear to replicate via an RNA intermediate (Lambowitz and Chiang 1995). Type IA plasmids are circular retroplasmids, and Type IB are linear retroplasmids further sub-divisible into those that have a "clothespin" like structure with a teleomere-like end and covalently closed terminus (Type IB1) and those having both ends closed covalently (Type IB2). Type III plasmids can be divided into two subclasses. Type IIIA includes linear plasmids encoding DNA and RNA polymerases with 5' terminal attached proteins. Type IIIB includes linear plasmids with covalently closed ends (hairpin elements), but the Type IIIB plasmids are poorly understood and they appear to lack both recognizable ORFs and terminal attached proteins. Type IIIB plasmids resemble linear retroplasmids that have covalently closed ends (Type IB2), but the lack of a reverse transcriptase-like ORF makes assignment difficult.

5.3.1 Mitochondrial retroplasmids (Type I)

The best known mRPs are the small, circular *mauriceville* (3.6 kb) and *varkud* (3.7 kb) plasmids of *Neurospora* species, which encode functional reverse transcriptases, and replicate via an RNA intermediate (Lambowitz and Chiang 1995; Mohr *et al.* 2000). A small (2.6 kb) circular mRP element has also been described from *Trichoderma harzianum*, designated the pThr1 element, which encodes a putative reverse transcriptase (Antal *et al.* 2002). Studies on mRPs found in *Neurospora* species suggest the following model for replication. The host mitochondrial RNA polymerase transcribes the mRP, thus generating a transcript that, after

RNA processing, yields a full length linear RNA transcript with a 3' tRNA-like structure that ends with two tandem CCA sequences (Chen and Lambowitz 1997). As replication intermediates, these transcripts are recognized by the mRP encoded reverse transcriptase at the 3' tRNA-like structure and/or at the 3' CCA sequence where *de novo* cDNA synthesis is initiated (Chen and Lambowitz 1997).

Serial transfer of *Neurospora* strains harboring the *mauriceville* or *varkud* retroplasmids frequently results in erratic colony growth, respiratory dysfunction, mtDNA rearrangement due to integration of these plasmids and, eventually, senescence (Griffith 1992; Bertrand 2000). Variant forms of both the *mauriceville* and *verkud* retroplasmids have been detected in *Neurospora* strains. These appear to have arisen by recombination events that result in deletions of plasmid sequences, and insertion of mtDNA segments or segments of plMEs (Arganoza and Akins 1995; Stevenson *et al.* 2000; Fox and Kennell 2001). These variant forms can induce senescence in *Neurospora crassa* by over- replicating and/or by inserting into the mtDNA, but the rate at which strains degenerate and the rate at which retroplasmids accumulate and/or insert, are highly variable and do not necessarily correspond with each other. This suggests that senescence and/or tolerance of retroplasmids could be host-specific and dependent on the nuclear background of a particular strain (Fox and Kennell 2001).

Linear mitochondrial plasmids encoding reverse transcriptases have been found in *Fusarium oxysporum* (Kistler *et al.* 1997) and in *Rhizoctonia solani* (Katsura *et al.* 2001); and a linear retroplasmid has been recorded in *Epichloe typhina* (Mogen *et al.* 1991). However, the sequence available for the latter Et2.0L element appears to be truncated, as only some of the N-terminal domains of a putative reverse transcriptase ORF can be identified. The *Rhizoctonia solani* linear mRP (pRS224) consists of 4 986 nucleotides, encodes an ORF for a putative reverse transcriptase, and both termini are covalently closed “hairpin-like” structures of 236 and 264 nucleotides (Katsura *et al.* 2001). This plasmid forms a single-stranded circle when denatured.

It has been suggested that the mRPs are possibly the ancestors of retroviruses, and perhaps related to the earliest DNA-based life forms that emerged at the time of transition from an RNA to a DNA world (Wang and Lambowitz 1993; Weiner and Maizels 1994). The genomic tag theory proposed by Maizels and Weiner (1993) suggests that retroelements such as the mRPs, could be the ancestors of modern day eukaryotic chromosomes. In light of these theories, the two recently characterized linear mRPs of *Fusarium oxysporum* (pFOXC2, and pFOXC3) are intriguing because these elements appear to replicate via their encoded reverse transcriptases and one of the linear ends of the mRPs has structural features that resemble eukaryotic telomeres (Walther and Kennell 1999). The *Fusarium* mRPs are1.9 kb in length and their structures can be described as “clothespin-like”. One terminus has a hairpin configuration (covalently closed) and the other terminus has a telomere-like iteration of a 5 bp sequence.

In the past, the *varkud* and *mauriceville* mRPs were thought to be related to mitochondrial mobile elements, in particular introns that encode reverse transcriptases, i.e. group II introns (Nargang *et al.* 1984; Akins *et al.* 1988). The *mauriceville* plasmid can integrate into the mitochondrial DNA (Akins *et al.* 1986), but there is no indication of integrated copies behaving as introns, and no structural RNA elements have been identified so far that are associated with either group I or group II introns. Based on comparisons of the conserved reverse transcriptase domains (Eickbush 1994; Nakamura and Cech 1998) from all (circular and linear versions) mRPs available, it is clear that the plasmid-encoded reverse transcriptases are members of a monophyletic family of reverse transcriptases (Walther and Kennell 1999; Eickbush and Malik 2002, Antal *et al.* 2002). Further, phylogenetic analysis has shown that mRPs and the group II intron-encoded reverse transcriptases share a distant ancestor (Eickbush and Malik 2002). Analysis of these two families of reverse transcriptases suggests that exchanges of ORFs between group II introns and mRPs has not occurred, nor were mRPs derived from renegade mitochondrial group II introns that had adopted a "plasmid-like" lifestyle. Instead, mRPs appear to have had their own evolutionary origin, and are probably neither ancestors of, nor derived from, group II introns.

5.3.2 Circular plasmids encoding a DNA polymerase (Type II)

The best characterized circular, autonomously-replicating mitochondrial elements are the *fiji* and *laBelle* plasmids of *Neurospora intermedia.* Each has a single ORF that encodes a B-family DNA polymerase (Stohl *et al.* 1982; Li and Nargang 1993). And while some regions of the DNA polymerase encoded by the *laBelle* plasmid are similar to some domains characteristic of reverse transcriptases (Pande *et al.* 1989; Schulte and Lambowitz 1991), these features are not conserved in the polymerase encoded by the *fiji* plasmid (Li and Nargang 1993). Furthermore, it has been demonstrated that the proteins encoded by the *laBelle* and *fiji* plasmids lack reverse-transcriptase activity (Schulte and Lambowitz 1991; Li and Nargang 1993). Neither the *labelle* plasmid nor the related *fiji* element have been found to integrate into mtDNA nor induce senescence in *Neurospora intermedia.*

The *harbin-1* plasmid of *Neurospora intermedia* has DNA homology, as detected by hybridization, to the *laBelle* plasmid (Yang and Griffiths 1993), but this element has not been characterized sufficiently to determine whether it encodes a DNA polymerase. Circular mitochondrial plasmids have also been described from *Cryphonectria parasitica,* pUG1 (Gobbi *et al.* 1997) and the closely related pCRY1 plasmid (Monteiro-Vitorello *et al.* 2000). The pUG1 and pCRY1 plasmids contain a long open reading frame that is transcribed and potentially encodes a 1214 amino acid B-family DNA polymerase similar to those encoded by the *laBelle* and *fiji* circular mitochondrial plasmids of *Neurospora* species. A comparison of isogenic, plasmid-free and plasmid-containing cultures of *C. parasitica,* indicates that pCRY1 is an infectious agent (can move horizontally) capable of reducing the pathogenicity of some, but not all, strains of this fungus (Monteiro-Vitorello *et al.* 2000; Baidyaroy *et al.* 2000a).

5.3.3 Linear Mitochondrial Plasmids (Type III)

Linear plasmids are present in both the cytoplasm and organelles of many lower and higher eukaryotes (Griffiths 1995, 1998; Kempken 1995a; Meinhardt *et al.* 1997). And within the ascomycetous yeasts the type of plasmid present can be distinguished readily as mitochondrial plasmids can be cured (eliminated) by treating cells with ethidium bromide, and cytoplasmic plasmids can be selectively eliminated by UV irradiation (Blaisonneau *et al.* 1999). The most common linear forms of mitochondrial plasmids in the filamentous fungi are invertron-like elements that encode a DNA and an RNA polymerase, have terminal inverted repeats (TIR) and have proteins covalently attached to both 5' ends. Phylogenetic analysis based on the ORFs of the mitochondrial linear plasmids suggests that these plasmids share a common ancestor with some phages and the adenovirus (Kempken 1995a). Terminal inverted repeats are important for the formation of replication intermediates. They also contain sequence motifs required for both transcription and replication (Meinhardt *et al.* 1997). As reported so far, linear plasmids with TIRs in the yeasts are located in the cytoplasm (Gunge 1995) and are more complex in organization and function then those in the filamentous fungi. In yeasts, linear cytoplasmic DNA elements encode a killer toxin and the protein that enables the host to tolerate the plasmid encoded toxin (Fukuhara 1995). The first true yeast mitochondrial plasmid has recently been reported from *Pichia kluyveri* (Blaisonneau *et al.* 1999). It is a 7.1 kbp element and is a typical member of the type IIIA group, it has 5' terminal bound proteins and it encodes a DNA and an RNA polymerase.

Although the current view is that linear plasmids do not produce obvious physiological or phenotypic effects in their hosts, the *kalilo* and *maranhar* plasmids of *Neurospora* species (Bertrand *et al.* 1986; Chan *et al.* 1991; Court *et al.* 1991) induce senescence by integrating into the mitochondrial chromosome. Plasmid integration appears to be a rare event, senescence presumably is the result of accumulation of suppressive defective mtDNAs, possibly generated by a single integration event (Chan *et al.* 1991). In contrast, the life span of the fungus *Podospora anserina* is prolonged substantially by the integration of the pAL2-1 linear plasmid into the mitochondrial chromosome (Hermanns *et al.* 1994). Some *Neurospora* species likely benefit from the presence of benign versions of the *kalilo* plasmid (i.e., they do not insert into mtDNA) such as the LA-*kalilo* form (Griffiths 1998). Here, the presence of the plasmid has been associated with an increased tolerance to higher temperatures, and an increase in fertility as measured by an increase in perithecial production (Bok and Griffiths 2000). Linear plasmids have been found in several plant pathogenic fungi such as *Glomerella musae*, *Tilletia* spp, *Fusarium oxysporum* f. sp. *cucurbitae*, *Cochliobolus heterostrophus*, *Gaeumannomyces graminis* var. *tritici*, and *Claviceps purpurea* (reviewed in Freeman *et al.* 1997; Meinhardt *et al.* 1997), but reports of the effect of these plasmids on virulence and pathogenicity are somewhat conflicting. This suggests that the interaction between plasmids and their hosts is complex, probably involving nuclear and mitochondrial factors as well as structural features of the plasmid and the proteins they encode.

In fungal linear plasmids, the presence of TIRs and terminal proteins bound to the 5' ends indicates that they likely replicate via a protein primed mechanism similar to that observed in adenovirus (Kornberg and Baker 1992). In the past it had been speculated that for the *kalilo* plasmids and other linear plasmids the 5' terminal proteins were plasmid encoded (Vierula *et al.* 1990; Kempken 1995a). Recently it has been shown in the linear *Pleurotus ostreatus* pMLP1 element that the 5' terminal protein is indeed encoded by the plasmid DNA polymerases gene (Kim *et al.* 2000).

A different type of mitochondrial linear plasmid has been noted in various vegetative incompatibility groups of *Rhizoctonia solani*. Unlike typical linear plasmids that are resistant to 5' exonuclease digestion (due to 5' terminally bound proteins) but susceptible to 3' digestion, a group of plasmids (pRS 64-1, pRS64-2, pRS64-3, pRS104, pRS188) from *R. solani* is resistant to both 3' and 5' exonuclease treatment. These plasmids have covalently closed termini (hairpin plasmids), and the hairpin loops differ in size (113 and 105 bp), shape and sequence (Katsura *et al.* 1997). In general these plasmids share regions of sequence homology, have hairpin-like structures that can be folded into cruciform base-paired regions. While no biologically significant ORFs have been detected, some of these plasmids contain a small putative ORF encoding a potential 68 or 91 amino acid peptide that has been implicated in vegetative incompatibility (Hongo *et al.* 1994). So far these plasmids have not been shown to be associated with pathogenicity. From a structural view point these hairpin plasmids share many features with the 4.98 kbp pRS224 hairpin retroplasmid that has been found in *R. solani*. It is therefore possible that in *R. solani* the ORF-less type hairpin plasmids are derived from linear retroplasmids.

6. CONCLUSIONS

Fungal mitochondria are warehouses of evolutionary relics such as catalytic RNA molecules, retroplasmids, and viral derived linear plasmids. Mitochondrial group I and group II introns offer a range of catalytic RNAs that could be of great value to biotechnology as ribozymes that can be designed to cleave RNA molecules, thus silencing nuclear genes or neutralizing virus gene expression. Homing endonucleases encoded by many group I introns also offer an almost untapped reserve of novel and rare cutting endonucleases. However, more biochemical work is needed to characterize these proteins and to examine their mode of target site recognition. Although only a relative small number of plasmids has been characterized from a limited number of fungal taxa, at least three major types of plasmids have been delineated. The biology of these plasmids, and their interaction with both the mitochondrial and nuclear genomes, is still poorly understood and needs to be addressed in order to truly understand the possible phenotypic effects these elements have on their hosts.

Techniques for the genetic manipulation of mitchondrial genomes in filamentous fungi are not yet available. The application of mitochondrial plasmids as vector systems requires considerably more basic research so that the biology (mode of transmission, replication, copy number control) of these elements can be understood. Nor is a transformation system that can target the mitochondrion yet available for filamentous fungi. Also more work is needed on potential selectable markers, efficient universal origins of replication, and promoters. The plMEs might offer an opportunity for defining origins of replication that might at least be species- specific however, plMEs in many instances are associated with mitochondrial instabilities and their inheritance patterns can be unpredictable.

Although the gene content of fungal mtDNAs is fairly consistent among the major classes of fungi, the nature of the intergenic regions, structural segments that are potentially mobile (MUSE segments, G+C-rich clusters etc.), are still poorly understood and must await further study and the availability of more mtDNA sequences from a greater variety of fungi. The availability of both the nuclear and mitochondrial genome sequences for several model system fungi (*S. cerevisiae*, *S. pombe*, *N. crassa*, and soon *Podospora* and *Aspergillus*) combined with modern techniques involving micro-arraying of genes on DNA chips that allows for simultaneous monitoring of all transcribed regions should greatly benefit our understanding of mitochondrial function and nucleo-mitochondrial interactions. With regards to plant pathogenic fungi, a better understanding of mitochondrial function offers an opportunity to screen natural strains or to generate mutants that have cytoplasmically contagious hypovirulence phenotypes associated with respiratory defects. This would provide an alternative control strategy for some plant pathogens.

Mitochondrial chromosomes and extrachromosomal elements are an exciting source of material for investigating diverse biological processes including evolution, RNA splicing, catalytic RNA, transposition, intracellular communication etc. Mitochondrial genomes also provide the basis for the development of molecular genetic tools for biotechnological and ecological applications.

Acknowledgments: I would like to thank Drs. D.A. Court, G.R. Klassen, J. Reid, and D.C. Bay and M. Young, for valuable discussion and comments. Critical comments by B. Franz Lang and two reviewers were also invaluable. I am also grateful to John C. Kennell for providing preprints of "in press" manuscripts. Funding from the National Sciences and Engineering Research Council of Canada (NSERC) is also gratefully acknowledged.

REFERENCES

Abu-Amero SN, Charter NW, Buck KW, and Brasier CM (1995). Nucleotide-sequence analysis indicates that a DNA plasmid in a diseased isolate of *Ophiostoma novo-ulmi* is derived by recombination between two long repeat sequences in the mitochondrial large subunit ribosomal RNA gene. Curr Genet 28:54-59.

Akins RA, Kelley RL, and Lambowitz AM (1986). Mitochondrial plasmids of *Neurospora*: integration into mitochondrial DNA and evidence for reverse transcription in mitochondria. Cell 47:505-516.

Akins RA, and Lambowitz AM (1987). A protein required for splicing group I introns in *Neurospora* mitochondria is mitochondrial tyrosyl-tRNA synthetase or derivative thereof. Cell 50:331-345.

Akins RA, Grant DM, Stohl LL, Bottorf DA, Nargang FE, and Lambowitz AM (1988). Nucleotide sequence of the Verkud mitochondrial plasmid of *Neurospora* and synthesis of a hybrid transcript with a 5' leader derived from mitochondrial DNA. J Mol Biol 204:1-25.

Albert B, and Sellem CH (2002). Dynamics of the mitochondrial genome during *Podospora anserina* aging. Curr Genet 40:365-373.

Almasan A, and Mishra NC (1990). Characterization of a novel plasmid-like element in *Neurospora crassa* derived mostly from the mitochondrial DNA. Nucleic Acids Res 18:5871-5877.

Antal Z, Manczinger L, Kredics L, Kevei F, and Nagy E (2002). Complete DNA sequence and analysis of a mitochondrial plasmid in the mycoparasitic *Trichoderma harzianum* strain T95. Plasmid 47:148-152.

Arganoza MT, Min J, Hu Z, and Akins RA (1994). Distribution of seven homology groups of mitochondrial plasmids in *Neurospora*: evidence for widespread mobility between species in nature. Curr Genet 26:62-73

Arganoza MT, and Akins RA (1995). Recombinant mitochondrial plasmids in *Neurospora* composed of Verkud and a new multimeric mitochondrial plasmid. Curr Genet 29:34-43.

Baidyaroy D, Glynn JM, and Bertrand H (2000a). Dynamics of asexual transmission of a mitochondrial plasmid in *Cryphonectria parasitica.* Curr Genet 37:257-267.

Baidyaroy D, Huber DH, Fulbright DW, and Bertrand H (2000b). Transmissible mitochondrial hypovirulence in a natural population of *Cryphonectria parasitica.* Mol Plant Microbe Interact 13:88-95.

Barroso G, and Labarère J (1997). Genetic evidence for nonrandom sorting of mitochondria in the basidiomycete *Agrocybe aegerita.* Appl Environ Microbiol 63:4686-4691.

Begel O, Boulay J, Albert B, Dufour E, and Sainsard-Chanet A (1999). Mitochondrial group II introns, cytochrome c oxidase, and senescence in *Podospora anserina.* Mol Cell Biol 19:4093-4100.

Belcour L, Rossignol M, Koll F, Sellem CH, and Oldani C (1997). Plasticity of the mitochondrial genome in *Podospora.* Polymorphisms for 15 optional sequences: group I, group II introns, intronic ORFs and an intergenic region. Curr Genet 31:308-317

Belfort M, and Roberts RJ (1997). Homing endonucleases: keeping the house in order. Nucleic Acids Res 25:3379-3388.

Belfort M, Derbyshire V, Parker MM, Cousineau B, and Lambowitz AM (2002). Mobile Introns: Pathways and Proteins. In: NL Craig, R Craigie, M Gellert, and AM Lambowitz, Ed. Mobile DNA II. Washington DC: ASM Press. pp 761-783.

Bendich AJ (1993). Reaching for the ring: the study of mitochondrial genome structure. Curr Genet. 24:279-290.

Bendich AJ (1996). Structural analysis of mitochondrial DNA molecules from fungi and plants using moving pictures and pulsed-field electrophoresis. J Mol Biol 255: 564-588.

Berger KH, and Yaffe MP (2000). Mitochondrial DNA inheritance in *Saccharomyces cerevisiae*. Trends Microbiol 8:508-513.

Bertrand H (1995). Senescence is coupled to induction of an oxidative phosphorylation stress response by mitochondrial DNA mutations in *Neurospora.* Can J Bot 73 (Suppl 1): S198-S204.

Bertrand H (2000). Role of mitochondrial DNA in the senescence and hypovirulence of fungi and potential for plant disease control. Annu Rev Phytopathol 38:397-422.

Bertrand H, Bridge P, Collins RA, Garriga G, and Lambowitz AM (1982). RNA splicing in *Neurospora* mitochondria. Characterization of new nuclear mutants with defects in splicing the mitochondrial large rRNA. Cell 29:517-526.

Bertrand H, Chan B S-S, and Griffiths AJF (1985). Insertion of a foreign nucleotide sequence into mitochondrial DNA causes senescence in *Neurospora intermedia.* Cell 41:877-884.

Bertrand H, Griffiths AJF, Court DA and Cheng CK (1986). An extrachromosomal plasmid is the etiological precursor of kalDNA insertion sequences in the mitochondrial chromosome of senescent *Neurospora.* Cell 47:829-837.

Birky Jr CW (2001). The inheritance of genes in mitochondria and chloroplasts: laws, mechanisms, and models. Annu Rev Genet 35:125-148.

Blaisonneau J, Nosek J, and Fukuhara H (1999). Linear DNA plasmid pPK2 of *Pichia klyveri*: Distinction between cytoplasmic and mitochondrial linear plasmids in yeasts. Yeast 15:781-1999.

Böckelmann B, and Esser K (1986). Plasmids of mitochondrial origin in senescent mycelia of *Podospora curvicolla.* Curr Genet 10:803-810.

Bok J-W, and Griffiths AHF (1999). Transfer of *Neurospora* kalilo plasmids between species and strains by introgression. Curr Genet 36:275-281.

Bok J-W, and Griffiths AHF (2000). Possible benefits of kalilo plasmids to their *Neurospora* hosts. Plasmid 43:176-180.

Boldoght IR, Yang H-C, and Pon LA (2001). Mitochondrial inheritance in budding yeast. Traffic 2:368-374.

Bonen L, and Vogel J (2002). The ins and outs of group II introns. Trends Genet 17:322-331.

Bullerwell CE, Burger G, and Lang BF (2000). A novel motif for identifying Rps3 homologs in fungal mitochondrial genomes. Trends Biochem Sci 25:363-365.

Burke JM, and RajBhandary UL (1982). Intron within the large rRNA gene of *N. crassa* mitochondria: A long open reading frame and a consensus sequence possibly important in splicing. Cell 31:509-520

Butow RA, Perlman PS, and Grossman LI (1985). The unusual *var1* gene of yeast mitochondrial DNA. Science 228:1496-1501.

Butow RA, and Zinn AR (1986). Mobile elements in the yeast mitochondrial genome. Basic Life Sci 40: 29-37.

Caprara MG, Mohr G, and Lambowitz AM (1996). A tyrosyl tRNA synthetase protein induces tertiary folding of the group I intron catalytic core. J Mol Biol 257:512-531

Carlile MJ, and Watkinson SC (1994). The Fungi. San Diego, CA.: Academic Press Inc.

Cavalier-Smith T (1998). A revised six-kingdom system of life. Biol Rev 73:203-266.

Cech TR (1990). Self-splicing of group-I introns. Annu Rev Biochem 55:599-629.

Chan BS, Court DA, Vierula PJ, and Bertrand H (1991). The kalilo linear senescence-inducing plasmid of *Neurospora* is an invertron and encodes DNA and RNA polymerases. Curr Genet. 20:225-237.

Charter NW, Buck KW, and Brasier CM (1993). De-novo generation of mitochondrial DNA plasmids following cytoplasmic transmission of a degenerative disease in *Ophiostoma novo-ulmi*. Curr Genet 24:505-514.

Chen B, and Lambowitz AM (1997). *De novo* and DNA primer-mediated initiation of cDNA synthesis by the Mauriceville retroplasmid reverse transcriptase involve recognition of a 3' CCA sequence. J Mol Biol 271:311-332.

Chevalier BS, and Stoddard BL (2001). Homing endonucleases: structural and functional insight into the catalysts of intron/intein mobility. Nucleic Acids Res 29: 3757-3774.

Chiang CC, Kennell JC, Wanner LA, Lambowitz AM (1994). A mitochondrial retroplasmid integrates into mitochondrial DNA by a novel mechanism involving the synthesis of a hybrid cDNA and homologous recombination. Mol Cell Biol 14:6419-6432.

Chung KR, Leuchtmann A, Schardl CL (1996). Inheritance of mitochondrial DNA and plasmids in the ascomycetous fungus, *Epichloe typhina*. Genetics. 142:259-265.

Clark-Walker GD. (1989). In vivo rearrangement of mitochondrial DNA in *Saccharomyces cerevisiae*. Proc Natl Acad Sci USA 86:8847-8851.

Clark-Walker GD (1992). Evolution of mitochondrial genomes in fungi. Int Rev Cytol 141:89-127.

Coenen A, Croft JH, Slakhorst M, Debets F, Hoekstra R (1996). Mitochondrial inheritance in *Aspergillus nidulans*. Genetic Research 67:93-100.

Collins RA (1993). *Neurospora crassa* laboratory strain 74-OR23-1A: mitochondrial genes, In: S. J. O'Brien, Ed. Genetic Maps, 6th ed. Cold Spring Harbor, NY: Cold Spring Harbor Laboratory Press, pp 3.19-3.21.

Collins RA, and Saville BJ (1990). Independent transfer of mitochondrial chromosomes and plasmids during unstable vegetative fusion in *Neurospora*. Nature 345:177-179

Contamine V, and Picard M (2000). Maintenance and integrity of the mitochondrial genome: a plethora of nuclear genes in the budding yeast. Micro Mol Biol Rev 64:281-315.

Court DA, Griffiths AJF, Kraus SR, Russell PJ, and Bertrand H (1991). A new senescence-inducing linear plasmid in field-isolated *Neurospora crassa* strains from India. Curr Genet 19:129-137.

Court DA, and Bertrand H (1992). Genetic organization and structural features of *maranhar*, a senescence-inducing linear mitochondrial plasmid of *Neurospora crassa*. Curr Genet 22:385-397.

Cummings DJ, Turker MS, and Domenico JM (1986). Mitochondrial excision-amplification plasmids in senescent and long-lived cultures of *Podospora anserina*. In: RB Wicker, A Hinnebusch, IC Gunsalus, AM Lambowitz and A Hollaender. Ed. Extrachromosomal Elements in Lower Eukaryotes. New York: Plenum Press, pp 129-146.

Cummings DJ, Michel F, and McNally KL (1989). DNA sequence analysis of the 24.5 kilobase pair cytochrome oxidase subunit I mitochondrial gene from *Podospora anserina*: a gene with sixteen intron. Curr Genet 16:381-406.

Cummings DJ, McNally KL, Domenico JM and Matsuura ET (1990). The complete DNA sequence of the mitochondrial genome of *Podospora anserina*. Curr Genet 17:375-402.

Dai L, Toor N, Olson R, Keeping A, and Zimmerly S (2003). Database for mobile group II introns. Nucleic Acids Res 31:424-426.

Dalgaard JZ, Klar AJ, Moser MJ, Holley WR, Chatterjee A, and Mian IS (1997). Statistical modeling and analysis of the LAGLIDADG family of site-specific endonucleases and identification of an intein that encodes a site-specific endonuclease of the HNH family. Nucleic Acids Res 25:4626-4638.

Deacon JW (1997). Modern mycology. 3rd. ed. Cambridge, UK: Blackwell Science Ltd. University Press.

Debets F, Yang X, and Griffiths AJF (1995). The dynamics of mitochondrial plasmids in a Hawaiian population of *Neurospora intermedia*. Curr Genet 29:44-49.

Delahodde A, Goguel V, Becam AM, Creusot F, Perea J, Banroques J, and Jacp C (1989). Site-specific DNA endonuclease and RNA maturase activities of two homologous intron-encoded proteins from yeast mitochondria. Cell 56:431-441.

De Vries H, Alzner-Deweerd B, Breitenberger CA, Chang DD, De Jonge JC, and Rajbhandary UL (1986). The E35 stopper mutant of *Neurospora crassa*: precise localization of deletion endpoints in mitochondrial DNA and evidence that the deleted DNA codes for a subunit of NADH dehydrogenase. EMBO J 5:779-785.

De Zamaroczy M, and Bernardi G (1985). Sequence organization of the mitochondrial genome of yeast - a review. Gene 41:1-17.

Dinouel N, Drissi R, Miyakawa I, Sor F, Rousset S, and Fukuhara H (1993). Linear mitochondrial DNAs of yeasts: closed-loop structure of the termini and possible linear-circular conversion mechanisms. Mol Cell Biol 13:2315-2323.
Doudna JA, and Cech TR (2002). The chemical repertoire of natural ribozymes. Nature 418:222-228.
Dujon B (1980). Sequence of the intron and flanking exons of the mitochondrial 21S rRNA gene of yeast strains having different alleles at the omega and *rib-1* loci. Cell 20:185-197.
Dujon B (1989). Group I introns as mobile genetic elements: facts and mechanistic speculation - a review. Gene 82:91-114
Dujon B, and Belcour L (1989). Mitochondrial DNA instabilities and rearrangements in yeasts and fungi. In: DE Berg and MM Howe Mobile DNA, Ed. Washington DC: AMS Press. pp 861-878.
Eickbush TH (1994). Origin and evolutionary relationships of retroelements. In: Morse SS, Ed. The evolutionary biology of viruses. New York: Raven Press. pp 121-157.
Eickbush TH (1997). Telomerase and retrotransposons: which came first? Science 277:911-912.
Eickbush TH (1999). Mobile introns: retrohoming by complete reverse splicing. Curr Biol 9:R11-R14.
Eickbush TH, and Malik HS (2002). Origins and evolution of retrotransposons. In: NL Craig, R Craigie, M Gellert, and AM Lambowitz, Ed. Mobile DNA II. Washington DC: ASM Press. pp 1111-1144.
Fangman WL, Henly JW, Churchill G, and Brewer B (1989). Stable maintenance of a 35-base-pair yeast mitochondrial genome. Mol Cell Biol 9:1917-1921.
Forget L, Ustinova J, Wang Z, Huss VAR, and Lang BF (2002). *Hyaloraphidium curvatum*: A linear mitochondrial genome, tRNA editing, and an evolutionary link to lower fungi. Mol Biol Evol 19:310-319.
Fox AN, and Kennell JC (2001). Association between variant plasmid formation and senescence in retroplasmid-containing strains of *Neurospora* spp. Curr Genet 39:92-100.
Freeman S, Redman RS, Grantham G, and Rodriguez RJ (1997). Characterization of a linear DNA plasmid from the filamentous fungal pathogen *Glomerella musae* [Anamorph: *Colletotrichum musae* (Berk. & Curt.) Arx.] Curr Genet 32:152-156.
Fukuhara H, Sor F, Drissi R, Dinouel N, Miyakawa I, Rousset S, and Viola AM (1993). Linear mitochondrial DNAs of yeasts: frequency of occurrence and general features. Mol Cell Biol 13:2309-2314.
Fukuhara H (1995). Linear DNA plasmids of yeasts. FEMS Microbiol Lett 131:1-9.
Garriga G, and Lambowitz AM (1986). Protein-dependent splicing of a group I intron in ribonucleoprotein particles and soluble fractions. Cell 46:669-680.
Gillham NM (1994). Organelle genes and genomes. New York: Oxford University Press,.
Gimble FS (2000). Invasion of a multitude of genetic niches by mobile endonuclease genes. FEMS Microbiol Let 185:99-107.
Gobbi E, Carpanelli A, Firrao G, and Locci R (1997). The *Cryphonectria parasitica* plasmid pUG1 contains a large ORF with motifs characteristic of family B DNA polymerases. Nucleic Acids Res. 25:3275-3280.
Goddard MR, and Burt A (1999). Recurrent invasion and extinction of a selfish gene. Proc Natl Acad Sci USA 96:13880-13885.
Goguel V, Delahodde A, and Jacq C (1992). Connections between RNA splicing and DNA intron mobility in yeast mitochondria: RNA maturase and DNA endonuclease switching experiments. Mol Cell Biol 12:696-705.
Gonzalez P, Barroso G, and Labarère J (1999). Molecular gene organization and secondary structure of the mitochondria large subunit ribosomal RNA form the cultivated Basidiomycota *Agrocybe aegerita*: a 13 kbp gene possessing six unusual nucleotide extension and eight introns. Nucleic Acids Res 27:1754-1761.
Gray MW, Burger G, and Lang BF (1999). Mitochondrial evolution. Nature 283:1476-1481.
Gray MW, Lang BF, Cedergren R, Golding GB, Lemieux C, Sankoff D, Turmel M, Brossard N , Delage E, Littlejohn TG, Plante I, Rioux P, Saint-Louis D, Zhu Y, and Burger G (1998). Genome structure and gene content in protist mitochondrial DNAs. Nucleic Acids Res 26:865-878.
Griffiths AJF (1992). Fungal senescence. Annu Rev Genet 26:351-372.
Griffiths AJF (1995). Natural plasmids of filamentous fungi. Microbiol Rev 59:673-685.
Griffiths AJF (1996). Miochondrial inheritance in filamentous fungi. J Genet 75:403-414.
Griffiths AJF (1998). The kalilo family of fungal plasmids. Bot Bull Acad Sin 39:147-152.
Griffiths AJF, and Bertrand H (1984). Unstable cytoplasms in Hawaiian strains of *Neurospora intermedia.* Curr Genet 8:387-398.
Griffiths AJF, Kraus SR, Barton R, Court DA, Meyers C J, and Bertrand H (1990). Heterokaryotic transmission of senescence plasmid DNA in *Neurospora.* Curr Genet 17:139-145.
Griffiths AJF, Collins RA, and Nargang FE (1995). Mitochondrial genetics of *Neurospora.* In: U Kück, Ed. The Mycota, Volume II: Genetics and Biotechnology. New York: Springer-Verlag. pp 93-108.
Grivell LA (1995). Nucleo-mitochondrial interactions in mitochondrial gene expression. Crit Rev Biochem Mol Biol 30:121-164.

Gross SR, Hsieh T, and Levine PH (1984). Intramolecular recombination as a source of mitochondrial chromosome heteromorphism in *Neurospora.* Cell 38: 233-239.
Gross SR, Mary A, and Levine PH (1989a). Change in chromosome number associated with a double deletion in the *Neurospora crassa* mitochondrial chromosome. Genetics 121:685-691.
Gross SR, Levine PH, Metzger S, and Glaser G (1989b). Recombination and replication of plasmid-like derivatives of a short section of the mitochondrial chromosome of *Neurospora crassa.* Genetics 121:693-701.
Gunge N, Takahashi S, Fukuda K, Ohnishi T, and Meinhardt F (1994). UV hypersensitivity of years linear plasmids. Curr Genet 26:369-373.
Gunge N (1995). Plasmid DNA and the killer phenomenon in *Kluyveromyces.* In: U Kück, Ed. The Mycota, Volume II: Genetics and Biotechnology. New York: Springer-Verlag. pp 189-209.
Guo H, Karberg M, Long M, Jones JP, Sullenger B, and Lambowitz AM (2000). Group II introns designed to insert into therapeutically relevant DNA target sites in human cells. Science 289:452-457.
Hamari Z, Juhász Á, Gácser A, Kucsera J, Pfeiffer I, and Kevei F (2001). Intron mobility results in rearrangement in mitochondrial DNAs of heterokaryon incompatible *Aspergillus japonicus* strains after protoplast fusion. Fung Genet Biol 33:83-95.
Hamari Z, Juhász Á, and Kevei F (2002). Role of mobile introns in mitochondrial genome diversity of fungi. Acta Micro Immun Hung 49:331-335.
Hausner G, Monteiro-Vitorello CB, Searles DB, Maland M, Fulbright DW, and Bertrand H (1999). A long open reading frame in the mitochondrial LSU rRNA group-I intron of *Cryphonectria parasitica* encodes a putative S5 ribosomal protein fused to a maturase. Curr Genetics 35:109-117.
Hausner G, Belkhiri A, and Klassen GR (2000). Phylogenetic analysis of the small subunit ribosomal RNA gene of the hyphochytrid *Rhizidiomyces apophysatus.* Can J Bot 78:124-128.
Henke MR, Butow RA, and Perlman PS (1995). Maturase and endonuclease functions depend on separate conserved domains of the bifunctional protein encoded by the group I intron aI4alpha of yeast mitochondrial DNA. EMBO J 14:5094-5099.
Hermanns J, and Osiewacz HD (1994). Three mitochondrial unassigned open reading frames of *Podospora anserina* represent remnants of a viral-type RNA polymerase gene. Curr Genet. 25:150-157.
Hermanns J, Asseburg A, and Osiewacz HD (1994). Evidence for a life span-prolonging effect of a linear plasmid in a longevity mutant of *Podospora anserina.* Mol Gen Genet 243:297-307.
Ho Y, Kim SJ, and Waring RB (1997). A protein encoded by a group I intron in *Aspergillus nidulans* directly assists RNA splicing and is a DNA endonuclease. Proc Natl Acad Sci USA 94:8994-8999.
Hongo M, Miyasaka A, Suzuki F, and Hashiba T (1994). Expression of the linear DNA plasmid pRS64 in the plant pathogenic fungus *Rhizoctonia solani.* Mol Gen Genet 245:265-271.
Hudspeth MES (1992). The fungal mitochondrial genome - a broader perspective. In: DK Arora, RP Elander, and KG Mukerji, Ed. Handbook of applied Mycology, Vol. 4: Fungal Biotechnology. New York: Marcel Dekker Inc. pp 213-242.
Jacquier A, and Dujon B (1983). The intron of the mitochondrial 21S rRNA gene: distribution in different yeast species and sequence comparison between *Kluyveromyces thermotolerans* and *Saccharomyces cerevisiae.* Mol Gen Genet 192:487-499.
Jacquier A (1990). Self-splicing group II introns and nuclear pre-mRNA introns: how similar are they? Trends Biochem Sci 15:351-354.
Jacquier A (1996). Group II introns: Elaborate ribozymes. Biochimie 78:474-487.
Jamet-Vierny C, Boulay J, Begel O, Silar P (1997a). Contribution of various classes of defective mitochondrial DNA molecules to senescence in *Podospora anserina.* Curr Genet 31: 171-178.
Jamet-Vierny C, Boulay J, and Briand J-F (1997b). Intramolecular cross-overs generate deleted mitochondrial DNA molecules in *Podospora anserina.* Curr Genet 31:162-170.
Jamet-Vierny C, Rossignol M, Haedens V, and Silar P (1999). What triggers senescence in *Podospora anserina*? Fung Genet Biol 27: 26-35.
Johansen S, Einvik C, Elde M, Haugen P, Vader A, and Haugli F (1997). Group I introns in biotechnology: prospects of application of ribozymes and rare-cutting homing endonucleases. In: MR El-Gewely, Ed. Biotechnology Annual Reviews. Vol. 3. Amsterdam: Elsevier Science BV. pp 111-150.
Jurica MS, and Stoddard BL (1999). Homing endonucleases: structure, function and evolution. Cell Mol Life Sci 55:1304-1326.
Kang D, and Hamasaki N (2002). Maintenance of mitochondrial DNA integrity: repair and degradation. Curr Genet 41:311-322.
Karlberg O, Canbäck B, Kurland CG, and Anderson SGE (2000). The dual origin of the yeast mitochondrial proteome. Yeast 17:170-187.

Katsura K, Suzuki F, Miyashita S-I, Nishi T, Hirochika H, and Hashiba T (1997). The complete nucleotide sequence and characterization of the linear DNA plasmid pRS64-2 from the plant pathogenic fungus *Rhizoctonia solani*. Curr Genet 32:431-435.

Katsura K, Sasaki A, Nagaska A, Fuji M, Miyake Y, and Hashiba T (2001). Complete nucleotide sequence of the linear DNA plasmid pRS224 with hairpin loops from *Rhizoctonia solani* and its unique transcriptional form. Curr Genet 40:195-202.

Keeling PJ, and Fast NM (2002). Microsporidia: Biology and evolution of highly reduced intracellular parasites. Ann Rev Microbiol 56:93-116.

Kempken F (1995a). Plasmid DNA in mycelial fungi. In: U Kück, Ed. The Mycota, Volume II: Genetics and Biotechnology. New York: Springer-Verlag. pp 169-188.

Kempken F (1995b). Horizontal transfer of a mitochondrial plasmid. Mol Gen Genet 248:89-94.

Kennell JC, and Cohen SM (2003). Fungal Mitochondria: Genomes, Genetic Elements and Gene Expression. IN: DK Arora, Ed. The Handbook of Fungal Biotechnology, 2nd Edition. New York: Marcel Dekker Inc.

Kim E-K, Jeong J-H, Youn HS, Koo YB, and Roe J-H (2000). The terminal protein of a linear mitochondrial plasmid is encoded in the N-terminus of the DNA polymerase gene in white-rot fungus *Pleurotus ostreatus*. Curr Genet 38:283-290.

Kistler HC, Benny U, and Powell WA (1997). Linear mitochondrial plasmids of *Fusarium oxysporum* contain genes with sequence similarity to genes encoding a reverse transcriptase from *Neurospora* spp. Appl Environ Microbiol 63:3311-3313.

Koll F, Boulay J, Belcour L, and d'Aubenton-Carafa Y (1996). Contribution of ultra-short invasive elements to the evolution of the mitochondrial genome in the genus *Podospora*. Nucleic Acids Res 24:1734-1741.

Kornberg A, and Baker TA (1992). DNA replication. New York: Freeman WH and Company.

Kubelik AR, Kennell JC, Akins RA, and Lambowitz AM (1990). Identification of *Neurospora* mitochondrial promoters and analysis of synthesis of the mitochondrial small rRNA in wild-type and the promoter mutant [poky]. J Biol Chem 265:4515-4526.

Lambowitz AM (1989). Infectious introns. Cell 56:323-326.

Lambowitz AM, and Belfort M (1993). Introns as mobile genetic elements. Annu Rev Biochem 62:587-622.

Lambowitz AM, and Perlman PS (1990). Involvement of aminoacyl-tRNA synthetases and other proteins in group I and group II intron splicing. Trends Biochem Sci 15:440-444.

Lambowitz AM, and Chiang C-C (1995). The Mauricville and Verkud plasmids: primitive retroelements found in *Neurospora* mitochondria. Can J Bot 73 (Suppl 1): S173-S179.

Lambowitz AM, Caprara MG, Zimmerly S, and Perlman PS (1999). Group I and group II ribozymes as RNPs: Clues to the past and guides to the future. In: RF Gesteland, TR Cech, and JF Atkins, Ed. The RNA World 2nd ed. Cold Spring Harbor NY: Cold Spring Harbor Laboratory Press. pp 451-485.

Landweber LF, Simon PJ, and Wagner TA (1998). Ribozyme engineering and early evolution. BioScience 48:94-103.

Lang BF (1984). The mitochondrial genome of the fission yeast *Schizosaccharomyces pombe*: highly homologous introns are inserted at the same position of the otherwise less conserved *cox1* genes in *Schizosaccharomyces pombe* and *Aspergillus nidulans*. EMBO J 3:2129-2136.

Lang BF, Ahne F, and Bonen L (1985). The mitochondrial genome of the fission yeast *Schizosaccharomyces pombe*. The cytochrome b gene has an intron closely related to the first two introns in the *Saccharomyces cerevisiae cox1* gene. J Mol Biol 184:353-366.

Lang BF, Seif E, Gray MW, O'Kelly CJ, and Burger G (1999). A comparative genomics approach to the evolution of eukaryotes and their mitochondria. J Eukaryot Microbiol 46:320–326.

LaPolla RJ, and Lambowitz AM (1981). Mitochondrial ribosome assembly in *Neurospora crassa*. J Biol Chem 256:7064-7067.

Lazarus CM, Earl AJ, Turner G, and Kuntzel H (1980). Amplification of a mitochondrial DNA sequence in the cytoplasmically inherited "ragged" mutant of *Aspergillus amstelodami*. Eur J Biochem. 106:633-641.

Lee SB and Taylor JW (1993). Uniparental inheritance and replacement of mitochondrial DNA in *Neurospora tetrasperma*. Genetics 134:1063-1075.

Leipe DD, Wainright PO, Gunderson JH, Porter D, Patterson DJ, Valois F, Himmerich S, and Sogin ML (1994). The stramenopiles from a molecular perspective:16S-like rRNA sequences from *Labyrinthuloides minuta* and *Cafeteria roenbergensis*. Phycologia 33:369-377.

Li Q, and Nargang FE (1993). Two *Neurospora* mitochondrial plasmids encode DNA polymerases containing motifs characteristic of family B DNA polymerases but lack the sequence asp-thr-asp. Proc Natl Acad Sci USA 90:4299-4303

Ling F, and Shibata T (2002). Recombination-dependent mtDNA partitioning: in vivo role of Mhr1p to promote pairing of homologous DNA. EMBO J 21:4730-4740.

Lockshon D, Zweifel SG, Freeman-Cook LL, Lorimer HE, Brewer BJ, and Fangman WL (1995). A role of recombination junctions in the segregation of mitochondrial DNA in yeast. Cell 81:947-955.

Lonergan KM, and Gray MW (1993). Editing of transfer RNAs in *Acanthamoeba castellanii* mitochondria. Science 259:812-816.
Mahanti N, Bertrand H, Monteiro-Vitorello C, and Fulbright DW (1993). Elevated mitochondrial alternative oxidase activity in dsRNA-free, hypovirulent isolates of *Cryphonectria parasitica.* Physiol Mol Plant Pathol 42: 455-463.
Maizels N, and Weiner AM (1993). The genomic tag hypothesis: modern viruses as molecular fossils of ancient strategies for genomic replication. In: RF Gesteland and JF Atkins, Eds. RNA World. Cold Spring Harbor NY: Cold Spring Harbor Laboratory Press. pp 577-602.
Maleszka R, Skelly PJ and Clark-Walker GD (1991). Rolling circle replication in yeast mitochondria. EMBO J 10:3923-3929.
Maleszka R, and Clark-Walker GD (1992). *In vivo* conformation of mitochondrial DNA in fungi and zoosporic moulds. Curr Genet 22:341-344.
Mannella CA, and Lambowitz AM (1979). Unidirectional gene conversion associated with two insertions in *Neurospora crassa* mitochondrial DNA. Genetics 93:645-654.
May G, and Taylor JW (1989). Independent transfer of mitochondrial plasmids in *Neurospora crassa.* Nature 359:320-322.
Meinhardt F, Schaffrath R, and Larsen M (1997). Microbial linear plasmids. Appl Microbiol Biotechnol 47:329-336.
Michel F, Jacquier A, Dujon B (1982). Comparison of fungal mitochondrial introns reveals extensive homologies in RNA secondary structure. Biochimie 64:867-881.
Michel F, and Cummings DJ (1985). Analysis of class I introns in a mitochondrial plasmid associated with senescence of *Podospora anserina* reveals extraordinary resemblance to the *Tetrahymena* ribosomal intron. Curr Genet 10: 69-79.
Michel F, and Lang BF (1985). Mitochondrial class II introns encode proteins related to the reverse transcriptases of retroviruses. Nature 316:641-643.
Michel F, Umesono K, and Ozeki H (1989). Comparative and functional anatomy of group II catalytic introns - a review. Gene 82:5-30.
Michel F, and Westhof E (1990). Modelling of the three-dimensional architecture of group I catalytic introns based on comparative sequence analysis. J Mol Biol 216:585-610.
Michel F, and Ferat JL (1995). Structure and activities of group II introns. Annu Rev Biochem 64:435-461.
Miyakawa I, Aoi H, Sando N, and Kuroiwa T (1984). Fluorescence microscopic studies of mitochondrial nucleoids during meiosis and sporulation in the yeast *Saccharomyces cerevisiae.* J Cell Sci 66:21-38.
Mogen KL, Siegel MR, and Schardl CL (1991). Linear DNA plasmids of the perennial ryegrass choke pathogen, *Epichloe typhina* (Clavicipitaceae). Curr Genet 20:519-526.
Monteiro-Vitorello CB, Bell JA, Fulbright DW, and Bertrand H (1995). A cytoplasmically transmissible hypovirulence phenotype associated with mitochondrial DNA mutations in the chestnut blight fungus *Cryphonectria parasistica.* Proc Natl Acad Sci USA 92:5935-5939.
Monteiro-Vitorello CB, Baidyaroy D, Bell JA, Hausner G, Fulbright DW, and Bertrand H (2000). A circular mitochondrial plasmid incites cytoplasmically-transmissible hypovirulence in some strains of *Cryphonectria parasitica.* Curr Genetics 37:242-256.
Mohr G, and Lambowitz AM (1991). Integration of a group I intron into a ribosomal RNA sequence promoted by a tyrosyl-tRNA synthetase. Nature 354:164-167
Mohr G, Perlman PS, and Lambowitz AM (1993). Evolutionary relationships among group II intron-encoded proteins and identification of a conserved domain that may be related to maturase function. Nucleic Acids Res 21:4991-4997.
Mohr G, Rennard R, Cherniack AD, Stryker J, and Lambowitz AM (2001). Function of the *Neurospora crassa* mitochondrial tyrosyl-tRNA synthetase in RNA splicing. Role of the idiosyncratic N-terminal extension and different modes of interaction with different group I introns. J Mol Biol 307:75-92.
Mohr S, Wanner LA, Bertrand H, and Lambowitz AM (2000). Characterization of an unusual tRNA-like sequence found inserted in a *Neurospora* retroplasmid. Nucleic Acids Res 28:1514-1524.
Mohr S, Stryker JM, and Lambowitz AM (2002). A DEAD-box protein functions as an ATP-dependent RNA chaperone in group I intron splicing. Cell 109:769-779.
Mota EM, and Collins RA (1988). Independent evolution of structural and coding regions in a *Neurospora* mitochondrial intron. Nature 332:654-656.
Murray HL, Mikheeva S, Coljee VW, Turczyk BM, Donahue WF, Bar-Shalom A, and Jarrell KA (2001). Excision of group II introns as circles. Mol Cell 8:201-211.
Nakamura TM, and Cech T (1998). Reversing time: origin of telomerase. Cell 92:587-590.
Nargang FE, Bell JB, Stohl LL, and Lambowitz AM (1984). The DNA sequence and genetic organization of a *Neurospora* mitochondrial plasmid suggest a relationship to introns and mobile elements. Cell 38:41-453.

Nargang FE, Pande S, Kennell JC, Akins RA, and Lambowitz AM (1992). Evidence that a 1.6 kb kilobase region of *Neurospora* mtDNA was derived by insertion of part of the Labelle mitochondrial plasmid. Nucleic Acids Res 20:1101-1108.

Naruse A, Yamamoto H, and Sekiguchi J (1993). Nucleotide sequence of the large mitochondrial rRNA gene of *Penicillium chrysogenum*. Biochimica Biophysica Acta 1172:353-356.

Neupert W (1997). Protein import into mitochondria. Annu Rev Biochem 66:683-717.

Nosek J, Dinouel N, Ková□ L, and Fukuhara H (1995). Linear mitochondrial DNAs from yeasts: telomeres with large tandem repetitions. Mol Gen Genet 247:61-72.

Nosek J, Tomáška L, Fukuhara H, Suyama Y, and Ková□ L (1998). Linear mitochondrial genomes: 30 years down the line. Trends Genet 14:184-188.

Oeser B, Rogmann-Backwinkel P, and Tudzynski P (1993). Interaction between mitochondrial plasmids in *Claviceps purpurea*: analysis of plasmid-homologous sequences upstream of the LrRNA gene. Curr Genet 23:315-322.

Oeser B, and Tudzynski P (1989). The linear mitochondrial plasmid pClK1 of the phytopathogenic fungus *Claviceps purpurea* may code for a DNA polymerase and an RNA polymerase. Mol Gen Genet 217:132-140.

Olson A, and Stenlid J (2001). Plant Pathogens: Mitochondrial control of fungal hybrid virulence. Nature 411:438.

Osiewacz HD, and Esser K (1984). The mitochondrial plasmid of *Podospora anserina*: a mobile intron of a mitochondrial gene. Curr Genet 8:299-305.

Osiewacz HD (2002). Mitochondrial functions and aging. Gene 286: 65-71.

Pande S, Lemire EG, and Nargang FE (1989). The mitochondrial plasmid from *Neurospora intermedia* strain LaBelle-1b contains a long open reading frame with blocks of amino acids characteristic of reverse transcriptases and related proteins. Nucleic Acids Res 17: 2023-2042.

Paquin B, Laforest M-J, and Lang F (1994). Interspecific transfer of mitochondrial genes in fungi and creation of a homologous hybrid gene. Proc Natl Acad Sci USA 91:11807-11810.

Paquin B, and Lang BF (1996). The mitochondrial DNA of *Allomyces macrogynus*: The complete genomic sequence from an ancestral fungus. J Mol Biol 255:688-701.

Paquin B, Laforest M-J, and Lang BF (2000). Double-hairpin elements in the mitochondrial DNA of *Allmoyces*: evidence for mobility. Mol Biol Evol 17:1760-1768.

Paquin B, Laforest M-J, Forget L, Roewer I, Wang Z, Longcore J, and Lang BF (1997). The fungal mitochondrial genome project: evolution of fungal mitochondrial genomes and their gene expression. Curr Genet 31:380-395.

Pel HJ, and Grivell LA (1993). The biology of yeast mitochondrial introns. Mol Biol Reports 18: 1-13.

Pietrokovski S (2001). Intein spread and extinction in evolution. Trends Genet 17:465-472.

Pyle AM (2000). New tricks from an itinerant intron. Nature Struct Biol 7:352-354.

Robison MM, Royer JC, and Horgen PA (1991). Homology between mitochondrial DNA of *Agaricus bisporus* and an internal portion of a linear mitochondrial plasmid of *Agaricus bitorquis*. Curr Genet 19:495-502.

Robison MM, and Horgen PA (1996). Plasmid RNA polymerase-like mitochondrial sequences in *Agaricus bitorquis*. Curr Genet 29:370-376.

Robison MM, Kerrigan RW, and Horgen PA (1997). Distribution of plasmids and plasmid-like mitochondrial sequence in the genus *Agaricus*. Mycologia 89:43-47.

Röhr H, Kües U, and Stahl U (1999). Recombination: Organelle DNA of plants and fungi: Inheritance and recombination. Progress in Botany 60:39-87.

Roman J, and Woodson SA (1995). Reverse splicing of the tetrahymena IVS: Evidence for multiple reaction sites in the 23S rRNA. RNA 1:478-490.

Rubidge T (1992). The structure and function of plasmids in filamentous fungi. In: DK Arora, RP Elander and KG Mukerji. Eds. Handbook of applied Mycology, Vol. 4: Fungal Biotechnology. New York: Marcel Dekker Inc. pp 243-258.

Saguez C, Lecellier G, and Koll F (2000). Intronic GIY-YIG endonuclease gene in the mitochondrial genome of *Podospora curvicolla*: evidence for mobility. Nucleic Acids Res 28:1299-1306.

Saldanha R, Mohr G, Belfort M, and Lambowitz AM (1993). Group I and group II introns. FASEB J 7:15-24.

Salvo JL, Rodeghier B, Rubin A, and Troischt T (1998). Optional introns in mitochondrial DNA of *Podospora anserina* are the primary source of observed size polymorphisms. Fung Genet Biol 23:162-168.

Schäfer B, Merlos-Lange AM, Anderl C, Welser F, Zimmer M, and Wolf K (1991). The mitochondrial genome of fission yeast: inability of all introns to splice autocatalytically, and construction and characterization of an intronless genome. Mol Gen Genet 225:158-167.

Schäfer B, Wilde B, Massardo DR, Manna F, Del Giudice L, Wolf K (1994). A mitochondrial group-I intron in fission yeast encodes a maturase and is mobile in crosses. Curr Genet 25:336-341.

Schmidt U, Sägebarth R, Schmelzer C, and Stahl U (1993). Self-splicing of a *Podospora* group IIA intron in vitro. J Mol Biol 231:559-568.
Schmidt WM, Schweyen RJ, Wolf K, and Mueller MW (1994). Transposable group II introns in fission and budding yeast. Site-specific genomic instabilities and formation of group II IVS plDNAs. J Mol Biol 243:157-166.
Schründer J, Gunge N, and Meinhardt F (1996). Extranuclear expression of the bacterial xylose-isomerase (*xylA*) and the UPD-glucose-dehydrogenase (*hasB*) genes in yeast using *Kluyveromyces lactis* linear plasmids as vectors. Curr Mirobiol 33:323-330.
Schulte U, and Lambowitz AM (1991). The LaBelle mitochondrial plasmid of *Neurospora intermedia* encodes a novel DNA polymerase that may be derived from a reverse transcriptase. Mol Cell Biol 11:1696-1706.
Sellem CH, d'Aubenton-Carafa Y, Rossignol M, and Belcour L (1996). Mitochondrial intronic open reading frames in *Podospora*: mobility and consecutive exonic sequence variations. Genetics 143:777-788.
Sellem CH, and Belcour L (1997). Intron open reading frames as mobile elements and evolution of a group I intron. Mol Biol Evol 14:518-526.
Sekito T, Okamoto K, Kitano H, and Yoshida K (1995). The complete mitochondrial DNA sequence of *Hansenula wingei* reveals new characteristics of yeast mitochondria. Curr Genet 28:39-53.
Seraphin B, Boulet A, Simon M, and Faye G (1987). Construction of a yeast strain devoid of mitochondrial introns and its use to screen nuclear genes involved in mitochondrial splicing. Proc Natl Acad Sci USA 84:6810-6814.
Shepherd HS (1992). Linear, non-mitochondrial plasmids of *Alternaria alternata*. Curr Genet 21:169-172.
Shimko N, Liu L, Lang BF, and Burger G (2001). GOBASE: The organella genome database. Nucleic Acids Res 29:128-132.
Shnyreva AV (1995). Mitochondrial introns in fungi and their evolutionary role. Russian Journal of Genetics 31:741-747.
Silar P, Koll F, and Rossignol M (1997). Cytosolic ribosomal mutations that abolish accumulation of circular intron in the mitochondria without preventing senescence of *Podospora anserina*. Genetics 145:697-705.
Silliker ME, and Cummings DJ (1990). A mitochondrial DNA rearrangement and three new mitochondrial plasmids from long-lived strains of *Podospora anserina*. Plasmid 24:37-44.
Silliker ME, Liotta MR, and Cummings DJ (1996). Elimination of mitochondrial mutations by sexual reproduction: two *Podospora anserina* mitochondrial mutants yield only wild-type progeny when mated. Curr Genet 30:318-324.
Steinhilber W, and Cummings DJ (1986). A DNA polymerase activity with characteristics of a reverse transcriptase in *Podospora anserina*. Curr Genet 10:389-392.
Stevenson CB, Fox AN, and Kennell JC (2000). Senescence associated with the over-replication of a mitochondrial retroplasmid in *Neurospora crassa*. Mol Gen Genet 263:433-444.
Suh S-Q, Jones KG, and Blackwell M (1999). A group I intron in the nuclear small subunit rRNA gene of *Cryptendoxyla hypophloia*, an ascomycetes fungus: Evidence for a new major class of group I introns. J Mol Evol 48:493-500.
Szczepanek T, and Lazowska J (1996). Replacement of two non-adjacent amino acids in the *S. cerevisiae* bi2 intron-encoded RNA maturase is sufficient to gain a homing-endonuclease activity. EMBO J 15:3758-3767.
Taylor JW (1986). Fungal evolutionary biology and mitochondrial DNA. Exp Mycol 10:259-269.
Toor N, Hausner G, and Zimmerly S (2001). Coevolution of the group II intron RNA structure with its intron-encoded reverse transcriptase. RNA 7:1142-1152.
Toor N, and Zimmerly S (2002). Identification of a family of group II introns encoding LAGLIDADG ORFs typical of group I introns. RNA 8:1373-1377.
Turcq B, Dobinson KF, Serizawa N, and Lambowitz AM (1992). A protein required for RNA processing and splicing in *Neurospora* mitochondria is related to gene products involved in cell cycle protein phosphatase functions. Proc Natl Acad Sci USA 89:1676-1680.
Turker MS, Domenico JM, and Cummings DJ (1987). Excision-amplification of mitochondrial DNA during senescence in *Podospora anserina*. J Mol Biol 198:171-185.
Van der Gaag M, Debets AJM, Osiewacz HD, and Hoekstra RF (1998). The dynamics of plAL2-1 homologous linear plasmids in *Podospora anserina*. Mol Gen Genet 258:521-529.
Van Dyck L, Neupert W, and Langer T (1998). The ATP-dependent PIM1 protease is required for the expression of intron_containing genes in mitochondria. Genes Dev 12:1515-1524.
van Open MJH, Catmull J, McDonald BJ, Hislop NR, Hagerman PJ, and Miller D (2000). The mitochondrial genome of *Acropora tenuis* (Cnidaria; Scleractinia) contains a large group I intron and a candidate control region. J Mol Evol 55:1-13.
Vierula PJ, Cheng CK, Court DA, Humphrey RW, Thomas DY, and Bertrand H (1990). The kalilo senescence plasmid of *Neurospora intermedia* has covalently-linked 5' terminal proteins. Curr Genet 17:195-201.

Wainright PO, Hinkle G, Sogin ML, and Stickel SK (1993). Monophyletic origins of the metazoa: an evolutionary link with fungi. Science 260:340-342.

Walther TC, and Kennell JC (1999). Linear mitochondrial plasmids of *F. oxysporum* are novel, telomere-like retroelements. Mol Cell 4:229-238.

Wang H and Lambowitz AM (1993). The Mauriceville plasmid reverse transcriptase can initiate cDNA synthesis *de novo* and may be related to the progenitor of reverse transcriptases and DNA polymerases. Cell 75:1071-1081.

Weiner AM, and Maizels N (1994). Unlocking the secretes of retroviral evolution. Curr Biol 4:560-563.

Wenzlau JM, Saldanha RJ, Butow RA, and Perlman PS (1989). A latent intron-encoded maturase is also an endonuclease needed for intron mobility. Cell 56:421-430.

Wenzlau JM and Perlman PS (1990). Mobility of two optional G + C-rich clusters of the var1 gene of yeast mitochondrial DNA. Genetics 126: 53-62.

Wesolowski M, and Fukuhara H (1981). Linear mitochondrial deoxyribonucleic acid from the yeast *Hansenula mrakii*. Mol Cell Biol 1:387-393.

Westermann B (2002). Merging mitochondria matters: Cellular role and molecular machinery of mitochondrial fusion. EMBO Rep 3:527-531.

Westermann B and Prokisch H (2002). Mitochondrial dynamics in filamentous fungi. Fung Genet Biol 36:91-97.

Wolf K (1994). Mitochondrial introns in yeast - mobile genetic elements. Endocytobiosis & Cell Res 10:55-63.

Wolf K (1995). Mitochondrial genetics of yeast. In: U Kück, Ed. The Mycota, Volume II: Genetics and Biotechnology. New York: Springer-Verlag pp 75-91.

Wolf K (1996). Mitochondrial genetics of *Saccharomyces cerevisiae*. In: CJ Bos, Ed. Fungal Genetics: Principles and practice. New York: Marcel Dekker Inc. pp 247-257.

Wolf K, and Del Giudice L (1988). The variable mitochondrial genome of ascomycetes: Organization, mutational alterations, and expression. Advances in Genetics 25:185-308

Wolff G, Burger G, Lang BF, and Kück U (1993). Mitochondrial genes in the colourless alga *Prototheca wickerhamii* resemble plant genes in their exons but fungal genes in their introns. Nucleic Acids Res 21:719-726.

Yaffe MP (1999). The machinery of mitochondrial inheritance and behavior. Science 283:1493-1497.

Yang X, and Griffiths AJF (1993). Plasmid diversity in senescent and nonsenescent strains of *Neurospora*. Mol Gen Genet 237:177-186.

Yin S, Heckman J, and Rajbhandary UL (1981). Highly conserved GC-rich palindromic DNA sequences flank tRNA genes in *Neurospora crassa* mitochondria. Cell 26:325-332.

Zimmerly S, Guo H, Perlman PS, and Lambowitz AM (1995a). Group II intron mobility occurs by target DNA-primed reverse transcription. Cell 82:545-554.

Zimmerly S, Guo H, Eskes R, Yang J, Perlman PS, and Lambowitz AM. (1995b). A group II intron RNA is a catalytic component of a DNA endonuclease involved in intron mobility. Cell 83:529-538.

Zimmerly S, Hausner G, and Wu X-C (2001). Phylogenetic analysis of group II intron ORFs. Nucleic Acids Res 29:1238-1250.

Zinn, AR, Pohlman JK, Perlman PS, and Butow RA (1988). *In vivo* double-stranded breaks occur at the recombinogenic G+C-rich sequences in the yeast mitochondrial genome. Proc Natl Acad Sci USA 85:2686-2690.

Applied Mycology & Biotechnology
An International Series. Volume 3. Fungal Genomics

Evolution of the Fungi and their Mitochondrial Genomes

Charles E. Bullerwell[2], Jessica Leigh[1], Elias Seif[1], Joyce E. Longcore[3] and B. Franz Lang[1]
Program in Evolutionary Biology, Canadian Institute for Advanced Research; [1]Département de Biochimie, Université de Montréal, 2900 Boul. Edouard-Montpetit, Montréal (Québec), H3T 1J4, Canada. (Franz.Lang@Umontreal.ca); [2]Department of Biochemistry and Molecular Biology, Dalhousie University, Halifax (Nova Scotia), B3H 4R2, Canada; [3]Department of Biological Sciences, University of Maine, Orono (Maine) 04469-5722, U.S.A.

Despite the importance of fungi as model eukaryotic organisms, fungal mitochondrial genomics has only recently received considerable attention. Over the past several years, the number of available, completely sequenced mitochondrial genomes from fungi has increased from just 3 to 22 sequences, including representatives of the four principle divisions of this kingdom: Ascomycota, Basidiomycota, Zygomycota and Chytridiomycota. This wealth of data from a wide range of diverse fungi has allowed a more complete understanding of the organization and content of their mitochondrial genomes. In addition, mtDNA-encoded protein sequences have proven invaluable for molecular phylogenetics, elucidating the phylogeny of the fungi and their relationship to other eukaryotes. Finally, in light of this phylogenetic framework, the comparison of fungal mitochondrial genome sequences has allowed for an appreciation of how mitochondrial genomes have evolved in terms of gene content, gene order, gene organization, gene expression and genome conformation. These advances will help us to better understand fungal biology, and therefore some of our most important eukaryotic model organisms.

1. INTRODUCTION

Fungi constitute some of the most well-studied and well-understood organisms in science. In particular, the "baker's yeast" *Saccharomyces cerevisiae* is perhaps the most frequently used eukaryotic model system in genetics, molecular biology, and biochemistry, as well as in several genomics disciplines. Other fungi, notably the fission yeast *Schizosaccharomyces pombe* and the filamentous euascomycetes *Neurospora crassa* and *Aspergillus nidulans*, also have proven to be of great utility in studies of such aspects of cell biology as the cell cycle, the genetics and regulation of nitrogen metabolism, and in a more general sense, as less derived and more gene-rich eukaryotic models, compared to *S. cerevisiae*.

The advantages of these fungal model systems are many-fold. In large part, they owe their popularity to the ease with which they can be grown and manipulated in the laboratory. Indeed, a wide variety of efficient molecular techniques are available for most of them (e.g., molecular transformation; genetic analyses of large numbers of colonies; easy inactivation, or replacement of nuclear genes) allowing experimentation at a genomics level. In addition, the modest size (relative to animals and plants) of their nuclear genomes permits complete genome sequencing with reasonably little effort and expenditure. The nuclear genome sequences of the ascomycetes *S. cerevisiae* (Goffeau *et al.* 1996) and *S. pombe* (Wood *et al.* 2002) have both recently been completed. The availability of this whole genome information is the basis for experiments on a genomics scale such as the exploration of gene expression using micro-arrays and the cataloguing of protein interactions. The ease of genomics experimentation in the case of *S. cerevisiae* is further due to the surprisingly low number of genes in this genome (~ 5,600; Goffeau *et al.* 1996), most of which are without introns. In *S. pombe*, the number of genes is similarly low (~ 4,940; Wood *et al.* 2002), although the number of introns is much higher in this species.

One aspect of fungal molecular biology that has only recently received substantial attention is the comparative analysis of fungal mitochondrial genomes. Despite their relatively small size, as recently as 1996 only three completely-sequenced fungal mitochondrial DNAs (mtDNAs) were available: *S. cerevisiae* (only a composite of several yeast genomes was available at that time; Foury *et al.* 1998), *S. pombe* (Lang *et al.* 1983; Lang 1993) and *Podospora anserina* (Cummings *et al.* 1990). In addition, the sequences of *N. crassa* and *A. nidulans* were near completion (the *A. nidulans* sequence remains unfinished at present). Because subsequent research projects tended to focus on additional members of Ascomycota, an understanding of fungal mitochondrial genome evolution was not possible. Only more recent, systematic sequencing of fungal mitochondrial genomes has produced mtDNA sequences from representatives of the four principle divisions of fungi: Ascomycota, Basidiomycota, Zygomycota and Chytridiomycota. The sum of these sequences now describes not only unexpected variation within and among fungal groups, but also has permitted the inference of a robust fungal phylogeny. This phylogeny provides a basis for interpreting changes in gene structure, expression and function of gene products, and changes in genome organization from a wider evolutionary perspective. In other words, a robust phylogenetic framework has allowed the mapping of genetic, biochemical, functional and genomics changes to a phylogenetic tree, and consequently permits interpretation of the origin and evolution of character changes. This type of study is termed "evolutionary genomics".

In this chapter, we will review studies that emphasize how the evolutionary genomics approach has revolutionized our understanding of fungal mitochondrial evolution. Although the fungal phylogeny based on concatenated mitochondrial protein sequence data has been addressed in a recent review (Leigh *et al.* 2003), we will revisit this topic in the light of more recent results. We will not, however, elaborate on mitochondrial genetics, gene composition, introns, plasmids, or a detailed description of mitochondrial biogenesis and functions from a biochemical standpoint. These topics have been covered in detail in several recent reviews (e.g., Paquin *et al.* 1997; Lang *et al.* 1999; Kennel and Cohen 2003; this volume, chapter by Hausner). Finally, we have attempted to include as much of the most recent information as possible, which has necessitated the occasional reference to publications "in press" or "unpublished". To facilitate access to the information in these forthcoming publications, we have created a website (http://megasun.bch.umontreal.ca/People/lang/FMGP/Reviews/) that will summarize relevant updates to these references, and supply links to new information.

2. TAXONOMY AND PHYLOGENY OF THE FUNGI

Just thirty years ago, when *The Fungi* series was issued, editor G.C. Ainsworth (Ainsworth 1973) expressed dissatisfaction with the classification of organisms into only two groups, the plants and the animals. Instead, Ainsworth supported use of Whittaker's (1969) Five-Kingdom System, wherein fungi were accorded kingdom status along with animals, plants, protists and bacteria. Since that time, classifications of organisms within the kingdom Fungi, and even definitions of what constitutes a member of this kingdom, have continued to change as a result of increasingly refined methods (Table 1). In this section we address the history of fungal taxonomy, from the morphological and ultrastructural features that have been extensively used to chart the interrelationships of the fungi, to the more recent developments in molecular phylogenetics that have complemented, but largely supplanted, these methods.

2.1 Classical Fungal Taxonomy

Members of Chytridiomycota are of particular evolutionary interest because they are believed to be a deep divergence of minimally-derived Fungi (Berbee and Taylor 1993). Nearly all members of this group produce flagellated, asexual reproductive spores, whereas flagella (and the basal bodies or centriole structures from which they arise) are lacking in the other fungal phyla. The ancestral quality of this trait is evidenced by the presence of the same microtubular substructure in the flagella of chytridiomycete spores as is found in the cilia of certain protists, animals and lower plants.

One of the most important alterations to fungal taxonomy was the removal of three groups of organisms, the oomycetes (e.g., *Saprolegnia, Phytophthora*), labyrinthulomycetes, and hyphochytriomycetes, from the fungal kingdom. These three groups of organisms were considered specifically related to the chytridiomycetes as part of the "Phycomycetes" (Sparrow 1943, 1960; later, both were classified in Mastigomycotina, Sparrow 1973) based on the presence of flagellated spores. However, information on cell wall composition, physiology of the lysine synthesis pathway and ultrastructural features of both mitochondria and zoospores led to the recognition that these three groups should be classified elsewhere in the eukaryotic tree. The presence of flagellated spores is now considered to be a convergent morphology in chytridiomycetes and these other groups, and molecular data has placed the oomycetes, labyrinthulomycetes and hyphochytriomycetes within the protist lineage Stramenopila with overwhelming support.

Groups within Chytridiomycota have also proven unstable, mostly because classical taxonomy depended on only a few, frequently non-homologous, morphological characters. Although others pioneered systematic ultrastructural studies of zoospores in the 1960's and 70's, it was Barr (1980) who formally used ultrastructural characters to segregate a new order (Spizellomycetales) from the Chytridiales. Further, he based genus descriptions in the new order on zoospore characters. In fact, the five orders in Chytridiomycota (Table 1) described on the basis of their zoosporic ultrastructure (Barr 1980, 2001) are consistent with current molecular phylogenies (see below), demonstrating the robustness of these characters. For example, the older, ordinal description of the Monoblepharidales based on thallus morphology was not inclusive, because some genera (Harpochytrium and Oedogoniomyces) lack the oogamous sexual reproduction and mycelial hyphae that characterized this order. In the morphology-based taxonomy, the order Harpochytriales described these morphologically simple fungi. However, zoosporic ultrastructural characters (e.g., Gauriloff *et al.* 1980), especially those associated with the kinetid (basal body and associated structures), are sufficient to classify the genera Harpochytrium and Oedogoniomyces within the Monoblepharidales. The order Harpochytriales was ultimately dropped. In another example, ultrastructural characters of zoospores revealed the need to reclassify the plant parasitic genus Physoderma from the Chytridiales to the Blastocladiales (Lange and Olson 1980), although the thallus morphology of *Physoderma* spp. strongly resembles that of the Chytridiales. Analyses of ultrastructural characters have also shown (Barr 1980) that several clades exist within the largest chytrid order, the Chytridiales. These same clades (Table 1) have been supported and extended by analysis of 18S rDNA sequences (James *et al.* 2000). Multiple-gene-based molecular phylogenies (see below) promise to resolve the more difficult questions of the branching order of the five chytridiomycete orders and higher groupings within these orders.

The placement of species into the other group of 'lower' fungi, Zygomycota, is based on the production of coenocytic thalli, the lack of motile spores at any

Table 1. Changes in higher-level classification of the Fungi from 1973 to present.

***The Fungi* IVB (Ainsworth 1973)**	***The Mycota* VII, Part A (McLaughlin *et al.* 2000)**	**Recent**
Eumycota	Chytridiomycota	Chytridiomycota[1]
Mastigomycotina	Chytridiomycetes	Chytridiomycetes
Chytridiomycetes	Blastocladiales	Blastocladiales
Blastocladiales	Monoblepharidales	Monoblepharidales
Harpochytriales	Neocallimastigales	Neocallimastigales
Monoblepharidales	Spizellomycetales	Spizellomycetales
Chytridiales	Chytridiales	Chytridiales[2]
Oomycetes		*Chytridium*-clade
Hyphochytriomycetes		*Rhizophydium*-clade
Plasmodiophoromycetes		*Nowakowskiella*-clade
		Lacustromyces-clade
Zygomycotina	Zygomycota	Zygomycota[3]
Zygomycetes	Zygomycetes	Zygomycetes
Trichomycetes	Trichomycetes	Trichomycetes
		Glomeromycota[4]
Ascomycotina	Ascomycota	Ascomycota[3, 5]
Hemiascomycetes	Saccharomycetes	Archiascomycetes[6]
Loculoascomycetes	Loculoascomycetes	Hemiascomycetes
Plectomycetes	Plectomycetes	Euascomycetes
Laboulbeniomycetes	Hymenoascomycetes	
Pyrenomycetes		
Discomycetes		
Basidiomycotina	Basidiomycota	Basidiomycota[3]
Teliomycetes	Urediniomycetes	Urediniomycetes
Hymenomycetes	Ustilaginomycetes	Ustilaginomycetes
Gasteromycetes	Heterobasidiomycetes	Hymenomycetes
	Homobasidiomycetes	

[1] Barr 2001; [2] Barr 1980, 2001; James *et al.* 2000; [3] Blackwell *et al.* 1996; [4] Schüßler et al 2001; [5] An *et al.* 2002; [6] Some authors consider this a grouping of early-branching ascomycetes. The results of analyses of mitochondrial data cast doubt on its existence.

developmental stage, and the lack of centrioles during mitosis. The classification of this group has not been particularly stable and the authors of several recent analyses of nuclear ribosomal sequences have even questioned the monophyly of Zygomycota, as well as that of Chytridiomycota. Ultrastructural evidence, which has been so valuable in hypothesizing relationships within the Chytridiomycota, has not proven definitive in resolving proposals for reclassification of this division based on molecular evidence. For example, it has been suggested that *Basidiobolus* (Zygomycota, Entomophthorales) might belong among the chytrids, an intriguing suggestion in light of the retention in *Basidiobolus* of a ring of microtubules in a centriole-like, nucleus-associated organelle (McKerracher and Heath 1985). This could indicate recent divergence from a flagellated ancestor. However, molecular studies have not resolved this question with adequate statistical support (Nagahama *et al.* 1995; Jensen *et al.* 1998). It has also been suggested that the Blastocladiales (Chytridiomycota) might group within Zygomycota (Bruns *et al.* 1992; Nagahama *et al.*

1995; James *et al.* 2000), a suggestion further supported by the position of the blastocladialean *Allomyces* in phylogenies based on mitochondrial data (Paquin *et al.* 1997), although not confirmed in subsequent analyses using more sophisticated inference methods (Leigh *et al.* 2003; Bullerwell *et al.* 2003b). Contradicting this suggestion is also the fact that the zoospore ultrastructure of blastocladialeans exhibits features typical for members of Chytridiomycota. Finally, Schüßler *et al.* (2001) has separated the Glomales (Glomerales), which contains the ecologically important arbuscular mycorrhizal fungi, into a new phylum, the Glomeromycota. The authors based this decision on the results of their analysis of SSU rDNA sequences, and hypothesize that the Glomeromycota probably share a common ancestry with the Ascomycota-Basidiomycota clade. Much more genomics data (mitochondrial and nuclear) from a broad selection of species will be required to address these issues.

In contrast to the historical and present difficulties in lower fungal classification, the placement of species into Ascomycota or Basidiomycota has not been altered to a large degree in recent years. The sexual characters on which the division of these phyla is based, which are visible by light microscopy, define groups that have remained stable through the advent of ultrastructural and molecular characters. Mycelia of both of these groups are regularly septate, both groups form dikaryotic cells before sexual reproduction, and some species in both phyla produce a yeast form of growth. The ascomycetes reproduce sexually by ascospores, which are produced within an ascus (sac-like cell within which karyogamy, meiosis and subsequent mitosis take place). Members of Basidiomycota reproduce sexually by basidiospores that are formed externally from a basidium (a cell within which karyogamy and meiosis takes place). Some aspects of higher fungal classification, however, have been greatly improved by newer technologies; for example, the sequencing of nuclear ribosomal genes has enabled many "deuteromycetes" (fungi of both phyla, but primarily ascomycetes, that are classified by their asexual reproductive structures) to be correlated with their sexually reproducing stage or relatives. In addition, whereas classical morphological, ultrastructural, and nuclear ribosomal sequences have yielded insufficient information to clarify deep divisions within these phlya with adequate support, mitochondrial protein sequence data have proven adept at resolving many of these relationships (see below).

2.2 The Promise of Molecular Phylogenetics

As molecular sequence data have become available, providing an unprecedented number of universal, phylogenetically strong characters that can easily be interpreted with computational methods, molecular phylogenetics has developed into the new taxonomic standard. Its promise is to provide a universal classification scheme that completes, complements, and if necessary, corrects the historical taxonomy based on morphological, biochemical, and ultrastructural characters.

2.2.1 Fungal phylogeny based on rRNA and nucleus-encoded proteins

The universally present (in all domains of life and in organelles) ribosomal RNA (rRNA) genes, particularly those encoding the small subunit (SSU) rRNA, quickly became the standard data set for molecular phylogenetics. In addition to being easily amplified by PCR (in contrast to protein-coding genes), these genes have a high level of sequence conservation. Two major databases of these sequences (as well as multiple alignments and secondary structure models) can be found at http://www.psb.rug.ac.be/rRNA/ and http://www.rna.icmb.utexas.edu. According to the former database, there are currently over 1550 publicly available fungal SSU rRNA sequences. Although the availability of sequence data from such a broad range of species makes rRNA molecules appealing for phylogenetic reconstruction, the use of these data has distinct limitations. In phylogenetic analyses, datasets containing more characters have more phylogenetic signal, and therefore result in better resolution of inter-taxa relationships. However, the amount of available sequence data in rRNA genes is practically limited to the SSU and large subunit (LSU). Both of these sequences contain a high proportion of nearly invariant positions, as well as a large number of highly variable sites that are difficult to align, leaving limited phylogenetically informative data with which to infer either very distant or very close evolutionary relationships. It has been suggested that, even if sequence data from both the LSU and SSU rRNA were available, there would be too little information to resolve deep fungal phylogenetic relationships with confidence (Berbee *et al.* 2000).

Although, initially, rRNA-based phylogenies appeared successful in resolving phylogenetic relationships in the fungi (e.g., Nishida and Sugiyama 1993; Bowman *et al.* 1992), it now seems that this success may be restricted to relationships within the four main fungal divisions. In addition, the robustness of published results may have been due to artifacts of the inference methods used. Since fungal species evolve at vastly different rates, long-branch attraction (LBA), which can cause quickly evolving species to branch together regardless of their true relationship, is a significant concern in phylogenetic reconstruction. Early rRNA-based analyses used uncorrected parsimony and distance-based methods, both of which are highly susceptible to LBA. The maximum likelihood-based reanalysis of one of these datasets by Leigh *et al.* (2003) produced a tree that was topologically different from the original published tree (Nishida and Sugiyama 1993). Yet, it also remained without significant support.

Other attempts have been made to analyze fungal phylogeny using amino acid sequences inferred from nuclear protein-coding genes (e.g., Liu *et al.* 1999; Keeling *et al.* 2000; Baldauf *et al.* 2000; Landvik *et al.* 2001). Amino acid sequences have several advantages over rRNA in phylogenetics. Because of the potentially large number of well-conserved and ubiquitous protein-coding genes, protein datasets can be much larger than rRNA datasets, resulting in greater signal and resolution. Additionally, whereas the mutations that lead to changes in protein sequence occur at the DNA level, selection pressure is on protein sequence and structure. Therefore, comparison of protein sequences provides a more realistic model of evolution at long evolutionary distance. Lastly, the use of protein sequence data in molecular phylogenetics allows additional complications associated with nucleotide sequences to be avoided (or at least reduced to a less significant level), such as nucleotide composition biases (overall or strand-specific).

Currently, the primary drawback of using nuclear-encoded protein sequences is the lack of available genomic data. At present, taxon sampling for whole genome studies of fungi is largely restricted to the ascomycetes, and is therefore of little use in evaluating the global fungal phylogeny. Two additional problems with nucleus-encoded proteins are gene duplications (paralogy) and lateral gene transfer (LGT), a process by which genes from one species are integrated into the genome of another species, and not inherited 'vertically'. While LGT is thought to be rampant in Eubacteria and Archaea (e.g., Doolittle 1999; Nesbo *et al.* 2001; Gogarten *et al.* 2002; Lawrence 2002), the prevalence of this phenomenon in eukaryotes is most likely minimal, although this conclusion currently relies on a small number of nuclear genome sequences that are not representative of eukaryotes in general. The result of LGT, as with paralogy, is that a tree based on a single gene does not necessarily match the tree representing the evolution of the species.

2.2.2 Advantages and disadvantages of mitochondrial protein sequences

A solution to some of the discussed problems is to reconstruct phylogenies using protein sequences encoded by mitochondrial genomes. Mitochondrial DNAs are often small with a high percentage of coding sequence, which reduces the effort involved in obtaining sequences from a wide range of organisms. Very little evidence exists for duplicated protein-coding genes or LGT in mitochondria (excluding lateral transfer of introns), one exception being the partial duplication of the *atp6* gene in *A. macrogynus* that exists as part of a mobile element (Paquin *et al.* 1994). In addition, the monophyletic origin of mitochondria from within the α-Proteobacteria is generally accepted, and no examples of mitochondria acquired by secondary endosymbiosis have been described. Consequently, the phylogeny determined by the analysis of mitochondrial sequences can be expected to reflect the phylogeny of eukaryotes, and the α-Proteobacteria can be used as an outgroup for these analyses. Finally, although the concatenation of nuclear proteins involved in different metabolic pathways may be problematic due to different selective pressures, the concatenation of the mitochondrial proteins commonly used for phylogenetics is justifiable: they are all involved in oxidative phosphorylation and can be considered to be under similar functional constraints.

One of the few disadvantages to using mitochondrial data in phylogenetic reconstruction is the limited amount of data available from highly reduced (with respect to their bacterial counterparts) mitochondrial genomes. At most, mitochondria have roughly 10% of the genes found in *Rickettsia prowazekii*, the α-proteobacterium that branches closest to mitochondria. For example, the mitochondrial genomes of most fungi contain a set of only fourteen protein-coding genes, and this set is further reduced in the fission yeasts (e.g., *Schizosaccharomyces pombe*) and in budding yeasts of the Saccharomyces genus. A further complication is the absence of mitochondrial DNA in some species, such as those of the order Neocallimasticales as well as members of Microsporidia. This limits the size and completeness of these data sets.

2.2.3 Fungal phylogeny based on concatenated mitochondrial proteins

Mitochondrial data have proven invaluable in phylogenetic reconstruction in general, and the resolution of fungal phylogeny using these data is unprecedented (e.g., Lang *et al.* 2002; Forget *et al.* 2002; Leigh *et al.* 2003; Bullerwell *et al.* 2003a,b). Figure 1 shows the phylogeny obtained from the maximum likelihood (ML) analysis of amino acid sequences inferred from twelve mitochondrial protein-coding genes (Atp6, Atp9, Cob, Cox1, Cox2, Cox3, Nad1, Nad2, Nad3, Nad4, Nad4L, Nad5). This analysis shows that the Fungi clearly form a monophyletic group (ML bootstrap of 100%; 100 resamplings were performed), as do the Holozoa (metazoa plus protists along the animal lineage; ML bootstrap of 100%). A highly supported monophyletic superset of these two groups (termed Ophistokonts, i.e., Fungi plus Metazoa) is also recovered (ML bootstrap of 98%). Bootstrap support for the internal branches among the Fungi is high, and the four fungal divisions are clearly defined, with the Ascomycota and Basidiomycota appearing as a monophyletic group, the Zygomycota branching prior to the ascomycete-basidiomycete divergence, and the Chytridiomycota branching at the base of the Fungi.

Although this topology is in accordance with the established taxonomy, certain peculiarities remain. First, there is little support (ML bootstrap of 64%) for the monophyly of Chytridiomycota including *Allomyces macrogynus* (discussed above). However, support for chytridiomycete paraphyly has dropped considerably with the advent of more sophisticated methods of phylogenetic analysis, and with the availability of data from additional species (support was 95% for the divergence of *A. macrogynus* from the branch leading to Zygomycota and 'higher' fungi in the analysis of Paquin *et al.* 1997). Another problematic issue is the exact placement of the Schizosaccharomyces genus within Ascomycota. According to SSU rRNA data this genus is a member of the archiascomycetes (formerly Taphrinomycotina according to NCBI), a group proposed to branch at the base of the ascomycetes (Nishida and Sugiyama 1994; Alexopolous *et al.* 1996). In contrast, recent

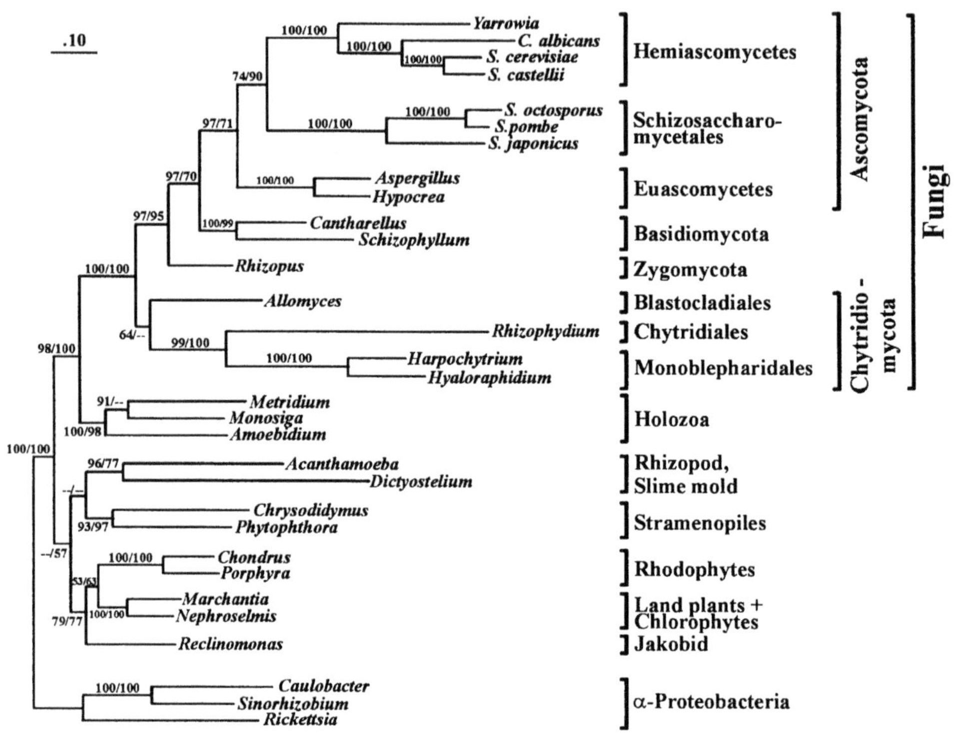

Fig. 1. Phylogenetic analysis based on concatenated mitochondrial proteins. The phylogenetic tree was constructed from unambiguously aligned portions of the concatenated protein sequences of Cox1, Cox2, Cox3, Cob, Atp6, Atp9, Nad1, Nad2, Nad3, Nad4L, Nad4, Nad5, a total of 2632 amino acid positions. The topology shown was initially inferred using ProML (Felsenstein 2002), and branch lengths were recalculated using CodeML (Yang 1997); the PMB model of protein evolution (Tillier, unpublished) was used with both programs. A value of 1.0 was chosen for the alpha factor. Maximum likelihood bootstrap support (%, first number) was calculated from 100 replicates using ProML. Distance bootstrap support (%, second number) was calculated from 1000 replicates using Tree-Puzzle (Strimmer and von Haeseler 1996), to generate pairwise distance tables, and trees inferred with Weighbor (Bruno *et al.* 2000). The WAG model of protein evolution (Whelan and Goldman 2001) was used in the distance approach. In addition, because distance methods are highly sensitive to missing data (whereas ML is not), Nad protein sequences were removed from the dataset, leaving a total of 1318 amino acids for distance tree inference. Relationships among fungi were similar in both distance and ML topologies, except that *Allomyces macrogynus* appears at the base of the chytridiomycetes in the ML tree (with 64% bootstrap support), whereas it branches at the base of the ascomycete-basidiomycete-zygomycete group in the distance tree (with 51% bootstrap support; value not shown, as it conflicts with the ML topology). Sequences were obtained from Genbank: *Yarrowia lipolytica* (NC002659), *Candida albicans* (NC002653), *Saccharomyces cerevisiae* (NC001224), *Saccharomyces castellii* (NC003920), *Schizosaccharomyces octosporus* (NC004312), *Schizosaccharomyces pombe* (NC001326), *Schizosaccharomyces japonicus* (NC004332), *Aspergillus nidulans* (ODAS1, CAA33481, AAA99207, AAA31737, CAA25707, AAA31736, CAA23994, X15442, P15956, CAA23995, CAA33116X00790, X15441, X06960, J01387, X01507), *Hypocrea jecorina* (NC003388), *Schizophyllum commune* (NC003049), *A. macrogynus* (NC001715), *Rhizophydium136* (NC003053), *Harpochytrium105* (AY182006), *Hyaloraphidium curvatum* (NC003048), *Metridium senile* (NC000933), *Monosiga brevicollis* (NC004309), *Amoebidium parasiticum* (AF538042-AF538052), *Acanthamoeba castellanii* (NC001637), *Dictyostelium discoideum* (NC000895), *Chrysodidymus synuroides* (NC002174), *Phytophthora infestans* (NC002387), *Chondrus crispus* (NC001677), *Porphyra purpurea* (NC002007), *Marchantia polymorpha* (NC001660), *Nephroselmis olivacea* (AF110138), *Reclinomonas americana* (NC001823), *Caulobacter crescentus* (NC002696), *Sinorhizobium meliloti* (NC003047), *Rickettsia prowazekii* (NC000963). Protein sequences of *Cantharellus cibarius* and *Rhizopus stolonifer* can be downloaded from http://megasun.bch.umontreal.ca/People/lang/FMGP/proteins/.

evidence indicates that the fission yeasts may not branch at the base of the ascomycetes, but form a monophyletic group with the budding yeasts (Fig. 1; Bullerwell *et al.* 2003a; Leigh *et al.* 2003), a scenario, however, that might result from LBA caused by the accelerated evolutionary rates in these lineages. Despite these unresolved issues, the current tree shows unparalleled resolution of the global fungal phylogeny, outperforming both rRNA data and the currently available nucleus-encoded protein sequence data.

3. EVOLUTIONARY GENOMICS: A TOOL FOR STUDYING GENE STRUCTURE AND EVOLUTION

Placing mitochondrial genome comparisons in an evolutionary context helps to reveal information that often cannot be predicted by simple genome comparisons, or directly from individual genome sequences. The principle of such inferences relies on the availability of a robust (well-supported) phylogenetic tree, a non-trivial demand for fungi due to their variable, and in most cases elevated, rates of sequence evolution. An evolutionary genomics analysis permits predictions about the presence or absence of features by analyzing their known distribution in lineages of descent. Here we present first a summary of the currently available completely-sequenced fungal mitochondrial genomes and describe them in terms of their gene complements, genome conformations and genome size variation. Second, we discuss particular features of fungal mitochondrial genes and genomes from an evolutionary perspective.

3.1 Completely Sequenced fungal Mitochondrial Genomes

The absence of sequence information has little effect on sequence-based phylogenetic inferences; however, numerous other predictions rely more heavily on the availability of complete mtDNA sequences. Among the extra information that can be extracted from complete genome data are variation in gene complement, gene order (including the direction of transcription of genes relative to each other) and the presence or absence of conserved regulatory elements in intergenic sequences. In addition, complete random sequencing of mtDNAs can reveal variations and peculiarities in genome conformation (e.g., the presence of genome variants in *Porphyra purpurea*; Burger *et al.* 1999; and the organization of the *Spizellomyces punctatus* genome in three distinct, circular-mapping DNAs; Laforest *et al.* 1997), as well as any form of polymorphic nucleotide positions or insertions/deletions that occasionally occur in mitochondrial DNA populations due to lack of genetic segregation (e.g., the presence of polymorphisms in mtDNAs of fungal specimens directly collected from nature, which have not undergone strict clonal selection in the laboratory; BFL, unpublished data). Finally, because there is so much variation in fungal mitochondrial genome size, gene order, and intron content, it is frequently more productive to directly sequence complete mtDNAs instead of using PCR amplification techniques, which have been productive in completely sequencing smaller mtDNAs with high gene order conservation, such as those of animals.

3.1.1 Gene complement

Nine mtDNAs have been added to the thirteen publicly-available, complete mitochondrial genome sequences that are listed in two recent reviews (Kennell and Cohen 2003; this volume, chapter by Hausner). New additions include the chytridiomycetes *Monoblepharella*15, *Harpochytrium*94 and *Harpochytrium*105 (Order Monoblepharidales; Bullerwell *et al.* 2003b), the ascomycetes *Schizosaccharomyces octosporus* and *Schizosaccharomyces japonicus* var. *japonicus* (Bullerwell *et al.* 2003a), and three zygomycetes (*Rhizopus stolonifer*, *Smittium culisetae* and *Mortierella verticillata*; BFL, submitted to GenBank with a release date of July 2003). In expectation of further releases of complete sequences, we invite interested readers to consult our webpage at http://megasun.bch.umontreal.ca/People/FMGP/Reviews/. To avoid repeating the information presented in the reviews cited above, we list the features of all above-mentioned genomes mainly to characterize new additions (Table 2,3). Otherwise, we will consider mtDNAs primarily from an evolutionary perspective.

Fungal mtDNAs are surprisingly constant in terms of the genetic information that they encode, considering the large amount of sequence and gene order divergence in this lineage. The basic fungal mitochondrial gene complement consists of genes encoding the large and small subunit ribosomal RNAs (*rnl* and *rns*), three subunits of the cytochrome oxidase complex (*cox*1, 2 and 3), apocytochrome b (*cob*), three subunits of the ATP-synthase complex (*atp6, 8 and 9*), seven subunits of the NADH dehydrogenase complex (*nad1, 2, 3, 4, 4L, 5* and *6*) and a variable number of transfer RNAs (discussed below). The genes for the small ribosomal subunit protein 3 (*rps3*; Bullerwell *et al.* 2000) and the gene encoding the RNA component of RNase P (*rnpB*) have a much more scattered distribution (Table 2). *nad* genes, and in one case *atp9*, are absent in some mtDNAs. Differences in fungal mtDNA coding content also include unidentified reading frames, introns, and certain plasmid-encoded polymerases, which are widespread in filamentous fungi (for recent reviews, see also Kennell and Cohen 2003; this volume, chapter by Hausner).

Despite enormous differences in genome size and structure, as well as the large evolutionary distances involved in the comparisons, the basic fungal mtDNA gene complement is almost the same as that in animal mitochondria. In contrast, a close unicellular relative of animals, the choanoflagellate *Monosiga brevicollis* (Lang *et al.* 2002) has a much larger gene complement, including many ribosomal protein genes (Burger *et al.* 2003). This indicates that an independent reduction of gene complement to a similar set of genes has occurred in both the fungal and animal lineages since their divergence from a common ancestor. It also raises the possibility that similarly gene-rich mtDNAs may remain to be identified in extant fungal groups, for instance, in unexplored members of Chytridiomycota. Also, protists that are phylogenetically close to the fungal/animal group might turn out to be part of the Fungi, and contain an extended mitochondrial gene set.

3.1.2 Variability of the gene complement and intron content

The seven genes coding for subunits of the NADH dehydrogenase complex are absent not only from the mtDNAs of the three known representatives of the genus Schizosaccharomyces, *S. pombe*, *S. octosporus* and *S. japonicus* (Fig.1, Table 2), but also from the nuclear DNA of *S. pombe* (Wood *et al.* 2002). In addition, although present in several hemiascomycete genera (e.g., Yarrowia, Pichia Candida) *nad* genes are absent in the mitochondrial genomes of *S. cerevisiae* and *Saccharomyces castellii*. This clearly demonstrates that *nad* genes have been lost independently from the mtDNA in the Saccharomyces and Schizosaccharomyces genera during fungal evolution; i.e., the similar reduction in mitochondrial gene content in these two genera cannot be used to support their decent from a common ancestor (in fact, mitochondrial gene content is among the least reliable characters in phylogenetic inferences).

Another case of variability in fungal mtDNAs is the complement of tRNAs. A complete tRNA complement, i.e., one that would be sufficient to decode all codons found in standard protein-coding genes, by applying mitochondrial "super wobble" rules of anticodon-codon decoding (~24-26 tRNAs), is present in all ascomycete, basidiomycete and zygomycete mtDNAs examined to date. A complete tRNA complement is also present in the chytridiomycete *Allomyces macrogynus* (Paquin and Lang 1996). In contrast, the six other examined chytridiomycete fungi (four from the taxonomic order Monoblepharidales, one from the Chytridiales and one from the Spizellomycetales) encode highly reduced sets of only 7-9 mitochondrial tRNAs (Laforest *et al.* 1997; Forget *et al.* 2002; Bullerwell *et al.* 2003b). In all six of these mtDNAs a common history of tRNA gene loss is evident as they all encode $tRNA^{fMet}$, $tRNA^{Pro}$, $tRNA^{Gln}$, $tRNA^{Trp}$ and $tRNA^{Tyr}$. Therefore, it seems likely that the bulk of tRNAs in the chytridiomycete lineage were lost in a common ancestor, followed by a few additional losses, duplications, and acquisitions in individual lineages (Fig. 2). Import from the cytoplasm is assumed for the 'missing' mitochondrial tRNAs. The tRNAs that remain in these systems presumably reflect tRNAs that might not be easily replaceable by their cytoplasmic counterparts, e.g., the methionine initiator tRNA, which is specific for the bacteria-like translation apparatus of mitochondria, and tRNAs that recognize codons with altered specificity such as those that recognize UGA as Trp and UAG as Leu (see below).

Table 2. Size and gene content of publicly-available complete fungal mitochondrial genomes.

Organismal Order Species	Size [1]	Genome Struct. [2]	*nad* Genes [3]	*atp9*	*rps3*	*rnpB*	ORFs [4]	tRNAs
Chytridiomycota								
Allomyces macrogynus	57.5	Circ-m	●	●	●	○	6i, 4f	25
*Harpochytrium*94	19.5	Circ-m	●	●	○	○	-	8
*Harpochytrium*105	24.2	Circ-m	●	●	○	○	-	8
*Monoblepharella*15	60.4	Circ-m	●	●	○	○	7i,5f	9
Hyaloraphidium curvatum	29.6	Linear	●	●	○	○	1i,3f	7
Spizellomyces punctatus	58.8; 1.4; 1.1	Circ-m (3)	●	●	○	○	4i,14f	8
*Rhizophydium*136	68.8	Circ-m	●	●	○	○	13i,7f	7
Zygomycota								
Mortierella verticillata	58.7	Circ-m	●	●	●	●	2i,7f	24
Rhizopus stolonifer	54.2	Circ-m	●	●	○	●	5i,4f	24
Smittium culisetae	58.7	Circ-m	●	●	●	●	12i,2f	24
Ascomycota								
Hypocrea jecorina *(Trichoderma reesei)*	42.1	Circ-m	●	●	●	○	1i,3f	26
Neurospora crassa	64.8	Circ-m	●	(●)[5]	●	○	8i,5f	27
Podospora anserina	100	Circ-m	●	○	●	○	39i,1f	27
Candida albicans	40.4	Circ-m	●	●	○	○	-	30
Saccharomyces castellii		Circ-m	○	●	●	●	1i	23
Saccharomyces cerevisiae	85.8		○	●	●	●	10i,4f	24
Pichia canadensis *(Hansenula wingei)*	27.7	Circ-m	●	●	●	●	1i	25
Yarrowia lipolytica	47.9	Circ-m	●	●	○	○	-	27
Schizosaccharomyces jap.	80.0	Circ-m	○	●	○	○	3i	25
Schizosaccharomyces oct.	44.2	Circ-m	○	●	●	●	6i	24
Schizosaccharomyces pom.	19.4	Circ-m	○	●	●	●	3i	25
Basidiomycota								
Schizophyllum commune	49.7	Circ-m	●	●	●	○	5f	24

[1]Size in kbp, rounded values;[2] Circ-m, circular mapping, and expected to be predominantly in form of long linear concatemer as shown in various eukaryotes (Bendich, 1996);[3] Ubiquitous genes in fungal mtDNAs are cob (apocytochrome b), cox1,2,3 (cytochrome oxidase subunits), atp6,8 (ATPase subunits), rns, rnl (small and large subunit rRNAs) and a various number of genes coding for tRNAs;[4] ORF length $\geq$ 100; intronic ORFs are labeled 'i', free-standing ORFs, 'f'; [5] the mtDNA-encoded atp9 gene of *N. crassa* is a pseudo-gene under vegetative growth conditions.

Proven or putative mobile elements have a widespread distribution in fungal mitochondrial genomes, and they represent one of the strongest sources of variability in mtDNAs. The most abundant proven mobile elements are introns, which are present in the four principle divisions of fungi, but are highly variable in terms of individual presence or absence. For example, they are absent from both *Harpochytrium* species and the basidiomycete *S. commune*, but they are frequent in their relatives, such as the chytridiomycete *Monoblepharella15* (Bullerwell *et al.* 2003b) and the basidiomycete *Microbotryum violaceum* (BFL, unpublished data). The abundance of introns in fungal mtDNAs is in contrast to the paucity of introns in animal mtDNAs. Further, in contrast to the situation in plant mitochondria, the vast majority of fungal mitochondrial introns is of group I, whereas the vast majority is of group II in plant mtDNAs.

Intron numbers can be remarkably high in fungal mtDNAs. For example, *Rhizophydium136* contains 37 mitochondrial introns, *Podospora anserina* contains 36, and *Allomyces macrogynus* contains 28 introns (Table 3; Forget *et al.* 2002; Cummings *et al.* 1990; Paquin and Lang 1996). There is a clear tendency for introns to be preferentially inserted in highly conserved regions of mitochondrial genes (Lang 1984), which is probably the reason for an elevated number of intron insertions in the highly conserved *cox1* gene (the next most intron-rich gene is *cob*). For instance, in *P. anserina*, the *cox1* gene extends over 24.5 kbp and contains 16 introns (Cummings *et al.* 1989). The wealth of data supporting intron mobility (e.g., Lambowitz and Belfort 1993; Belfort and Perlman 1995) will not be discussed here. However, it can be surmised that introns were present in the mitochondrial DNA of an early ancestor of the fungal lineage, and that they have been considerably shuffled within the fungal lineage by lateral transfer events, an ongoing, frequent process as we can learn from comparative studies of *S. pombe* isolates (see below).

3.1.3 Genome conformation

In vivo, several fungal mitochondrial genomes have been shown to consist predominantly of linear, multimeric head-to-tail concatamers (Bendich 1993, 1996), not - as widely assumed -of monomeric circles. It should be noted that this observation is not in contradiction to the fact that most fungal mtDNAs examined to date map and assemble as circular molecules, both in restriction and sequence analyses. The only demonstrated examples of true monomeric molecules (i.e., comprising the main proportion of DNA molecules within a population) are linear mtDNAs. This conformation has been described in several members of the hemiascomycete genera Candida, Pichia and Williopsis (Fukuhara *et al.* 1993; Nosek *et al.* 1998), interspersed with species containing circular-mapping genomes, and in the chytridiomycete *Hyaloraphidium curvatum* (Forget *et al.* 2002). In fungi, linear genome conformations have a scattered distribution, and have likely emerged independently several times, a distribution pattern also observed for distantly related protists (see Lang *et al.* 1999). Determining how true linear genomes are maintained, and how they evolved from their circular-mapping counterparts, may reveal insights into the replication and maintenance of fungal mitochondrial genomes as a whole. It is interesting that some fungal mtDNAs include plasmid components, which sometimes encode polymerases involved in their maintenance (see Kennell and Cohen 2003; this volume, chapter by Hausner). Certain types of mitochondrial plasmids are maintained as monomeric, linear molecules with inverted repeats at their termini, organized much like linear mtDNAs. It is known that these plasmids occasionally insert into mtDNAs, leaving behind polymerase genes. It is tempting to speculate that a plasmid-derived DNA polymerase might confer the capacity for replication of linear mtDNA molecules, considering that such DNA polymerase genes are found in certain linear mtDNAs (e.g., in *Ochromonas danica*; Chesnick *et al.* 2000; and in the jakobid flagellate *Jakoba libera*; BFL, unpublished).

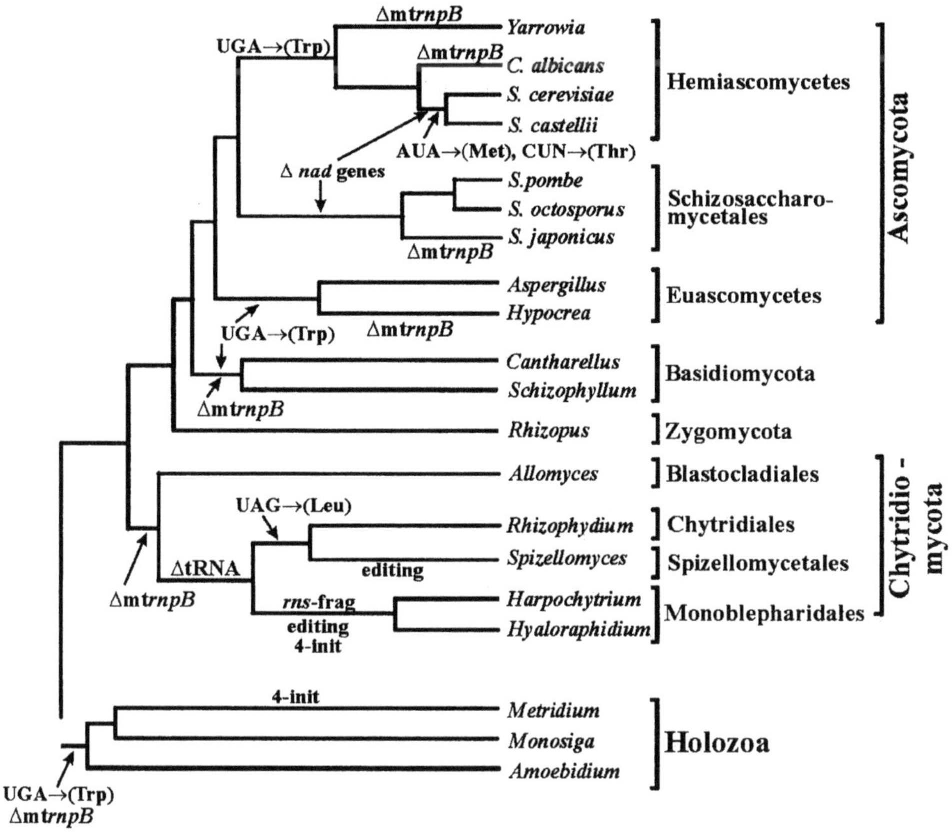

Fig. 2. Occurrence of features in fungal mtDNAs. editing, 5'tRNA editing; *rns*-frag, fragmentation of SSU rRNA gene; 4-init, quartet initiation codons plus modified initiator tRNAs; ΔtRNA, loss of several tRNA genes; Δ *nad* genes, loss of all genes of NADH dehyrogenase, both in mtDNA and nuclear DNA; Δmt*rnpB*; loss of mitochondrial gene (*rnpB*) encoding RNA subunit of RNase P. For abbreviations of species names, see Fig. 1.

Other, less frequently observed, circular-mapping mitochondrial plasmids are likely genuine parts of multi-chromosome mitochondrial genomes, because they encode regular mitochondrial genes (such plasmids are common in flowering plants). Two such mitochondrial plasmids are present in *S. punctatus* (one encoding an *atp9* gene, the other carrying a conserved repeat region characteristic for this mitochondrial genome; Laforest *et al.* 1997). A highly unusual variant of multi-chromosome mtDNA is present in a unicellular relative of animals, *Amoebidium parasiticum*, which possesses several hundred linear mitochondrial chromosomes, all of which share a terminus-specific sequence pattern (Burger *et al.* 2003). Evidently, we are still far from understanding how linear mitochondrial genomes replicate, and how their structure evolves.

3.1.4 Genome size variation

mtDNA size is highly variable in fungi. This is sometimes due to variation in intron content, which can result in large differences in genome size even between closely-related species. For instance, genome size variation from 17.4 to 24.4 kbp in naturally-occurring strains of *S. pombe* is due to the presence or absence of introns (Zimmer *et al.* 1987). These elements are likely superfluous in mitochondrial systems, and, in fact, they have been experimentally eliminated from *S. pombe* mtDNA without affecting cellular survival (Schäfer *et al.* 1991). The length of 'non-coding' intergenic regions may also vary considerably, even within a lineage. For instance, the mtDNA sizes within the genus Schizosaccharomyces vary from 17 to 80 kbp (Bullerwell *et al.* 2003a), and those of the Monoblepharidales vary from 19 to 60 kbp (Bullerwell *et al.* 2003b). In both of these examples, intergenic regions are responsible for most of the variation in genome size.

Table 3. Features of publicly available complete fungal mitochondrial genomes.

Organismal Order Species	**Trans. Code** [1]	**Introns** [2]	**tRNAs** [3]	**tRNA Editing**	***rns* in pieces**	**Acc. # and References**
Chytridiomycota						
Allomyces macrogynus	universal	28	25			NC001715
Harpochytrium94	universal	-	8	■	■	AY182005
Harpochytrium105	universal	-	8	■	■	AY182006
Monoblepharella15	universal	8	9	■	■	AY182007
Hyaloraphidium curvatum	universal	1	7	■	■	NC003048
Spizellomyces punctatus	UAG(L)	12	8	■		NC003052, 60, 61
Rhizophydium136	UAG(L)	37	7			NC003053
Zygomycota						
Mortierella verticillata	universal	4	24			[5]
Rhizopus stolonifer	universal	9	24			[5]
Smittium culisetae	universal	14	24			[5]
Ascomycota						
Hypocrea jecorina (Trichoderma reesei)	UGA(W)	10	26			NC003388
Neurospora crassa	UGA(W)	16	27			Whitehead Inst.
Podospora anserina	UGA(W)	36	27			NC001329
Candida albicans	UGA(W)	6	30			NC002653
Saccharomyces castellii	UGA(W), AUA(M), CUN(T)	2	23			NC003920
Saccharomyces cerevisiae	UGA(W), AUA(M), CUN(T)	13	24			NC001224
Pichia canadensis (Hansenula wingei)	UGA(W)	2	25			NC001762
Yarrowia lipolytica	UGA(W)	17	27			NC002659
Schizosaccharomyces jap.	universal	2	25			NC004332
Schizosaccharomyces oct.	universal	6	24			NC004312
Schizosaccharomyces pom.	universal [4]	3	25			NC001326
Basidiomycota						
Schizophyllum commune	UGA(W)	-	24			NC003049

[1]Deviations from the standard translation code are indicated in bold; [2] Total number of introns, the two mitochondrial intron classes I and II are not distinguished; [3] Includes duplicated genes; [4] In *S. pombe*, one UGA(W) is present in *rps3* and two in intronic ORFs; [5] For accession numbers (to be released July 2003) see http://megasun.bch.umontreal.ca/People/FMGP/Reviews/.

It is of considerable interest that fungal mtDNAs are much less stringently selected for compactness than their animal counterparts. Some fungal groups have a tendency towards compactness; however, others seem to rapidly accumulate non-coding, often repetitive and A+T-rich sequences (e.g., in the mtDNAs of isolates of the basidiomycete *Microbotryum violaceum*; BFL, unpublished data). Whether genome size is regulated to any extent, or whether seemingly uncontrolled genome expansion only leads to inefficiency and increased extinction rates, remains a tantalizing puzzle. Sequencing of many more fungal mtDNAs is required to provide the data for a more systematic analysis and understanding of both genome expansion and contraction processes.

3.2 Mapping the Origin and Evolution of Mitochondrial Gene Expression Features

The combination of complete genome sequence data from a diverse range of fungi coupled to the robust phylogenetic framework has allowed a more comprehensive understanding of overall changes in gene expression features, including deviations from the translation code, the presence or absence of RNA editing, codon usage biases, and compositional nucleotide or amino acid bias. In this section, we discuss several features of mitochondrial genomes that serve to demonstrate additional insights that the evolutionary genomics approach can offer.

3.2.1 Genetic code variation

Many of the currently known fungal mitochondrial genomes do not use the universal translation code (Table 3). The most common deviation from the standard code observed in these genomes is the reassignment of UGA 'stop' codons as tryptophan. This modification has been described in ascomycetes (e.g., *S. cerevisiae* and *C. albicans*; Fox *et al.* 1979; Anderson *et al.* 2001), basidiomycetes (*Schizophyllum commune* and *Cantharellus cibarius*; Paquin *et al.* 1997; BFL, unpublished data) and zygomycetes. Changes in the genetic code, particularly the deviation UGA(Trp), are, in fact, quite common in mitochondrial systems (Gray *et al.* 1998). Interestingly, UGA codons are also greatly preferred over UGG in the eubacterial genus Mycoplasma (Yamao *et al.* 1985) and its close relatives, attesting to the widely dispersed occurrence of this feature. It is clear that this deviation from the universal code has emerged many times independently. The deviation of UAG coding for leucine instead of termination is found in the mtDNAs of the chytridiomycetes *S. punctatus* and *Rhizophydium136*. The same change has also been observed in mitochondria of certain chlorophycean algae (Hayashi-Ishimaru 1996; note, however, that UAG codes for alanine or termination in other chlorophycean algal systems). Finally, AUA codes for methionine instead of isoleucine, and CUN for threonine instead of leucine, in hemiascomycetes of the genus Saccharomyces (although not in mitochondria of other hemiascomycetes, as suggested in some GenBank records).

Codons are sometimes completely unused in mitochondrial genomes. The relatively large number of unassigned codons in the mtDNAs of the Monoblepharidales (up to 14 in *Harpochytrium* species) shows that this phenomenon can be quite extensive (Forget *et al.* 2002; Bullerwell *et al.* 2003b). The disuse of a codon may represent the first step in changing the genetic code. Once components of the translation machinery that were previously required to recognize these codons (i.e., tRNAs, or release factors in the case of termination codons) are eliminated, modified or novel tRNAs might take over the decoding of these codons. This may be particularly straightforward in mitochondrial systems, where coding rules are more relaxed. For example, most tRNAs containing an unmodified uridine in the wobble position of the anticodon are able to decode all four nucleotides in third codon positions. Thus, the reduced mitochondrial anticodon-codon specificity (compared to most nuclear or bacterial translation systems) may facilitate the transition from one codon identity to another.

In the standard protein-coding genes of the basidiomycete *S. commune*, over 20% of tryptophan residues are specified by UGA (Paquin *et al.* 1997). Mitochondrial and eubacterial systems that use this non-standard genetic code generally encode a $tRNA^{Trp}$ with the anticodon sequence 5'-UCA-3', capable of forming base-pairs with UGA as well as the standard UGG. Yet, the $tRNA^{Trp}$ encoded by the *S. commune* mtDNA has the anticodon sequence CCA, normally recognizing only UGG codons in mRNAs. It has been proposed (Paquin *et al.* 1997) that this tRNA is in fact able to inefficiently decode UGA codons in addition to UGG codons, suggesting that the translation system can tolerate a certain threshold of UGAs. A similar proposition has been made for the mtDNA of *S. pombe* (Bullerwell *et al.* 2003a), a mitochondrial system that contains one UGA codon in the *rps3* reading frame, the product of which is critical for cell survival (Neu *et al.* 1998). Remarkably, in another basidiomycete *Cantharellus cibarius,* for which more than 50% of tryptophan codons are encoded by UGA in the mtDNA, there is a duplicate *trnW*. It contains two nucleotide differences from its duplicate, one of which is the expected alteration of the CCA anticodon sequence to UCA (BFL, unpublished data). This observation lends support to the role of tRNA gene duplication and divergence in reassignment of mitochondrial codon identity.

A possible intermediate in the process of codon reassignment can be observed in three independent examples: the ascomycete *S. octosporus*, the basidiomycete *S. commune* and the zygomycete *R. stolonifer*. In these mitochondrial systems, $tRNA^{Ile}$(cau) (cytosine modified to lysidine), the tRNA capable of decoding AUA codons in mitochondrial (and bacterial) transcripts, is absent. Similarly, ATA codons are absent in standard protein-coding genes in these three mtDNAs (Bullerwell *et al.* 2003a; BFL, unpublished). This suggests that selective pressure promotes the retention of a full complement of tRNAs in these mitochondrial systems only until the necessity for a tRNA (i.e., the presence of the cognate codon) disappears. At that point, the tRNA can be eliminated without compromising the survival of the organism. The unused AUA codon is then available for a new assignment, to AUA(Met), for instance, which is relatively frequent in mitochondria (e.g., in *Saccharomyces* and animals).

3.2.2 Promoters and RNA processing

Whereas comparisons of fungal mtDNAs from a broad selection of species have given us an overall understanding of the state of these genomes, comparisons of genomes at shorter evolutionary distances have identified functional sequence elements that, because of their low degree of conservation, are not detectable at larger evolutionary distances. One particular instance in which comparisons of mtDNAs from closely related organisms can be of great utility is the identification of promoters.

In the yeast *S. cerevisiae*, a highly conserved, consensus nonanucleotide (5'-ATATAAGTA-3') was identified immediately upstream of proven or putative transcription initiation sites (Osinga *et al.* 1982; Christianson and Rabinowitz 1983). Examination of the mtDNA sequences from two closely related budding yeasts, *Kluyveromyces lactis* and *Torulopsis glabrata*, revealed that the same consensus nonanucleotide was also present (Clark-Walker *et al.* 1985). However, when more distantly related organisms are compared, less conservation is observed. For example, in the fission yeast *S. pombe*, transcription initiates from two promoters (5'-ATATATGTA-3' and 5'-ATATGTGA-3') that show only negligible similarity to the yeast consensus. In the mtDNAs of two further members of *Schizosaccharomyces*, *S. octosporus* and *S. japonicus*, similar promoters have not been identified (Bullerwell *et al.* 2003a). Recently, the involvement in transcription initiation of a short, conserved sequence in *H. curvatum* was suggested, based on its orientation and distribution (Forget *et al.* 2002). A comparison of the *H. curvatum* mtDNA with those of its relatives *Monoblepharella15*, *Harpochytrium94* and *Harpochytrium105* (Bullerwell *et al.* 2003b) did not reveal similar conserved sequences, which is not surprising in light of the considerable evolutionary divergence between these organisms. However, other strongly conserved motifs were identified in the three latter genomes. The consensus sequence 5'-TTATAGGAAAT-3' was identified between 15 and 26 nucleotides upstream of the start codons of the *atp6*, *atp8*, *nad3*, *nad4L*, *nad5* and *nad6* genes. Another conserved sequence, 5'-AGAGTGTANTNNAAT-3', was identified 8 to 13 nucleotides upstream of the *cox3* gene in all three species. Initially, it was predicted that some or all of these sequences might serve as mitochondrial promoters in these three species. Primer extension experiments were performed to test this hypothesis, and contrary to expectations, the 5'-ends of these RNAs were found to map 1 to 6 nucleotides upstream of the 5'-end of the consensus sequence (i.e., these sequence motifs are located within the transcripts), thus arguing against their role as promoters. It is likely that these sequences function in translation regulation (Bullerwell *et al.* 2003b), as do certain structures that are situated in the 5'-untranslated regions of mitochondrial transcripts in yeast (Costanzo and Fox 1988; Mulero and Fox 1993; Steele *et al.* 1996). Unfortunately, the currently available data leave us without predictions about potential mitochondrial promoters in these and any other chytridiomycete fungi.

Similarly, mechanisms for transcript processing have been difficult to predict, except in closely related species. In yeast mitochondria, relatively short primary transcripts are produced from several promoters, and a dodecanucleotide consensus sequence appears to be involved in signaling the 3' processing of these precursor RNAs (Clark-Walker *et al.* 1985). In contrast, the fission yeast *S. pombe* and the euascomycetes *N. crassa* and *A. nidulans* appear to transcribe long primary transcripts. In these instances, RNA processing proceeds in a different way, relying on tRNA processing to release mRNAs (Lang *et al.* 1983; de Vries *et al.* 1985; Burger *et al.* 1985; Dyson *et al.* 1989). In the three species of *Schizosaccharomyces*, an additional conserved C-rich RNA motif has been identified close to the 3' region of most genes. S1 nuclease protection experiments in *S. pombe* demonstrated the presence of S1 nuclease protection signals directly adjacent (downstream) to these motifs (Lang *et al.* 1983; Trinkl *et al.* 1989). The presence of these motifs downstream of all genes in *S. octosporus* and *S. japonicus* (Bullerwell *et al.* 2003a) predict that they too are involved in RNA 3'-end transcript processing. Again, at this time nothing is known about transcript processing in non-ascomycete fungal systems.

3.2.3 Fragmentation of the *rns* gene

Fragmentation of ribosomal RNAs has been observed in a wide variety of mitochondrial systems such as *Tetrahymena pyriformis* (Schnare *et al.* 1986), certain green algae (e.g., Boer and Gray 1988; Turmel *et al.* 1999; Nedelcu *et al.* 2000), *Plasmodium falciparum* (Gillespie *et al.* 1999) and *Theileria parva* (Kairo *et al.* 1994). Although rRNA genes appear to be contiguous in all zygomycete, basidiomycete and ascomycete mtDNAs examined to date, fragmentation of the mitochondrial small subunit rRNA gene (*rns*) has been described for all four characterized members of the order Monoblepharidales (*H. curvatum*, *Monoblepharella15*, *Harpochytrium94* and *Harpochytrium105*; Forget *et al.* 2002; Bullerwell *et al.* 2003b). Modeling of the secondary structures of these four rRNAs predicts that the break point is located in the same variable region corresponding to nucleotides 590-649 of the *E. coli rns* gene, and that the two mitochondrial rRNA pieces have the potential to assemble by intermolecular base-pairing. The common location of break points suggests that a single event gave rise to the fragmentation in these four species. The three other characterized chytridiomycete mtDNAs (*A. macrogynus*, *S. punctatus* and *Rhizophydium136*) encode intact ribosomal RNA genes (Table 3), thus the event that gave rise to the fragmentation likely occurred in a common ancestor of these four monoblepharidalean fungi, as indicated in Figure 2.

3.2.4 Mitochondrial RNase P-RNA: gene discovery and prediction of RNA secondary structure

RNase P is a ribonuclease in prokaryotes, eukaryotes and eukaryotic organelles that is involved in the removal of 5' leader sequences from tRNA precursors (Frank and Pace 1998). In *E. coli* and other eubacteria, RNase P is composed of one RNA subunit and one protein subunit. The RNA subunit of RNase P (P-RNA) is essential for enzymatic activity, both *in vivo* and *in vitro* (Stark *et al.* 1978; Guerrier-Takada *et al.* 1983; Gardiner *et al.* 1985). The single eubacterial protein component of RNase P is only essential for the enzymatic activity *in vivo*, and its participation in the formation of an active site architecture, and substrate interaction has been suggested (True and Celander 1998; Crary *et al.* 1998). Further, it may also increase the catalytic activity of RNase P by acting as an electrostatic shield between negatively charged P-RNA and tRNA molecules (Guerrier-Takada *et al.* 1983; Gardiner *et al.* 1985).

Mitochondrial RNase P (mtRNaseP) has been characterized in much less detail and only in a few species. The most detailed information on the biochemical and genetic properties of mtRNaseP is available for *S. cerevisiae*. The gene encoding its RNA subunit was first identified by analyzing yeast mitochondrial mutants deficient in mitochondrial tRNA processing and protein synthesis (Underbrink-Lyon *et al.* 1983; Miller and Martin 1983). Its protein subunit is unusually large (105 kDa), and is encoded in the nucleus (Morales *et al.* 1992; Dang and Martin 1993). Unlike in *S. cerevisiae*, seven polypeptides were co-purified with the mtRNaseP activity in the euascomycete *A. nidulans* (Lee *et al.* 1996b). The RNA subunit from both *S. cerevisiae* and *A. nidulans* is essential for catalytic activity, as in eubacteria.

Biochemical studies on the structure and function of catalytic and structural RNAs requires rigorous testing, demanding large experimental efforts. A highly efficient, complementary approach is the comparison of RNA molecules from a wide range of phylogenetically diverse organisms, an approach that has been taken to identify new genes encoding mitochondrial P-RNAs (mtP-RNAs), and to explore their potential RNA secondary structure. These RNAs are difficult to identify due to significant reductions in their secondary structure in mitochondria, in comparison to those of eubacteria. Searching mtDNA sequences for universally-conserved sequence features of RNase P RNAs, followed by comparative modeling of RNA structures from closely related species, has proven successful in the identification of several new fungal mitochondrial homologs (Table 2, Fig. 3).

The least-derived (most eubacterial-like) mtP-RNA identified to date is that of the jakobid *R. americana* (Figure 3; Lang *et al.* 1997). Its features have since served as a guide to remodel the secondary structures of more derived mtP-RNAs. For instance, a new secondary structure model has been proposed for the mitochondrion-encoded *A. nidulans* mtP-RNA, to better reflect the eubacterial consensus (Martin and Lang 1997) than did a previously published model (Lee *et al.* 1996a). Similarly, the folding of P-RNAs encoded in the mtDNAs of the green alga *Nephroselmis olivacea* (Turmel *et al.* 1999) and the ascomycete *Taphrina deformans* (ES and BFL, unpublished data) has been modeled using this approach. Because hemiascomycete and fission yeast mtP- RNAs are extremely rich in A and U, however, RNA secondary structure prediction has been more problematic, and has required comparisons among mitochondrial RNA molecules of closely related species, such as *S. pombe* and *S. octosporus* (ES and BFL, unpublished data). In addition, hemiascomycete mtP-RNAs exhibit further size reductions and variations that obscure the identification of conserved helical regions in the RNA secondary structure, even at short evolutionary distances (Wise and Martin 1991). Despite these difficulties, by comparative modeling of fourteen fungal mtP-RNA sequences, three types of structures have been identified: (i) the most bacteria-like structures occur in zygomycetes and the ascomycetes *T. deformans*, and *A. nidulans*; (ii) structures of intermediate similarity to the eubacterial P-RNAs occur in *S. pombe* and *S. octosporus*; and (iii) highly derived structures occur in almost all hemiascomycete mitochondria.

The failure to identify a mitochondrial gene that encodes an mtP-RNA has several possible explanations: (i) a protein-only enzyme may substitute its function (similar to the situation in spinach chloroplast, where the RNase P has biochemical and physical properties consistent with the presence of a protein-only enzyme; Thomas *et al.* 1995, 2000); (ii) the RNA is encoded by a mitochondrial gene that remains unidentified due to its highly divergent sequence; and (iii) the RNA subunit may be nucleus-encoded, and imported into mitochondria. We currently favor the idea that the mitochondrial *rnpB* gene is absent from many mtDNAs, i.e., either that their mtP-RNA is imported from the cytoplasm, or that it has become a protein-only enzyme. Strong support for this notion comes from the two *Harpochytrium* species, wherein unidentified intergenic regions are simply too short to accommodate a gene of this size.

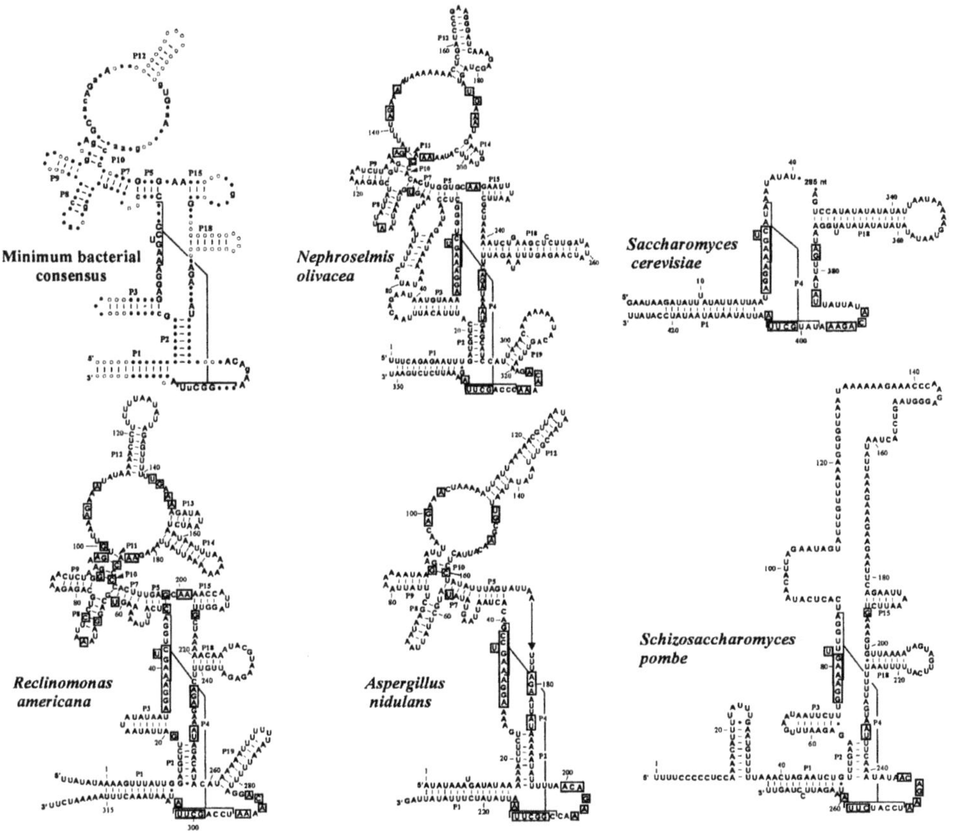

Fig. 3. Secondary structure predictions of mitochondrial RNase P RNAs from *R. americana* (Lang *et al.* 1997), *N. olivacea* (Turmel *et al.*, 1999), *A. nidulans* (Martin and Lang 1997), *S. pombe*, and *S. cerevisiae* (ES and BFL, unpublished). The phylogenetic minimum bacterial consensus structure (Brown 1998) is shown for comparison. The key to the bacterial consensus is as follows: capital letter 100% conserved; lower case 90% conserved; filled circle (nucleotide present in all RNAs); open circle (nucleotide present in 90% of the RNAs). Boxed residues in the mitochondrial RNA structures denote conservation of minimum bacterial consensus nucleotides.

Given the large variation in RNA secondary structure among mtP-RNAs, it will be of great interest to examine the actual composition of RNase P ribonucleoproteins from a variety of fungal mitochondrial systems to determine the number and size of protein subunits . Based on the idea that loss of RNA structure elements is likely compensated by proteins, we expect to find an inverse correlation between the degree of complexity of mtP-RNA structure and size and number of RNase P proteins.

3.2.5 Identification of genes encoding ribosomal protein Rps3

In 1979, Var1 (a protein encoded by hemiascomycete mitochondrial DNAs) was shown to be a stoichiometric component of yeast mitochondrial ribosomes (Groot *et al.* 1979; Terpstra *et al.* 1979). In the same year, in the euascomycete *N. crassa*, another ribosomal protein, called S5, was also found to be a stoichiometric component of mitochondrial ribosomes (Lambowitz *et al.* 1979). Although speculation abounded (Butow *et al.* 1985; de Zamaroczy and Bernardi 1987) as to the origins of these two unusual ribosomal proteins (they bear little sequence similarity to known ribosomal proteins), they remained a mystery for over twenty years.

During that time, mitochondrial genome data accumulated, revealing that mtDNAs encode a variable number of ribosomal proteins, ranging from 27 in the minimally-derived mtDNA of the jakobid flagellate *R. americana*, to none in animal mtDNAs. However, it was not until 1996 that a homolog of the small subunit ribosomal protein 3 gene (*rps3*) was discovered in the mtDNA of the little-derived chytridiomycete *A. macrogynus* (Paquin and Lang 1996). RNA transcripts of this gene were not detectable, and, therefore, it was possible that it represented a pseudogene. Nevertheless, this observation suggested that other mitochondrial genomes might encode ribosomal proteins. Indeed, an *rps3* homolog was subsequently identified in the zygomycetes *M. verticillata* and *S. culisetae*, based on comparison of these sequences to *rps3* from eubacteria and to that of the *A. macrogynus* mtDNA. A link between these genes and those encoding Var1 and 5S in the ascomycetes was then possible, based largely on a conserved sequence motif in the carboxy-terminal region of these highly divergent protein sequences (Bullerwell *et al.* 2000).

The *rps3* homologs identified in fungal mtDNAs are very different from the more conservative ones encoded in prokaryotes, eukaryotic nuclei, and plant chloroplasts. This difference is likely due to the high evolutionary rate in fungal mitochondria, coupled with relaxed functional constraints on the Rps3 protein. It is interesting that only one region, the carboxy-terminal region, is conserved to any appreciable degree in fungal mitochondrial Rps3 homologs. The Rps3 carboxy-terminal motif was also identified in Orf227, a protein encoded by the mtDNA of *S. pombe*. Mutations in the carboxy-terminal region of this protein have been shown to be responsible for a mitochondrial mutator phenotype (Zimmer *et al.* 1991; Neu *et al.* 1998), a phenotype possibly due to impaired mtDNA repair (Ahne *et al.* 1988). In fact, the involvement of Rps3 in the repair of oxidative DNA damage has also been show in *Drosophila* (Wilson *et al.* 1994; Sandigursky *et al.* 1997) and in mammals (Kim *et al.* 1995; Wool *et al.* 1996). The conservation of this region in fungal Rps3 homologs raises the intriguing possibility that it may be involved not only in the assembly of the small ribosomal subunit, but also in the DNA repair process.

The *rps3* gene has a scattered distribution in fungal mitochondria (Table 2). It is not present in the mtDNA of the zygomycete *R. stolonifer* (although it is present in its relatives *M. verticillata* and *S. culisetae*), nor is it present in chytridiomycete mtDNAs, with the exception of *A. macrogynus*. Similarly, animal mtDNAs lack *rps3*, whereas this gene (along with many other ribosomal protein genes) is encoded in the mtDNAs of the choanoflagellate *M. brevicollis*, and the ichthyosporean *A. parasiticum* (Burger *et al.* 2003), both relatives of the animals. Independent loss of the *rps3* gene from the mtDNA has seemingly occurred numerous times in diverse lineages (see Lang *et al.* 1999). It is nevertheless likely, because of its important role in mitochondrial function, that it is replaced by a nuclear DNA-encoded Rps3 homolog imported from the cytoplasm.

3.2.6 The origin of 5'tRNA editing in Chytridiomycota

RNA editing has been reported in the chytridiomycete fungus *S. punctatus* (Laforest *et al.* 1997). Modeling of the secondary structures of the mtDNA-encoded tRNAs in this fungus reveals potential mispairing in the first three base pairs of the acceptor stems of all eight tRNAs. This mispairing is of such an extensive nature that orthodox tRNA folding and function is impossible. Sequencing of tRNA molecules showed that mismatches were repaired at the RNA level (Laforest *et al.* 1997; M.-J. Laforest and B.F.L., unpublished data). The postulated biochemical activity of 5'tRNA editing involves substitution of the first three 5'-nucleotides of the tRNA molecule with nucleotides that reconstitute standard Watson-Crick base pairs. A similar type of editing had previously been described in *Acanthamoeba castellanii* (Lonergan and Gray 1993), an amoeboid protist with no specific phylogenetic link to chytridiomycete fungi. These observations raised the intriguing possibility that a similar type of activity had evolved independently in two distant lineages.

Subsequent mtDNA sequence data from chytridiomycetes of the taxonomic order Monoblepharidales has revealed that 5'tRNA editing is also present in the mitochondria of *Monoblepharella15*, *Harpochytrium94* and *Harpochytrium105* (Laforest and BFL, unpublished data). Modeling evidence has also been presented for this type of editing in *H. curvatum* (Forget *et al.* 2002), although the mismatching in tRNA acceptor stems in this organism is much less extensive than in its close relatives or in *S. punctatus*. Combined with the absence of tRNA editing in the mtDNA-encoded tRNAs of *A. macrogynus* and any other examined fungus, this data might suggest that tRNA editing emerged only once, at the base of the chytridiomycetes, subsequent to the divergence of *A. macrogynus*. Contrary to this prediction, however, no evidence for 5' tRNA editing has been found in *Rhizophydium136* (Laforest and Lang, unpublished data), a member of the taxonomic order Chytridiales, which branches specifically with *S. punctatus* (Bullerwell *et al.* 2003b) to the exclusion of the Monoblepharidales (see also Fig. 1). It is therefore plausible that this feature was acquired independently in two chytridiomycete orders, as indicated in Figure 2.

The presence of 5'tRNA editing has only been tested directly in *A. castellanii*, *S. punctatus*, *Monoblepharella15*, *Harpochytrium94* and *Harpochytrium105* systems. However, mismatching in tRNA acceptor stems has also been reported in the cellular slime mold *Dictyostelium discoideum* (Ogawa *et al.* 2000), and the heterolobosean *Naegleria gruberi* (M.W. Gray *et al.*, unpublished observation). The same type of editing has therefore most likely emerged several times in distant lineages, at least once in Heterolobosea, once in Amoebozoa (which includes *A. castellanii* and *D. discoideum*; Cavalier-Smith, 1998) and twice in chytridiomycete fungi. According to our current hypothesis, 5'tRNA editing evolved independently by modification of an existing enzyme system, rather than by several (otherwise rare) horizontal transfers among species.

If these activities are indeed of independent origin, there might be mechanistic differences between systems. Although some work has been done to characterize the biochemistry of the editing activity in *A. castellanii* mitochondria (Price and Gray 1999), similar studies have not been undertaken in other systems. The work of Price and Gray (1999) established that the *A. castellanii* activity is composed of at least two components: (1) a nuclease that removes the first three 5'-nucleotides from tRNA acceptor stems; and (2) a nucleotidyltransferase that adds nucleotides sequentially in a 3'-to-5' direction (contrary to all other known polymerases, which add in a 5'-to-3' direction), using the 3'-half of the acceptor stem as a guide. It will be necessary to test all the systems in which editing has been postulated to be sure that editing actually occurs, as well as to determine whether different enzymes account for the observed editing events.

3.2.7 Translation initiation in the Monoblepharidales

Comparative analysis of the four available monoblepharidalean mitochondrial DNAs revealed an unusual feature: almost every ATG and GTG start codon in these genomes is immediately preceded on the 5'-side by a guanosine residue (only two exceptions were noted; Bullerwell *et al.* 2003b). Further analysis revealed that the initiator tRNAs^{fMet} encoded by these genomes also have non-standard features: they contain non-Watson-Crick base pairs at the base of their anticodon stems (potentially enlarging the anticodon loops to nine instead of the usual seven nucleotides). In addition, the nucleotide at position 37 (immediately 3' to the anticodon) is occupied by a cytidine residue, whereas a purine residue (usually modified) is found at the corresponding position in the vast majority of tRNAs. Taken together, the unorthodox features of the initiator tRNAs and start codons strongly suggest a four base-pair interaction between the extended CAUC anticodon of the initiator tRNAs, and the quartet GAUG/GGUG start codons. The proposed translation initiation mechanism in Monoblepharidales mitochondria would provide a precise choice of translation initiation sites, analogous to Shine-Dalgarno sequences in eubacteria. Indeed, when comparing predicted mitochondrial protein sequences from the four Monoblepharidales with each other and with other eukaryotes, the most consistent translation initiation sites are exclusively located at the quartet GAUG/GGUG start codons in the Monoblepharidales, further supporting the proposed hypothesis (Bullerwell *et al.* 2003b).

Intriguingly, a similar situation apparently exists in the mitochondria of the sea anemone *Metridium senile* (Bullerwell *et al.* 2003b; Beagley *et al.* 1998). This mtDNA also encodes a tRNA^{fMet} with a cytidine residue at position 37. Further inspection of its mtDNA reveals that five of thirteen protein-coding genes have a guanosine residue immediately preceding the predicted initiation codons. In addition, three other genes have in-frame GAUG codons within 12 codons of the predicted start codon, which adds up to a significant proportion of potential GAUG start codons (8/13). As there is no evidence to support a specific phylogenetic relationship between the Monoblepharidales and *M. senile*, this extended anticodon-codon interaction, coupled to changes of the initiator tRNA structure, clearly evolved more than once independently.

4. CONCLUSIONS

Fungal mitochondrial genomes represent a microcosm of genomes in general. Analysis of these sequences reveals that mitochondria are among nature's most advanced evolutionary laboratories, and provide scientists with opportunities to discover novel molecular principles, which may be more easily recognized in small, well-defined systems such as these. Moreover, in many instances, principles discovered in mitochondrial systems have subsequently been detected in bacterial or nuclear genomes (e.g., deviations from the 'universal' translation code, group I and group II introns, various forms of RNA editing, genes in pieces, etc.). Thus, the study of mitochondrial genomes has implications far beyond organelle biology.

Analysis of complete mitochondrial genome sequences from a broad range of fungal species from all major lineages, in combination with the robust fungal phylogeny inferred using concatenated mitochondrial protein sequences, have allowed for the first time a detailed description of the evolution of fungi and their mitochondria. Our current knowledge is a giant leap from the situation less than a decade ago, when complete genome sequences were only available from a small number of representatives from Ascomycota. However, to truly understand fungal mitochondrial evolution, more work remains to be done. Mitochondrial sequencing projects must continue, focused on fungal groups where little data is currently available (such as Chytridiomycota and Zygomycota) as well as on groups where phylogenetic relationships have not been resolved with adequate support (such as the branching order within Ascomycota). In concert with these sequencing efforts, further molecular, biochemical, and genetic experimentation will be necessary to test hypotheses of gene expression and the function of gene products. Through these combined efforts, a deep understanding of fungal mitochondria and their genomes will be reached.

Acknowledgement: We would like to acknowledge M.-J. Laforest for sharing unpublished data. This work was supported by the 'Canadian Institute of Health Research' (CIHR). BFL is Imasco fellow in the program of Evolutionary Biology of the Canadian Institute for Advanced Research (CIAR), whom we thank for salary and interaction support. JEL was supported by NSF grants IBN-9977063 and DEB-9978094.

REFERENCES

Ahne A., Muller-Derlich J, Merlos-Lange A.M, Kanbay F, Wolf K, and Lang BF (1988). Two distinct mechanisms for deletion in mitochondrial DNA of *Schizosaccharomyces pombe* mutator strains. Slipped mispairing mediated by direct repeats and erroneous intron splicing. J Mol Biol 202:725-734.

Ainsworth GC (1973). Introduction and keys to higher taxa. In: GC Ainsworth, FK Sparrow, and AS Sussman, eds. The Fungi IVB. New York: Academic Press. pp 1-7.

Alexopolous CJ, Mims CW, and Blackwell M (1996) Introductory Mycology 4th ed. New York: John Wiley and Sons.

Anderson JB, Wickens C, Khan M, Cowen LE, Federspiel N, Jones T, and Kohn LM (2001). Infrequent genetic exchange and recombination in the mitochondrial genome of Candida albicans. J Bacteriol 183:865-872.

An KD, Nishida H, Miura Y, and Yokota A (2002). Aminoadipate reductase gene: a new fungal-specific gene from comparative evolutionary analyses. BMC Evolutionary Biology 2: 6-9.

Baldauf SL, Roger AJ, Wenk-Siefert I, and Doolittle WF (2000). A kingdom-level phylogeny of eukaryotes based on combined protein data. Science 290:972-977.

Barr DJ (1980). An outline for the reclassification of the Chytridiales, and for a new order, the Spizellomycetales. Can J Bot 58:2380-2394.

Barr DJ (2001). Chytridiomycota. In: DJ McLaughlin, EG McLaughlin, and PA,Lemke, eds. The Mycota VIIA. Berlin, Heidelberg: Springer-Verlag. pp 93-112.

Beagley CT, Okimoto R, and Wolstenholme DR (1998). The mitochondrial genome of the sea anemone *Metridium senile* (Cnidaria): introns, a paucity of tRNA genes, and a near-standard genetic code. Genetics 148:1091-1108.

Belfort M and Perlman PS (1995). Mechanisms of intron mobility. J Biol Chem 270:30237-30240.

Bendich AJ (1993). Reaching for the ring: the study of mitochondrial genome structure. Curr Genet 24:279-290.

Bendich AJ (1996). Structural analysis of mitochondrial DNA molecules from fungi and plants using moving pictures and pulsed-field gel electrophoresis. J Mol Biol 255:564-588.

Berbee ML, and Taylor JW (1993). Dating the evolutionary radiations of the true fungi. Can J Bot 71:1114-1127.

Berbee ML, Carmean DA, and Winka K (2000). Ribosomal DNA and resolution of branching order among the ascomycota: how many nucleotides are enough? Mol Phylogenet Evol 17:337-344.

Blackwell M, Vilgalys R, and Taylor JW (1996). Fungi. In: DR Maddison, coord and ed. The Tree of Life Web Project (http://tolweb.org/tree/)

Boer PH, and Gray MW (1988). Scrambled ribosomal RNA gene pieces in *Chlamydomonas reinhardtii* mitochondrial DNA. Cell 55:399-411.

Bowman BH, Taylor JW, Brownlee AG, Lee J, Lu SD, and White TJ (1992). Molecular evolution of the fungi: relationship of the Basidiomycetes, Ascomycetes, and Chytridiomycetes. Mol Biol Evol 9:285-296.

Brown JW (1998). The Ribonuclease P database. Nucleic Acids Res 27:314.

Bruno WJ, Socci ND, and Halpern AL (2000). Weighted neighbor joining: a likelihood-based approach to distance-based phylogeny reconstruction. Mol Biol Evol 17:189-197.

Bruns TD, Vilgalys R, Barns SM, Gonzalez D, Hibbett DS, Lane DJ, Simon L, Stickel S, Szaro TM, Weisburg WG, and Sogin ML (1992). Evolutionary relationships within the fungi: analyses of nuclear small subunit rRNA sequences. Mol Phylogenet Evol 1:231-241.

Bullerwell CE, Burger G, and Lang BF (2000). A novel motif for identifying Rps3 homologs in fungal mitochondrial genomes. Trends Biochem Sci 25:363-365.

Bullerwell CE, Leigh J, Forget L, and Lang BF (2003a). A comparison of three fission yeast mitochondrial genomes. Nucleic Acids Res 31:759-768.

Bullerwell CE, Forget L, and Lang BF (2003b). Evolution of monoblepharidalean fungi based on complete mitochondrial genome sequences. Nucleic Acids Res, in press.

Burger G, Helmer Citterich M, Nelson MA, Werner S, and Macino G (1985). RNA processing in *Neurospora crassa*: transfer RNAs punctuate a large precursor transcript. EMBO J 4:197-204.
Burger G, Saint-Louis D, Gray MW, and Lang, BF (1999). Complete sequence of the mitochondrial DNA of the red alga *Porphyra purpurea*: cyanobacterial introns and shared ancestry of red and green algae. Plant Cell 11:1675-1694.
Burger G, Forget L, Zhu Y, Gray MW, and Lang BF (2003). Unique mitochondrial genome architecture in unicellular relatives of animals. Proc Natl Acad Sci USA., in press.
Butow RA, Perlman PS, and Grossman LI (1985). The unusual var1 gene of yeast mitochondrial DNA. Science 228:1496-1501.
Cavalier-Smith T (1998). A revised six-kingdom system of life. Biol Rev Camb Philos Soc 73:203-266.
Chesnick JM, Goff M, Graham J, Ocampo C, Lang BF, and Burger G (2000). The mitochondrial genome of the stramenopile alga *Chrysodidymus synuroideus*. Complete sequence, gene content and genome organization. Nucleic Acids Res 28:2512-2518.
Christianson T and Rabinowitz M (1983). Identification of multiple transcriptional initiation sites on the yeast mitochondrial genome by *in vitro* capping with guanylyltransferase. J Biol Chem 258:14025-14033.
Clark-Walker GD, McArthur CR, and Sriprakash KS (1985). Location of transcriptional control signals and transfer RNA sequences in *Torulopsis glabrata* mitochondrial DNA. EMBO J 4:465-473.
Crary SM, Niranjanakumari S, and Fierke CA (1998). The protein component of *Bacillus subtilis* ribonuclease P increases catalytic efficiency by enhancing interactions with the 5' leader sequence of pre-tRNA Asp. Biochemistry 37:9409-9416.
Costanzo MC, and Fox TD (1988). Specific translational activation by nuclear gene products occurs in the 5'-untranslated leader of a yeast mitochondrial mRNA. Proc Natl Acad Sci USA. 85:2677-2681.
Cummings DJ, McNally KL, Domenico JM, and Matsuura ET (1990). The complete DNA sequence of the mitochondrial genome of *Podospora anserina*. Curr Genet 17:375-402.
Cummings DJ, Michel F, and McNally KL (1989). DNA sequence analysis of the 24.5 kilobase pair cytochrome oxidase subunit I mitochondrial gene from *Podospora anserina*: a gene with sixteen introns. Curr Genet 16:381-406.
Dang YL and Martin NC (1993). Yeast mitochondrial RNase P. Sequence of the RPM2 gene and demonstration that its product is a protein subunit of the enzyme. J Biol Chem 268:19791-19796.
de Vries H, Haima P, Brinker M, and de Jonge JC (1985). The *Neurospora* mitochondrial genome: the region coding for the polycistronic cytochrome oxidase subunit 1 transcript is preceded by a transfer RNA gene. FEBS Lett 179:337-342.
de Zamaroczy M and Bernardi G (1987). The AT spacers and the var1 genes from the mitochondrial genomes of *Saccharomyces cerevisiae* and *Torulopsis glabrata*: evolutionary origin and mechanism of formation. Gene 54:1-22.
Doolittle WF (1999). Lateral genomics. Trends Cell Biol 9:M5-M8.
Dyson NJ, Brown TA, Ray JA, Waring RB, Scazzocchio C, and Davies RW (1989). Processing of mitochondrial RNA in *Aspergillus nidulans*. J Mol Biol 208:587-599.
Felsenstein J (2002). Phylip (Phylogeny Inference Package) Version 3.6 a 2.1. Distributed by the author. Seattle:University of Washington.
Forget L, Ustinova J, Wang Z, Huss VAR, and Lang BF (2002). *Hyaloraphidium curvatum*: A linear mitochondrial genome, tRNA editing, and an evolutionary link to lower fungi. Mol Biol Evol 19:310-319.
Foury F, Roganti T, Lecrenier N, and Purnelle B (1998). The complete sequence of the mitochondrial genome of *Saccharomyces cerevisiae*. FEBS Lett 440:325-331.
Fox TD (1979). Five TGA "stop" codons occur within the translated sequence of the yeast mitochondrial gene for cytochrome c oxidase subunit II. Proc Natl Acad Sci USA. 76:6534-6538.
Frank DN and Pace NR (1998). Ribonuclease P: unity and diversity in a tRNA processing ribozyme. Annu Rev Biochem 67:153-180.
Fukuhara H, Sor F, Drissi R, Dinouel N, Miyakawa I, Rousset S, and Viola AM (1993). Linear mitochondrial DNAs of yeasts: frequency of occurrence and general features. Mol Cell Biol 13:2309-2314.
Gardiner KJ, Marsh TL, and Pace NR (1985). Ion dependence of the *Bacillus subtilis* RNase P reaction. J Biol Chem 260:5415-5419.
Gauriloff LP, Delay RJ, and Fuller MS (1980). Comparative ultrastructure and biochemistry of chytridiomycetous fungi and the future of the Harpochytriales. Can J Bot 58: 2098-2109.
Gillespie DE, Salazar NA, Rehkopf DH, and Feagin JE, (1999). The fragmented mitochondrial ribosomal RNAs of *Plasmodium falciparum* have short A tails. Nucleic Acids Res 27:2416-2422.
Goffeau A, Barrell BG, Bussey H, Davis RW, Dujon B *et al.* (1996). Life with 6000 genes. Science 274:546-567.
Gogarten JP, Doolittle WF, Lawrence JG (2002). Prokaryotic evolution in light of gene transfer. Mol Biol Evol 19:2226-2238.

Gray MW, Lang BF, Cedergren R, Golding GB, Lemieux C, Sankoff D, Turmel M, Brossard N, Delage E, Littlejohn TG, Plante I, Rioux P, Saint-Louis D, Zhu Y, and Burger G (1998). Genome structure and gene content in protist mitochondrial DNAs. Nucleic Acids Res 26:865-878.

Groot GS, Mason TL, and Van Harten-Loosbroek N (1979). Var1 is associated with the small ribosomal subunit of mitochondrial ribosomes in yeast. Mol Gen Genet 174:339-342.

Guerrier-Takada C, Gardiner K, Marsh T, Pace N, and Altman S, (1983). The RNA Moiety of ribonuclease P is the

catalytic subunit of the enzyme. Cell 35:849-857.

Hayashi-Ishimaru Y, Ohama T, Kawatsu Y, Nakamura K, and Osawa S (1996). UAG is a sense codon in several chlorophycean algae. Curr Genet 30:29-33.

James TY, Porter D, Leander CA, Vilgalys R, Longcore JE (2000). Molecular phylogenetics of the Chytridiomycota supports the utility of ultrastructural data in chytrid systematics. Can J Bot 78:336-350.

Jensen AB, Gargas A, and Eilenberg J (1998). Relationships of the insect-pathogenic order Entomopthorales (Zygomycota, Fungi) based on phylogenetic analyses of nuclear small subunit ribosomal DNA sequences (SSU rDNA). Fungal Genet Biol 24:325-334.

Kairo A., Fairlamb AH, Gobright E, and Nene V (1994). A 7.1 kb linear DNA molecule of *Theileria parva* has scrambled rDNA sequences and open reading frames for mitochondrially encoded proteins. EMBO J 13: 898-905.

Keeling PJ, Luker MA, and Palmer JD (2000). Evidence from beta-tubulin phylogeny that microsporidia evolved from within the fungi. Mol Biol Evol 17:23-31.

Kennell JC and Cohen SM (2003). Fungal mitochondria: genomes, genetic elements, and gene expression. In: D Arora, ed. Handbook of fungal biotechnology. 2nd ed. New York: Marcel Dekker Inc., in press.

Kim J, Chubatsu LS, Admon A, Stahl J, Fellous R, and Linn S (1995). Implication of mammalian ribosomal protein S3 in the processing of DNA damage. J Biol Chem 270:13620-132629.

Laforest MJ, Roewer I, and Lang BF (1997). Mitochondrial tRNAs in the lower fungus *Spizellomyces punctatus*: tRNA editing and UAG 'stop' codons recognized as leucine. Nucleic Acids Res 25:626-632.

Lambowitz AM, LaPolla RJ, and Collins RA (1979). Mitochondrial ribosome assembly in *Neurospora.* Two dimensional gel electrophoresis analysis of mitochondrial ribosomal proteins. J Cell Biol 82:17-31.

Lambowitz AM and Belfort M (1993). Introns as mobile genetic elements. Annu Rev Biochem 62: 587-622.

Landvik S, Eriksson OE, and Berbee ML (2001). Neolecta -- a fungal dinosaur? Evidence from beta-tubulin amino acid sequences. Mycologia 93:1151-1163.

Lang BF (1993). The mitochondrial genome of *Schizosaccharomyces pombe*. In: SJ O'Brien, ed. Genetic Maps. 6th ed. Cold Spring Harbor: Cold Spring Harbor Laboratory Press.

Lang BF (1984). The mitochondrial genome of the fission yeast *Schizosaccharomyces pombe*: highly homologous introns are inserted at the same position of the otherwise less conserved cox1 genes in *Schizosaccharomyces pombe* and *Aspergillus nidulans*. EMBO J 3:2129-2136.

Lang BF, Ahne F, Distler S, Trinkl H, Kaudewitz F, and Wolf K (1983). Sequence of the mitochondrial DNA, arrangement of genes and processing of their transcripts in *Schizosaccharomyces pombe*. In: A Nasim, P Young, and BF Johnston, eds. Molecular Biology of the Fission Yeast. San Diego:Academic Press.

Lang BF, Burger G, O=Kelly CJ, Cedergren R, Golding GB, Lemieux C, Sankoff D, Turmel M, and Gray MW (1997). An ancestral mitochondrial DNA resembling a eubacterial genome in miniature. Nature 387:493-497.

Lang BF, Gray MW, and Burger G (1999). Mitochondrial genome evolution and the origin of the eukaryotes. Annu Rev Genet 33:351-397.

Lang BF, O'Kelly C, Nerad T, Gray MW, and Burger G (2002). The closest unicellular relatives of the animals. Curr Biol 12:1773-1778.

Lange L and Olson LW (1980). Transfer of the Physodermataceae from the Chytridiales to the Blastocladiales. Trans Brit Mycol Soc 74:449-457.

Lawrence JG (2001). Gene transfer in bacteria: speciation without species? Theor Popul Biol 61:449-460

Lee YC, Lee BJ, Hwang DS, and Kang HS (1996 a). Purification and characterization of mitochondrial ribonuclease P from *Aspergillus nidulans*. Eur J Biochem 235:289-296.

Lee YC, Lee BJ, and Kang HS (1996 b). The RNA component of mitochondrial ribonuclease P from *Aspergillus nidulans*. Eur J Biochem 235:297-303.

Leigh J, Seif E, Rodriguez N, Jacob Y, and Lang BF (2003). Fungal evolution meets fungal genomics. In: D Arora, ed. Handbook of Fungal Biotechnology. 2nd ed. New York: Marcel Dekker Inc., in press.

Liu YJ, Whelen S, and Hall BD (1999). Phylogenetic relationships among ascomycetes: evidence from an RNA polymerase II subunit. Mol Biol Evol 16:1799-1808.

Lonergan KM and Gray MW (1993). Editing of transfer RNAs in *Acanthamoeba castellanii* mitochondria. Science 259:812-816.

Martin CA and Lang BF (1997). Mitochondrial RNase P: the RNA family grows. Nucleic Acids Symp Ser 36:42-44.
McKerracher LJ, and Heath IB (1985). The structure and cycle of the nucleus-associated organelle in two species of *Basidiobolus*. Mycologia 77:412-417.
McLaughlin DJ (2000). Volume Preface. Pp. XI-XIV in The Mycota VIIA. DJ McLaughlin, EG McLaughlin, and PA Lemke, eds. Berlin, Heidelberg: Springer-Verlag, ppXI-XIV.
Miller DL and Martin NC (1983). Characterization of the yeast mitochondrial locus necessary for tRNA biosynthesis: DNA sequence analysis and identification of a new transcript. Cell 34:911-917.
Morales MJ, Dang YL, Lou YC, Sulo P, and Martin NC (1992). A 105-kDa protein is required for yeast mitochondrial RNase P activity. Proc Natl Acad Sci USA. 89:9875-9879.
Mulero JJ and Fox TD (1993). PET111 acts in the 5'-leader of the *Saccharomyces cerevisiae* mitochondrial cox2 mRNA to promote its translation. Genetics 133:509-516.
Nagahama T, Sato H, Shimazu M, and Sugiyama J (1995). Phylogenetic divergence of the entomophthoralean fungi: evidence from nuclear 18S ribosomal RNA gene sequences. Mycologia 87:203-209.
Nedelcu AM, Lee RW, Lemieux C, Gray MW, and Burger G (2000). The complete mitochondrial DNA sequence of *Scenedesmus obliquus* reflects an intermediate stage in the evolution of the green algal mitochondrial genome. Genome Res 10:819-831.
Nesbo CL, L'Haridon S, Stetter KO, Doolittle WF (2001). Phylogenetic analyses of two "archaeal" genes in *Thermotoga maritima* reveal multiple transfers between archaea and bacteria. Mol Biol Evol 18:362-375.
Neu R, Goffart S, Wolf K, and Schäfer B (1998). Relocation of urf a from the mitochondrion to the nucleus cures the mitochondrial mutator phenotype in the fission yeast *Schizosaccharomyces pombe*. Mol Gen Genet 258:389-396.
Nishida H and Sugiyama J (1994). Archiascomycetes: detection of a major new lineage within the Ascomycota. Mycoscience 35:361-366.
Nishida H and Sugiyama J (1993). Phylogenetic relationships among Taphrina, Saitoella, and other higher fungi. Mol Biol Evol 10:431-436.
Nosek J, Tomaska L, Fukuhara H., Suyama Y, and Kovac L (1998). Linear mitochondrial genomes: 30 years down the line. Trends Genet 14:184-188.
Ogawa S, Yoshino R, Angata K, Iwamoto M, Pi M, Kuroe K, Matsuo K, Morio T, Urushihara H, Yanagisawa K, and Tanaka Y (2000). The mitochondrial DNA of *Dictyostelium discoideum*: complete sequence, gene content and genome organization. Mol Gen Genet 263:514-519.
Osinga KA, De Haan M, Christianson T, and Tabak HF (1982). A nonanucleotide sequence involved in promotion of ribosomal RNA synthesis and RNA priming of DNA replication in yeast mitochondria. Nucleic Acids Res 10:7993-8006.
Paquin B, Laforest M-J, Forget L, Roewer I, Wang Z, Longcore J. and Lang BF (1997). The fungal mitochondrial genome project: evolution of fungal mitochondrial genomes and their gene expression. Curr Genet 31: 380-395.
Paquin B, Laforest M-J, and Lang BF (1994). Interspecific transfer of mitochondrial genes in fungi and creation of a homologous hybrid gene. Proc Natl Acad Sci USA 91:11807-11810.
Paquin B and Lang BF (1996). The mitochondrial DNA of *Allomyces macrogynus*: the complete genomic sequence from an ancestral fungus. J Mol Biol 255:688-701.
Price DH, and Gray MW (1999). A novel nucleotide incorporation activity implicated in the editing of mitochondrial transfer RNAs in *Acanthamoeba castellanii*. RNA 5:302-317.
Sandigursky M, Yacoub A, Kelley MR, Deutsch WA, and Franklin WA (1997). The Drosophila ribosomal protein S3 contains a DNA deoxyribophosphodiesterase (dRpase) activity. J Biol Chem 272:17480-17484.
Schäfer B, Merlos-Lange AM, Anderl C, Welser F, Zimmer M, and Wolf K (1991). The mitochondrial genome of the fission yeast: inability of all introns to splice autocatalytically, and construction and characterization of an intronless genome. Mol Gen Genet 225:158-167.
Schnare MN, Heinonen TY, Young PG, and Gray MW (1986). A discontinuous small subunit ribosomal RNA in *Tetrahymena pyriformis* mitochondria. J Biol Chem 261: 5187-5193.
Schüßler A, Schwarzott D, and Walker C (2001). A new fungal phylum, the Glomeromycota: phylogeny and evolution. Mycol Res 105:1413-1421.
Sparrow FK (1943). Aquatic Phycomycetes. Ann Arbor: University of Michigan Press.
Sparrow FK (1960). Aquatic Phycoycetes, 2nd rev. ed. Ann Arbor: University of Michigan Press.
Sparrow FK (1973). Chytridiomycetes, Hyphochytridiomycetes In: GC Ainsworth, FKSparrow, and AS Sussman, eds. The Fungi IV B. New York: Academic Press, pp85-110.
Stark BC, Kole R, Bowman EJ, and Altman S (1978). Ribonuclease P: an enzyme with an essential RNA component. Proc Natl Acad Sci USA 75:3717-3721.

Steele DF, Butler CA, and Fox TD (1996). Expression of a recoded nuclear gene inserted into yeast mitochondrial DNA is limited by mRNA-specific translational activation. Proc Natl Acad Sci USA 93:5253-5257.

Strimmer K and von Haeseler A (1996). Quartet puzzling: a quartet maximum-likelihood method for reconstructing tree topologies. Mol Biol Evol 13:964-969.

Terpstra P, Zanders E, and Butow RA (1979). The association of var1 with the 38 S mitochondrial ribosomal subunit in yeast. J Biol Chem 254:12653-12661.

Thomas BC, Gao L, Stomp D, Li X, and Gegenheimer PA (1995). Spinach chloroplast RNase P: a putative protein enzyme. Nucleic Acids Symp Ser 33:95-98.

Thomas BC, Li X, Gegenheimer P (2000). Chloroplast ribonuclease P does not utilize the ribozyme-type pre-tRNA cleavage mechanism. RNA 6:545-553.

Trinkl H, Lang BF, and Wolf K (1989). Nucleotide sequence of the gene encoding the small ribosomal RNA in the mitochondrial genome of the fission yeast *Schizosaccharomyces pombe*. Nucleic Acids Res 17:6730.

True HL and Celander DW (1998). Protein components contribute to active site architecture for eukaryotic ribonuclease P. J Biol Chem 273:7193-7196.

Turmel M, Lemieux C, Burger G, Lang BF, Otis C, Plante I, and Gray MW (1999). The complete mitochondrial DNA sequences of *Nephroselmis olivacea* and *Pedinomonas minor*: two radically different evolutionary patterns within green algae. Plant Cell 11:1717-1729.

Underbrink-Lyon K, Miller DL, Ross NA, Fukuhara H, and Martin NC (1983). Characterization of a yeast mitochondrial locus necessary for tRNA biosynthesis. Deletion mapping and restriction mapping studies. Mol Gen Genet 191:512-518.

Whelan S and Goldman N (2001). A general empirical model of protein evolution derived from multiple protein families using a maximum-likelihood approach. Mol Biol Evol 18:691-699

Whittaker RH, (1969). New concepts of kingdoms of organisms. Science 163:150-160.

Wilson DM 3rd, Deutsch WA, and Kelley MR (1994). Drosophila ribosomal protein S3 contains an activity that cleaves DNA at apurinic/apyrimidinic sites. J Biol Chem 269:25359-25364.

Wise CA and Martin NC (1991). Dramatic size variation of yeast mitochondrial RNAs suggests that RNase P RNAs can be quite small. J Biol Chem 266:19154-19157.

Wood V, Gwilliam R, Rajaendream MA, Lyne M, Lyne R *et al.* (2002). The genome sequence of *Schizosaccharomyces pombe*. Nature 415:871-880.

Wool I (1996). Extraribosomal functions of ribosomal proteins. Trends Biochem Sci 21:164-165.

Yamao F, Muto A, Kawauchi Y, Iwami M, Iwagami S, Azumi Y, and Osawa S (1985). UGA is read as tryptophan in *Mycoplasma capricolum*. Proc Natl Acad Sci USA 82:2306-2309.

Yang Z (1997). PAML: A program package for phylogenetic analysis by maximum likelihood. Comput Appl Biosci 13:555-556.

Zimmer M, Krabusch M, and Wolf K (1991). Characterization of a novel open reading frame, urf a, in the mitochondrial genome of fission yeast: Correlation of urf a mutations with a mitochondrial mutator phenotype and a possible role for frameshifting in urf a expression. Curr Genet 19:95-102.

Zimmer M, Welser F, Oraler G, and Wolf K (1987). Distribution of mitochondrial introns in the species *Schizosaccharomyces pombe* and the origin of the group II intron in the gene encoding apocytochrome b. Curr Genet 12:329-336.

Applied Mycology & Biotechnology
An International Series. Volume 3. Fungal Genomics

8

Ribosome Biogenesis in Yeast: rRNA Processing and Quality Control

Ross N. Nazar
Department of Molecular Biology and Genetics, University of Guelph, Guelph, Ontario, Canada N1G 2W1 (rnnazar@uoguelph.ca).

The ribosome, being a cell's factory for protein synthesis, represents a cellular substructure that is critical, both for cell growth and survival. In the eukaryotes, the cytoplasmic ribosome is synthesized, almost exclusively, in a specialized subnuclear compartment called the nucleolus. Biogenesis begins with the transcription of rRNA precursors (pre-rRNAs) from rRNA genes (rDNA) that are localized in the nucleolus and ends with the transport of pre-ribosomal subunits to the cytoplasm where the final steps in the maturation process occur. In the course of ribosome biogenesis, the pre-rRNAs are cleaved and covalently modified while assembling with ribosomal proteins to form the mature subunits. These assembly and maturation processes are dependent on both cis-acting elements and many non-ribosomal protein or RNA trans-acting factors. The mature ribosome is the result of numerous macromolecule interactions involving many RNA and protein molecules which are actual ribosomal constituents and other molecules which ultimately are excluded from the mature particle. Together, the maturation processes provide a cell, and perhaps biotechnologists, with a means to control cell growth and even a mechanism for quality control.

1. INTRODUCTION

A great variety of genetic and biochemical approaches has been used extensively to study ribosome biogenesis over a 40 year period. In general, the maturation of the pre-rRNA and the basic processes in the biogenesis of the ribosomal subunits appear to be largely conserved among the eukaryotes, but differences in many species-specific details also have been reported. In all the organisms that have been examined, the nascent pre-rRNA is first assembled into an 80-90S nucleolar particle. Structural rearrangements and nucleotide modifications occur as the ribosomal proteins are recruited, followed with cleavages which ultimately result in the mature cytoplasmic 40S and 60S ribosomal subunits. Since yeasts can be manipulated easily both genetically and biochemically, a majority of these studies has occurred in yeast cells, initially in *Sacharomyces cerevisiae* (Grivell and Planta, 1990) but, more recently, also in *Schizosacharomyces pombe*. Despite extensive progress in the elucidation of the biosynthetic pathways, surprisingly little is still understood about the role or need for the extensive changes which occur during ribosome maturation, including the need for the very large number of trans-acting factors which have been identified. As a number of recent reviews have extensively considered the maturation pathways and trans-acting factors, this manuscript is somewhat focused at providing some insight into the overall function of the maturation processes including the possibility that they represent, at least in part, quality control mechanisms.

2. PROCESSING OF rRNA PRECURSORS

In the cytoplasmic ribosomes of eukaryotic cells the smaller 40S ribosomal subunit contains a single 18 S rRNA and 30-48 ribosomal proteins. In contrast, the larger 60S subunit contains at least three discrete rRNA components (5S, 5.8S and 25-28S rRNAs) as well as 40-50 ribosomal proteins (see Raue and Planta, 1991; Woolford and Warner; 1991; Planta and Mager, 1998). Ribosome biogenesis begins with the transcription of the ribosomal RNAs. All of the rRNAs, except the smallest 5S rRNA molecule (approximately 120 nucleotides in length) are transcribed in the nucleolus as a single large precursor (pre-rRNA) by the DNA dependent RNA polymerase I (Pol I). The 5S rRNA is transcribed independently by RNA polymerase III (Pol III). In *S. cerevisiae*, the genes for the 5 S rRNA are linked with rDNA as a large repeating unit; in *S. pombe*, as in most eukaryotic cells, the 5S rRNA encoding genes are not linked and are localized elsewhere in the nucleus often in large clusters of repeating transcriptional units. In all known examples, the ribosomal 5S rRNA also is transcribed as a precursor molecules with a short extension at the 3' end. To sustain rapid cell growth the ribosomal RNA genes are always present in many copies and represent moderately repeated sequence families. In *S. cerevisiae* the 9.1 kb rDNA transcriptional unit is repeated 100 to 200 times on the long arm of chromosome XII and in *S. pombe* the somewhat longer 10.8 kb unit has been estimated to have a copy number which exceeds 115 units. As might be anticipated, a variety of studies (see Mager and Planta, 1991; Warner, 1999) indicate that the synthesis of rRNA is coordinated with the synthesis of the ribosomal proteins to ensure an efficient assembly of pre-ribosomal particles. The tandemly arranged rDNA transcriptional units act as a nucleolar organizer (NOR), a circular area or center in electronmicrographs of a nucleolus (see Busch and Smetana, 1970; Hadjiolov, 1985) surrounded by electron dense filaments collectively called the pars fibrosa. The filaments are made up of newly transcribed pre-rRNA. The nascent RNA then interacts with proteins and other RNA molecules to form the nucleolar pre-ribosomal particles which accumulate and mature in the pars granulosa, the more granular outer region of the nucleolus, before being exported to the cytoplasm via the nuclear pores.

In all eukaryotic pre-rRNAs, the mature rRNA sequences are separated by at least two internal transcribed non conserved spacer (ITS) sequences, ITS1 and ITS2, and are flanked by two external transcribed non conserved spacer (ETS) sequences, the 5' ETS and the 3' ETS. Precursors vary greatly in both composition and size, primarily because of species-specific differences in the transcribed spacers as well as the variable regions within the mature ribosomal RNAs, themselves. The yeasts contain some of the shorter precursor molecules (35S nRNA in *S. cerevisiae* and *S. pombe*) while the human transcript is among the longest (45S nRNA). Only the core sequence elements within the mature RNAs are highly conserved. In *S. cerevisiae*, the initial transcript is more then 9000 nucleotides in length but after processing only 5100 nucleotides remain as the mature RNA. Similarly, in S. pombe, approximately 5500 nucleotides of a somewhat longer transcript remain as ribosomal RNA. This loss is even greater in higher eukaryotes, an observation which raises interesting questions about the need for such large and variable spacer regions.

As illustrated in Figure 1, the maturation of the 25S pre-rRNA in yeast is complex, apparently consisting of at least 10 distinct processing sites in a multi step pathway. With transcription first completed, the precursor first is assembled into an 80-90S nucleolar pre-ribosomal particle before cleavage begins in the 5' ETS and ends with the separation of the 5.8S and 25S rRNA sequences and a final trimming to form the mature ribosomal RNAs.

While most of the processing occurs in the nucleolus, at least in yeast, the final maturation of the pre-40S ribosomal subunit actually takes place in the cytoplasm when the 20S pre-rRNA is finally trimmed to the 18S rRNA molecule (Udem and Warner, 1973; Valasek *et al.*, 2001; Vanrobays *et al.*, 2001). The final maturation of the pre-60S ribosomal subunit also occurs in the cytoplasm with the addition of some ribosomal proteins and possibly some structural rearrangements (Venema and Tollervey, 1999; Senger et al., 2001).

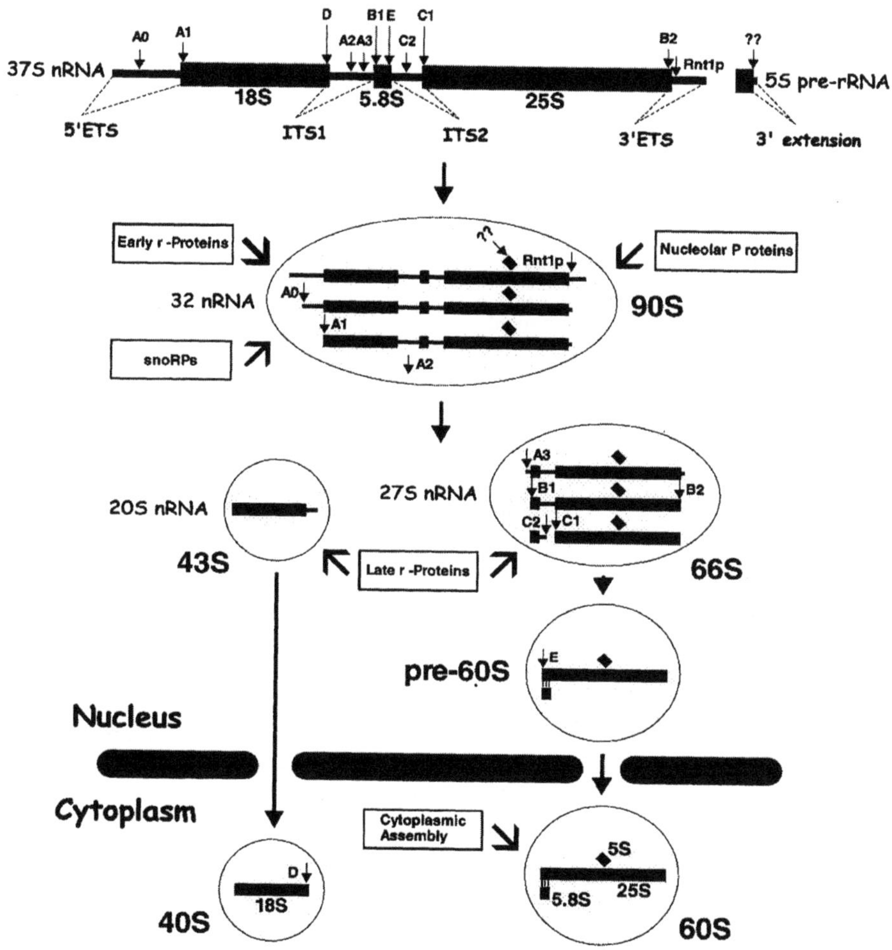

Fig. 1. Pre-rRNA processing and the ribosomal subunit assembly pathway in *S. cerevisiae*.
The rRNA is transcribed as 37S pre-rRNA and 5S pre-rRNA molecules which are modified and assembled with ribosomal proteins (r-Protein) and trans-acting factors (nucleollar proteins and snoRPs) into a 90S nucleolar pre-ribosomal particle before being cleaved into the 43S and 66S pre-ribosomal subunits and exported to the cytoplasm for final maturation to the 40S and 60S ribosomal subunits. The major cleavage sites, A0-C2, which produce the nucleolar 35S, 32S, 27S and 20S pre-rRNA intermediates, as well as the mature 18S, 5.8S and 25S rRNAs are indicated with the small arrows. The Rnt1p cleavage site which produces the first detectable 35S nRNA pre-rRNA and the unknown enzyme activity which removes the 3' extension on the 5S pre-rRNA also are indicated.

Initially, many of the ribosomal proteins are rapidly integrated in the formation of the nucleolar pre-ribosomal particle but some assemble at later steps including the few which are added in the cytoplasm (Kruiswijk *et al.*, 1978). Despite a general similarity in the cleavage pathways, some details regarding individual steps in different organisms have been surprisingly variable. In *S. pombe*, for example, the results for 5' and 3'ETS transcript mapping suggest that the processing of the external spacers may be more complex with additional intermediate cleavage sites and cleavage sites C1 and C2 are exchanged as the cleavage in the ITS2 region appears to precede maturation at the 5'end of the 25S rRNA (Good *et al.*, 1997). I a second example, Raue (2002) recently reported a novel pathway for ITS1 processing in yeast that bypasses both the A2 and A3 sites. While the 5S rRNA is synthesized independently of the other ribosomal RNAs by an alternate RNA polymerase (Pol III), this RNA again is synthesized as a longer precursor with a short extension (7 to 13 nucleotides) at the 3' end. This extension subsequently is removed by endo or exonuclease cleavage, apparently also by alternate mechanisms which are species-specific (see Lee and Nazar, 1997).

In addition to RNA cleavage, the nucleolar rRNA precursor undergoes numerous other changes in the form of covalent modifications (see Maden, 1990). These include isomerization of some uridine bases to pseudouridine (Ψ) by base rotation, methylation of the 2'-hydroxyl group of specific sugar residues (2'-O-ribose methylation), and a smaller number of base methylations to form the minor base constituents. Most of these changes occur shortly after transcription, even before RNA cleavage. A few of the changes occur later; apparently some of these even occur in the cytoplasm. In all the eukaryotes, including the yeasts, the majority of these modifications are dependent on small nucleolar RNAs (snoRNAs) and associated nucleolar proteins which form ribonucleoprotein complexes (snoRPs) that interact with the modified sites to facilitate the modifications.

In describing the many changes which occur in the course of rRNA maturation and ribosome biogenesis, it seems important to note that the cleavages or modifications in the course of rRNA processing do not take place while the RNA is being transcribed or on "naked" nascent RNA. Instead, the rRNA transcript initially interacts with many nucleolar and ribosomal protein molecules as well as the very significant number of snoRNAs. Together, they form a large 80-90S nucleolar ribonucleoprotein particle that is acted upon by all of the processing and maturation steps to form the nearly mature ribosomal subunits which ultimately are released to the cytoplasm. A variety of early biochemical analyses in several different organisms as well as elegant electron micrographs have conclusively demonstrated (see Busch and Smetana, 1970) the formation of such particles containing intact nascent rRNA transcripts. Despite the seeming independence of the individual cleavage or modification steps, initiation of the maturation processes is critically dependent on the formation of this particle. While a great deal is now known about the detailed crystal structure of the ribosomes from several species (Ban *et al.*, 2000; Yusupov *et al.*, 2001; Harms *et al.*, 2001), after three decades of research, remarkably little is known about the structure or essential features of this nucleolar ribonucleoprotein complex.

3. TRANS-ACTING FACTORS IN RIBOSOME BIOGENESIS

Although the structure of the nucleolar pre-ribosomal ribonucleoprotein complex remains unclear, many of the trans-acting factors which make up or interact with this complex in the course of ribosome maturation have been identified. Based on genetic analyses and some biochemical characterizations, to date more than 100 protein factors have been identified in *S. cerevisiae* alone, including proteins which interact with snoRNAs to form small nucleolar ribonucleoprotein complexes (snoRPs). Based on genomic databases as well as direct analyses, most of the snoRNAs that participate in rRNA modification also have been identified in *S. cerevisiae* and many of the analogous RNAs have been putatively identified in *S. pombe*.

To simplify the complexity of this long list of factors, based on function or location, the proteins often are divided into at least six groups and the RNAs can be placed in three categories. The proteins include, ribonucleases, ATP-dependent RNA-like helicases, snoRNA-associated proteins, other nucleolus-associated proteins, transport factors and unknown proteins which mediate genetic defects. The RNAs include Box C/D snoRNAs which are associated with rRNA methylation, Box H/ACA snoRNAs which are associated with pseudouridylation and other snoRNAs, associated with RNA cleavage or the initiation of rRNA processing. The complex pattern of interactions among these many factors has been carefully reviewed by Kressler, Linder and de LA Cruz (1999) and Venema and Tollervey (1999); readers should refer to their articles if this type of detail is required. While most of the factors initially were identified in *S. cerevisiae*, computer-based analyses of genomic databases have identified or putatively identified many of the equivalent trans-acting proteins in *S. pombe* as well. These have been categorized as Nucleolus GO:0005730 or RRNA Processing GO:0006364. For these lists the reader is referred to the *S. pombe* genome sequencing project database at the Sanger Center, UK (www.sanger.ac.uk/ Projects/S_pombe/).

3.1 Ribonuclease Activities Associated with rRNA Processing

Studies of the cleavage pathways and the termini of intermediate precursor molecules indicate both exo- and endonuclease activities are essential for rRNA maturation. In the yeasts, at least four specific nuclease activities have been characterized with respect rRNA processing: RNase MRP, an RNA dependent nuclease which has been associated with the maturation of the 5.8 S rRNA (Schmitt and Clayton, 1993; Chu *et al.*, 1994), an RNase III-like analog or endonuclease activity associated with processing at the 3' end of the pre-rRNA transcripts and possibly the 5' ETS (Abou Elela *et al.*, 1996; Rotondo and Frendewey, 1996), a 3' to 5' exonuclease activity associated with the exosome (Mitchell *et al.*, 1997) that appears to be essential for trimming at the 3' extended termini of the 5.8 S and 18 S rRNAs, and a 5' to 3' exonuclease activity (Heyer *et al.*, 1995; Johnson, 1997) which appears to trim the 5.8S rRNA and perhaps other 5'-extended termini.

In *Escherichia coli*, RNase III, a double-strand specific endonuclease is responsible for the initial cleavages which separate the mature rRNA sequences. This enzyme is specific for and cuts helical structure that forms between the termini of both the 16S and 23S rRNAs. Although the termini do not interact in a similar fashion in the eukaryotic pre-rRNAs, helical structures are present in the junction regions and RNase III-like nucleases have been observed in eukaryotic cells, including Rnt1p in *S. cerevisiae* (Abou Elela *et al.*, 1996) and the Pac1 nuclease in *S. pombe* (Rotondo *et al.*,1996). Both these enzymes have been shown, *in vitro*, to cleave at known *in vivo* intermediate termini in the 3' ETS region. A *Δrnt1* strain also has been shown to display of very slow growth phenotype (Chanfreau *et al.*, 1998) and the *rnt1-1* mutant strain has been shown to accumulate the primary transcripts, in vivo (Abou Elela *et al.*, 1996).

While Rnt1p also has been shown (Abou Elela *et al.*, 1996) to cleave a 5' ETS substrate at site A0, *in vitro*, a more recent study (Kufel *et al.*, 1999) has shown that, although functionally delayed, cleavage at this site still occurs in the *Δrnt1* mutant strain. Therefore, if Rnt1p is involved in the cleavage at site A0, this function can be replaced by another enzyme activity. Differences or inconsistencies in this and other examples of rRNA cleavage ultimately may be explained by the fact that nuclease activities such as that of Rnt1p are not restricted to the rRNA precursors or their specificity may be modified by additional factors which influence the structure of the RNA substrate or even the enzyme itself. For example, Rnt1p also is involved in the processing of some snRNA or snoRNA precursors such as the dicistronic snR190-U14 precursor (Abou Elela and Ares, 1998; Chanfreau *et al.*, 1998). Since RNAs such as snoU14 also have been implicated in rRNA processing (e.g., Li *et al.*, 1990), some effects could be indirect. Alternatively, in *S. pombe*, a recent study has shown that protein factors can influence both the efficiency and specificity of Pac1 nuclease cleavages in the 3' ETS (Spasov *et al.*, 2002); similar influences by nucleolar protein factors also may explain the differences in the Rnt1p studies.

In contrast to the Rnt1p protein, the role of RNase MRP in the cleavage of the A3 site has been well established, both *in vivo* and *in vitro* (Schmitt and Clayton, 1993; Chu *et al.*, 1994; Lygerou *et al.*, 1996). The snoRNA constituent of RNase MRP is structurally related to RNase P but is unusual because, unlike other snoRNAs, it does not have a complementary sequence which permits base pairing with the pre-rRNA. This nuclease also contains nine protein subunits, eight of which (Pop1p, Pop3p, Pop4p, Pop5p, Pop6p, Pop7p/Rpp2p, Pop8p and Rpp1p) are shared with RNase P (Chamberlain *et al.*, 1998), and a unique protein, Snm1p (Schmitt and Clayton, 1994). Mutations in these proteins result in an inhibition of cleavage at the A3 site and an increased population of 60 S subunits containing 5.8S rRNAs with 5'-extended termini. Since the cut at the A3 site is not essential for cell viability but all known components of the RNase MRP are essential (Henry *et al.*, 1994), other RNase MRP cleavages in the pre rRNA or other RNA substrates are anticipated. In this respect the endonuclease(s) responsible for cleavage at A1, A2 and D have not been identified, again raising the possibility that additional protein factors may influence the specificity of the RNase MRP or the Rnt1p nuclease with respect to other sites.

As indicated initially, a variety of studies show that exonuclease activities also play important roles in rRNA processing. For example, the formation of the 3' end of the 5.8S rRNA from its 7S precursor involves a protein complex called the exosome (Mitchell *et al.*, 1996; 1997) which removes the extended terminal by a 3' to 5' exonucleolytic mechanism. This exosome activity is associated with at least 14 protein constituents, three of which (Rrp4p, Rr41p/Ski6 P, and our Rp44p) have been demonstrated to be 3' to 5' exonucleases and the remaining eight (Csi14p, Mtr3p, Rrp40p, Rrp42p, Rrp43p, Rrp45p, Rrp46p, and Rrp6p), identified as putative exonucleases. Mutations in or deletions of any one of most of the components impairs the normal synthesis of the 5.8S rRNA, and lead to the accumulation of 3'-extended forms of the molecule (Allmang *et al.*, 1999).

Processive 5' to 3' exonuclease activities also have been identified in *S. cerevisiae* and at least one protein with this activity has been linked to rRNA maturation. The essential Rat1p (Amberg *et al.*, 1992) and the nonessential Xrn1p (Larimer and Stevens, 1990), which appear to be functionally equivalent proteins, have been shown to have processive 5' to 3'endonuclease activity, *in vitro*, and temperature sensitive or disrupted gene mutations have been observed to result in the accumulation of 5'-extended 5.8S rRNAs (Henry *et al.*, 1994). These enzymes are not restricted to effects on rRNA processing and also participate in a variety of cleavage reactions including mRNA degradation (Caponigro *et al.*, 1996), snoRNA processing (Chanfreau *et al.*, 1998) and even DNA strand exchange (Johnson, 1997).

Studies on the maturation of the pre-5S rRNA in *S. cerevisiae* also indicate that a 3' to 5' processive nuclease activity plays an important role in the removal of the extended 3' end. Early experiments suggested that a single endonucleolytic cleavage, catalyzed by the product of the RNA82 gene, may be required (Piper *et al.*, 1983). However, more recent studies, *in vitro*, indicate that a 3' to 5' endonucleolytic activity in nuclear extracts can effectively trim the pre-5S rRNA to the mature form (Lee and Nazar, 1997). In this case the activity appears limited by the secondary structure (helix I) formed between the interacting termini of the mature 5S rRNA. The mature RNA is then stabilized with respect to further digestion by the formation of a ribonucleoprotein complex with the 5S RNA binding protein (YL3). In this respect, it also has been interesting to note that a timely termination of RNA transcription, followed by 3' end processing and protein binding, is essential for the efficient integration of the 5S rRNA into mature ribosomes. Mutations which interfere with a timely termination, the efficient processing of the 3' end extension or the binding of the ribosomal protein severely affect the incorporation of the 5 S RNA molecule into mature ribosomes (Lee *et al.*, 1995). Studies on structural features which underlie the protein binding suggest that protection from random endonucleolytic cleavages is essential for 5S rRNA integration and that a compromised protein binding by RNA mutation (Van Ryk *et al.*, 1992) results in a ribosomal subunit imbalance and essentially the elimination of mutant RNAs which are unable to effectively interact with the ribosomal protein (YL3). Equally, a depletion of the ribosomal protein results in unstable, newly synthesized 5S rRNA and severe effects on the assembly of 60S ribosomal subunits (Deshmukh *et al.*, 1993).

Although, at least some of these ribonuclease activities have been recognized for many years, in general it has been difficult to demonstrate the formation of mature termini or to reproduce the cleavage pathways, *in vitro*. This probably does not simply reflect a lack of appropriate conditions but more likely reflects the complex interaction of factors with precursor structure which occurs within the 80-90 S nucleolar ribonucleoprotein pre-ribosomal particle. For example, in *S. cerevisiae* the RNase MRP activity is associated (Schmitt and Clayton, 1994; Chamberlain *et al.*, 1996) with at least five trans-acting factors or subunits (Pop1p, Pop3p, Pop4-Pop7p, Pop8p and Snm1p) and the exosome activity (Allmang *et al.*, 1999) is known to be associated with at least 11 protein constituents (Rrp4p, Rrp40p, Cs14p, Rrp41/Ski6p, Rrp42,-43,-45 and-46p, Mtr3p, Rrp44p/Dis3p and Rrp6p). Furthermore, other nucleolar interactions may alter the cleavage site structure. For example RAC protein recently was shown to dramatically altered the efficiency and specificity of the Pac1 RNase-III like nuclease in *S. pombe* (Spasov *et al.*, 2002).

One area of study which illustrates the complexities of pre-rRNA cleavage is the processing of the 5' ETS and the initiation of such early events. While there are as many as 100 snoRNAs in yeast, most play a role in rRNA modification (Balakin *et al.*, 1996; Cavaille *et al.*, 1996). The function of these modifications remains unclear and the individual snoRNAs which are essential for the modifications are nonessential for cell viability. In contrast, a small group of snoRNAs including U3, U8, U14, U22, snR10, snR30 and MRP, have been linked with rRNA cleavage reactions in yeast or other organisms (Tollervey, 1987; Kass *et al.*, 1990; Savin and Gerbie, 1990; Li *et al.*, 1990; Hughes and Ares, 1991; Schmitt and Clayton, 1993; Mougey *et al.*, 1993; Peculis and Steitz, 1993; Tycowski *et al.*, 1994) and can be essential for cell viability. As described above, MRP RNA constitutes the RNA component of RNase MRP and is a unique snoRNA because it does not contain complementary sequences that permit a pairing with the pre-rRNA. U3 RNA, which is the most abundant and best studied snoRNA, and U14 RNA are both C/D-box snoRNAs which base-pair with the pre-rRNA; in *S. cerevisiae* U3 base-pairs with the 5' ETS (Beltrame and Tollervey, 1995; Sharma *et al.*, 1999) and U14 base-pairs with 18S rRNA sequences (Liang and Fortier, 1995). Depletion of these RNAs in *S. cerevisiae* results in an inhibition of cleavage at the A0, A1 and A2 or A1 and A2 sites, respectively. Surprisingly, neither RNA pairs with the pre-rRNA at or adjacent to the cleavage sites. A crude *in vitro* system appears to cleave vertebrate 5' ETS faithfully in a U3 snRNA-dependent fashion (Kass *et al.* , 1990; Mougey *et al.*, 1993) but the enzymes responsible have not been isolated. As described above, in yeast at least the A0 site has been cleaved with Rnt1p(Abou Elela and Ares,1996) but an alternate nuclease activity has been detected in the *Δrnt1* mutant strain (Kufel *et al.*, 1999). The snR10 and snR30 RNAs are both H/ACA-box snoRNAs; depletion of these RNAs also leads to the inhibition of cleavage at sites A1 and A2 as well as a depletion of the 18 S rRNA.

In addition to the snoRNAs, these early cleavages in the pre-ribosomal RNA precursor also have been linked to a number of trans-acting proteins including Dbp4p, Dbp8p, Fal1p, Rok1p, Rrp3p, Dim1p and Rrp5p and the snoRNAs, themselves, are known to recruit other proteins to form snRPs (for reviews see Kressler *et al.*, 1999; Venema and Tollervey, 1999). For example, the U3 snoRP is known to contain 3 proteins (Nop1p, Nop56p and Nop58p) which probably form the core of all C/D-box snoRPs but the U3 snRNA also is associated with a number of other unique nucleolar proteins such as Sof1p (Jansen *et al.*, 1993), a protein with seven WD repeats which could mediate protein-protein interactions.

While nuclease activity has only been detected in crude vertebrate extracts (Mougey *et al.*, 1993), it is clear that most if not all eukaryotes, including yeasts, do form a 5' ETS processing complex of snoRNAs and trans-acting protein factors which are essential for the initiation of rRNA processing and the early cleavages at A0, A1 and A2. It remains unclear if the cleavage activities are directly associated with one or more of the snoRNAs and further experimentation remains essential for a clarification of these important initial events in a ribosome biogenesis.

3.2 Putative ATP-dependent RNA Helicases

A surprisingly large number of trans-acting factors which have been linked to ribosomes biogenesis are actually putative RNA helicases of the DEAD-box and related families. Characterized by evolutionarily conserved motifs, these types of protein also have been linked to mRNA splicing, initiation of translation and RNA degradation (see de la Cruz *et al.*, 1999). In general, many examples of such proteins possess RNA dependent ATPase activity but, with respect to ribosome biogenesis, very few actually have been shown to have this activity and, as a result, most of the proteins often are referred to as putative ATP-dependent RNA helicases. At least three different functions are envisaged for these RNA helicase-like proteins: they may function to make critical RNA sites accessible to nuclease cleavage, they may alter the RNA structure in order to permit the binding of specific trans-acting factors or ribosomal proteins or, they may even be required to establish or dissociate snoRNA-pre rRNA base-pairing required for RNA cleavage or modification reactions. A variety of different defects in rRNA processing or ribosomes biogenesis already have been linked to this family of proteins. For example, disruption of genes for Dbp3p (Weaver *et al.*, 1997) and Dbp7p (Daugeron and Linder, 1998) or the deletion of Dbp6p (Kressler *et al.*, 1998) results in a defective assembly of 60S subunits and aberrant rRNA processing; deletion of Dbp8p (Daugeron and Linder, 2001) results in similar effects on the 18 S rRNA and Dbp4p can influence some known effects of U14 snoRNA mutants (Liang *et al.*, 1997). Fal1p (Kressler *et al.*, 1997), Rok1p (Venema *et al.*, 1997), and Rrp3p (O'day *et al.*, 1996), are important in early cleavages at A0, A1 or A2. Similarly, nine other helicase-like proteins affect processing and assembly of the large subunit. In this respect Dbp3p affects cleavage at A3 (Weaver *et al.*, 1997); Dbp7p (Weaver *et al.*, 1997) and Dbp9p (Dougeron *et al.*, 2001) lead to reductions of the 27 S pre-rRNA; Dbp10p (Burger *et al.*, 2000), Dob1p/Mtr4p (de la Cruz *et al.*, 1998a), Drs1p (Ripmaster *et al.*, 1993), and Spb4p(de la Cruz *et al.*, 1998b) result in various defects in the assembly of the 60 S subunits; and, Sen1p appears to act indirectly to delay 35 S pre-rRNA processing (Ursic *et al.*, 1997).

3.3 Other Trans-acting Protein Factors

In addition to nucleases and RNA helicases at least two other categories of trans-acting factors have been identified and for some others the function is not simply unknown. Some of these additional factors act early in the ribosome assembly process and others act much later in the movement of the nearly mature subunits to the cytoplasm or during the final maturation steps in the cytoplasm. While most of the ribosomal proteins are rapidly integrated with ribosomal RNA during the formation of the 80-90 S nucleolar pre- ribosomal particle (Kruiswyk at al., 1978), little is actually known about factors which may influence these processes. Despite numerous attempts, the lack of *in vitro* reconstitution assays for eukaryotic ribosomes greatly limits our understanding of the assembly process. In fact, this may be the best evidence that trans-acting factors are critical in this phase of ribosome biogenesis. A recent study by Woolford and coworkers (Harnpicharnchai *et al.*, 2001), based on affinities for a single nucleolar protein, Nop7p, effectively illustrates the complexity of the interactions between the trans-acting factors and ribosomal proteins with more then eight proteins identified. In addition, the ribosomal proteins themselves may act as trans-acting factors in rRNA processing or subunit assembly (e.g., Vilardell and Warner, 1977). At least one trans-acting factors, Rp7p, has been linked specifically with the assembly of the 40S ribosomal subunit. Deletion of this protein leads not only to an inhibition of pre-rRNA cleavage beginning with A0 to A2, but also to a reduced ratio between the 40S and 60S ribosomal subunits (Baudin-Baillieu *et al.*, 1997). Similarly, mutations in or the depletion of trans-acting factors which have been linked with the assembly of 60S ribosomal subunits often lead to an abortive assembly or degradation of pre-60S ribosomal particles. For example, the depletion of Nop4p or Nop8p, as well as either of the putative RNA helicases Dbp6p and Dbp9p leads to decreased levels of the 27 S pre-rRNA and a decreased accumulation of 60S ribosomal subunits (Sun and Woolford, 1994; Kressler *et al.*,1998). Similarly, Nip7p, an apparent multifunctional protein, which is found both in the nucleus and nucleolus but also is associated with free 60S ribosomal subunits in the cytoplasm, is essential for 60S subunit accumulation(Zanchin and Goldfarb, 1999).

Proteins which are required for the nuclear export of almost complete pre-ribosomal subunits also can have striking effects on ribosome biogenesis. The nuclear export of the pre-60S ribosomal subunit which involves the Ran-cycle (Hurt *et al.*, 1999; Stage-Zimmermann *et al.*, 2000) requires the export factor Xpo1/Crm1, which binds to the ribosomal protein Rpl10p via Nmd3p (Ho *et al.*, 2000; Gadal *et al.*, 2001). Similarly, the export of the pre-40S subunit also depends on the Ran-cycle (Moy and Silver, 1999), but no specific export factors have been identified. Recently a nucleolar protein in *Caenorhabditis elegans*, RBD-1, which was essential for 40S ribosomal subunit maturation, was found to accompany the subunits throughout its transport to the cytoplasm but, as yet, no role in the export process has been established (Bjork *et al.*, 2002).

Two other factors, Nop3p and Nsrlp, also have been advanced as mediators of nucleocytoplasmic transport although other functions also are possible. Pre-rRNA processing defects are observed upon Nop3p deletion and the protein has been observed to shuttle between the nucleus and the cytoplasm (Lee *et al.*, 1996) but a role in the nuclear import of a ribosomal protein or trans-acting factor has not been eliminated. Similarly, Nsr1p is a nucleolar protein which is structurally and functionally similar to mammalian nucleolin (Lee *et al.*, 1991) that also shuttles between the nucleus and cytoplasm, recognizes nuclear localization signals, and participates in pre-rRNA processing (see Tuteja and Tuteja, 1998). Analyses to date suggest that Nsr1p plays a role in pre-rRNA processing and/or ribosome assembly or participates in the import of ribosomal proteins or other important components (Lee *et al.*, 1992).

In addition to protein factors that mediate the export of the maturing subunits other protein factors appeared to be important in intranuclear trafficking. The ubiquitous La protein (Lhp1p in *S. cerevisiae*), for example, associates with the 3' termini of many newly synthesized nuclear small RNAs, including all nascent transcripts made by RNA polymerase III. (see Wolin and Cedervall, 2002). Genetic and biochemical analyses indicate that binding by the La protein protects the 3' ends of these RNAs from exonucleases and may chaperone the RNAs through multiple steps in their maturation pathways. In the case of the 5S rRNA, the nascent transcript transiently associates with the La protein before being bound by its cognate ribosomal binding protein (Rpl5p/YL3 in *S. cerevisiae*) and integrated into ribosomal structure (Steitz *et al.*, 1988). The association of the 5S RNA with the La protein may, itself, be chaperoned by one or more of the Lsm proteins which recently were reported to facilitate RNA-protein interactions and structural changes required during ribosomal subunit assembly (Kufel *et al.*, 2002). Finally, a range of miscellaneous proteins which preferentially affect ribosome biogenesis have been isolated by specific genetic screens. These include the drs, spb and mak mutants (Sachs and Davis, 1989; Ripmaster *et al.*, 1993; Ohtake and Wickner,1995). Some specifically affect 60S ribosomal subunit production and others affect 40S ribosomal subunit biosynthesis, sometimes raising the possibility of intriguing relationships with seemingly unrelated proteins or functions such as a poly (A) binding protein (Sachs and Davis, 1989) or plasma membrane ATPases (Siegmund *et al.*, 1998). In some cases such proteins appear to have dual functions. For example, eIF3j/Hcr1p, a protein associated with eIF3 that has been shown to bind to, and stabilize, the multifactor complex containing eIFs 1, 2, 3 and 5, and Met-tRNA(i)(Met) is also required for the rapid processing of the 20S pre-rRNA to 18 S rRNA (Valasek *et al.*, 2001).

3.4 rRNA Modified and the snoRNAs

As noted in the introduction, in the course of rRNA maturation and ribosome biogenesis, a number of nucleotides are covalently modified, either with the addition of methyl groups to specific nucleotides or the conversion of some uridines to pseudouridines (see Maden, 1990; Ofengand and Bakin, 1997). In eukaryotic cells the vast majority of methylations are at the 2'-O-position of the ribose moiety. A few are directed at specific bases but, in general, little is actually known about base methylation in eukaryotes. Although first characterized more than three decades ago (Singh and Lane, 1964; Lane and Tamaoki, 1967), a functional role for the majority of these modifications remains surprisingly unclear and largely speculative. A number of specific roles have been advanced. These include an effect on ribosome maturation (Vaughan *et al.*, 1967; Sirum-Connolly and Mason, 1993), influences on rRNA conformation (Rottman *et al.*, 1974; Nazar *et al.*, 1983), and even a role in peptide bond formation (Lane *et al.*, 1992; Lane *et al.*, 1995). In bacteria, where base methylation is more common, some modifications have been shown to confer resistance to toxic antibiotics (reviewed in Cundliffe, 1989) and even appear to be a factor in the heat shock response (Bugl *et al.*, 2000). Early occurring modifications also might fine-tune the folding of nascent pre-rRNA or modulate ribosomal protein binding in the course of ribosome assembly (Caboche and Bachellerie, 1977; Tollervey *et al.*, 1993). Indeed, altered processing has been observed in response to an inhibition of methylation, but this was not critical for either rRNA maturation or ribosomes biogenesis (reviewed in Bachellerie and Cavaille, 1997).

Until recently, little also was known about the modification processes in eukaryotic cells. In bacteria, specific methyl transferases and pseudouridine synthetases have been shown to carry out the modification, but comparable activities were not observed in eukaryotic cells. Kinetic analyses indicate that most modifications occur in the nascent pre-rRNA, usually before the transcript is completed. Since all appear to be limited to the universally conserved core regions of the mature are rRNAs, a number of researchers have speculated that the modifications must be involved in some fundamental aspects of ribosome biogenesis or function.

An important clue regarding the unusual nature of the modification processes has come from studies of the many small nucleolar RNAs (snoRNAs) which are found in the nuclei of eukaryotic cells. Initially, sequence complementarities between a family of fibrillarin-associated snoRNAs and the mature rRNA sequences were noted and then demonstrated to be essential for rRNA methylation (Kiss-Laszlo *et al.*, 1996; Nicoloso *et al.*, 1996). Subsequently, complementarity between the second family of snoRNAs and modified regions in the mature RNAs also was shown to be essential for the site-specific synthesis of pseudouridines (Ni *et al.*, 1997; Ganot *et al.*, 1997). Together, the studies have provided the basis for our current understanding of rRNA modification in the eukaryotic cell.

The snoRNAs which participate in rRNA modifications are present in relatively small amounts with estimated concentrations of 10^3 to 10^4 molecules per cell (Ro-Choi, 1997). Based on conserved sequence elements and function, they have been divided into two families, containing either C/D boxes or H/ACA boxes. In the first family, the core element for box C is UGAUGA/U and, for box D, it is CUGA (Smith and Steitz, 1997). In the second, box H (ANANNA) is located at a hinge region between the stand-loop structures and box ACA is on the 3' side of a stem-loop structure (Balakin *et al.*, 1996).

Yeasts contain 75 to 100 different snoRNAs and mammals have been reported to have as many as 200. Some are transcribed from independent promoters as single or polycistronic transcripts. Others are included within the introns of protein genes, involved in ribosome synthesis or function. In *S. cerevisiae*, no fewer than 63 different snoRNAs have been identified as participating in specific rRNA modifications. Twenty of these are members of the H/ACA family and the remainder contain the C/D boxes. Together, they have been implicated in at least 54 different rRNA methylations and 23 different pseudouridine conversions. For updated lists readers are referred to online databases (e.g. for *S. cerevisiae* go to www.bio.umass.edu/biochem/rna-sequence/Yeast_snoRNA_Database/ snoRNA_DataBase.html. The rRNAs of the higher eukaryotes have been observed to contain even more modified nucleotides. In mammals there appear to be 100 to 104 methylated sugars, approximately 95 pseudouridines and even 9 to 10 methylated bases (Maden, 1990).

The discovery of a relationship between snoRNAs and rRNA modification has led to more detailed studies on the modification mechanisms and the recognition of the "guide" role which these RNAs play. A variety of studies have shown clearly that the observed stretches of complementarity to conserved sequences in the rRNA are essential, but the snoRNAs actually appear not to modify the RNA. Instead, they form ribonucleoprotein complexes with fibrillarin and the other nucleolar proteins and presumably guide the modification of the rRNA by activities associated with the protein complex. Nearly all C/D snoRNAs are associated with the 34 kD nucleolar fibrillarin protein or Nop1 in *S. cerevisiae* (Schimmang *et al.*, 1989). In yeast, the box C/D snoRNAs also interact with Nop5p/58p (Wu *et al.*, 1998) which is critical for the normal processing and Nop56p (Gautier *et al.*, 1997) which is not an essential protein but can delay rRNA processing. These three core proteins of box C/D snoRPs are all critical for the methylation of rRNA but none of these proteins actually resemble methyltransferases. Proteins which have been predicted to have rRNA methyltransferase activity, Nop2p, Ncl1p and Spb1p (Hang *et al.*, 1997; Wu *et al.*, 1998; Kressler *et al.*, 1999) have not been linked to the C/D snoRNAs.

Equally, Gar1 protein is bound to snoRNAs which are involved in pseudouridylation (Bousquet-Antonelli, 1997). This protein is essential for viability in yeast and temperature sensitive mutants do show lower levels of pseudouridines although, again, there is no sequence similarity with known pseudouridylases. At least three other proteins, Cbf5p (Lafontaine *et al.*, 1998), Nhp2p and Nop10 (Henras *et al.*, 1998) also have been shown to associate with this class of snoRNA. Cbf5p shows some sequence similarities to *E. coli* pseudouridyl synthetase but no enzymatic activity has been detected (Koonin, 1996). Since fibrillarin or the other well characterized snoRNA binding proteins do not resemble methyltranferases or pseudouridylases, the actual source of enzyme activity remains unclear and a ribozyme function has not been entirely eliminated.

The only base methylation which has been studied in yeast is a dimethylation at two adjacent adenosines ($m^{6,2}A_{1779}$ and $m^{6,2}A_{1780}$) at the 3' end of the 18 S rRNA of *S. cerevisiae*, an evolutionarily conserved modification which also is present in bacterial rRNAs. These do not require a snoRNA but are dependent on the essential protein Dim1p that also is required for processing at sites A1 and A2 to form the 20 S pre-rRNA (Lafontaine *et al.*, 1994; Lafontaine, 1995). The dimethylation actually occurs on the 20S pre-rRNA so the modification itself is not required for rRNA cleavage. A *dim1-2* mutant strain indicates that dimethylation also is not essential for cell viability but translation, *in vitro*, is compromised suggesting that under certain physiological conditions the modification may fine tune ribosome function, *in vivo* (Lafontaine *et al.*, 1998).

4. ROLE OF CIS-ACTING SEQUENCE ELEMENTS IN rRNA PROCESSING

While many trans-acting factors have been implicated in the processing of the eukaryotic pre-rRNAs, the role of cis-acting elements remains controversial and somewhat unclear. Recombinant DNA methods have permitted the cloning and sequencing of rDNAs from a very large number of diverse eukaryotic organisms. Although a comparative approach has been very useful with studies on the structure of the mature ribosomal RNAs, comparative studies have not lead to generally accepted models for the nonconserved transcribed spacers sequences. While the core structure of the mature RNAs is very similar in size and nucleotide sequence, a common secondary structure or consensus sequence elements in the spacer regions have been very difficult to identify as these regions differ greatly with respect to length and show very little sequence homology (see Nazar *et al.*, 1987). Such attempts have caused some workers to speculate that the processing mechanisms may vary widely with phylogeny or that the spacers may not even play a role in rRNA processing (Subrahmanyan *et al.*, 1982; Bachellerie *et al.*, 1983; Tague and Gerbi, 1984). Such a conclusion was eliminated when Planta and co-workers first expressed plasmid-associated "tagged" rDNA transcriptional units in *S. cerevisiae* (Musters *et al.*, 1989; 1990). The insertion of tags, which did not interfere with processing or the accumulation of mature rRNAs, permitted the detection of plasmid-derived transcripts with mutations in spacer regions. Large deletions in the 5' ETS or either of the internal transcribed spacers were found to be detrimental and demonstrated that the spacers are necessary for normal rRNA maturation (Musters *et al.*, 1990; van der Sande *et al.*, 1992). Clearly the spacers contain elements which control their own removal.

Subsequent analysis of systematic mutations in the internal transcribed spacers permitted several more specific conclusions (van Nues *et al.* 1994). A division of the ITS1 region into five structural domains, as suggested by an estimate of the secondary structure (Yeh *et al.*, 1990), indicated that the domains IV and V were dispensable for 18S rRNA maturation while each of these domain appeared to be individually sufficient for 25S rRNA production. Based on these studies the researchers suggested that ITS1 is organized into two functionally and structurally distinct halves. Similarly, the ITS2 region was divided into five structural domains based on the secondary structure model of Yeh and Lee (1990). Subsequent mutational analyses indicated that the cis-acting elements required for the correct and efficient processing of ITS2 were found to be structurally conserved features of domains II,

III, IV and V in *S. cerevisiae* and other closely related yeasts (van Nues *et al.*, 1995a). Mutations in each of these elements severely reduced the production of the mature 25S rRNA but 18 S rRNA was still detected. Deletion of the variable segments in domains IV, V and VI had little or no effect on processing, although reduced growth rates were observed which suggested a direct role in the assembly of fully functional 60S ribosomal subunits.

In addition to the spacer regions, changes in the mature rRNA sequences also appear to have significant effects on the formation as well as function of the ribosomal subunits. For example, in *S. cerevisiae*, a large insertion or replacement of helix 6 in variable region VI of the 18 S rRNA abolished almost all 18 S rRNA formation with a concomitant accumulation of the 37S pre-rRNA and an aberrant 23S intermediate precursor (van Nues *et al.*, 1997). Similarly, the insertion of an IVS (intervening sequence) or ribozyme sequence from Tetrahymena rDNA into its analogous position in the *S. pombe* 25S rDNA also had dramatic effects on rRNA production (Good *et al.*, 1994). Subsequent analyses of transformants expressing the mutant gene indicated that the IVS was excised normally from the recombinant transcripts but, after the excision of the ribozyme, the spliced pre-rRNAs did not mature normally. Both of these examples indicate that critical sequence elements or features also must be present in the mature rRNAs which can affect rRNA processing and accumulation, either directly or indirectly.

A number of sequence analyses also have been focused on the external transcribed spacers. In *S. cerevisiae*, for example, large deletions of 200 to 400 bp of the 5' ETS disrupted the processing of the 18 S rRNA (Musters *et al.*, 1990) and smaller scale deletions at proposed U3 snoRNA binding sites in the 5' ETS indicate that the -470 element is necessary for 18S rRNA production, but the -665 element is dispensable (Beltrame and Tollervey, 1992; Beltrame *et al.*, 1994). More recent systematic analysis of the proximal region in the 5' ETS of the genes encoding the rRNAs of *S. pombe* was focused on a crucifix-like estimate of the secondary structure predicted by computer modeling and nuclease digestion analyses (Intine *et al.*, 1999). A series of large and small mutations in this structure indicated strong correlations with known or putative events in rRNA maturation. Changes associated with an intermediate cleavage site or with the putative U3 snoRNA binding site were critical to 18S rRNA production and also reduced the production of the large ribosomal subunit RNA by as much as 60%. Smaller or compensatory changes in the sequence also indicated that the effects were sequence dependent and not simply the result of disrupted structure.

The actual recognition signals that define the cleavage at the 5' end of the yeast 18S rRNA (site A1), however, appeared not to be dependent on the U3 snoRNA binding site. Instead, Tollervey and coworkers (Venema *et al.*, 1995) have concluded that the maturation of the 5' end sequence is defined by two partially independent features, one involving a small cluster of phylogenetically conserved nucleotides, immediately upstream of the A1 cleavage site, and a second, defined by a fixed distance of three nucleotides from a 5' stem-loop/pseudoknot structure within the 18 S rRNA sequence. In *S. cerevisiae*, the cleavage at A0 appeared to be dependent on an RNase III-like enzyme activity with specificity for double stranded RNA (Abou Elela *et al.*, 1996), but in the vertebrates, a single stranded configuration appears to be essential (Craig *et al.*, 1991) and a more recent study in *S. cerevisiae* (Kufel *et al.*, 1999) reports contrary results. Clearly, further studies are required to resolve the many interesting questions which have been raised by past observations.

Although the 3' ETS initially received relatively little attention, it also contains cis-acting sequence elements and structural features that are essential for rRNA processing. Melekhovets and coworkers (1994) first demonstrated that a large conserved hairpin loop structure in the 3'ETS of *S. pombe* is essential in the maturation of the 3'end of the 25S rRNA as well as the ITS2 region. This structure, which is located just downstream of the mature 25S rRNA sequence, also has been described in many other pre-rRNAs (e.g., Kempers-Veenstra *et al.*, 1986; Walker *et al.*, 1990). Subsequent genetic and biochemical analyses have indicated that a critical sequence within the upper end of the hairpin structure serves as a binding site for nucleolar protein (Hitchen *et al.*, 1997) and this protein binding appears to mediate the deleterious effects of sequence mutations in this region. Furthermore, studies on the RNase III-like, Pac1 nucleases in *S. pombe* and the analogous protein, Rnt1p in *S. cerevisiae* indicate that both enzymes cleave efficiently in their cognate 3' external transcribed spacer sequences. *In vitro* analyses have shown that in *S. cerevisiae* (Abou Elela *et al.*, 1996), a single cut is observed at a natural intermediate cleavage site (+ 21) and in *S. pombe* (Rotondo and Frendewey, 1996), two staggered cuts also are introduced at known intermediate cleavage sites (+ 41 and + 83), both in the highly conserved hairpin structure. Neither enzyme produces mature termini with only a purified RNA substrate, but until recently, and unlike the role of RNase III in the maturation of the bacterial rRNAs, the significance of this activity in the biogenesis of the eukaryotic ribosomes remained unclear.

In the recent studies on spacer binding proteins in *S. pombe*, a large protein complex of 20 or more polypeptides was a isolated using the ITS1 rRNA sequence (Lalev *et al.*, 2000). It was putatively called the ribosome assembly chaperone (RAC) in recognition of its affinity for the alternate spacers in rDNA, including the 3' ETS region which, previously had been shown to interact with unknown cellular protein (Hitchen *et al.*, 1997). This complex exhibited no nuclease activity but mutations in the RAC protein binding site within the 3' ETS also were known to disrupt rRNA processing. In view of this, Pac1 nuclease cleavage was reexamined in the presence of RAC protein (Spasov *et al.*, 2002) and found to be altered dramatically with respect to both cleavage efficiency and specificity. In the presence of the protein complex, the Pac1 enzyme was found to cleave at the 3' end of the 25S rRNA sequence, leading to the complete removal of the 3' ETS. Taken together, these observations suggest that, at least the 3' ETS, contains two important types of sequence element which are essential to rRNA maturation: appropriate structural features which are recognized by the cleaving enzyme activity, and a protein binding site which serves to organize or guide the cleavage process. Additional studies on the other spacers regions in the *S. pombe* rDNA further indicate that similar elements may exist in all four spacers and that there may be a common theme in the removal all the spacers even though alternate enzymatic activities are utilized. As noted earlier, many sequence comparisons of the spacer regions in diverse organisms (see Nazar, 1987) have noted a very strong divergence in these regions both with respect to the actual nucleotide sequence and the length of the spacers. Some studies have focused on the structure of these elements in specific organisms (e.g., Yeh and Lee, 1990) but, in view of the strong sequence divergence, models generally have been restricted to closely related examples (e.g., Hershkovitz *et al.*, 1996; Michot *et al.*, 1999). A "biological spring" hypothesis has been advanced (Nazar *et al.*, 1987) to explain, at least in part, why the sequences can be so diverse and yet play a critical role in rRNA maturation. The hypothesis was based on the fact that, despite the great diversity in the secondary structure estimates for the internal transcribed spacers, one feature was common, namely the fact that in each case the secondary structure brought the maturing termini into relatively close proximity. In this sense the secondary structure plays the role of a kind of spring to assemble the termini together for cleavage.

More recent comparisons of the internal transcribed spacers in *S. pombe* with those of very divergent organisms have recognized some generally conserved core structure and at least one additional common feature namely, RAC protein binding sites. Computer modeling and probes for nuclease protection of the ITS1 region in *S. pombe* (Lalev and Nazar; 1998) suggest a highly organized structure consisting of a central extended hairpin with smaller hairpin branches immediately adjacent to the maturing termini. These features are retained in divergent examples, even when they are much longer or very short. Protein binding studies further indicate that, as observed in the 3' ETS region, the extended hairpin is not only a site of intermediate RNA cleavage during rRNA processing, but also a site for interaction with RAC protein (Hitchen *et al.*, 1997; Spasov *et al.*, 2002). Similarly, with that same analytical methods, the structure of the *S. pombe* ITS2 region has been re-evaluated with respect to greatly divergent organisms containing much longer and shorter spacers sequences (Lalev and Nazar, 1999). Again, a simple core structure was recognized, consisting of a single extended hairpin in which diversity between the organisms was the result of changes in hairpin length as well as additional branched helices. Of greater probable significance is the fact that the extended hairpin is again a site for intermediate RNA cleavage during rRNA processing, as well as a binding site for RAC protein (Lalev and Nazar, 1999; Abeyrathne *et al.*, 2002). Together, all of these analyses suggest structural equivalence in the transcribed spacers in which the spacer structure serves to organize the termini for cleavage as well as to bind and organize trans-acting protein factors which ultimately carry out RNA cleavages to produce the mature RNA termini. Subsequent mutational analyses (Spasov *et al.*, 2002; Abeyrathne *et al.*, 2002) continue to support this model but more study clearly is required to fully verify the suggestions and provide the necessary details.

An additional question which has been raised by the protein binding studies with spacer sequences from *S. pombe* is the possibility that the spacers may act in ways that are analogous to the roles of snRNAs in mRNA splicing. In a comparison of the factor binding sites in ITS1 and the 3' ETS, sequence similarities were noted in the distal helical regions which represent the core of these binding sites and also were found to share some features with known protein binding sites in the U1 snRNA (Lalev and Nazar, 1999). These same regions were observed to contain even greater similarities when ITS1 and ITS2 were compared, suggesting that the spacers may act in a manner which is analogous with that of free snRNAs when they organize a spliceosomal complex in the course of mRNA maturation. Such an intriguing possibility awaits further experimentation.

5. QUALITY CONTROL AND RIBOSOME BIOGENESIS

The ribosomes of both bacteria and eukaryotic cells utilize essentially identical mechanisms to synthesize proteins and contain core structures which are highly conserved. Despite these great similarities eukaryotic cells have a more complex scheme for ribosome biogenesis with many more protein and RNA factors required for RNA processing and ribosome assembly. As noted earlier, the pre-rRNA must be fully transcribed and intergrated into the nucleolar 80-90S pre-ribosomal particle before cleavage is even initiated. The need for such a complex maturation pathway poses intriguing questions both with respect to ribosome biogenesis and the evolution of the eukaryotic cell.

5.1 Interdependencies in Ribosome Biogenesis

As indicated earlier, the pathways for rRNA processing and ribosome maturation appear to consist of many individual steps which, at first approximation, might be expected to proceed independently of each other and even without the completion of transcript expression. Certainly, in bacteria, nascent mRNA transcripts are utilized efficiently for protein synthesis before RNA synthesis is complete. Indeed, studies of rRNA processing in *S. cerevisiae* have reported an independent maturation of the two ribosomal subunits (Musters *et al.*, 1990; van der Sande *et al.*, 1992). With the use of a low copy expression vector and hybridization analyses to detect the vector-derived RNA molecules, the disruption of processing which led to the sysnthesis of the large subunit RNA did not prevent the synthesis of the small subunit and, equally, disruption of the small subunit did not prevent the incorporation of RNA into the large subunit. In fact, when cells are forced genetically to utilize plasmid-associated rRNA genes, ribosomes can be synthesized in trans with the large and small subunit RNAs being expressed from alternate plasmid-associated templates (Liang and Fournier, 1997). Such studies raise important questions about the need for the nucleolar 80-90 S ribonucleoprotein particle and why rRNA transcripts are completed before being cleaved and modified.

A series of more recent studies on rRNA maturation in *S. pombe* appear to offer an explanation. In these studies a high copy plasmid was selected (Abou Elela *et al.*, 1994;1995) to express the rDNA transciptional unit into which mutations were introduced systematically. Unlike the earlier studies in *S. cerevisiae*, where the mutant RNA represented a very small portion of the transcripts (Musters *et al.*, 1989) or the entire population (Nogi *et al.*, 1991), in these analyses approximately equal amounts of mutant and normal RNA were transcribed in the *S. pombe* cells and quantitative comparisons could be accurately undertaken. In such an analysis on the removal of the 3' ETS in the pre-rRNA of *S. pombe*, mutations in a highly conserved extended hairpin structure that closely follows the mature 25S rRNA sequence were shown to inhibit not only the removal of the 3' ETS sequence, but also to inhibit completely the removal of the ITS2 spacer region, located some 3000 nucleotides upstream of the 3 'ETS (Melekhovets *et al.*, 1994). Furthermore, subsequent analysis of the plasmid-derived 18 S rRNA indicated that the incorporation of this RNA into mature ribosomes continued but also was reduced severely and only 15 to 20 percent of normal levels were observed. Similar analyses with mutations in the ITS2 region have shown that such changes can not only inhibit the maturation of the 5.8S and 25S rRNAs but, again, also can severely reduce levels of plasmid-derived 18 S rRNA in the mature ribosomes (Good *et al.*, 1997b). Finally, mutations in the 5' ETS, which have been shown to inhibit the maturation of the 18S rRNA, also have been shown to inhibit severely the incorporation of the cognate 5.8S and 25S rRNAs into the mature ribosomes (Intine *et al.*, 1999). Taken together, all of these analyses indicate that, at least in *S. pombe*, there are significant interdependences in the cleavage steps which can dramatically influence the incorporation of the rRNAs into mature ribosomes. While changes in one subunit may not critically affect the formation of the other subunit (Van Nues *et al.*, 1995b) and, indeed, ribosomal subunits can be produced "in trans" (Liang and Fournier, 1997), the interdependencies clearly influence the yield and, as a result, could affect the survival of an organism in a competitive environment.

5.2 RNA Processing as a Quality Control Mechanism

In response to questions regarding the interdependences and the need for the nucleolar ribonucleoprotein complex, a suggestion has been made that, taken together, these features represent a "quality control" function in ribosome biogenesis which, at least in part, helps to insure that only functional ribosomes are synthesized. The need for such control may reflect an adaptation to an important basic characteristic of protein synthesis. In all cells, protein synthesis is carried out by polyribosomes often consisting of many ribosomes, efficiently translating a messenger RNA in a sequential fashion. Presumably such an efficient use of mRNA enables large amounts of protein to be made when required for rapid cell growth in a competitive environment. In such a circumstance, a defective ribosome not only produces less or no protein itself, but also might inhibit the movement of other ribosomes which follow with much more severe consequences. As a result, it seems especially important that, at least for long-term survival, cells should avoid the production of defective ribosomes by some sort of quality control mechanism. The formation of an 80 to 90 S nucleolar ribonucleoprotein complex, and the interdependencies which are incorporated into the complex may serve such a function. Figure 2 shows a model for the assembly of the nucleolar pre-ribosomal particle which was first suggested to explain structural features in the internal transcribed spacers

(Nazar *et al.*, 1987) as well as the distant interdependencies in rRNA processing (Nazar *et al.*, 1996). The model proposes that as ribosomal proteins assemble on the mature ribosomal RNA sequences, nucleolar constituents simultaneously assemble on the spacer regions basically forming a nucleolar pre-ribosomal particle consisting of three domains, two of which correspond with the ribosomal subunits and one domain composed of the spacer sequences, nucleolar proteins and RNAs, comprising a common processing domain. Essentially the process of fitting everything correctly into the large particle for efficient maturation may be acting as a kind of a "checklist" to ensure that everything is normal in all regions of the precursor. Failing this check makes the pre-rRNA susceptible to "housekeeping" enzymes which rapidly destroy it and prevent its incorporation into mature subunits.

In search of structural features that underlie the interdependencies in rRNA processing, interactions between the space regions and soluble cellular constituents have been characterized and attempts made to isolate proteins which specifically interact with the individual transcribed spacers sequences (Hitchen *et al.*, 1997; Lalev *et al.*, 1998; Lalev *et al.*, 1999). Initially, using the ITS1 spacer as a ligand for affinity chromatography, a large protein complex of 20 or more polypeptides was isolated (Lalev *et al.*, 2000). The protein components ranged in size from 20 to 200 kDa. Although no nuclease activity could be demonstrated, peptide mapping by Maldi-Toff mass spectroscopy identified eight hypothetical RNA binding proteins which included four different RNA binding motifs. Similar attempts, again using affinity chromatography, to isolate proteins which interacted with the ITS2 or 3' ETS spacer yielded essentially the same complex of proteins (Lalev and Nazar, 2001). Equally important, at least one protein binding site has been mapped in each of the 3 spacer regions and disrupting mutations in the site not only affect protein binding but have been shown to inhibit rRNA processing and essentially eliminate the mutant RNA from the mature ribosome population (Hitchen *et al.*, 1997; Lalev *et al.*, 2000; Abeyrathne *et al.*, 2002). Furthermore, the same protein complex has been shown to interact with multiple binding sites in the 5' ETS (Spasov *et al.*, unpublished results) with mutations in these binding sites also adversely affecting rRNA processing.

Because nuclease activity could not be demonstrated with this protein complex but it's binding to the spacer regions was critical to rRNA maturation, it was tentatively called a ribosome assembly chaperone (RAC), presumably acting as a kind of rack on which critical structure is organized for rRNA processing and perhaps ribosome biogenesis, in general. Recently the notion that this complex may have a chaperone function was supported experimentally when, also in *S. pombe*, the RAC protein complex was observed to direct the complete removal of the 3' ETS by the Pac1 nuclease (Spasov *et al.*, 2002). In the presence of the RAC protein/3' ETS complex, cleavage by this RNase III-like homologue is not restricted to known intermediate sites but also was directed at the 3' end of the 25S rRNA.

The observations that the RAC protein binding sites in the spacer regions are critical to rRNA processing and can be critical to interdependences in rRNA processing are fully consistent with the model raised in Figure 2. They also provide further argument for the notion that the function of the 80-90S nucleolar pre-ribosomal particle together with a very complex maturation pathway represents, at least in part, a quality control mechanism which helps avoid the formation of defective ribosomes.

5.3 Modification of rRNA as a Quality Control Mechanism

As already noted, many roles have been suggested for rRNA modifications, including influences on rRNA conformation, protein binding, RNA stability and even ribosome function. Despite this, almost nothing is known about the function of the specific methylations or pseudouridylations. In mammalian cells, a few of the modifications are partial and appeared to correlate with cellular differentiation (Nazar *et al.*, 1975; Munholland *et al.*, 1987). In the case of the human 5.8S rRNA, this type of modulated modification has been shown to alter the confirmation of the 5.8S rRNA and influence the stability of its interaction with the 28S rRNA (Nazar *et al.*, 1983). To date, however, similar examples have not been observed in yeast. Much more surprising is the observation that cells which lack most of the snoRNAs or are depleted of core binding proteins can grow normally, and even pseudouridine residues which appeared to be good candidates for a direct involvement in peptidyl transferase function (Lane *et al.*, 1992), again are not essential (Ni *et al.*, 1997).

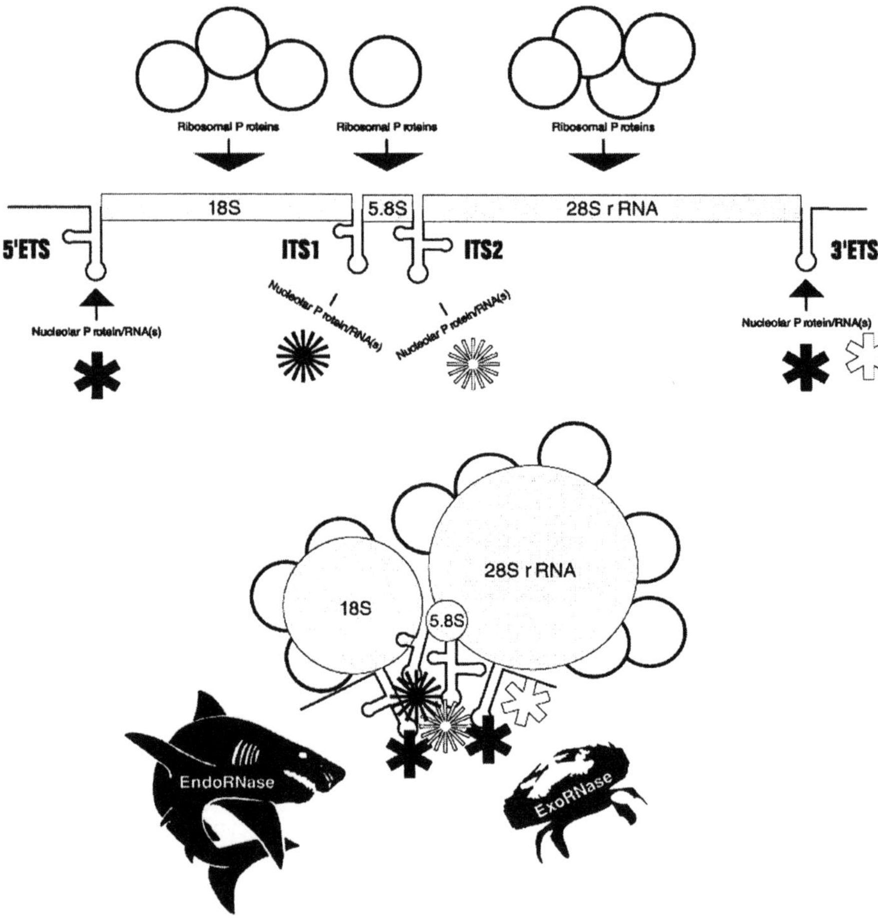

Fig. 2. A model for the assembly of a nucleolar 80-90S pre-ribosomal particle. Ribosomal proteins assemble on the rRNA sequences to form the ribosomal subunits whereas the nucleolar spacer-associated factors assemble on the transcribed non conserved spacers to form a common processing domain that is acted on by nucleases and related factors.

In the experiments which led to these conclusion (Maxwell and Fournier, 1995; Balakin *et al.*, 1996; Cavaille *et al.*, 1996), targeted mutations were introduced into specific snoRNAs or other mutations were introduced with the aim of disrupting the modification process(es). In each case, the absence of specific methylations or pseudouridylations was clearly demonstrated but the cells remained viable and grew normally.

At least one explanation for this unexpected and fascinating result was advanced recently when modified positions in the rRNAs were altered with respect to the nucleotide sequence. An efficiently expressed rDNA plasmid again was used to effectively express mutant rRNA which would have to compete with normal molecules. Under these conditions, a quantitative analysis was used to assess the effect of such nucleotide changes within the peptidyl transferase center of the 25S rRNA of *S. pombe* cells. The results show that, unlike normal RNA, in the same cell and relative to a less conserved but modified position outside the center, these mutant RNAs were highly unstable and rapidly degraded with little or no effect on cell growth (Song *et al.*, 2002). The results provided direct evidence that positions of modification can be critical sites for nuclease attack, either because they are directly susceptible to nucleases or because protein binding to the site is compromised.

Whichever the case, the results raised the possibility that, as is observed for rRNA processing, rRNA modification also may represent a quality control mechanism which helps insure that only functional rRNA is incorporated into mature ribosomes particles. As discussed earlier, the site specific rRNA modifications are dependent on snoRNAs that share long sequence complementarities with the modified sequences and pair directly with the rRNA in order to guide the modification processes. In a sense, such interactions effectively proof-read portions of the newly transcribed rRNA sequence. Altered sequences would not be expected to pair as effectively and, therefore, a modification process is likely to be disrupted. In turn, based on the observations with altered nucleotides at the modified positions (Song and Nazar, 2002), the unmodified RNA would be unstable or less stable and fully or, at least, partially eliminated from the mature ribosomal population. Since the modified nucleotides are largely localized in active centers of the ribosomes, this proofreading would be especially helpful in eliminating mutant rRNA sequences and would tend to avoid any deleterious or lethal effects of sequence changes in the functional regions. The experiments to date are preliminary in nature but further studies under appropriately competitive conditions ultimately may resolve these important questions about rRNA modifications and their roles in ribosome biogenesis.

6. CONCLUSIONS

The last decade has witnessed dramatic advances in our understanding of the ribosome and its biogenesis, including the discovery and characterization of many trans-acting protein factors, the unanticipated role of the many snoRNAs in rRNA methylation and pseudouridylation, as well as the critical cis-acting elements and interdependencies in rRNA processing. Together these findings have provided a fundamental understanding of the basic steps in ribosome biogenesis as well as the complexities, which surround each step. Despite this very significant progress and the many intriguing observations surrounding each step, many details regarding the mechanisms for ribosome biogenesis remained unclear including details of the RNA cleavage mechanisms at most sites in the maturation pathways and the role of the many modification events. Mechanisms of ribosome assembly remain especially unclear and though models for quality-control in ribosome biogenesis have been advanced experimentation must continue at very significant levels if the remaining questions are to be answered.

Acknowledgement: Thanks to Dr. E.J. Robb for her comments regarding the manuscript.

REFERENCES

Abeyrathne PD, Lalev AI, and Nazar RN (2002). A RAC protein-binding site in the internal transcribed spacer 2 of Pre-rRNA transcripts from *Schizosaccharomyces pombe*. J Biol Chem 277:21291-21299.

Abou Elela S, Good L, Melekhovets YF, and Nazar RN (1994). Inhibition of protein synthesis by an efficiently expressed mutation in the yeast 5.8S ribosomal RNA. Nucleic Acids Res 22:686-693.

Abou Elela S, Good L, and Nazar RN (1995). An efficiently expressed 5.8S rRNA 'tag' for in vivo studies of yeast rRNA biosynthesis and function. Biochim Biophys Acta 1262:164-167.

Abou Elela S, Igel H, and Ares M Jr (1996). RNase III cleaves eukaryotic preribosomal RNA at a U3 snoRNP-dependent site. Cell 5:115-124

Abou Elela S, and Ares M Jr (1998). Depletion of yeast RNase III blocks correct U2 3' end formation and results in polyadenylated but functional U2 snRNA. EMBO J 17:3738-3746.

Allmang C, Kufel J, Chanfreau G, Mitchell P, Petfalski E, and Tollervey D (1999) Functions of the exosome in rRNA, snoRNA and snRNA synthesis. EMBO J 18:5399-5410.

Amberg DC, Goldstein AL, and Cole CN (1992). Isolation and characterization of RAT1: an essential gene of *Saccharomyces cerevisiae* required for the efficient nucleocytoplasmic trafficking of mRNA. Genes Dev 6:1173-1189.

Balakin AG, Smith L, and Fournier MJ (1996). The RNA world of the nucleolus: two major families of small RNAs defined by different box elements with related functions. Cell 86:823-834.

Bachellerie JP, Michot B, and Raynal F (1983). Recognition signals for mouse pre-rRNA processing. A potential role for U3 nuclec,ar RNA. Mol Biol Rep 9:79-86.

Bachellerie JP, and Cavaille J (1997). Guiding ribose methylation of rRNA. Trends Biochem Sci 22:257-261.

Ban N, Nissen P, Hansen J, Moore PB, and Steitz TA (2000). The complete atomic structure of the large ribosomal subunit at 2.4 A resolution. Science 289:905-920.

Baudin-Baillieu A, Tollervey D, Cullin C, and Lacroute F (1997). Functional analysis of Rrp7p, an essential yeast protein involved in pre-rRNA processing and ribosome assembly. Mol Cell Biol 17:5023-5032.

Beltrame M, and Tollervey D (1992). Identification and functional analysis of two U3 binding sites on yeast pre-ribosomal RNA. EMBO J 11:1531-1542.

Beltrame M, Henry Y, and Tollervey D (1994). Mutational analysis of an essential binding site for the U3 snoRNA in the 5' external transcribed spacer of yeast pre-rRNA. Nucleic Acids Res 22:5139-5147.

Beltrame M, and Tollervey D (1995). Base pairing between U3 and the pre-ribosomal RNA is required for 18S rRNA synthesis. EMBO J 14:4350-4356.

Bjork P, Bauren G, Jin S, Tong YG, Burglin TR, Hellman U, and Wieslander L (2002). A Novel Conserved RNA-binding Domain Protein, RBD-1, Is Essential For Ribosome Biogenesis. Mol Biol Cell 13:3683-3695.

Bousquet-Antonelli C, Henry Y, G'elugne JP, Caizergues-Ferrer M, and Kiss T (1997). A small nucleolar RNP protein is required for pseudouridylation of eukaryotic ribosomal RNAs. EMBO J 16:4770-4776.

Bugl H, Fauman EB, Staker BL, Zheng F, Kushner SR, Saper MA, Bardwell JC, and Jakob U (2000). RNA methylation under heat shock control. Mol Cell 6:349-60.

Burger F, Daugeron MC, and Linder P (2000). Dbp10p, a putative RNA helicase from *Saccharomyces cerevisiae*, is required for ribosome biogenesis. Nucleic Acids Res 28:2315-2323.

Busch H, and Smetana K (1970). The Nucleolus. New York: Academic Press.

Caboche M, and Bachellerie JP (1977). RNA methylation and control of eukaryotic RNA biosynthesis. Effects of cycloleucine, a specific inhibitor of methylation, on ribosomal RNA maturation. Eur J Biochem 74:19-29.

Caponigro G, and Parker R (1996). Mechanisms and control of mRNA turnover in *Saccharomyces cerevisiae*. Microbiol Rev 60:233-249.

Cavaille J, Nicoloso M, and Bachellerie JP (1996). Targeted ribose methylation of RNA in vivo directed by tailored antisense RNA guides. Nature 383:732-735.

Chamberlain JR, Pagan-Ramos, Kindelberger DW, and Engelke DR (1996). An RNase P RNA subunit mutation affects ribosomal RNA processing. Nucleic Acids Res 24:3158-3166.

Chamberlain JR, Lee Y, Lane WS, and Engelke DR (1998). Purification and characterization of the nuclear RNase P holoenzyme complex reveals extensive subunit overlap with RNase MRP. Genes Dev 12:1678-1690.

Chanfreau G, Rotondo G, Legrain P, and Jacquier A (1998). Processing of a dicistronic small nucleolar RNA precursor by the RNA endonuclease Rnt1. EMBO J 17:3726-3737.

Chu S, Archer RH, Zengel JM, and Lindahl L (1994). The RNA of RNase MRP is required for normal processing of ribosomal RNA. Proc Natl Acad Sci U S A 91:659-663.

Craig N, Kass S, and Sollner-Webb B (1991). Sequence organization and RNA structural motifs directing the mouse primary rRNA-processing event. Mol Cell Biol 11:458-467.

Cundliffe E (1989). How antibiotic-producing organisms avoid suicide. Annu Rev Microbiol 43:207-233.

Daugeron MC, and Linder P (1998). Dbp7p, a putative ATP-dependent RNA helicase from *Saccharomyces cerevisiae*, is required for 60S ribosomal subunit assembly. RNA 4:566-4581.

Daugeron MC, and Linder P (2001). Characterization and mutational analysis of yeast Dbp8p, a putative RNA helicase involved in ribosome biogenesis. Nucleic Acids Res 29:1144-1155.

Daugeron MC, Kressler D, and Linder P (2001). Dbp9p, a putative ATP-dependent RNA helicase involved in 60S-ribosomal-subunit biogenesis, functionally interacts with Dbp6p. RNA 7:1317-1334.

de la Cruz J, Kressler D, Tollervey D, and Linder P (1998a). Dob1p (Mtr4p) is a putative ATP-dependent RNA helicase required for the 3' end formation of 5.8S rRNA in *Saccharomyces cerevisiae*. EMBO J 17:1128-1140.

de la Cruz J, Kressler D, Rojo M, Tollervey D, and Linder P (1998b). Spb4p, an essential putative RNA helicase, is required for a late step in the assembly of 60S ribosomal subunits in *Saccharomyces cerevisiae*. RNA 4:1268-1281.

de la Cruz J, Kressler D, and Linder P (1999). Unwinding RNA in *Saccharomyces cerevisiae*: DEAD-box proteins and related families. Trends Biochem Sci 24:192-198.

Deshmukh M, Tsay YF, Paulovich AG, and Woolford JL Jr (1993). Yeast ribosomal protein L1 is required for the stability of newly synthesized 5S rRNA and the assembly of 60S ribosomal subunits. Mol Cell Biol 13:2835-2845.

Gadal O, Strauss D, Kessl J, Trumpower B, Tollervey D, and Hurt E (2001). Nuclear export of 60s ribosomal subunits depends on Xpo1p and requires a nuclear export sequence-containing factor, Nmd3p, that associates with the large subunit protein Rpl10p. Mol Cell Biol 21:3405-3415.

Ganot P, Bortolin ML, and Kiss T (1997). Site-specific pseudouridine formation in preribosomal RNA is guided by small nucleolar RNAs. Cell 89:799-809.

Gautier T, Berges T, Tollervey D, and Hurt E (1997). Nucleolar KKE/D repeat proteins Nop56p and Nop58p interact with Nop1p and are required for ribosome biogenesis. Mol Cell Biol 17:7088-7098.

Good L, Elela SA, and Nazar RN (1994). Tetrahymena ribozyme disrupts rRNA processing in yeast. J Biol Chem 269:22169-22172.

Good L, Intine RV, and Nazar RN (1997). The ribosomal-RNA-processing pathway in *Schizosaccharomyces pombe*. Eur J Biochem 247:314-321.

Good L, Intine RV, and Nazar RN (1997). Interdependence in the processing of ribosomal RNAs in *Schizosaccharomyces pombe*. J Mol Biol 273:782-788.

Grivell LA, and Planta RJ (1990). Yeast: the model 'eukaryote'? Trends Biotechnol 8:241-243.

Hadjiolov, A (1985). The Nucleolus and Ribosome Biogenesis. New York: Springer-Verlag.

Harms J, Schluenzen F, Zarivach R, Bashan A, Gat S, Agmon I, Bartels H, Franceschi F, and Yonath (2001). A high resolution structure of the large ribosomal subunit from a mesophilic eubacterium. Cell 107:679-688.

Harnpicharnchai P, Jakovljevic J, Horsey E, Miles T, Roman J, Rout M, Meagher D, Imai B, Guo Y, Brame CJ, Shabanowitz J, Hunt DF, and Woolford JL Jr (2001). Composition and functional characterization of yeast 66S ribosome assembly intermediates. Mol Cell 8:505-515.

Henras A, Henry Y, Bousquet-Antonelli C, Noaillac-Depeyre J, Gelugne JP, and Caizergues-Ferrer M (1998) Nhp2p and Nop10p are essential for the function of H/ACA snoRNPs. EMBO J 17:7078-7090.

Henry Y, Wood H, Morrissey JP, Petfalski E, Kearsey S, and Tollervey D (1994). The 5' end of yeast 5.8S rRNA is generated by exonucleases from an upstream cleavage site. EMBO J 13:2452-2463.

Hershkovitz MA, and Zimmer EA (1996). Conservation patterns in angiosperm rDNA ITS2 sequences. Nucleic Acids Res 24:2857-2867.

Heyer WD, Johnson AW, Reinhart U, and Kolodner RD (1995). Regulation and intracellular localization of *Saccharomyces cerevisiae* strand exchange protein 1 (Sep1/Xrn1/Kem1), a multifunctional exonuclease. Mol Cell Biol 15:2728-2736.

Hitchen J, Ivakine E, Melekhovets YF, Lalev A, and Nazar RN (1997). Structural features in the 3' external transcribed spacer affecting intragenic processing of yeast rRNA. J Mol Biol 274:481-490.

Ho JH, Kallstrom G, and Johnson AW (2000). Nmd3p is a Crm1p-dependent adapter protein for nuclear export of the large ribosomal subunit. J Cell Biol. 151:1057-1066.

Hong B, Brockenbrough JS, Wu P,and Aris JP (1997). Nop2p is required for pre-rRNA processing and 60S ribosome subunit synthesis in yeast. Mol Cell Biol 17:378-388.

Hughes JM, and Ares M Jr (1991). Depletion of U3 small nucleolar RNA inhibits cleavage in the 5' external transcribed spacer of yeast pre-ribosomal RNA and impairs formation of 18S ribosomal RNA. EMBO J 10:4231-4239.

Hurt E, Hannus S, Schmelzl B, Lau D, Tollervey D, and Simos G (1999). A novel in vivo assay reveals inhibition of ribosomal nuclear export in ran-cycle and nucleoporin mutants. J Cell Biol 144:389-401.

Intine RV, Good L, and Nazar RN (1999). Essential structural features in the *Schizosaccharomyces pombe* pre-rRNA 5' external transcribed spacer. J Mol Biol 286:695-708.

Jansen R, Tollervey D, and Hurt EC (1993). A U3 snoRNP protein with homology to splicing factor PRP4 and G beta domains is required for ribosomal RNA processing. EMBO J 12:2549-2558.

Johnson AW (1997). Rat1p and Xrn1p are functionally interchangeable exoribonucleases that are restricted to and required in the nucleus and cytoplasm, respectively. Mol Cell Biol 17:6122-6130.

Kass S, Tyc K, Steitz JA, and Sollner-Webb B (1990). The U3 small nucleolar ribonucleoprotein functions in the first step of preribosomal RNA processing. Cell 60:897-908.

Kempers-Veenstra AE, Oliemans J, Offenberg H, Dekker AF, Piper PW, Planta RJ, and Klootwijk J (1986). 3'-End formation of transcripts from the yeast rRNA operon. EMBO J 5:2703-2710.

Kiss-Laszlo Z, Henry Y, Bachellerie JP, Caizergues-Ferrer M, and Kiss T (1996). Site-specific ribose methylation of preribosomal RNA: a novel function for small nucleolar RNAs. Cell 85:1077-1088.

Koonin EV (1996). Pseudouridine synthases: four families of enzymes containing a putative uridine-binding motif also conserved in dUTPases and dCTP deaminases. Nucleic Acids Res 24:2411-2415.

Kressler D, de la Cruz J, Rojo M, and Linder P (1997). Fal1p is an essential DEAD-box protein involved in 40S-ribosomal-subunit biogenesis in *Saccharomyces cerevisiae*. Mol Cell Biol 17:7283-7294.

Kressler D, de la Cruz J, Rojo M, and Linder P (1998). Dbp6p is an essential putative ATP-dependent RNA helicase required for 60S-ribosomal-subunit assembly in *Saccharomyces cerevisiae*. Mol Cell Biol 18:1855-1865.

Kressler D, Linder P, and de La Cruz J (1999). Protein trans-acting factors involved in ribosome biogenesis in *Saccharomyces cerevisiae*. Mol Cell Biol 19:7897-7912.

Kressler D, Rojo M, Linder P, and de la Cruz J (1999). Spb1p is a putative methyltransferase required for 60S ribosomal subunit biogenesis in *Saccharomyces cerevisiae*. Nucleic Acids Res 27:4598-5608.

Kruiswijk T, Planta RJ, and Krop JM (1978). The course of the assembly of ribosomal subunits in yeast. Biochim Biophys Acta 517:378-389.

Kufel J, Dichtl B, and Tollervey D (1999) Yeast Rnt1p is required for cleavage of the pre-ribosomal RNA in the 3' ETS but not the 5' ETS. RNA 5:909-1017.

Kufel J, Allmang C, Petfalski E, Beggs J, and Tollervey D (2002). Lsm proteins are required for normal processing and stability of ribosomal RNAs. J Biol Chem in press.

Lafontaine D, Delcour J, Glasser AL, Desgres J, and Vandenhaute J (1994). The DIM1 gene responsible for the conserved m6(2)Am6(2)A dimethylation in the 3'-terminal loop of 18 S rRNA is essential in yeast. J Mol Biol 241:492-497.

Lafontaine D, Vandenhaute J, and Tollervey D (1995). The 18S rRNA dimethylase Dim1p is required for pre-ribosomal RNA processing in yeast. Genes Dev 9:2470-2481.

Lafontaine DL, Preiss T, and Tollervey D (1998). Yeast 18S rRNA dimethylase Dim1p: a quality control mechanism in ribosome synthesis? Mol Cell Biol 18:2360-2370.

Lalev AI, and Nazar RN (1998). Conserved core structure in the internal transcribed spacer 1 of the *Schizosaccharomyces pombe* precursor ribosomal RNA. J Mol Biol 284:1341-1351.

Lalev AI, and Nazar RN (1999). Structural equivalence in the transcribed spacers of pre-rRNA transcripts in *Schizosaccharomyces pombe*. Nucleic Acids Res 1999 27:3071-3078.

Lalev AI, and Nazar RN (2001). A chaperone for ribosome maturation. J Biol Chem 276:16655-16659.

Lalev AI, Abeyrathne PD, and Nazar RN (2000). Ribosomal RNA maturation in *Schizosaccharomyces pombe* is dependent on a large ribonucleoprotein complex of the internal transcribed spacer 1. J Mol Biol 302:65-77.

Lane BG, and Tamaoki T (1967). Studies of the chain termini and alkali-stable dinucleotide sequences in 16 s and 28 s ribosomal RNA from L cells. J Mol Biol 27:335-348.

Lane BG, Ofengand J, and Gray MW (1992). Pseudouridine in the large-subunit (23 S-like) ribosomal RNA. The site of peptidyl transfer in the ribosome? FEBS Lett 302:1-4.

Lane BG, Ofengand J, and Gray MW (1995). Pseudouridine and O2'-methylated nucleosides. Significance of their selective occurrence in rRNA domains that function in ribosome-catalyzed synthesis of the peptide bonds in proteins. Biochimie 77:7-15.

Larimer FW, and Stevens A (1990). Disruption of the gene XRN1, coding for a 5'→3' exoribonuclease, restricts yeast cell growth. Gene 95:85-90.

Lee MS, Henry M, and Silver PA (1996). A protein that shuttles between the nucleus and the cytoplasm is an important mediator of RNA export. Genes Dev 10:1233-1246.

Lee WC, Xue ZX, and Melese T (1991). The NSR1 gene encodes a protein that specifically binds nuclear localization sequences and has two RNA recognition motifs. J Cell Biol 113:1-12.

Lee WC, Zabetakis D, and Melese T (1992). NSR1 is required for pre-rRNA processing and for the proper maintenance of steady-state levels of ribosomal subunits. Mol Cell Biol 12:3865-3871.

Lee Y, Melekhovets YF, and Nazar RN (1995). Termination as a factor in "quality control" during ribosome biogenesis. J Biol Chem 270:28003-28005.

Lee Y, and Nazar RN (1997). Ribosomal 5 S rRNA maturation in *Saccharomyces cerevisiae*. J Biol Chem 272:15206-15212.

Li HV, Zagorski J, and Fournier MJ (1990). Depletion of U14 small nuclear RNA (snR128) disrupts production of 18S rRNA in *Saccharomyces cerevisiae*. Mol Cell Biol 10:1145-1152.
Liang WQ, and Fournier MJ (1995). U14 base-pairs with 18S rRNA: a novel snoRNA interaction required for rRNA processing. Genes Dev 9:2433-2443.
Liang WQ, Clark JA, and Fournier MJ (1997). The rRNA-processing function of the yeast U14 small nucleolar RNA can be rescued by a conserved RNA helicase-like protein. Mol Cell Biol 17:4124-4132.
Lygerou Z, Allmang C, Tollervey D, and Seraphin B (1996). Accurate processing of a eukaryotic precursor ribosomal RNA by ribonuclease MRP in vitro. Science 272:268-270.
Maden BE (1990). The numerous modified nucleotides in eukaryotic ribosomal RNA. Prog Nucleic Acid Res Mol Biol 39:241-303.
Mager WH, and Planta RJ (1991). Coordinate expression of ribosomal protein genes in yeast as a function of cellular growth rate. Mol Cell Biochem. 104:181-187.
Maxwell ES, and Fournier MJ (1995). The small nucleolar RNAs. Annu Rev Biochem 64:897-934.
Melekhovets YF, Good L, Elela SA, and Nazar RN (1994). Intragenic processing in yeast rRNA is dependent on the 3' external transcribed spacer. J Mol Biol 239:170-180.
Michot B, Joseph N, Mazan S, and Bachellerie JP (1999). Evolutionarily conserved structural features in the ITS2 of mammalian pre-rRNAs and potential interactions with the snoRNA U8 detected by comparative analysis of new mouse sequences. Nucleic Acids Res 27:2271-2282.
Mitchell P, Petfalski E, Shevchenko A, Mann M, and Tollervey D (1997). The exosome: a conserved eukaryotic RNA processing complex containing multiple 3'→5' exoribonucleases. Cell 91:457-466.
Mitchell P, Petfalski E, and Tollervey D (1996). The 3' end of yeast 5.8S rRNA is generated by an exonuclease processing mechanism. Genes Dev 10:502-513.
Mougey EB, O'Reilly M, Osheim Y, Miller OL Jr, Beyer A, and Sollner-Webb B (1993). The terminal balls characteristic of eukaryotic rRNA transcription units in chromatin spreads are rRNA processing complexes.
Moy TI, and Silver PA (1999). Nuclear export of the small ribosomal subunit requires the ran-GTPase cycle and certain nucleoporins. Genes Dev. 13:2118-2133.
Munholland JM, and Nazar RN (1987). Methylation of ribosomal RNA as a possible factor in cell differentiation. Cancer Res 47:169-172.
Musters W, Boon K, van der Sande CA, van Heerikhuizen H, and Planta RJ (1990). Functional analysis of transcribed spacers of yeast ribosomal DNA. EMBO J 9:3989-3996.
Musters W, Venema J, van der Linden G, van Heerikhuizen H, Klootwijk J, and Planta RJ (1989). A system for the analysis of yeast ribosomal DNA mutations. Mol Cell Biol 9:551-559.
Genes Dev 7:1609-1619.
Nazar RN, Sitz TO, and Busch H (1975). Tissue specific differences in the 2'-O-methylation of eukaryotic 5.8S ribosomal RNA. FEBS Lett 59:83-87.
Nazar RN, Lo AC, Wildeman AG, and Sitz TO (1983). Effect of 2'-O-methylation on the structure of mammalian 5.8S rRNAs and the 5.8S-28S rRNA junction. Nucleic Acids Res 11:5989-6001.
Nazar RN, Wong WM, and Abrahamson JL (1987) Nucleotide sequence of the 18-25 S ribosomal RNA intergenic region from a thermophile, *Thermomyces lanuginosus*. J Biol Chem 262:7523-7527.
Nazar RN, Good L, Intine RVA, Lee Y, and Melekhovets YF (1996). RNA processing as a "quality control" factor in ribosome biogenesis. Abstracts, RNA'96, First Annual Meeting of the RNA Society, Madison, pp503-503.
Ni J, Tien AL, and Fournier MJ (1997). Small nucleolar RNAs direct site-specific synthesis of pseudouridine in ribosomal RNA. Cell 89:565-573.
Nicoloso M, Qu LH, Michot B, and Bachellerie JP (1996). Intron-encoded, antisense small nucleolar RNAs: the characterization of nine novel species points to their direct role as guides for the 2'-O-ribose methylation of rRNAs. J Mol Biol 260:178-195.
Nogi Y, Yano R, and Nomura M (1991). Synthesis of large rRNAs by RNA polymerase II in mutants of Saccharomyces cerevisiae defective in RNA polymerase I. Proc Natl Acad Sci U S A 88:3962-3966.
O'Day CL, Chavanikamannil F, and Abelson J (1996). 18S rRNA processing requires the RNA helicase-like protein Rrp3. Nucleic Acids Res 24:3201-3207.
Ofengand J, and Bakin A (1997). Mapping to nucleotide resolution of pseudouridine residues in large subunit ribosomal RNAs from representative eukaryotes, prokaryotes, archaebacteria, mitochondria and chloroplasts. J Mol Biol 266:246-268.
Ohtake Y, and Wickner RB (1995). Yeast virus propagation depends critically on free 60S ribosomal subunit concentration. Mol Cell Biol 15:2772-2781.
Peculis BA, and Steitz JA (1993). Disruption of U8 nucleolar snRNA inhibits 5.8S and 28S rRNA processing in the Xenopus oocyte. Cell 73:1233-1245.

Piper PW, Bellatin JA, and Lockheart A (1983). Altered maturation of sequences at the 3' terminus of 5S gene transcripts in a *Saccharomyces cerevisiae* mutant that lacks a RNA processing endonuclease. EMBO J 2:353-359.
Planta RJ, and Mager WH (1998). The list of cytoplasmic ribosomal proteins of *Saccharomyces cerevisiae*. Yeast 14:471-477.
Raue HA, and Planta RJ (1991) Ribosome biogenesis in yeast. Prog Nucleic Acid Res Mol Biol 41:89-129.
Raue HA (2002). A novel pathway for ITS1 processing in yeast that bypasses both the A2 and A3 sites. Abstracts, The Dynamics of Ribosome Structure and Function: Ribosome Meeting 2002, Queenstown, pp 79-79.
Ripmaster TL, Vaughn GP, and Woolford JL Jr (1993). DRS1 to DRS7, novel genes required for ribosome assembly and function in *Saccharomyces cerevisiae*. Mol Cell Biol 13:7901-7912.
Ro-Choi TS (1997). Nucleolar snoRNA and ribosome production. Mol Cells 7:451-467.
Rotondo G, and Frendewey D (1996). Purification and characterization of the Pac1 ribonuclease of *Schizosaccharomyces pombe*. Nucleic Acids Res 24:2377-2386.
Rottman F, Friderici K, Comstock P, and Khan MK (1974). Influence of 2'-O-alkylation on the structure of single-stranded polynucleotides and the stability of 2'-O-alkylated polynucleotide complexes. Biochemistry 13:2762-2771.
Sachs AB, and Davis RW (1989). The poly(A) binding protein is required for poly(A) shortening and 60S ribosomal subunit-dependent translation initiation. Cell 58:857-867.
Savino R, and Gerbi SA (1990). In vivo disruption of *Xenopus* U3 snRNA affects ribosomal RNA processing. EMBO J 9:2299-2308.
Schimmang T, Tollervey D, Kern H, Frank R, and Hurt EC (1989). A yeast nucleolar protein related to mammalian fibrillarin is associated with small nucleolar RNA and is essential for viability. EMBO J 8:4015-4024.
Schmitt ME, and Clayton DA (1993). Nuclear RNase MRP is required for correct processing of pre-5.8S rRNA in *Saccharomyces cerevisiae*. Mol Cell Biol 13:7935-7941.
Schmitt ME, and Clayton DA (1994). Characterization of a unique protein component of yeast RNase MRP: an RNA-binding protein with a zinc-cluster domain. Genes Dev 8:2617-2628.
Senger B, Lafontaine DL, Graindorge JS, Gadal O, Camasses A, Sanni A, Garnier JM, Breitenbach M, Hurt E, and Fasiolo F (2001). The nucle(ol)ar Tif6p and Efl1p are required for a late cytoplasmic step of ribosome synthesis.
Mol Cell 8:1363-1373.
Sharma K, Venema J, and Tollervey D. (1999). The 5' end of the 18S rRNA can be positioned from within the mature rRNA. RNA 5:678-86.
Siegmund A, Grant A, Angeletti C, Malone L, Nichols JW, and Rudolph HK (1998). Loss of Drs2p does not abolish transfer of fluorescence-labeled phospholipids across the plasma membrane of *Saccharomyces cerevisiae*. J Biol Chem 273:34399-34405.
Singh H, and Lane BG (1964). The separation, estimation, and characterization of alkali-stable oligonucleotides derived from commercial ribonucleate preparations. Can J Biochem 42:87-93.
Sirum-Connolly K, and Mason TL (1993). Functional requirement of a site-specific ribose methylation in ribosomal RNA. Science 262:1886-1889.
Smith CM, and Steitz JA (1997). Sno storm in the nucleolus: new roles for myriad small RNPs. Cell 89:669-672.
Song X, and Nazar RN (2002). Modification of rRNA as a 'quality control mechanism' in ribosome biogenesis. FEBS Lett 523:182-186.
Spasov K, Perdomo LI, Evakine E, and Nazar RN (2002). RAC protein directs the complete removal of the 3' external transcribed spacer by the Pac1 nuclease. Mol Cell 9:433-437.
Stage-Zimmermann T, Schmidt U, and Silver PA (2000). Factors affecting nuclear export of the 60S ribosomal subunit *in vivo*. Mol Biol Cell 11:3777-3789.
Subrahmanyam CS, Cassidy B, Busch H, and Rothblum LI (1982). Nucleotide sequence of the region between the 18S rRNA sequence and the 28S rRNA sequence of rat ribosomal DNA. Nucleic Acids Res 10:3667-3680.
Sun C, and Woolford JL Jr (1994). The yeast NOP4 gene product is an essential nucleolar protein required for pre-rRNA processing and accumulation of 60S ribosomal subunits. EMBO J 13:3127-3135.
Tague BW, and Gerbi SA (1984). Processing of the large rRNA precursor: two proposed categories of RNA-RNA interactions in eukaryotes. J Mol Evol 20:362-367.
Tollervey D (1987). A yeast small nuclear RNA is required for normal processing of pre-ribosomal RNA. EMBO J 6:4169-4175.

Tollervey D, Lehtonen H, Jansen R, Kern H, and Hurt EC (1993). Temperature-sensitive mutations demonstrate roles for yeast fibrillarin in pre-rRNA processing, pre-rRNA methylation, and ribosome assembly. Cell 72:443-457.

Tuteja R, and Tuteja N (1998). Nucleolin: a multifunctional major nucleolar phosphoprotein. Crit Rev Biochem Mol Biol 33:407-436.

Tycowski KT, Shu MD, and Steitz JA (1994). Requirement for intron-encoded U22 small nucleolar RNA in 18S ribosomal RNA maturation. Science 266:1558-1561.

Udem SA, and Warner JR. (1973). The cytoplasmic maturation of a ribosomal precursor ribonucleic acid in yeast. J Biol Chem 248:1412-1416.

Ursic D, Himmel KL, Gurley KA, Webb F, and Culbertson MR (1997). The yeast SEN1 gene is required for the processing of diverse RNA classes. Nucleic Acids Res. 25:4778-4785.

Valasek L, Hasek J, Nielsen KH,and Hinnebusch AG(2001). Dual function of eIF3j/Hcr1p in processing 20 S pre-rRNA and translation initiation. J Biol Chem 276:43351-43360.

van der Sande CA, Kwa M, van Nues RW, van Heerikhuizen H, Raue HA, and Planta RJ (1992). Functional analysis of internal transcribed spacer 2 of *Saccharomyces cerevisiae* ribosomal DNA. J Mol Biol 223:899-910.

van Nues RW, Rientjes JM, van der Sande CA, Zerp SF, Sluiter C, Venema J, Planta RJ, and Raue HA (1994). Separate structural elements within internal transcribed spacer 1 of *Saccharomyces cerevisiae* precursor ribosomal RNA direct the formation of 17S and 26S rRNA. Nucleic Acids Res 22:912-919.

van Nues RW, Rientjes JM, Morre SA, Mollee E, Planta RJ, Venema J, and Raue HA (1995a). Evolutionarily conserved structural elements are critical for processing of Internal Transcribed Spacer 2 from *Saccharomyces cerevisiae* precursor ribosomal RNA. J Mol Biol 250:24-36.

van Nues RW, Venema J, Rientjes JM, Dirks-Mulder A, and Raue HA (1995b). Processing of eukaryotic pre-rRNA: the role of the transcribed spacers. Biochem Cell Biol 73:789-801.

van Nues RW, Venema J, Planta RJ, and Raue HA (1997). Variable region V1 of *Saccharomyces cerevisiae* 18S rRNA participates in biogenesis and function of the small ribosomal subunit. Chromosoma 105:523-531.

Vanrobays E, Gleizes PE, Bousquet-Antonelli C, Noaillac-Depeyre J, Caizergues-Ferrer M, and Gelugne JP (2001). Processing of 20S pre-rRNA to 18S ribosomal RNA in yeast requires Rrp10p, an essential non-ribosomal cytoplasmic protein. EMBO J 20:4204-4213.

Van Ryk DI, Lee Y, Nazar RN (1992). Unbalanced ribosome assembly in *Saccharomyces cerevisiae* expressing mutant 5 S rRNAs. J Biol Chem 267:16177-16181

Vaughan MH Jr, Soeiro R, Warner JR, and Darnell JE Jr (1967). The effects of methionine deprivation on ribosome synthesis in HeLa cells. Proc Natl Acad Sci U S A 58:1527-1534.

Venema J, Henry Y, and Tollervey D (1995). Two distinct recognition signals define the site of endonucleolytic cleavage at the 5'-end of yeast 18S rRNA. EMBO J 14:4883-4892.

Venema J, Bousquet-Antonelli C, Gelugne JP, Caizergues-Ferrer M, and Tollervey D (1997). Rok1p is a putative RNA helicase required for rRNA processing. Mol Cell Biol 17:3398-3407.

Venema J, and Tollervey D (1999). Ribosome synthesis in *Saccharomyces cerevisiae*. Annu Rev Genet 33:261-311.

Vilardell J, and Warner JR (1997). Ribosomal protein L32 of Saccharomyces cerevisiae influences both the splicing of its own transcript and the processing of rRNA. Mol Cell Biol 17:1959-1965.

Walker K, Wong WM, and Nazar RN (1990). Termination region in rRNA genes from a eucaryotic thermophile, Thermomyces lanuginosus. Mol Cell Biol 10:377-381.

Warner JR (1999). The economics of ribosome biosynthesis in yeast. Trends Biochem Sci 24:437-440.

Weaver PL, Sun C, and Chang TH (1997). Dbp3p, a putative RNA helicase in *Saccharomyces cerevisiae*, is required for efficient pre-rRNA processing predominantly at site A3. Mol Cell Biol 17:1354-1365.

Wolin SL, and Cedervall T (2000). The la protein. Annu Rev Biochem 71:375-403.

Woolford JL, and Warner JR (1991). The ribosome and its synthesis. P.587-626. In J.R. Broach, JR Pringle and EW Jones (ed.) The molecular and cellular biology of the yeast Saccharomyces: genome dynamics, protein synthesis, and energetics. Vol. 1. Cold Spring Harbor Laboratory Press. Cold Spring Harbor, N.Y.

Wu P, Brockenbrough JS, Paddy MR, and Aris JP (1998). NCL1, a novel gene for a non-essential nuclear protein in *Saccharomyces cerevisiae*. Gene 220:109-117.

Yeh LC, and Lee JC (1990). Structural analysis of the internal transcribed spacer 2 of the precursor ribosomal RNA from *Saccharomyces cerevisiae*. J Mol Biol 211:699-712.

Yeh LC, Thweatt R, and Lee JC (1990). Internal transcribed spacer 1 of the yeast precursor ribosomal RNA. Higher order structure and common structural motifs. Biochemistry 29:5911-5918.

Yusupov MM, Yusupova GZ, Baucom A, Lieberman K, Earnest TN, Cate JH, and Noller HF (2000). Crystal structure of the ribosome at 5.5 A resolution. Science 292:883-896.

Zanchin NI, and Goldfarb DS (1999). Nip7p interacts with Nop8p, an essential nucleolar protein required for 60S ribosome biogenesis, and the exosome subunit Rrp43p. Mol Cell Biol 19:1518-1525.

Applied Mycology & Biotechnology
An International Series. Volume 3. Fungal Genomics

9

Fungal Pathogenicity Genes

Paul Tudzynski[1] and Amir Sharon[2]
[1]Institut für Botanik, Schlossgarten 3, D-48149 Münster, Germany (tudzyns@uni-muenster.de)
[2]Department of Plant Sciences, Tel Aviv University, Tel Aviv 69978, Israel.

Molecular genetic tools in recent years allowed the identification and detailed functional analysis of genes involved in the interplay of pathogenic fungi and their host plants. In the focus of interest today are genes involved in signaling events which accompany and control all stages in the infection and colonization processes. From the view point of developing chemical control strategies, specially the early events in the interaction (i.e. the surface-bound events) are of interest. Further milestones are genes controlling the different life styles of fungi (bio-/necrotroph) and genes involved in overcoming/suppressing the host's defense. In addition, "black box" approaches based on genomic data have provided sets of new genes obviously involved in the interaction, but where the details of function have yet to be worked out.

1. INTRODUCTION

The factors influencing the interaction of pathogenic fungi and their hosts have been a major research topic in the fungal community in recent years. These detailed investigations have been fuelled by the necessity to develop new strategies for the control of these economically highly important organisms; inspite of strong efforts to develop and introduce new fungicides and resistant plant varieties, losses due to fungal diseases especially in agriculture are a growing stimulus for basic research in this field. An essential cue in this ongoing battle is the search for fungicide targets via the identification of pathogenicity determinants, encoded by pathogenicity genes. The definitions of pathogenicity genes are manifold; we will follow here the definition of Idnurm and Howlett (2001), which described pathogenicity genes "as genes necessary for desease development, but not essential for the pathogen to complete its lifecycle *in vitro*". We are aware of the problem to apply this definition on biotrophic (non-culturable) fungi. We definitely will not deal here with basic "housekeeping" genes (e.g. aminoacid metabolism, etc.), though for practical purposes (definition of targets for fungicides) also basic genes might be of high importance. We will not differentiate between pathogenicity genes (yes or no) and virulence genes (modulating desease severity), because these can depend on the host variety, age and type of tissue involved and on external conditions. And, finally, we will deal mainly with phytopathogens, according to our expertise.

We will focus on genes, which have been proved to have impact on pathogenicity, and this usually involves functional analysis by molecular techniques (disruption or, alternatively, enhanced expression upon contact with host or in infection structure). We will not review the literature on AVR genes and other fungal mechanisms that suppress disease, although such genes may be regarded as pathogenicity factors. The standard approach in the past to identify pathogenicity genes was forward genetics; i.e. indirect evidence for a factor being a pathogenicity determinant (biochemical and/or genetic data) was tested via the isolation and deletion of the corresponding gene. In parallel, insertional mutagenesis approaches (REMI, transposon tagging, T-DNA) have yielded a wealth of pathogenicity mutants in several systems and have led to several new, partly unexpected pathogenicity factors. The genomics approaches give additional, seemingly unlimited perspectives (Soanes *et al.* 2002); comparative genomic analyses can help to find answers to the old question, what is the difference between a saprophyte and a parasite. EST-analyses allow the comparison of gene sets expressed during pathogenesis and this can help to identify genes common to groups of pathogens (same host, same organ, etc.). Such comparative studies also can help to overcome one of the major obstacles of genomic data: the high percentage of ORFs, i.e. putative genes that show no homology to any annotated gene. ORFs that are present (and timely expressed) in more than one pathogen could be worth to be studied functionally.

One major problem for unequivocal identification of pathogenicity determinants is function redundancy, which means that the function of a gene could be taken over by others, even if the analysed gene seems to be a single copy one without apparent paralogs in the genome. Therefore attention has been focused recently on transcription factors and signal chain components that controll whole sets of pathogenicity genes. Important information can also be obtained by comparison of the different life-style of fungi, e.g. the role of homologous genes in biotrophic and necrotrophic fungi (see e.g. the AOS data below), and more important, the differential expression of genes in biotrophic and necrotrophic phases of the same fungus. The same holds true for the comparative analysis of mutualistic interactions (mycorrhiza, some endophytes), pinpointing the differences and the homologies to pathogenic systems. In this chapter we will try to present an overview of this rapidly expanding field. We do not intend to provide a complete list of pathogenicity genes, since new genes are added monthly. Instead, we will mention some genes in each category, update versus previous reviews, and discuss in some more details genes for which there is more information that can demonstrate general trends. For comprehensive listing of fungal pathogenicity elements readers are referred to recent reviews on the subject (Gold *et al.* 2001; Oliver and Osbourn 1995; Idnurm and Howlett 2001; Tudzynski and Tudzynski 1999, 2001; Yoder and Turgeon, 1996; Yoder and Turgeon 2001) as well as to relevant chapters in this volume.

2. SIGNALING

The success of a fungal pathogen depends to a high degree on its ability to perceive and to respond to signals generated by the plant, especially in the very early stages of infection (recognition), but also in later stages involving different cell types/tissues. Despite the rapidly increasing number of cloned signal component genes, the initial events of sensing of extracellular signals and transduction into an intracellular signal are still poorly understood. The binding of signal ligands to cell-surface receptors triggers a conformational change of receptors, e.g. in the case of heterotrimeric G proteins by dissociation of the Gα subunit from the ß- subunits which activates or inhibits appropriate target effectors such as protein kinases, adenylate cyclases, phospholipases, and ion channels (Kronstadt 1997). Components of signal chains have been studied recently in several pathogenic fungi, focusing on MAP kinase cascades, the classical cAMP pathway (heterotrimeric G-proteins, adenylate cyclase, cAMP-dependent proteinkinase A), and the crosstalk between them. Only in very few model systems these investigations have followed up whole signal chains (including receptors and downstream components), in most cases single components were functionally analysed to pinpoint pathogenicity – related cascades. The data obtained so far allow some general conclusions: (1) There are several examples for signal pathways involved only in pathogenicity; i.e. deletions of the corresponding genes do not effect vegetative properties *in vitro*; (2) Single components (like the different Gα subunits or MAPK) are highly conserved, even highly homologous to mammalian systems (3) The components of a given signal chain might differ considerably between fungi; (4)The same (or highly homologous) components can be members of cascades regulating different downstream components (see the results of MAPK knockouts in several fungi). In the following a few selected aspects of this field of research, which has developed rapidly into one of the major foci of molecular phytopathology, will be presented.

2.1 Receptors

Fungi undergo specific differentiation and developmental processes in response to distinct physical and chemical environmental signals. All these events start with an initial "recognition phase" in which specific receptors play an important role by detecting surface components or other ligands and transmitting this information to one or more downstream signaling pathways. So far only one fungal gene encoding a pathogenicity-related transmembrane receptor protein, *PTH11*, was described. It was identified by a REMI-mutagenesis approach in the rice blast fungus *Magnaporthe grisea* (De Zwaan *et al.* 1999). The REMI mutant tagged in the *PTH11* gene was almost fully apathogenic (drastic reduction of appressoria formation). The predicted secondary structure of Pth11p suggested that it is an integral membrane protein; this was confirmed by *in situ* localization experiments using a PTH11-GFP-gene fusion. Eukaryotic serpentine receptors have typically seven transmembrane domains (Bockaert and Pin 1999), whereas Pth11p appears to have nine, suggesting an atypical structure. Exogenous cAMP suppressed defects associated with *pth11* mutants, suggesting that Pth11p mediates cellular response through the cAMP pathway.

2.2 Heterotrimeric GTP-binding proteins (G-Proteins)

The importance of heterotrimeric G proteins in regulating diverse processes such as differentiation, mating, and pathogenicity has been demonstrated in a number of phytopathogenic fungi (a recent compilation in Tudzynski and Tudzynski 2001 lists 7 species). In most cases two or more Gα subunit genes were detected, only one of which had significant influence on pathogenicity e.g., *CPG-1* from the chesnut blight fungus *Cryphonectria parasitica* (Choi *et al.* 1995) and *MAGB* from *M. grisea* (Liu and Dean 1997). The defects linked to Gα knockouts are manifold: e.g., in *ctg-1* mutants of *Colletotrichum trifolii* the conidia fail to germinate, demonstrating the requirement of this Gα- subunit for a very early stage in the life cycle of this pathogen (Truesdell *et al.* 2000). *cga1* mutants of the northern corn leaf blight fungus *Cochliobolus heterostrophus* show reduced ability to form appressoria on glass surfaces and corn leaves, but nevertheless are able to induce lesions; in addition, CGA1 appears to be involved in mating: mutants are female sterile (Horwitz *et al.* 1999). In the gray mould fungus *Botrytis cinerea*, two Gα genes, *BCG1* and *BCG2*, were functionally characterized (Tudzynski *et al.* 2000). Both genes are expressed *in planta* at very early stages of infection. Knock-out-mutants of both genes caused wild type-like primary necrotic lesions in the first hours of infection on bean and tomato leaves. However, after two days, no further development was observed for the lesions caused by the *bcg1* mutants, whereas *bcg2*-mutants produced spreading secondary lesions, albeit retarded.

Several of the pathogenesis-related Gα mutants also show defects in vegetative parameters like reduced growth rate and altered colony morphology. Recently it was shown that an *Aspergillus nidulans* Gα subunit, *FADA*, is involved in the regulation of chitin content and porosity of the cell wall and in susceptibility to osmotin (Coca *et al.* 2000). This could explain the pleiomorphic phenotypes of various Gα mutants. Heterotrimeric G proteins can be connected with MAP kinase cascades or the cAMP pathway. Interestingly, the pathogenesis-related Gα subunits all belong to the $G\alpha_i$ class of the mammalian classification system, which in mammalian cells act as inhibitors of adenylate cyclase activity. In fungi, however, most of them have a stimulatory effect on adenylate cyclase: external cAMP led to reversion of appressorium development in a *M. grisea magb* mutant (Liu and Dean 1997) and fully recovered the wild-type colony morphology in *B. cinerea bcg1* mutants (Tudzynski *et al.* 2000). The only exception so far is the induction of appressoria formation in the cereal eyespot disease fungus *Tapesia yallundae*, which is elicited by mechanical pressure; signal transfer obviously is mediated by heterotrimeric G proteins and involves a reduced cAMP level (Bowyer, pers.communic.); a functional analysis (knock out) in this system is necessary to confirm these data.

2.3 cAMP Signaling Pathways

The cAMP signaling pathway in phytopathogenic fungi has been analyzed in detail in the past years. It has been shown that it plays a crucial role during pathogenic development in all systems analyzed so far. Fungal strains in which cAMP signaling is blocked at different levels are disturbed at distinct stages of the infection process *in planta* (see Tudzynski and Tudzynski 1999, 2001). It appears that especially the early infection stages such as spore germination, appressorium formation and penetration require an intact cAMP signaling pathway. In *M. grisea* cAMP-dependent protein kinase (PKA) activity increases during germination of condia and appressorium formation on hydrophobic surfaces. In *cpka* mutants (lacking the catalytic subunit of a PKA) appressoria formation is impaired (Kang *et al.* 1999). It could be shown that compartmentalization and rapid degradation of storage carbohydrate (glycogen) and lipid reserves, processes which are essential for the generation of turgor in appressoria, are controlled by the *CPKA/SUM1*-encoded PKA (Thines *et al.* 2000). Also, mutations in the other cAMP pathway components have drastic effect on early infection processes (Tudzynski and Tudzynski 2001).

In the corn smut fungus *Ustilago maydis* the cAMP pathway is needed not only for the early stages of infection, but also for subsequent fungal development *in planta* (Krüger *et al.* 2000). The components of the cAMP pathway involved in pathogenicity are an activating Gα subunit, *GPA3*, an adenylate cyclase, *UAC1*, and regulatory and catalytic subunits of a protein kinase A, *UBC1*, and *ADR1*, respectively.

Also biotrophic pathogens become now accessible for signalling studies: A catalytic subunit of PKA was recently cloned as an expressed sequence tag from the causal agent of barley powdery mildew, the obligate biotroph *Erysiphe* (*Blumeria*) *graminis* f. sp. *hordei* (Hall *et al.* 1999). In contrast to *M. grisea* and *C. trifolii*, appressoria differentiation is not induced by a single cAMP-mediated signal, such as contact with a hydrophobic surface (Lee and Dean 1993) or host cutin-derived compounds (Gilbert *et al.* 1996), but requires a complex series of external signals (Hall and Gurr 2000). Both, cAMP and 8-Br-cAMP are able to activate and inactivate PKA activity during appressoria differentiation demonstrating different requirements for cAMP signaling during the differentiation process. So far functional analyses by targeted inactivation cannot be performed in this strictly biotrophic fungus; however, the *B. graminis PKA-C* gene can complement the *cpkA* mutation of *M.grisea*, strongly suggesting a common function (Bindsev *et al.* 2001).

2.4 MAP Kinases

Mitogen-activated protein (MAP) kinases, a special family of serine/threonine protein kinases, are known to mediate the adjustment of intracellular activities of eukaryotic cells to environmental changes; they are activated by a MAP-kinase cascade, which has been shown to be highly conserved in a wide variety of eukaryotic organisms (Schaeffer and Webber

1999). In *Saccharomyces cerevisiae* five MAP kinase mediated signal transduction pathways have been identified, three of which have been shown to be also active in filamentous fungi: Fus3/Kss1 is involved (in *S. cerevisiae*) in mating responses and filamentous growth, Slt2 is responsible for cell integrity, and Hog1 for stress response (especially high osmotic pressure). Homologues of these 3 MAP kinases have been described in *M. grisea* (summarized in Xu 2000): *PMK1* (*Fus3*), *OSM1* (*Hog1*), and *MPS1* (*Slt2*). Only *PMK1* and *MPS1* have been shown to be involved in pathogenicity of *M. grisea*: *pmk1* mutants are impaired in appressorial formation, penetration and invasive growth; *mps1* mutants are defective in penetration but are able to invade the host's tissue after wounding. The sequences of MAPK genes are now available from a large number of fungi, which mostly can be grouped into three clades corresponding to the 3 *M. grisea* enyzmes (see Fig. 1). Many of the corresponding genes of phytopathogens have been functionally characterized (table 1); *PMK1* homologues were identified in many fungi and were shown to be essential for the infection process in several foliar pathogens (*Cochliobolus heterostrophus, Colletotrichum lagenarium, Pyrenophora teres* ; Lev *et al.* 1999; Takano *et al.* 2000; Ruiz-Roldan *et al.* 2001), as well as the non-specialized necrotrophic fungus *Botrytis cinerea* (Zheng *et al.* 2000), the vascular wilt pathogen *Fusarium oxysporum* (Di Pietro *et al.* 2001)

Table 1: Fungal MAPkinases involved in pathogenicity (also see Fig. 1).

Gene	Fungus	Effect of inactivation	References
(a) Fus3 homologues			
PMK1	*Magnaporthe grisea*	no appressoria; no penetration; no invasive growth; female sterile; normal veget. properties	Xu and Hamer 1996
CHK1	*Cochliobolus heterostrophus*	no appressoria, but penetration; reduced virulence; reduced invasive growth; no sporulation, autolysis of culture	Lev *et al.* 1999
UBC3/KPP2	*Usitlago maydis*	reduced filamentous growth; virulence and response to pheromones; mating deficiency	Mayorga and Gold 1999; Müller *et al.* 1999
CMK1	*Colletotrichum lagenarium*	no appressoria; apathogenic; reduced spore germination; no invasive growth	Takano *et al.* 2000
BMP1	*Botrytis cinerea*	apathogenic; no penetration; no invasive growth; normal sporulataion; reduced growth on rich media	Zheng *et al.* 2000
FMK1	*Fusarium oxysporum f.sp. lycopersici*	apathogenic; no invasive growth; normal vegetation properties	Di Pietro *et al.* 2001
PMK1	*Pyrenophora teres*	apathogenic; no appressoria; no conidiation	Ruiz-Roldan *et al.* 2001
CPMK1	*Claviceps purpurea*	apthogenic; no invasive growth; much vegetaion properties	Mey *et al.* 2002a
(b) SLT2 homologues			
MPS1	*Magnaporthe grisea*	apathogenic; no penetration; cell-wall defect; osmo-sensitive, reduced conidation	Xu *et al.* 1998
CPMK2	*Claviceps purpurea*	apathogenic; reduced penetration; cell-wall defect; reduced conidation	Mey *et al.* 2002b
MGV1	*Fusarium graminearum*	reduced virulence; female sterile; cell-wall defect	Hou *et al.* 2002

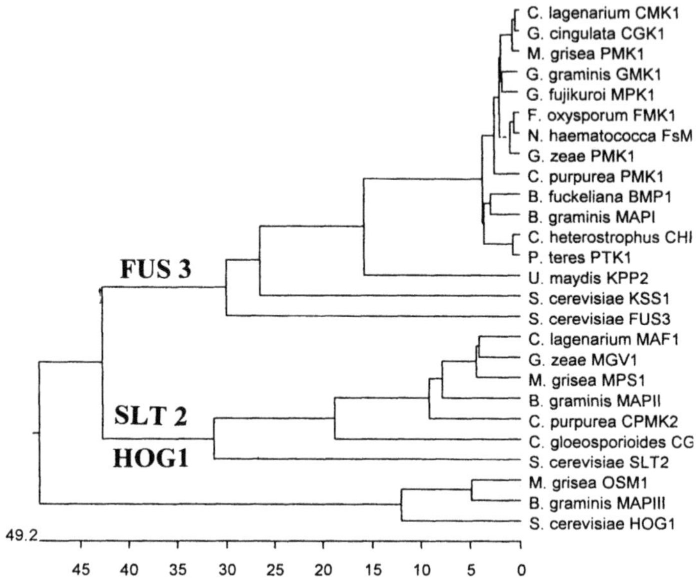

Fig. 1: Phylogram of MAPK sequences from phytopathogenic fungi; *S. cerevisiae* sequences are included for comparison. Accession numbers of sequences used for this analysis: *C. lagenarium* CMK1 (AAD50496), MAF1 (AAL50116); *G. cingulata* CGK1 (BAB21569); *M. grisea* PMK1 (AAC49521), MPS1 (AAC63682), OSM1 (AAF09475); *G. graminis* GMK1 (AAG44657); *G. fujikuroi* MPK1 (CAC36428); *F. oxysporum* FMK1 (AAG01162); *N. haematococca* FsMAPK (AAB72017); *G. zeae* PMK1 (AAL73403), MGV1 (AAM13670); *C. purpurea* CPMK1 (CAC47939), CPMK2 (CAC87145); *B. fuckeliana* BMP1 (AAG23132); *B. graminis* MAPI (AAG53654), MAPII (AAG53655), MAPIII (AAL83917); *C. heterostrophus* CHL (AAF05913); *P. teres* PTK1 (AAK52840); *U. maydis* KPP2 (AAF15528); *S. cerevisiae* KSS1 (NP011554), FUS3 (AAA34613), SLT2 (CAA41954), HOG1 (AAA34680); *C. gloeosporioides* CGMAP (AAN32906).

and the biotrophic grass pathogen *Claviceps purpurea* (Mey *et al.* 2002a). The phenotypes of knock out mutants of the MAP kinase type are highly variable, with respect to parasitic properties [from completely apathogenic (e.g *B. cinerea*) to reduced lesions (*C. heterostrophus*)] as well as vegetative parameters (some mutants are impaired in sporulation and/or growth) (see table 1). The *U. maydis* MAPK UBC3/KPP2 appears to be functionally closely related to Fus3/Kss1, as it is also required for mating and filamentous growth. A mutant strain lacking this MAP kinase shows only a slight reduction of virulence, indicating that it does not play an essential role in pathogenicity (Mayorga and Gold 1999, Müller *et al.* 1999). However, recently in *U. maydis* a second *PMK1*-homologous MAP kinase gene was identified, *KPP6*. Double mutants defective in both *KPP2* and *KPP6* show a strong reduction in virulence, indicating that this second MAP kinase could at least partially complement the loss of the Kpp2p enzyme (A. Brachmann, P. Müller, J. Schirawski, R. Kahmann, pers. communication). Thus, all (functionally analysed) PMK1-related MAP kinases from phytopathogenic fungi constitute a phylogenetic group of pathogenicity determinants, being essential for fungi with completely different pathogenic strategies, for a foliar, appressoria-forming pathogen like *M. grisea*, as well as for a biotrophic, non-appressoria forming flower pathogen like *C. purpurea*.

The number of yeast *Slt2/MPS1* homologues identified in filamentous fungi is considerably smaller (Fig. 1). Only in a few phytopathogens a corresponding gene has been detected: in the biotroph *B. graminis* (Zhang and Gurr 2001), in *F. graminearum* (Hou *et al.*

2002), and in *C. purpurea* (Mey *et al.* 2002b). So far, only the last two (to our knowledge) were functionally analysed. In *F. graminearum*, a *mgv1* mutant showed a pleiomorphic phenotype: conidiation was normal (in contrast to Δ*mps1*), but – like in Δ*mps1* – virulence was significantly reduced, female fertility was affected, and sensitivity to cell wall degrading enzymes increased (indicating a modified cell wall). In addition, heterokaryon formation and the ability to accumulate mycotoxins on inoculated wheat were impaired. This indicates a rather broad role of this signalling pathway in developmental processes in this fungus (Hou *et al.* 2002). A knock out in the *CPMK2* gene of *C. purpurea* resulted in complete apathogenic isolates that. failed to penetrate and to cause disease symptoms. *In vitro*, the mutants showed severely reduced sporulation, a modified cell wall (sensitivity to lytic enzymes) and impaired growth (hyperbranching, "curly" hyphae). These symptoms are comparable to those observed in the *M. grisea* Δ*mps1* of mutants. Interestingly, complementation experiments showed that the heterologous expression of *CPMK2* under the control of its own regulatory regions almost fully restored sporulation, differentiation of infection hyphae, and pathogenicity of the *M. grisea mps1* mutant. Comparable complementation experiments have been successful with the *Fus3/PMK1* orthologues; the *C. purpurea CPMK1*–gene complements the *pmk1* deficiency in *M. grisea* and restores both appressoria differentiation and pathogenicity (Mey *et al.* 2002a). The same was shown for *GMK1* from the take all disease of cereals *Gaeumanomyces graminis* (Dufresne and Osbourn 2001); vice versa, the *MPS1* gene was shown to complement the corresponding mutant of the tomato wilt pathogens *Fusarium oxysporum* (Di Pietro *et al.* 2001). In contrast, *PIM1* from the (non-pathogenic) yeast *Pichia pastoris* is not able to complement the deficiencies linked to the deletion of its homologue *Slt2* in *S. cerevisiae* (Cosano *et al.* 2001), indicating that members of this phylogenetic group of MAP kinases are not always functionally conserved. The fact that the *C. purpurea CPMK2* MAPK can replace *MPS1* in *M. grisea* indicates that the MAP kinase cascades in these two pathogenic fungi are highly conserved, despite their completely different lifestyle. It appears that the Slt2p/MPS1 MAP kinase cascade represents a second important common signalling pathway in phytopathogenic fungi. It will be interesting to see the functions of orthologous genes in other important plant pathogens.

In contrast to the wealth of data on MAPK genes, the analysis of the other MAP kinase cascade components in filamentous fungi is still in its infancy. Among the few upstream elements of the tripartite Fus3/Kss1-like pathways known in filamentous fungi are the *U. maydis* MAPKK *UBC5* (*Fuz7*) and the MAPKKK *UBC4* (Andrews *et al.* 2000), and the *C. gloeosporioides* MAPKK *CgMEK1* (Kim *et al.* 2001).

Knowledge about the downstream components of MAPK cascades in phytopathogenic fungi is also limited. In *U. maydis* a transcription factor (*PRF1*) has been identified, which - under joint control of a MAPK cascade and the cAMP pathway induces genes responsible for pathogenic development and hyphal growth (Hartmann *et al.* 1996; Kahmann *et al.* 1999). In order to identify downstream components of *PMK1* in *M. grisea*, *MST12*, a homologue of the yeast transcription factor *Ste12* (which is under control of *Fus3*) was cloned and functionally characterized. Deletion mutants were able to form apothecia (in contrast to Δ*pmk1*), but were unable to penetrate and colonize the host tissues. This indicates that *MST12* may function downstream of *PMK1*; but there must be additional downstream factors responsible for appressoria development (Park *et al.* 2002). In several other systems first data are coming up on putative target genes of MAPK. In *C. lagenarium* the temporal transcription pattern of three melanin biosynthetic genes is affected in the non-germinating conidia of the *cmk1* mutant (Takano *et al.* 2000), and mRNA level of an endopectate lyase is greatly reduced in the Δ*fmk1* mutant from *F. oxysporum* (Di Pietro *et al.* 2001). Interestingly, the expression of an endopolygalacturonase gene in *S. cerevisiae* depends on the Kss1 pathway (Madhani *et al.* 1999), suggesting that the transcriptional control of genes coding for secreted enzymes may involve comparable regulatory pathways in yeast and in pathogenic filamentous fungi. In *M. grisea* a subtractive library approach yielded two *PMK1*-controlled genes, *GAS1* and *GAS2*; both are highly expressed in appressoria, and deletion mutants are impared in appressorial penetration and lesion development (Xue *et al.* 2002). The function of these genes is open; they encode small proteins homologous to the *B. graminis gEgh16* gene, which is also highly expressed *in planta* and has no known homologues in other organisms. These genes could therefore be representatives of a novel, long-looked for, class of fungal-specific pathogenicity factors. Two *in planta* induced cellulase-encoding genes that are under control of the Chk1p MAP kinase have been recently discovered in *C. heterostrophus* (B. Horwitz, unpublished). Gene expression is delayed in Δ*chk1* background but not abolished. Therefore it seems that the *CHK1* is necessary for proper control of these cellulolytic enzymes but there are additional regulators that can activate these genes, maybe not during pathogenesis.

2.5 Miscellaneous

Since Ca^{++} has been *shown* to have severe impact on polar growth and differentiation in fungi (Hyde 1998) it is not suprising that Ca^{++} signaling is also involved in pathogenesis. In *M. grisea* recently a cyclophilin gene (*CYP1*) was identified during a screening for genes with high *in planta* expression level; deletion of *CYP1* led to impaired penetration peg formation and appressorium turgor generation. The *CYP1* encoded cyclophilin was shown to be the targed of the immosuppressive drug cyclosporine in *M. grisea;* since cyclosporine acts mainly on calcineurin (dependent on a complex formed with cyclophilin), these data strongly suggest that Ca-signalling is involved in early infection processes in *M. grisea* (Viaud *et al.* 2002).

Several other (non-mitogen-activated) proteinkinases have been correlated to pathogenicity in plant pathogens, e.g. serin/threonine kinases: *CLK1* from the bean pathogen *Colletotrichum lindemuthianum* (Dufresne *et al.* 1998), which is involved in colonization of host tissue, and *UKC*1 from *U. mayis* (Dürrenberger and Kronstadt 1999). A homologue of the yeast SNF1 kinase, *ccSNF1* from *C. carbonum*, was shown to be involved in control of (biosynthesis and) secretion of cell wall degrading enzymes; a Δ*ccsnf1* mutant displays significantly reduced pathogenicity.

In conclusion, research on signaling in plant-pathogenic fungi has yielded valuable insights into the complex pathways moderating pathogenesis-related processes; still, the status of knowledge is unsatisfying, and research in this area has to (and certainly will) be emphasized in the next years, as well in detail on model systems like *M. grisea* and *U. maydis*, but also horizontally in many important pathogens, to get an impression of the evolution of these regulatory pathways and to be able to use this knowledge for the development of broad control strategies.

3. EARLY EVENTS

Under this category we include genes that affect the processes of adhesion, surface sensing, and appressorium differentiation and function. Some aspects have been discussed under the signalling section and will be only briefly mentioned here.

3.1 Adhesion

Spores and later the germ tubes and appressoria that they differentiate are firmly glued to the host surface by adhesive materials. The specific composition of these adhesives differs among species and even between organs. Typically they include various water insoluble proteins and glycoproteins, lipids and polysaccharides (Nicholson and Epstein 1991; Nicholson and Kunoh 1994; Sugai *et al.* 1998; Tucker and Talbot 2001; Xiao *et al.* 1994). Fungi have different adhesion strategies (for a recent review see Tucker and Talbot 2001).

Spores of many fungi secrete adhesives called the "spore tip mucilage" that anchor the spore to the host surface immediately upon first contact. In *M. grisea,* extrusion of spore tip mucilage is induced when spores are hydrated. These adhesives are preformed in the dormant spores, and their rapid release is a passive process that does not require synthesis of new proteins. In *C. graminicola* spore adhesion includes an initial similar passive step, however biosynthesis of new materials is later necessary for maximum adhesion (Mercure *et al.* 1994; 1995; Sugai *et al.* 1998). Genes that encode for adhesive materials or for enzymes involved in their biosynthesis have not been identified. However, spore and appressorium adherents are clearly essential for successful infection. Therefore, in this case the active molecules are known, but the corresponding genes are missing (see table 2 for novel and missing genes).

After spore germination different mucilages are secreted and assist in anchoring of the germ tube and appressoria to the host surface. In addition, several groups of proteins have been suggested to assist in germ tube and appressorium attachment and to mediate the exchange of early signalling between the fungus and the plant. Evidences exist for involvement of cutinases, hydrophobins, lectins, and integrins in these processes. Cuticle degrading enzymes are embedded in the spore matrix or secreted by spores upon contact with the host surface (Deising *et al.* 1992; Schäfer 1994). It has been suggested that by degrading the cuticular waxes, these enzymes help removing the lypophylic waxes that coat plant organs thereby making them more receptive to water-coated fungal organs. Several lines of evidence have shown that cutinase activity is essential for spore attachment and pathogenicity. Spores of the bean rust *Uromyces viciae-fabae* contain a cutinase and other serine esterases that are localized on the spore surface. Treatment of spore with a serine esterase inhibitor, or washing the enzymes off the spore surface greately reduced spore attachment (Deising *et al.* 1986). Autoclaved spores failed to adhere to bean cuticle, but adhesion of autoclaved spore was restored when an active enzymatic fraction was added to dead spores. These and other results strongly suggest that cutinases are important for spore attachment in some cases. A *U. viciae-fabae* cutinase gene has been isolated, but since transformation is not possible in this obligate parasite null mutants have not been generated. It should be pointed out that cutinase genes have been isolated from many other fungi, in which the presumable role of cutinases has been to assist in fungal penetration (Dickman and Kolattkdy 1989; van Kan *et al.*. 1997; Yao and Koller 1995). However, when null mutants were generated they were usually still pathogenic (e.g. Oeser and Yoder 1994; Stahl *et al.* 1994; Sweigard *et al.* 1992; van Kan *et al.* 1997). The lack of a clear pathogenicity phenotype has been attributed to the presence of multiple isozymes whith overlap activity (Yao and Koller 1995;). Thus, although the genes and their products have been characterized, molecular evidence that will help elucidating the role of cutinses in plant diseases are still missing. The recent introduction of multiple gene disruption and RNAi technologies may help resolving such situations that involve multiple, highly similar gene families.

Hydrophobins are small, hydrophobic proteins produced only by fungi. They are heterogenous in structure except for eight cysteine residues in conserved positions (Wessels 1996). Hydrophobins compose a large percentage of the proteins that cover spores and hyphae surfaces and probably mediate the interaction of fungi with hydrophobic surfaces (Wösten 2001). There are two major groups of hydrophobins, class I and class II that are defined according to biochemical and molecular characteristics (Ebbole, 1997; Kershaw and Talbot 1997; Wessels 1996; Wösten and de Vocht 2000). Numerous hydrophobin genes have been isolated, but a role in pathogenesis has been demonstrated only in two cases. *MPG1*, a class I hydrophobin gene from *M. grisea* was isolated by a screen for genes that are differentially expressed during infection (Talbot *et al.* 1993). Δ*mpg1* deletion mutants were defected in apperessoria and lesion formation. The development phenotypes *in vitro* of Δ*mpg1* resembled the development of the wild type strain when grown on surfaces that do not induce appressorium formation, suggesting that the Mpg1p hydrophobin may contribute in some way to surface recognition and and/or perception. It is important to keep in mind that this may not be the only, or even the primary role of *MPG1*. Indeed, □*mpg1* mutants are also defected in conidial formation (Talbot *et al.* 1996). Both defects can be recovered by growing mutants in the presence of cAMP suggesting that Mpg1p is involved in perception and transduction of cAMP-mediated signalling that occurs down stream of *MPG1*. Another hydrophobin gene for which there are molecular evidences that indicate possible involvement in early pathogenesis is *CU* (encoding for cerato-ulmin), a class II hydrophobin from the Dutch elm disease fungus *Ophiostoma ulmi* (Takai 1974; Takai and Hiratsuka 1980). Although Δ*cu* mutants do not show reduced virulence, they are defected in attachment to bark beetles, which are the vectors that carry and transmit the disease (Temple *et al.* 1997). Like Mpg1p and many other hydrophobins, CU has additional roles and the possible effect on pathogenicity is probably a side effect rather than a primary role. These two examples show that although hydrophobins may be used is some instances as disease mediators, they have not originally evolved for this purpose, and their primary roles are probably not directly related to pathogenicity (Wösten 2001).

Two additional adhesion factors are lectins and integrins. While there are no clear molecular evidences for involvement of these factors in pathogenicity to plants, there are convincing biochemical and physiological evidences to suggest it (Correa and Hoch 1995; Correa *et al.* 1996; Hostetter 1999). *INT1*, a surface protein with similarity to vertebrate integrins was cloned from the human pathogenic yeast *Candida albicans* (Gale *et al.* 1998). Disruption of *INT1* suppressed hyphal growth and adhesion to epithelial cells and reduced the virulence of the mutants to mice indicating that the Int1p protein is essential for adhesion and disease development in *C. albicans*. A single copy *INT1* homologue is present in the *M. grisea* genome sequence. Generation of *M. grisea* null mutants of this gene and hetrologous expression in other fungi should provide molecular proofs for possible involvement of the gene in pathogenicity to plants. The drastic effects of *INT1* in *C. albicans* suggest that integrins may turn out to be novel pathogenicity factors that have a primary role in early stages of fungal-host recognition.

3.2 Surface Sensing

Concomitant with attachment of spores to a new surface, fungi are exposed to a variety of signals. While most fungi (although not all) require free water for germination, host-specific pathogens commonly need additional chemical or physical signals that are characteristics of the favorable hosts. Spore germination and appressorium formation in *M. grisea* are induced by hard, hydrophobic surface, while germination and appressoria formation in some *Colletotrichum* sp. can be induced by host waxes or ethylene, and are enhanced by contact with a hard surface (Kolattukudy *et al.* 1995; Robinson and Sharon 1999). Germ tubes and germlings often exhibit another level of sensing and respond to chemical signals as well as to the topographical (thigmotropism) architecture of the surface. Using plastic membranes, Wynn (Wynn 1976) showed that germ tube growth in rust fungi is solely directed by thigmotropism and does not involve any chemicals. There are evidences to suggest involvement of cell wall proteins and specific ion channels in sensing, but no genes have been cloned so far (Epstein and Nicholson 1997; Zhou *et al.* 1991). Hard surface has been known for many years to be necessary for induction of appressoria formation in many other species but the molecular basis of this requirement is yet unknown. *PTH11*, the putative

Table 2: Novel and missing genes.

Function	Gene	Homology	Fungus	References
1. Early events				
(a) Sensing				
Surface sensing	INT1	Integrin	*C. albicans*	Gale *et al.* 1998
	CHIP6	sterol glycosyl transferases	*C. gloeosporioides*	Kim *et al.* 2002
Thigmotropism Sensing	*Missing*	Ion channels, cell-wall protein	*Rust, Powdery mildew*	Epstein & Nicholson 1997; Zhou *et al.* 1991
(b) Penetration				
Appressorium formation	*CBP1*	chitin-binding protein	*M. grisea*	Kamakura *et al.* 2002
Appressorium penetration	GAS1, GAS2	none	*M. grisea*	Xue *et al.* 2002
	CYP1	Cyclophylin	*M. grisea*	Viaud *et al.* 2002
	gEgh16	none	*E. graminis*	Justesen *et al.* 1996
Penetration peg formation	*PLS1*	tetraspanin	*M. grisea*	Clergeot *et al.* 2001
Infectious hyphae formation	*PDE1*	aminophospholipid translocase	*M. grisea*	Balhadre *et al.* 1999
2. Colonization of host tissue				
(a) Biotrophic growth				
Oxydative stress response	*CPTf1*	Δ-ZIP transcription factor	*C. purpurea*	Tudzynski unpublished
Prevention of elicitor formation	*GIPs*	Glucanase inhibitors	*P. sojae*	Rose *et al.* 2002
Suppression of HR during biotrophy	*CGDN3*	cell-wall-associate receptor kinase	*C. gloeosporioides*	Stephanson *et al.* 2000
Activation of biotrophic genes	*CLTA1*	GAL4-like transcription factor	*C. lindemuthianum*	Dufresne *et al.* 2000
Extracellular matrix of biotrophic phase	*CIH1*	cell-wall proteins	*C. lindemuthianum*	Perfect *et al.* 2000
(b) Switch in growth modus, necrotrophic growth				
Host colonization	*FOW1*	mt carrier	*F. oxysporum*	Inoue *et al.* 2002
3. General pathogenicity factors				
Endophyte/pathogenic transition	*PATh-1*	unknown	*C. magma*	Freeman and Rodriguez, 1993
Biotrophic/ necrotrophic	*Missing*	-	-	-
General pathogenicity factor	*CPS1*	nonribosomal peptide synthetase	*C. heterostrophus*	Yoder 1998

membrane receptor identified in *M. grisea* (see section on signalling), may be involved in surface sensing and transmission of the signal through a Gα-cAMP signalling cascade that regulates appressorium formation (DeZwaan *et al.* 1999). Δ*pth11* mutants form appressoria at a reduced rate (15% compared to wild type), indicating the Pth11p protein is not required for appressorium formation but is involved in host surface recognition. It is appealing to speculate that Pth11p is involved in perception and transduction of signals mediated by the *M. grisea* Mpg1p hydrophobin, which also affects appressoria formation in a similar way (as discussed earlier). Recently a hard surface contact-induced gene has been isolated from *C. gloeosporioides* cv. avocado (Kim *et al.* 2002). The gene, named *CHIP6,* is induced in conidia upon contact with a hard surface. It encodes a protein with homology to sterol glycosyl transferases and catalyses *in vitro* transfer of glucosyl from UDP-glucose to cholesterol. Δ*chip6* mutants have normal growth phenotype and form normal appressoria, but have a significant reduction in virulence on the natural host avocado. The mechanism by which this enzyme affects pathogenicity is unclear.

3.3 Appressoria

A relatively large number of genes have been isolated that affect appressoria differentiation and function. Most of these genes have been isolated from the large appressorium- forming species *M. grisea* and *Colletotrichum* sp. Appressoirum-related genes can be divided into three sub-groups: 1) genes that operate prior to and are necessary for appressorium formation, 2) genes that are uniquely expressed in appressoria or contribute significantly to specific features of appressoria structure and may be regarded as appressorium-specific, and 3) genes that control and affect appressorium germination (formation of the penetration peg) and host penetration. Group 1of genes includes the *M. grisea MPG1,* and *PTH11,* and the *C. gloeosporioide CHIP6* that have been previously mentioned. General signalling components such as the alpha subunit of heterotrimeric G proteins, adenylate cyclase, and protein kinase A regulatory and catalytic subunits may also affect appressoria differentiation, however these conserved elements are involved in many other processes and have been discussed separately. An additional gene that affects appressorium formation in *M. grisea* and may be involved in recognition of physical factors on solid surfaces has been recently reported. This gene, named *CBP1,* encodes a chitin-binding protein and is specifically expressed in germ tubes before appressoria formation (Kamakura *et al.* 2002). Δ*cbp1* mutants produce normal appressoria on leaves and are fully pathogenic, but on artificial surfaces they produce abnormal appressoria. This indicates that the Cbp1p protein may be involved in surface sensing.

Groups 2 and 3 of genes include melanin biosynthesis genes, which are needed for functional appressoria, and apperssorium-specific genes that are expressed only in the appressorium. Melanin biosynthesis genes have been isolated from *C. lagenarium* (*PKS1, SCD1, THR1*) and from *M. grisea* (*RSY, BUF*). Mutant strains in either species are albino and unable to infect the host plants (Perpetua *et al.* 1996). In addition, transformation of *M. grisea* melanin-deficient mutants with *Alternaria alternata* melanin biosynthesis genes restored full pathogenicity (Kawamura *et al.* 1997). It should be pointed out, that albino strains of *C. heterostrophus* are fully pathogenic. In this genus appressoria are small and less developed compared to the appressoria formed by *Magnaporthe* and *Colletotrichum* and are not considered essential for pathogenicity. Taken together, these molecular analyses show that melanin is specifically essential for the function of appressoria, but has no effect on other, pathogenicity unrelated functions and therefore it is a true pathogenicity factor in those species that use large appressoria to penetrate the host.

There are not many known genes that can be classified under group 3, which are genes that affect appressorium germination and penetration-peg formation. The two *M. grisea* appressorium-specific genes, *GAS1* and *GAS2* have been previously mentioned (section on signalling). Both genes are expressed exclusively in appressoria and are localized to the cytoplasm. Deletion mutants of either gene had normal growth and conidiation and formed normal appressoria, but were reduced in appressorial penetration and lesion formation (Xue *et al.* 2002). These phenotypes classify the *M. grisea GAS* genes as specific pathogenicity factors that probably affect appressorial penetration. Together with the *E. graminis gEgh16* they may represent a novel class of fungal pathogenicity elements.

At least two additional appressorium specific genes from *M. grisea* were isolated by REMI. In the first mutant, appressoria had cellular structure and glycogen content similar to those of wild type before host penetration, but they were unable to differentiate penetration pegs (Clergeot *et al.* 2001). The gene, *PLS1*, encodes a putative integral membrane protein with homology to tetraspanin proteins, which are part of membrane signaling complexes in animals. The cytological, morphological and structural analyses suggest that *PLS1* is essential for the differentiation of the appressorium penetration peg (Clergeot *et al.* 2001). Another gene, *PDE1*, was isolated from a mutant that was impaired in its ability to elaborate penetration hyphae (Balhadère *et al.* 1999). *PDE1* has homology to aminophospholipid translocase group of P-type ATPase, it is expressed in germinating conidia and developing appressoria. The expression pattern and phenotype of the mutants suggest that *PDE1* is essential for development of penetration hyphae and subsequent proliferation of the fungus beyond colonization of the first epidermal cell.

Several transcripts designated *CAP* genes that are expressed in appressoria and during host invasion have been isolated from *C. gloeosporioides* (Hwang *et al.* 1995; Kolattukudy *et al.* 1995). Two of these peptides (CAP20 and CAP22) show homology to cell wall proteins and may be part of the appressoria wall. Deletion mutants of *CAP20* are non-pathogenic while mutations in some other *CAP* genes had no effect on pathogenicity. Since mutation of *CAP20* affects processes other than disease, it may not directly influence pathogenic development and therefore the *CAP* genes might not be considered pathogenicity factors until further examination. Additional genes that may participate in the early events of fungal pathogenesis are covered in recent reviews (Kahmann and Basse 2001; Idnurm and Howlett 2001; Tucker and Talbot 2001).

4. NECROTROPHIC GROWTH: GENES INVOLVED IN DEGRADATION OF HOST STRUCTURE AND METABOLISM

Apart from the few strictly biotrophic pathogens, the life cycle of most phytopathogens involves a necrotrophic stage, in which the fungus kills the plant cells, destroys its structure and lives from the "dead" organic material. The factors involved in these processes are among the first which have been studied in several interaction systems; a large number of "necrotrophic" genes have been functionally analyzed, but only a few of them – due to the complexity of the system – have been unequivocally shown to be essential for these processes. Here we will discuss the current status of research on cell wall-degrading enzymes (CWDE), toxins, and active-oxygen-species (AOS)-generating and -scavenging systems.

4.1 Cell Wall Degrading Enzymes

As in detail documented in a recent review (ten Have *et al.* 2002), so far only few functional analyses of cell wall-degrading enzymes have indicated an important role of single enzymes in pathogenicity, probably due to the complexity and redundancy of these enzymes. Interestingly, only pectin-degrading enzymes have been shown to be important, though they normally also represent very complex systems, and though pectin does not seem to be the most recalcitrant cell wall component. An interesting example is an endo-polygalacturonase (PG) gene from *B. cinerea*, the deletion of which reduced virulence on tomato, though altogether at least 5 endo PG genes are present (ten Have *et al.* 1998). In an elegant control experiment this result could be substantiated by modifying the plant partner: expression of a polygalacturonase inhibitory protein (from pear) resulted in a comparable reduction of virulence of *B. cinerea* on these transgenic plants as had been observed with PG mutants on wild type tomato (Powell *et al.* 2000).

The important role of pectin degradation has been further confirmed in other systems: Yakoby *et al.* (2001) showed that heterologous expression of a pectate lyase from *Colletotrichum gloeosprioides* in *Colletotrichum magna* led to increased virulence of transformants on watermelon (Yakoby *et al.* 2001). In *C. purpurea* replacement of two closely linked polygalacturonase genes resulted in drastic reduction of pathogenicity on rye (Oeser *et al.* 2002a); this drastic effect was unexpected, since in this system the disruption of other CWDE genes (celluloses, xylanases) did not effect virulence significantly (Oeser *et al.*

2002b), and since the pectin content in cereal tissues is extremely low. On the other hand - just as an example for several papers – a double mutant of *C. carbonum* lacking the two major extracellular polygalacturonases (having less than 1 % of total wild-type PG activity) displayed normal virulence on maize (Scott-Craig *et al.* 1998).

In the most thoroughly investigated system regarding CWDE, *C. carbonum*, John Walton's group finally (after a large set of knock-outs showing no effect on virulence; see e.g. Kim *et al.* 2001) took an alternative approach to determine the role of CWDEs: they cloned an orthologue of the yeast *SNF1* gene (encoding a protein kinase involved in carbon-catabolite repression), *ccNF1*. Disruption of this gene resulted in a significant reduction of expression of several CWDE-genes (coding e.g. for glucanases, xylanases, pectinases, and an arabinosidase) and in a significantly reduced number of spreading lesions. This interesting result now allows the inverse functional approach: increasing the expression of single genes in these mutants can define the role of specific enzymes/enzyme groups (Tonukari *et al.* 2000).

4.2 Toxins

The production of toxic substances, which weaken or kill plant cells in advance of the growing hyphae (or already prior to penetration) is a widespread phenomenon in plant pathogenic fungi. These toxins can be effective on several, unrelated host-plants (non-specific toxins"), or their effect can be restricted to a certain species or even a single variety/genotype (host-specific toxin). Non-specific and host-specific toxins were among the first pathogenicity determinants confirmed by molecular genetics (see the excellent reviews/monographs by Hohn 1997; Yoder *et al.* 1997; Kohmoto and Yoder 1998). Especially the molecular genetics of the host-specific toxins of *Cochliobolus* species was studied in detail. Race 1 ($Tox2^+$) isolates of *C. carbonum* produce the so-called HT-toxin (derived from the alternative name *Helminthosporium carbonum*). Detailed genetic and functional analyses have provided evidence that this toxin is a virulence and specificity factor in the maize/*C. carbonum* interaction (Walton 1996). In this pathogen the toxin genes are clustered, forming a giant locus (*TOX2*) spanning 600 kb and repeats of at least 7 genes involved in HC-toxin biosynthesis, export and regulation (Ahn *et al.* 2002). This clustering of genes involved in a toxin biosynthetic pathway seems to be the rule in fungi; this raises interesting questions about the evolution and conservation of such gene clusters (Walton 2000). This aspect was addressed also by Turgeon and Berbee (1998) and by Yoder (1998). Highly virulent (= toxin-producing) isolates of *C. heterostrophus*, *C. carbonum*, and *C. victoriae* arose suddenly in the field. Genes required for host-specific toxin biosynthesis by *C. heterostrophus* and *C. carbonum* obviously are unique to those races producing toxins, suggesting horizontal gene transfer.

Also different pathotypes of *Alternaria alternata* produce host-specific toxins; their role as pathogenicity factors on pear (Tanaka *et al.* 1999) and on apple (Johnson *et al.* 2000) has been confirmed by generation of knock out mutants. In *Pyrenophora tritici-repentis* a small proteinaceous host specific toxin was identified as a pathogenicity factor on (sensitive) wheat cultivars (Ciuffetti *et al.* 1997). An interesting aspect of the host-specific toxins is that their effects in several systems are similar to these of avirulence factors, i.e. a clear-cut distinction often is not possible (Wolpert *et al.* 2002).

Among the large number of unspecific toxins the trichothecenes produced by various *Fusarium* sp. have been analysed in detail. Also here the genes involved in the biosynthesis are clustered, at least in *F. sporotrichoides* and *F. graminearum* (Brown *et al.* 2001). Inactivation of a gene controlling the first step of trichotecenes biosynthesis (tox5/tri5) resulted in reduction of virulence of *Gibberella pulicaris* on parsnip (but not on potato) and of *G. zeae* on wheat (Desjardins *et al.* 1992; Proctor *et al.* 1995).

4.3 Active Oxygen Species: Generation vs Detoxification

One of the earliest defense reactions of plants against pathogens is the transient formation of active (or reactive) oxygen species (A/ROS). In analogy to mammalian systems, this reaction is termed oxidative burst (Lamb and Dixon 1997). H_2O_2 (and O_2^-) have been shown to cause very rapid stiffening of the cell walls by cross-linking of proteins and lignification reactions (Otte and Barz 1996). In incompatible interactions this oxidative burst triggers the induction of the so-called hypersensitive response (HR), and H_2O_2 might also induce further defense reactions in the surrounding tissue. The mechanism of this H_2O_2/O_2^- formation and their impact on defense reactions has been studied in detail in several plant systems. However, direct impact of this oxidative burst (and other AOS which are formed normally in differentiating tissue, lignin formation, etc.) on the pathogen is still neglected. In the last years, some groups have started investigation of the fungal part of the AOS story.

Interestingly, two different strategies of pathogenic fungi with respect to AOS become obvious: in the *B. cinerea*, a necrotroph, the formation of AOS *in planta* is directly correlated with agressiveness of fungal isolate (von Tiedemann 1997). Cytological analysis showed that *B. cinerea* produces H_2O_2 in axenic culture and *in planta* (K.B. Tenberge, unpubl. data), indicating that the fungus contributes to (and causes?) enhanced AOS formation by the plant, leading to killing of plant tissue and therewith obviously facilitating fungal growth. So far, three potential H_2O_2-generating systems have been described in *B. cinerea*: a toxin, botrydial, which decomposes under light to yield H_2O_2 (I. Gonzalez Collado, unpubl.), a glucose oxidase (*GOD,* Liu *et al.* 1998), and a superoxidedismutase (SOD). Functional analyses by targeted gene interaction showed that the GOD does not contribute significantly to the AOS generation (and virulence), whereas knock out of *CPSOD1* encoding a Cn/Zu SOD leads to significantly reduced H_2O_2 production *in vitro* and reduced lesion size on bean plants (Y. Rolke, K.-M. Weltring, B. Williamson, P. Tudzynski, unpubl.). Deletion of a secreted catalase has no impact on virulence in this system (Schouten *et al.* 2002). Interestingly, in the model system *A. thaliana*, Govrin and Levine (2000) showed that the hypersensitive response facilitates plant infection by *B. cinerea*, supporting this hypothesis. On the other hand, in more balanced systems such as the (hemi-) biotrophic *C. purpurea*, the fungus obviously needs to overcome the oxidative stress by secreting AOS-detoxifying enzymes. Deletion of genes encoding the major extracellular catalase and SOD, respectively, had no impact on virulence of *C. purpurea* on rye (Garre *et al.* 1998; Moore *et al.* 2002). However, inactivation of an H_2O_2 -induced transcription factor (*CPTF1*) controlling all catalase genes of the fungus has significant influence on virulence, and – unexpectedly – induces an oxidative burst in the rye ovarian tissue (which is never observed in wild-type infected plants). The deletion mutant secretes more H_2O_2 in axenic culture, indicating that this increased H_2O_2 level (due to complete lack of catalase activity) induces the plant response (S. Joshi, E. Nathues, B. Oeser, P. Tudzynski, unpubl.).The available data are preliminary, but the impact of the capability of a fungus to deal with the oxidative stress it faces *in planta* could very well contribute to its pathogenic potential.

5. SUPPRESSION OF HOST DEFENSE

Perhaps the main difference between pathogens and saprophytes is the ability of live substrates (the host) to recognize and respond to the presence of microorganisms by a series of so-called "defense mechanisms" that help preventing invasion and spread of microorganisms. These defenses are the main challenge that pathogens have to cope with. Only after overcoming the host defenses can pathogenic microorganism benefit from the host nutrients, by degrading of host high molecular weight constituents and absorbing low molecular weight metabolites. During the millions of years of co-evolution of fungi and plants, there has been a consistent arms race, in which new fungal pathogenicity factors are counter acted by new defense elements of the plant and visa versa (Stahl and Bishop 2000). These processes created different kinds of systems and strategies that assist pathogens to overcome plant defenses and successfully colonize their hosts, and at the same time provided plants with antifungal mechanisms that help restricting pathogen development. Thus, the fungal and plant factors that mediate specific interactions represent a snapshot of a dynamic situation in a specific system. This may result in some inconsistencies of effects of certain types of elements in different systems, depending on the current status of the arms race between the fungus and the host. For the sake of this discussion we divided these mechanisms and the associated genes into two sub-categories, mechanisms that specifically counteract and dismantle plant defense components, and mechanisms that help pathogens to avoid or suppress activated plant defenses.

5.1 Coping with Antifungal Plant Substances

Plants produce a vast array of secondary metabolites, many of which have *in vitro* antimicrobial activity. Antimicrobial compounds include preformed substances (phytoanticipins) and compounds that are synthesized upon a microbial challenge (phytoalexins). Phytoanticipins and phytoalexins have been implicated as phytoprotectants for many years, however, direct biochemical and molecular evidence for such a role have been obtained only in a limited number cases (Dixon 2001; Hammerschmidt 1999; Thomma *et al.* 1999). As may be expected, some pathogens have evolved hydrolytic enzymes that can degrade toxic phytoprotectans thereby allowing the fungus to overcome specific chemical barriers. One such example is saponins. These are widely occurring, preformed glycosylated molecules, many of which with antimicrobial activity *in vitro*. Presence or absence of saponins has been found in correlation with disease development in several plant-fungus interactions (Osbourn 1996; 1999). Infection of oats and wheat by the take all fungus *Gaeumannomyces graminis* is correlated with production of the triterpene saponin avenacin A in the host, and the avenacin A degrading enzyme avenacinase (Ava1) in the fungus. *ava1* mutant strains are unable to detoxify avenacin, they are non-pathogenic on avenacin A producing oats, but are fully pathogenic on wheat, which does not naturally produce saponins, and on oats mutants defective in saponin production (Papadopoulou *et al.* 1999). Additional genes that encode saponin-degrading enzymes have been isolated from a number of plant pathogenic fungi, but gene disruption had no clear effect on pathogenicity (Martin-Hernandez *et al.* 2000; Melton *et al.* 1998). Effect of phytoalexin-degrading enzymes on pathogenicity has been unequivocally demonstrated only in the interaction between *Nectria haematococca* and chickpea. Disruption of the *MAK1* gene led to inability of the fungus to detoxify the chickpea phytoalexin maackiain and to reduced virulence of the fungus. The effect was incomplete, suggesting involvement of additional defense factors. Overall, the available data show that phytoprotectants-degrading enzymes can affect pathogenesis in some systems, but not all, and therefore each case needs specific examination.

Another class of plant protecting molecules is antifungal proteins, also known as PR (pathogenesis related) proteins. As for the secondary metabolites, there are both preformed and induced antifungal proteins that are assumed to play a role in plant protection by direct toxicity to the fungus and by release of fungal elicitors that may activate plant defenses (Selitrennikoff 2001). As for the antifungal phytochemical degrading fungal enzymes, it is intuitive to predict that fungi would have evolved protection mechanisms also against antifungal proteins. A recent report by Rose *et al.* (2002) provided the first molecular evidence that fungi may indeed produce such molecules. A class of glucanase inhibitor proteins (collectively called GIPs) has been characterized in the soybean oomycete pathogen *Phytophthora sojae*. The GIPs are homologous with the trypsin class of serine proteases, but lack proteolytic activity. Structural motifs of protein-protein interaction found in GIPS, and high affinity binding of the soybean endoglucanases EGaseA with GIP1 suggest that they may act as glucanase specific inhibitors. GIP1 was also found to inhibit the release of glucan elicitors by EgaseA from *P. sojae* cell walls. Thus the GIPs may represent a novel class of fungal counter defensive proteins that suppress plant defense responses. Their discovery may encourage more intensive search of similar elements in other fungi.

Another protection mechanism against antifungal phytochemicals is secretion. Fungi possess vast array of transporters that mediate secretion and import of a variety of compounds. The majority of transporters are part of the homeostasis, however some are specifically functioning during plant infection. A specific group of transporters belonging to yeast ATP-binding cassette (ABC) transporters is implicated in multi drug resistance. Several such genes form fungal pathogens have been cloned and analyzed. The *PEP5* gene is part of a pathogenicity gene cluster located on a 1.6 Mb dispensible chromosome in *N. haematococca* that also includes the maackiain detoxifying gene *MAK1,* and *PDA1*, a gene for detoxification of the pea phytoalexin pisatin. Disruption of *PDA1* had only slight effect on pathogenicity of *N. haematococca* on pea, but transformation of isolates lacking the entire dispensable chromosome with three *PEP* genes (*PEP1,2,* and *5*) increased pathogenicity of these isolates (Han *et al.* 2001). While *PEP1* and *PEP2* have no database homologues, *PEP5* shows homology to multi drug facilitator transporters and may be involved in pisatin excretion. In *M. grisea, ABC1* has been shown essential for pathogenesis on rice. The Abc1p protein has homology to yeast ABC transporters and the *ABC1* transcript is inducible by toxic drugs and by the rice phytoalexin (Urban *et al.* 1999). Δ*abc1* deletion mutants arrest growth and die shortly after host penetration, indicating that *ABC1* is essential for *M. grisea* pathogenicity. However, the Δ*abc1* mutants were not hypersensitive to various antifungal compounds including the rice phytoalexin and therefore the specific mode of action through which Abc1p affects pathogenicity remains unsolved. Somewhat opposite result was obtained for the *BCATRB* gene that encodes an ABC transporter in *B. cinerea.* Disruption of the gene increased sensitivity of *B. cinerea* to antibiotics and fungicides, and to the grapevine phytoalexin resverastrol, but had almost no effect on fungal pathogenicity (Schoonbeek *et al.* 2001). This may suggest that additional mechanisms, which do not operate *in vitro* are functional *in planta* and compensate for lack of this transporter in the mutants. *GPABC1,* an ABC transporter with homology to the *M. grisea ABC1* has been found necessary for tolerance of the potatato tuber pathogen *Gibberella pulicaris* to the potato phytoalexin rishitin (Fleißner *et al.* 2002). In this case Δ*gpabc* mutant strains were still able to detoxify rishitin *in vitro,* but they lost their tolerance to the phytoalexin and were avirulent on potatoes. Collectively, these results show that multi drug resistant transporters can protect fungi from antifungal phyotoxins, thereby providing another level of pathogenicity mechanism.

5.2 Suppression of Active Plant Defenses

Plant defense systems include a range of responses that can be activated in a spatial and temporal manner in response to pathogen invasion. In most cases, when defense responses are timely coordinated with the intrusion event, disease development is arrested at the very early stages and there is no extensive damage to the attacked plant. Thus, early detection is the name of the game in many cases. This suggests that successful pathogens, and especially obligate parasites that prosper on live host, should have evolved mechanisms that will help them avoiding the plant detection system and prevent induction of the plant defenses, or that they developed ways to suppress the manifestation of these responses. Surprisingly however, there is relatively little information on this class of genes and molecules, maybe because discovery of such genes requires *in planta* screening and analyses, which are more difficult than *in vitro* work. One kind of such genes might be the *Phytophthora GIPs* that have been previously described, which may prevent the release of fungal elicitors and thus prevent activation of the plant defense system (Rose *et al.* 2002). Another example is the HC toxin produced by *C. carbonum* (see section on CWDE and toxins). Unlike other host-specific toxins, the cyclic tetra peptide HC-toxin does not cause hypersensitive response and there is no evidence that it induces plant defense responses (Wolpert *et al.* 2002). Evidences in fact indicate that HC-toxin inhibits the host histone deacetylase, thereby distorting the proper regulation of defense gene activation (Walton 1996; Wolpert *et al.* 2002).

The *CGDN3* gene has been isolated by screening for genes that are expressed in the early stages of *C. gloeosporioides* infection. It encodes a small, secreted protein with low homology to plant cell wall-associate receptor kinases. Mutants in *CGDN3* have normal growth and form normal appressoria but are unable to cause disease on the natural host *Stylosanthes guianensis* (Stephanson *et al.* 2000). Microscopic analysis of plants inoculated with the mutant strain revealed small necrotic spots that included few host cells underneath the inoculation site, suggesting that the mutant elicited a localized, host hypersensitive-like response. The mutants were able to grow necrotrophically and caused disease when conidia were inoculated directly onto wound site. Based on these and other observations the authors suggested that the Cgdn3p protein is associated with the biotrophic phase of primary infection and may be involved in suppression of elicitation of a hypersensive response in the compatible host.

6. *IN PLANTA* EXPRESSED GENES

In this group we included genes that are involved in establishment and maintenance of infection but that are not directly associated with acquisition of nutrients from the host. They include elements that control *in planta* gene expression and genes that affect disease development after the initial contact was established.

Different sets of fungal genes operate during pathogenesis. After penetration and inoculation of the first cells new genes are needed for the following phases. In necrotrophic fungi activation of toxin and CWDE are induced. Specific transcription factors such as the *C. carbonum ccSNF1* that controls i*n planta* expression of CWDE are involved in the regulation of such genes. In hemibiotrophic and biotrophic pathogens, genes that regulate the development of infection structures (e.g. infection vesicles and haustoria), prevent elicitation of host defenses etc. are required. Several such genes have been isolated from hemibiotrophic *Colletotrichum* sp. *CLTA1,* a GAL4-like gene belonging to fungal zinc cluster family of transcriptional activators was isolated from a non-pathogenic insertional mutant of the hemi biotroph *C. lindemuthianum.* The mutant isolate was able to induce small, hypersensitive-like necroses and was blocked in transition from biotrophic to necrotrophic phase (Dufresne *et al.* 2000). The *CLTA1* sequence data, and the phenotype of *catl1* mutants suggest that it is a specific transcription factor that activates biotrophic-specific genes in *C. lindemuthianum* – bean interaction. An example of such a genes may be the biotrophy related gene *CIH1* (Perfect *et al.* 2000). The Cih1p glycoprotein is proline rich and is embedded in the extracelluair matrix that separates the fungal cell wall from the host plasma membrane (Mendgen and Hahn 2002). Significantly, the Cih1p glycoprotein was shown to be present uniquely at the interface of the extra cellular matrix and its expression was switched off at the onset of necrotrophic development (Perfect *et al.* 2000). It will be interesting to learn if *CIH1* expression is modified in *clta1* mutants. Another mutant defective in the switch from bitrophic to necrotrophic phase was reported in the maize anthracnose pathogens *C. graminicola.* Mutants were fully capable of penetrating and colonizing host cells during the biotrophic phase, but their growth was arrested before transition to necrotrophy (Thon *et al.* 2002). The mutated gene, named *CPR1* shows similarity to a family of genes that encode a subunit of eukayotic microsomal signal peptide. Although the function of *CPR1* is yet unclear, the authors suggested that the *cpr1* mutant might be impaired in ability to secrete sufficient quantities of hydrolytic enzymes to support the transition to necrotrophy.

A cluster of five, highly conserved genes (*MIG2*-1 to 5*)* that are specifically expressed during the biotrophic phase has been reported in *U. maydis* (Basse *et al.* 2002). The *MIG2* genes do not show any sequence homology to known genes. They are secreted to the extra cellular space but their putative function is unknown. Deletion of *MIG2-1* had no effect on pathogenicity, possibly due to overlap in function of the other *MIG2* genes. *FOW1*, which encodes a protein with strong similarity to mitochondrial carrier proteins from yeast, has been isolated from the wilt pathogen *F. oxysporum.* Δ*fow1* deletion mutants of *F. oxysporum* f. sp. *melonis* and f. sp. *lycopersici* had normal growth and conidiation in culture, but were defected in ability to colonize the host plants (Inoue *et al.* 2002). These results suggest that the Fow1p protein is specifically required for host colonization.

7. GENERAL PATHOGENICITY GENES OF UNKNOWN FUNCTION

Although many genes can be classified under this definition, we will only mention two examples in which single genes have been found to have drastic effects on pathogenicity in a wide range of species. UV and insertional mutagenesis in *Colletotrichum magna* generated non-pathogenic isolates with several different pathogenic defects (Freeman and Rodriguez 1993; Redman *et al.* 1999). The UV *path-1* mutant was non pathogenic but retained the ability to colonize and reproduce within the host without any symptoms. DNA was isolated from one of the insertional mutants and the phenotype was reproduced by targeted gene disruption. Homologues of the disrupted locus, designated pGMR1, have been found in additional species and when disrupted produced similar phenotypes. Thus, the *path-1* class of genes represents elements that control the expression of virulent genes that are necessary for disease outbreak. A general pathogenicity factor was reported in *C. heterostrophus.* The gene, *CPS1* encodes a protein with similarity to nonribosomal peptide synthetasese and is conserved among plant and human fungal pathogens (Yoder 1998). Disruption of *CPS1* in *C. heterostrophus* and in three other species drastically reduced pathogenicity of the mutants. Thus, *CPS1* is a general pathogenicity factor in pathogenic fungi, perhaps a moderator of other virulence factors.

8. CONCLUSIONS

The last decade has been characterized by significant bio-technological developments that greatly influenced biological research. The main achievements are availability of genomic sequences and the development of high throughput data gathering and sophisticated bioinformatic methods. The relatively small genomes of filamentous fungi on one hand, and their biological complexity and biodiversity on the other, have made fungi attractive targets of sequencing initiatives. Two years ago, by the end of the last millennium, only the yeast genome was publicly available. Today, only two years later, the complete sequences of six filamentous fungi including human and plant pathogens are publicly available. The genomes of several other pathogenic fungi have already been sequenced by the private sector (Yoder and Turgeon 2001), and sequences of additional plant pathogens will be publicly available by the end of this year. In comparison, only one complete plant full genomic sequence (*A. thaliana*) and one near completion genomic sequence (*O. sativus*) are currently publicly available. The consequences of these recent developments are yet to come, but they will undoubtedly help filling in many gaps in our knowledge and understanding of fungal pathogenesis. There is already a great deal of information on specific processes and genes, but much more is still unknown. We anticipate that within a few years a large portion of the pathogenicity genes will be identified in several species in which intensive genomic studies have already been initiated e.g., *M. grisea, U. maydis, C. heterostrophus, A. gossypii*. Large-scale mutagenesis and functional analyses will provide information on genes' regulation and function. Whole genome comparisons between related pathogens and between pathogens and saprophytes will help defining the genetic information required for basic pathogenicity as well as host specificity. One of the greatest challenges will be in unraveling the complex molecular networks that regulate and control fungal pathogenicity. All these exciting developments are expected to expedite the development of new means to control human and plant fungal diseases, and for better utilization of fungi in agriculture and industry.

REFERENCES

Ahn J-H, Cheng Y-Q, and Walton JD (2002). An extended physical map of the TOX2 locus of *Cochliobolus carbonum* required for biosynthesis of HC-toxin. Fung Gen Biol 35:31-38.

Andrews DL, Egan JD, Mayorga ME, and Gold SE (2000). The *Ustilago maydis* ubc4 and ubc5 genes encode members of a MAP kinase cascade required for filamentous growth. Mol Plant-Microbe Interact 13:781-786.

Balhadère PV, Foster AJ, and Talbot NJ (1999). Identification of pathogenicity mutants of the rice blast fungus Magnaporthe grisea by insertinal mutagenesis. Mol Plant-Microbe Interact 12:129-142.

Basse CW, Kolb S, and Kahmann R (2002). A maize-specifically expressed gene cluster in *Ustilago maydis*. Molec Microbiol 43:75-93.

Bindsev L, Kershaw M.J, Talbot, NJ, and Oliver RP (2001). Complementation of the *Magnaporthe grisea cpkA* mutation by the *Blumeria graminis PKA*-c gene: functional genetic analysis of an obligate plant pathogen. Mol Plant-Microbe Interact 14:1368-1375.

Bockaert J, and Pin JP (1999). Molecular tinkering of G protein-coupled receptors: an evolutionary success. EMBO J 18:1723-1729.

Brown DW, McCormick SP, Alexander NJ, ProctorRH, and Desjardins AE (2001). A genetic and biochemical approach to study trichothecene diversity in *Fusarium sporotrichioides* and *Fusarium graminearum*. Fungal Gen and Biol 32:121-133.

Choi GH, Chen BS, and Nuss DL (1995). Virus mediated or transgenic suppression of a G protein alpha subunit and attenuation of fungal virulence. Proc Natl Acad Sci USA 92:305-309.

Ciuffetti LM, Tuori RP, and Gaventa JM (1997). A single gene encodes a selective toxin causal to the development of tan spot of wheat. The Plant Cell 9:135-144.

Clergeot PH, Gourgues M, Cots J, Laurans F, Latorse MP, Pepin R, Thatteau D, Notteghem JL, and Lebrun MH (2001). PLS1, a gene encoding a terraspanin-like protein, is required for penetration of rice leaf by the fungal pathogen Magnaporthe grisea. Proc Natl Acad Sci USA 98:6963-6968.

Coca MA, Damsz B, Yun, D-J, Hasegawa PM, Bressan RA, and Narasimhan ML (2000). Heterotrimeric G-proteins of a filamentous fungus regulate cell wall composition and susceptibility to a plant PR-5 protein. The Plant J 22:61-69.

Correa A Jr, and Hoch HC (1995). Identification of thigmoresponsive loci for cell differentiation in *Uromyces germlings*. Protoplasma 186:34-40.

Correa A Jr, Staples RC, and Hoch HC (1996). Inhibition of thigmostimulated cell differentiation with RGD-peptides in *Uromyces germlings*. Protoplasma 194:91-102.

Cosano IC, Martin H, Flandez M, Nombela C, and Molina M (2001). Pim1, a MAP Kinase involved in cell wall integrity in *Pichia pastoris*. Mol. Genet Genomics 265:604-614.

De Zwaan TM, Carroll AM, Valent B, and Sweigard JA (1999). *Magnaporthe grisea* Pth11p is a novel plasma membrane protein that mediates appressorium differentiation in response to inductive substrate cues. Plant Cell 11:2013-2030.

Deising H, Nicholson RL, Hug M, Howard RJ, and Mendgen K (1992). Adhesion pad formation and the involvement of cutinase and esterases in the attachment of uredospores to the host cuticle. Plant Cell 4:1101-1111.

Deising H, Zuckerman SH, and Andonov-Roland MM (1986). Isolation of a *Fusarium solani* mutant reduced in cutinase activity and virulence. J Bacteriol 168:911-916.

Desjardins AE, Gardner HW, and Weltring K-M (1992). Detoxification of sesquiterpene phytoalexins by *Gibberella pulicaris (Fusarium sambucinum)* and its importance for virulence on potato tubers. J of Industrial Microbiol 9:201-211.

DeZwaan TM, Carroll AM, Valent B, and Seigard JA (1999). *Magnaporthe grisea* Pth11p is a novel plasma membrane protein that mediates appressorium differentiation in response to inductive surface cues. Plant Cell 11:2013-2030.

Di Pietro F, García-Maceira I, Méglecz E, and Roncero MIG (2001). A MAP kinase of the vascular wilt fungus *Fusarium oxysporum* is essential for root penetration and pathogenesis. Mol Microbiol 39:1140-1152.

Dickman MB, and Kolattukudy PE (1989). Insertion of cutinase gene into a wound pathogen enables it to infect intact host. Nature 343:446-448.
Dixon RA (2001). Natural products and plant disease resistance. Nature 411:843-847.
Dufresne M, and Osbourn AE (2001). Definition of tissue-specific and general requirements for plant infection in a phytopathogenic fungus. Mol Plant-Microbe Interact 14:300-397.
Dufresne M, Bailey JA, Dron M, and Langin T (1998). Clk1, a serine/threonine protein kinase-encoding gene, is involved in pathogenicity of *Colletotrichum lindemuthianum* on common bean. Mol Plant-Microbe Interact 11:99-108.
Dufresne M, Perfect S, Pellier A-L, Bailey JA, and Langin T (2000). A GAL4-like protein is involved in the switch between biotrophic and necrotrophic phases of the infection process of *Colletotrichum linemuthianum* on common bean. The Plant Cell 12:1579-1589.
Dürrenberger F, and Kronstad J (1999). The ukc1 gene encodes a protein kinase involved in morphogenesis, pathogenicity and pigment formation in *Ustilago maydis*. Mol Gen Genet 261:281-289.
Ebbole DJ (1997). Hydrophobins and fungal infections of plants and animals. Trends. Microbiol 5:405-408.
Epstein L, and Nicholson RN (1997). Adhesion of spores and hyphae to plant surfaces. In The Mycota. V. Plant relationships, Pt. A. ed. GC Carroll, P. Tudzynski, Berlin/Heidelberg, Springer-Verlag, pp 11-25.
Fleißner A, Sopalla C, and Weltring K-M (2002). An ATP-binding cassette multidrug-resistance transporter is necessary for tolerance of *Gibberella pulicaris* to phytoalexins and virulence on potato tubers. Molecular Plant-Microbe Interactions 15:102-108.
Freeman S, and Rodriguez JR (1993). Genetic conversion of fungal plant pathogen to a nonpathogenic, endophytic mutualist. Science 260:75-78.
Gale CA, Bendel CM, McClellan M, Hauser M, Becker JM, Berman J, and Hostetter MK (1998). Linkage of adhesion, filamentous growth, and virulence in *Candida albicans* to a single gene, INT1. Science 279:1355-1358.
Garre V, Müller U, and Tudzynski P (1998). Cloning, characterization and targeted disruption of cpcat1, coding for an in planta secreted catalase of *Claviceps purpurea*. Mol Plant-Microbe Interact 11: 772-783.
Gilbert RD, Johnson AM, and Dean RA (1996). Chemical signals responsibible for appressorium formation in the rice blast fungus *Magnaporthe grisea*. Physiol Mol Plant Pathol 48:335-346.
Gold SE, García-Pedrajas M, and Martínez-Espinoza AD (2001). New (and used approaches to the study of fungal pathogenicity. Annu Rev Phytopathol 39:337-365.Govrin EM, and Levine A (2000). The hypersensitive response facilitates plant infection by the necrotrophic pathogen *Botrytis cinerea*. Curr Biol 10:751-757.
Hall AA, and Gurr SJ (2000). Initiation of appressorial germ tube differentiation and appressorial hooking: distinct morphological events regulated by cAMP signalling in *Blumeria graminis* f.sp. *horde*i. Physiol Mol Plant Pathol 56:39-46.
Hall AA, Bindslev L, Rouster J, Rasmussen SW, Oliver RP, and Gurr SJ (1999). Involvement of cAMP and protein kinase A in conidial differentiation by *Erysiphe graminis* f.sp. hordei. Mol Plant-Microbe Interact 12:960-968.
Hammerschmidt R (1999). Phytoalexins: What have we learned after 60 years? Ann Rev Phytopathol 37:285-306.
Han Y, Liu X, Benny U, Kistler HC, and VanEtten HD (2001). Genes determining pathogenicity to pea are clustered on a supernumerary chromosome in the fungal plant pathogen Nectria haematococca. Plant J 25:305-314.
Hartmann H, Kahmann R, and Bölker M (1996). The pheromone response factor coordinates filamentous growth and pathogenicity in *Ustilago maydis*. EMBO J 15:16-32-1641.
Hohn TH (1997). Fungal phytotoxins: biosynthesis and activity. In: Carroll GC, Tudzynski P (eds) The Mycota Vol V A Plant Relationships, Springer Verlag, Berlin pp 129-144.
Horwitz BA, Sharon A, Shun-Wen L, Ritter V, Sandrock TM, Yoder, OC, and Turgeon BG (1999). A G protein Alpha subunit from *Cochliobolus heterostrophus* involved in mating and appressorium formation. Fungal Gen Biol 26:19-32.
Hostetter MK (1999). Intergrin-like proteins in *Candida albicans* spp. and other microorganisms. Fung Genet Biol 28:135-145.
Hou Z, Xue C, Peng Y, Katan T, Kistler HC, and Xu J-R (2002). A mitrogen-activated protein kinase gene (MGV1) *in Fusarium graminearum* is required for female fertility heterokaryon formation, and plant infection. Mol Plant-Microb Interact 15:1119-1127.
Hwang CS, Flaishamn MA, and Kolattukudy PE (1995). Cloning of a gene expressed during appressorium formation by *Colletotrichum gloeosporioides* and a marked decrease in virulence by disruption of this gene. Plant Cell 7:183-193.
Hyde G (1998). Calcium imaging: a primer for mycologists. Fungal Genet Biol 24:14–23.
Idnurm A, and Howlett BJ (2001). Pathogenicity genes of phytopathogenic fungi. Mol Plant Pathol 2:241-255.

Inoue I, Namiki F, and Tsuge T (2002). Planf colonization by the vascular wilt fungus *Fusarium oxysporum* requires FOW1, a gene encoding a mitochondrial protein. Plant Cell 14:1869-1883

Johnson RD, Johnson L, Itoh Y, Kodama M, Otani H, and Kohmoto K (2000). Cloning and characterization of a cyclic peptide synthestase gene from *Alternaria alternata* apple pathotype whose product is involved in AM-toxin synthesis and pathogenicity. Mol.Plant-Microbe Interact. 13:742-753.

Justesen A, Somerville S, Christiansen S, and Giese H (1996). Isolation and characterization of two novel genes expressed in germinating conidia of the obligate biotroph *Erysiphe graminis* f.sp. *hordei*. Gene 170:131-135.

Kahmann R, Basse C, and Feldbrügge M (1999). Fungal-plant signaling in the *Ustilago maydis*-maize pathosystem. Curr Opin Microbiol 2:647-650.

Kahmann R, and Basse C (2001). Fungal gene expression during pathogenesis-related development and host plant colonization. Curr Opin Microbiol 2001,4:374–380

Kamakura T, Yamaguchi S, Saitoh K-I, Teraoka T, and Yamaguchi I (2002). A novel genes, CBP1, encoding a putative extracellular chitin-binding protein, may play an important role in the hydrophobic surface sensing of Magnaporthe grisea during appressorium diffrentiation. Molec Plant-Microbe Interact 15:437-444.

Kang SH, Khang CH, and Lee YH (1999). Regulation of cAMP-dependent protein kinase during appressorium formation in *Magnaporthe grisea*. FEMS Microb Letters 170:419-423.

Kawamura C, Moriwaki J, Kimura N, Fujita Y, Fuji SI, Hirano T, Koizumi S, and Tsuge T (1997). The melanin biosynthesis genes of Alternaria alternata can restore pathogenicity of the melanin-deficient mutants of *Magnaporthe grisea*. Mol Plant-Microbe Interact 10:446-453.

Kershaw MJ, and Talbot NJ (1997). Hydrophobins and repellents: proteins with fundamental roles in fungal morphogenesis. Fung Genet Biol 23:18-33.

Kim H, Ahn J-H, Görlach JM, Caprari C, Scott-Craig JS, and Walton JD (2001). Mutational analysis of β-glucanase genes from the plant-pathogenic fungus *Cochliobolus carbonum*. Mol Plant-Microbe Interact 14:1436-1443.

Kim DJ, Baek J-M, Uribe P, Kenerley CM, and Cook DR (2002). Cloning and characterization of multiple glycosyl hydrolase genes from *Trichoderma virens*. Curr Genet. 40:374-384.

Kim Y-K, Wang Y, Liu Z, and Kolattukudy PE (2002). Identification of a hard surface contact-induced gene in *Colletotrichum gloeosporioides* conidia as a sterol glycosyl tranferase, a novel fungal virulence factor. Plant J 30:177-187.

Kohmoto K, and Yoder OC (eds) (1998). Molecular genetics of host-specific toxins in plant diseases. Kluwer Academic Publ, Dordrecht.

Kolattukudy PE, Rogers LM, Li D, Hwang C-S, and Flaishman MA (1995). Surface signaling in pathogenesis. Proc Natl Acad Sci USA 92:4080-4087.

Kronstadt JW (1997). Virulence and cAMP in smuts, blasts and blights. Trends Plant Sci 2:193-199.

Krüger J, Loubradou G, Wanner G, Regenfelder E, Feldbrügge M, and Kahmann, R (2000). Activation of the cAMP pathway in *Ustilago maydis* reduces fungal proliferation and teliospore formation in plant tumors. Mol Plant Microb Interact 13:1034-1040.

Lamb C, and Dixon RA (1997). The oxidative burst in plant disease resistance. Annu Rev Plant Physiol Plant Mol Biol 48:251-275.

Lee HY, and Dean RA (1993). cAMP regulates infection structure formation in the plant pathogenic fungus *Magnaporthe grisea*. Plant Cell 5:693-700.

Lev S, Sharon A, Hadar R, Ma H, and Horwitz BA (1999). A mitogen-activated protein kinase of the corn leaf pathogen *Cochliobolus heterostrophus* is involved in conidiation, appressorium formation, and pathogenicity: Diverse roles for mitogen-activated protein kinase homologs in foliar pathogens. Proc Natl Acad Sci USA 96:13542-13547.

Liu S, and Dean RA (1997). G protein α subunit genes control growth, development and pathogenicity of *Magnaporthe grisea*. Mol Plant Microb Interact 10:1075-1086.

Liu S, Oeljeklaus S, Gerhardt B, and Tudzynski B (1998). Purification and characterization of glucose oxidase of *Botrytis cinerea*. Physiol Mol Plant Pathol 53:123-132.

Madhani HD, Galitski T, Lander ES, and Fink GR (1999). Effectors of a developmental mitogen-activated protein kinase cascade revealed by expression signatures of signaling mutants. Proc Natl Acad Sci USA 96:12530-12535.

Martin-Hernandez AM, Dufresne M, Hugouvieux V, Melton R, and OsbournAE (2000) Effects of targeted replacemment of the tomatinase gene on the interaction of *Septoria lycopersici* with tomato plants. Anonymous. Anonymous. Mol Plant-Microbe Interact. 13:1301-1311.

Mayorga ME, and Gold SE (1999). A MAP kinase encoded by the *ubc3* gene of Ustilago maydis is required for filamentous growth and full virulence. Mol Microbiol 34:485-497.

Melton RE, Flegg LM, Brown JKM, Oliver RP, Daniels MJ, and Osbourn AE (1998). Heterologous expression of Seproia lycopersici tomatinase in *Cladosporium fulvum*: Effects on compatible and incompatible interactions with tomato seedlings. Mol Plant-Microbe Interact 11:228-236.

Mendgen K, and Hahn M (2002). Plant infection and the establishment of fungal biotrophy. Tren Plant Sci 7:352-356.
Mercure EW, Leite B, and Nicholson RL (1994). Adhesion of ungerminated conidia of *Colletotrichum graminicola* to artificial hydrophobic surfaces. Physiol Mol Plant Pathol 45:421-440.
Mercure EW, Kunoh H, and Nicholson RL (1995). Visualisation of materials released from adhered, ungerminated conidia of *Colletotrichum graminicola*. Physiol Mol Plant Pathol 461:121-135.
Mey G, Oeser B, Lebrun MH, and Tudzynski P (2002a). The biotrophic, non-appressoria forming grass pathogen Claviceps purpurea needs a Fus3/Pmk1 homologous MAP kinase for colonization of rye ovarian tissue. Molec Plant Microbe-Interact 15: 303-312.
Mey G, and Tudzynski P (2002b) CPMK2, an Slt2-homologous MAP-kinase is essential for pathogenesis of *Claviceps purpurea* on rye: evidence for a second conserved pathogenesis-related MAP-kinase cascade in phytopathogenic fungi. (submitted).
Moore S, de Vries OMH, andTudzynski P (2002). The major Cu,Zn SOD of the phytopathogen *Claviceps purpurea* is not essential for pathogenicity. Mol Plant Pathol 3:9-22.
Müller P, Aichinger C, Feldbrügge M, and Kahmann R (1999). The MAP kinase Kpp2 regulates mating and pathogenic development in Ustilago maydis. Mol Microbiol 34:1007-1017
Nicholson RL, and Epstein L (1991). Adhesion of fungi to the plant surface: prerequisite for pathogenesis. In The Fungal Spore and Disease Initiation in Plants and Animals. Ed GT Cole, HC Hoch, New York, Plenum, pp 3-23.
Nicholson, AL, and Kunoh H (1994). Early interactions, adhesion, and establishment of the infection court by *Erysiphe graminis*. Can J Bot 73v(Suppl 1):609-615.
Oeser B, and Yoder OC (1994). Pathogenesis by Cochliobolus heterostrophus transformants expressing a cutinase-encoding gene from *Nectria haematococca*. Mol Plant-Microbe Interact 7:282-288.
Oeser B, Heidrich P, Müller U, Tudzynski P, and Tenberge KB (2002a). Polygalacturonase is a pathogenicity factor in the *Claviceps purpurea*/rye interaction. Fungal Genet Biol 36:176-186.
Oeser B, Tenberge KB, Moore S, Mihlan M, Heidrich PM, and Tudzynski P (2002b). Pathogenic development of Claviceps purpurea. In: Osiewacz, H. (ed.) Molecular Biology of Fungal Development. Marcel Dekker, New York, pp 419-455.
Oliver R, and Osbourn AE (1995). Molecular dissection of fungal phytopathogenicity. Microbiology 141:1-9.
Osbourn AE (1996). Saponins and plant defence- a soap story. Tren Plant Sci 1:4-8.
Osbourn AE (1999). Antimicrobial phytoprotectants and fungal pathogens: a commentary. Fung Genet Biol 26:163-168.
Otte O, and Barz W (1996). The elicitor-induced oxidative burst in cultured chickpea cells drives the rapid insolubilization of two cell wall structural proteins. Planta 200:238-246.
Papadopoulou K, Melton RE, Leggett M, Daniels MJ, and Osbourn AE (1999). Compromised disease resistance in saponin-deficient plants. Proc Natl Acad Sci 96:12923-12928.
Park G, Xue GY, Zheng L, Lam S, and Xu JR (2002). MST12 regulates infectious growth but not appressorium formation in the rice blast fungus Magnaporthe grisea. Mol Plant-Microbe Interact 15:183-192.
Perfect SE, Pixton KL, O'Connell RJ, and Green JR (2000). The distribution and expression of a biotrophy-related gene, CIH1, within the genus Colletotrichum. Mol.Plant Pathol. 1:213-221.
Perpetua NS, Kubo Y, Yasuda N, Takano Y, and Furusawa I (1996). Cloning and characterization of a melanin biosythesis THR1 reductase gene essential for appressorial penetration of Colletotrichum lagenarium. Mol Plant-Microbe Interact 8:593-601.
Powell ALT, van Kan JAL, ten have A, Visser J, Greve LC, Bennett AB, and Labavitch JM (2000). Transgenic expression of pear PGIP in tomato limits colonization. Mol Plant Microbe Interact 13:942-950.
Proctor RH, Hohn TM, and McCormick SP (1995). Reduced virulence of *Gibberella zeae* caused by disruption of a trichothecene toxin biosynthetic gene. Mol Plant-Microbe Interact 8:593-601.
Redman RS, Ranson JC, and Rodriguez RJ (1999). Conversion of the pathogenic fungus Colletotrichum magna to a nonpathogenic endophytic mutualist by gene disruption. Mol Plant Microbe Interact 12:969-975.
Robinson M., and Sharon A (1999). Transformation of the bioherbicide *Colletotrichum gloeosporioides* f. sp. *aeschynomene* by electroporation of germinated conidia. Curr Genet 36:98-104
Rose JKC, Ham K-S, Darvill AG, and Albersheim P (2002). Molecular cloning and characterization of glucanase inhibitor proteins: coevolution of a counterdefense mechanism by plant pathogens. Plant Cell 14;1329-1345.
Ruiz-Roldan MC, Maier F J and Schäfer W (2001). PTK1, a mitogen-activated-protein kinase gene, is required for conidiation, appressorium formation, and pathogenicity of *Pyrenophora teres* on Barley. Mol Plant-Microbe Interact 14:116-125.
Schaeffer HJ, and Webber MJ (1999). Mitogen-activated protein kinases: Specific messages from ubiquitous messengers. Mol Cell Biol 19:2435-2444.
Schäfer W (1994). Molecular mechanisms of fungal pathogenicity to plants. Ann Rev Phytopathol 32:461-477.

Schoonbeek H, Del Sorbo G, and De Waard MA (2001). The ABC transporter BcatrB affects the sensitivity of Botrytis cinerea to the phytoalexin resveratrol and the fungicide fenpiclonil. Mol Plant-Microb Interact 14:562-571.
Schouten A, Wagemakers L, Stefanato FL, van der Kaaij RM, and van Kan JAL (2002). Resveratrol acts as a natural produngicide and indueces self-intoxication by a specific laccase. Mol Microbiol 43:883-894.
Scott-Craig JS, Cheng YQ, Cervone F, De Lorenzo G, Pitkin JW, and Walton JD (1998). Targeted mutants of *Cochliobolus carbonum* lacking the two major extracellular polygalacturoneses. Appl Environ Microbiol 64:1497-1503.
Selitrennikoff CP (2001). Antifungal proteins. AppL Environ Microbiol 67:2883-2894.
Soanes DM, Skiner W, Keon J, Hargreaves J, and Talbor NJ (2002). Genomics of phytopathogenic fungi and the development of bioinformatic resources. Mol Plant-Microbe Interact 15:421-427.
Stahl DJ, Theuerkauf A, Heitefuss R, and Scafer W. (1994). Cutinase of *Nectria haematococca (Fusarium solani* f. sp. *pisi)* is not required for fungal virulence or organ specificity on pea. Mol Plant-Microbe Interact 7:713-725.
Stahl EA, and Bishop JG (2000). Plant-pathogen arms races at the molecular level. Curr Opin Plant Biol 3:299-304.
Stephenson SA, Hatfield J, Rusu AG, Maclean DJ, and Manner JM (2000). cgDN3: An essential pathogenicity gene of *Colletotrichum gloeosporioides* necessary to avert a hypersensitive-like response in the host *Stylosanthes guianesis*. Mol Plant-Microbe Interact 13:929-941.
Sugai JA, Leite B, and Nicholson RL (1998). Partial characterization of the extracellular matrix released onto hydrophobic surfaces by conidia and conidial germlings of *Colletotrichum graminicola*. Physiol Mol Plant Pathol 52:411-425.
Sweigard J, Chumley FG, and Valent B (1992). Cloning and analysis of *Cut*1, a cutinase gene from *Magnaporthe grisea*. Mol Gen Genet 232:174-182.
Sweigard JA, Chumley FG, and Valent B (1992). Disruption of a *Magnaporthe grisea* cutinase gene. Mol Gen Genet 232:183-190.
Takai S (1974). Pathogenicity and ceratoulmin production in *Ceratocystis ulmi*. Nature 252:124-126.
Takai S, and Hiratsuka Y (1980). Accumulation of the material contining the toxin cerato-ulmin on the hyphal surface of *Creatocystis ulmi*. Can J Bot 58:663-668.
Takano Y, Kikuchi T, Kubo Y, Hamer JE, Mise K, and Furusawa I (2000). The *Colletotrichum lagenarium* MAP kinase gene CMK1 regulates diverse aspects of fungal pathogenesis. Mol Plant-Microbe Interact 13:374-383.
Talbot NJ (1999). Fungal biology – coming up for air and sporulation. Nature 398:295-296.
Talbot NJ, Ebbole DJ, and Hamer JE (1993). Identification and characterization of MPG1 a gene involved in pathogenicity from the rice blast fungus *Magnaporthe grisea*. Plant Cell 5:1575-1590.
Talbot NJ, Kershaw MJ, Wakley GE, de Vries OMH, Wessels JGH, and Hamer JE (1996). MPG1 encodes a fungal hydrophobin involved in surface interactions during infection-related development of *Magnaporthe grisea*. Plant Cell 8:985-999.
Tanaka A, Shiotani H, Yamamoto M, and Tsuge T (1999). Insertional mutagenesis and cloning of the genes required for biosynthesis of the host-specific AK-toxin in the Japanese pear pathotype of *Alternaria alternata*. Mol Plant-Microbe Interact 12:691-702.
Temple B, Horgen PA, Bernier L, and Hintz WE (1997). Cerato-ulmin, a hydrophobin secreted by the causal agents of Dutch elm disease, is a parasitic fitness factor. Fung Genet Biol 22:39-53.
ten Have A, Mulder W, Visser J, and van Kan JAL (1998). The endopolygalacturonase gene *Bcpg*1 is required for full virulence of Botrytis cinerea. Molec Plant-Microbe Interact 11:1009-1016.
ten Have A, Tenberge KB, Benen IAE, Tudzynski P, Visser J, and van Kan JAL (2002) The contribution of cellwall degrading enzymes to pathogenesis of fungal plant pathogens. In: The Mycota, Vol. XI "Application in Agriculture" Kempken F (ed), Springer, Berlin, pp 341-358.
Thines E, Weber RWS, and Talbot NJ (2000). MAP kinase and protein kinase A-dependent mobilization of triacylglycerol and glycogen during appressorium tugor generation by *Magnaporthe grisea*. Plant Cell 12:1703-1718
Thomma BPHJ, Nelissen I, Eggermont K, and Broekaert WF (1999). Deficiency in phytoalexin production causes enhanced susceptibility of *Arabidopsis thaliana* to the fungus *Alternaria brassicicola*. Plant Journal 19:163-171.
Thon MR, Nuckles EM, Takach JE, and Vaillancourt LJ (2002). CPR1: A gene encoding a putative signal peptidase that fucntions in pahtogenicty of *Colletotrichum graminicola* to maize. Mol Plant Microbe Interactions 15:120-128.
Tonukari NJ, Scott-Craig JS, and Walton JD (2000). The *Cochliobolus carbonum* SNF1 gene is required for cell wall-degrading enzyme expression and virulence in maize. Plant Cell 12:237-247.

Truesdell GM, Zhonghui Y, and Dickman MB (2000). A Gα subunit gene from the phytopathogenic fungus *Colletotrichum trifolii* is required for conidial germination. Physiol Mol Plant Pathol 56:131-140.
Tucker SL and Talbot NJ. 2001. Surface Attachment and pre-penetration stage development by plant pathogenic fungi. Ann Rev Phytopathol 39:385-417
Tudzynski P, and Tudzynski B (1999). Phytopathogenic fungi: genetic aspects of host-pathogen interaction. Prog Bot 61:119-147.
Tudzynski B, and Tudzynski P (2001). Pathogenicity factors and signal transduction in plant-pathogenic fungi. Prog Bot 63:163-188.
Tudzynski B, Schulze Gronover C, Klimpel A, and Kasulke D (2000). Signaling and pathogenicity in the gray mold Botrytis cinerea. XIIth International Botrytis Symposium Reims, July 3-7, Abstr. L6.
Turgeon BG, and Berbee ML (1998). Evolution of pathogenic and reproductive strategies in *Cochliobolus* and related genera. In: K Kohmoto , OC Yoder eds. Molecular genetics of host-specific toxins in plant diseases. Dordrecht: Kluwer Academic Publ., Vol. 13 pp 153-163.
Urban M, Bhargava T, and Hamer JE (1999). An ATP-driven efflux pump is a novel pathogenicity factor in rice blast disease. EMBO J 18:512-521.
van Kan JAL, van t' Klooster JW, Wagemakers CAM, Dees DCT, and van der Vlugt-Bergmans CJB (1997). Cutinase A of *Botrytis cinerea* is expressed, but not essential, during penetration of gerbera and tomato. Mol Plant-Microbe Interact 10:30-38.
Viaud MC, Balhadere PV, and Talbot NJ (2002). A Magnaporthe grisea cyclophilin acts as a virulence determinant during plant infection. Plant Cell 14:917-930.
von Tiedemann A (1997). Evidence for a primary role of active oxygen species in induction of host cell death during infection of bean leaves with *Botrytis cinerea*. Physiol Molec Plant Pathol 50:151-166.
Walton JD (1996). Host-selective tToxins: Agents of compatibility. The Plant Cell 8:1723-1733.
Walton JD (2000). Horizontal gene transfer and the evolution of secondary metabolite gene clusters in fungi: an hypothesis. Fungal and genetics and biology 30:167-171.
Wessels JGH (1996). Fungal hydrophobins: proteins that function at an interface. Trends Plant Sci 1: 9–15.
Wolpert TJ, Dunkle LD, and Ciuffetti LM (2002). Host-selective toxins and avirulence determinants: wath's in a Name*. Annu Rev Phytopathol 40:251-285.
Wösten HAB (2001). Hydrophobins: Multipurpose proteins. Ann rev Microbiol 55:625-646.
Wösten HAB, and de Vocht ML (2000). Hydrophobins, the fungal coat unraveled. Biochim Biophys Acta 1469:79-86.
Wynn WK (1976). Appressorium formation over stomates by the bean rust fungus: response to the surface contact stimulus. Phytopathology 66:136-146.
Xu J-R, and Hamer JE (1996). MAP kinase and cAMP singnaling regulate infection structure formation and pathogenic growth in the rice blast fungus *Magnaporthe grisea*. Genes Dev 10:2696-2706.
Xu JR, Staiger CJ, and Hamer JE (1998). Inactivation of the mitogen-activated protein kinase Mps1 from the rice blast fungus prevents penetration of host cells but allows activation of plant defense responses. Proc Natl Acad Sci USA 95:12713-12718.
Xu JR (2000). MAP kinases in fungal pathogens. Fung Genet Biol 31:137-152.
Xue CY, Park G, Choi WB, Zeng L, Dean RA, and Xu JR (2002). Two novel fungal virulence genes specifically expressed in appressoria of the rice blast fungus. Plant Cell 14:2107-2119.
Yakoby N, Beno-Moualem D, Keen DT, Dinoor A, Pines O, and Prusky D (2001). Colletotrichum gloeosporioides pelB, encoding pectate lyase, is a key gene in fungal-fruit interactions. Mol Plant-Microbe Interact 14:988-995.
Yao C, and Koller W (1995). Diversity of cutinases from plant pathogenic fungi: Different cutinases are expressed during saprophytic and pathogenic stages of *Alternaria brassicicola*. Molec Plant-Microbe Interact 8:122-130.
Yoder OC and Turgeon BG. 1996. Molecular-genetic evaluation of fungal molecules for roles in pathogenesis to plants. J Genet 75:425-440
Yoder OC (1998). Polyketides and peptides as determinants of general or specific fungal virulence to plants. 6. Intern. Mycol. Congress Jerusalem, August 23.-28, Abstr. p 131.
Yoder OC, and Turgeon BG (2001). Fungal genomics and pathogenicity. Curr Opinion in Plant Biol 4:315-321.
Yoder OC, Macko V, Wolpert T, and Turgeon BG (1997). *Cochliobolus* spp. and their host-specific toxins. In:Carroll GC, Tudzynski P (eds) The Mycota Vol V A Plant Relationships, Springer Verlag, Berlin pp 145-166.

Zhang Z, and Gurr SJ (2001). Expression and sequence analysis of the *Blumeria graminis* mitogen-activated protein kinase genes, mpk1 and mpk2. Gene 266:57-65.

Zheng L, Campbell M, Murphy J, Lam S, Xu J-R (2000). The BMP1 gene is essential for pathogenicity in the gray mold fungus *Botrytis cinerea.* Mol Plant-Microb Interact 13:724-732.

Zhou XL Stumpf RC, Hoch HC, and Kung C (1991). A mechano-sensitive channel in whole cells and in membrane patches of the fungus *Uromyces*. Science 253:1415-1417.

Applied Mycology & Biotechnology
An International Series. Volume 3. Fungal Genomics

10

Genetic Improvement of Baker's Yeasts

Paul V. Attfield and Philip J. L. Bell
Microbiogen Pty Ltd, c/- Department of Biological Sciences, Macquarie University, Sydney, NSW 2109, Australia (pattfiel@rna.bio.mq.edu.au).

Yeasts have been used for many thousands of years to produce leavened bread. Nowadays the production of baker's yeast biomass represents a highly competitive multi-billion dollar global industry. The environmental conditions that prevail during manufacture and application of baker's yeasts, coupled with the sheer variety of bread making processes and recipes used around the world, place considerable demands on yeasts. These demands translate into technological and economic challenges for producers of baker's yeasts. One way to meet these challenges is to improve the physiological attributes of yeasts so that they are better suited to the complex requirements of the modern baking industry. Improvement or modification of yeast performance can be achieved to some extent by modifying the parameters of their growth and downstream processing during production. However, the potential of a yeast strain's performance is dictated in the first place, by the genetic makeup of that strain. The emerging knowledge of yeast genomics and proteomics promises to deliver important strategies for improving the genetic potential of strains of baker's yeasts. Genetic modification of baker's yeast can be achieved by classical or molecular procedures, or a combination of both approaches. However, given the general negativity surrounding GMO's, we contend that classical strategies remain the most practical approach to developing strains for commercial applications. Nevertheless, genomics and molecular techniques remain important for determining key genes, pathways and associated physiological functions that need to be enhanced in novel strains of baker's yeast.

1. INTRODUCTION

The topic of baker's yeast, its biology and technology has been ably dealt with in several books, chapters and reviews by Burrows (1970), Johnston and Oberman (1979), Oura *et al.* (1983), Spencer and Spencer (1983), Chen and Chiger (1985), Trivedi *et al.* (1986), Beudeker *et al.* (1990), Evans (1990), Nagodawithana and Trivedi (1990), and Reed and Nagodawithana, (1991). Strains of *Saccharomyces cerevisiae* represent almost all of the yeast that is produced for baking applications. In short, baker's yeast is produced by aerobic, fed-batch cultivation, most commonly using raw substrates such as sugar beet or cane molasses. Biomass is harvested, processed and transported to bakeries as either a suspension (cream), compressed blocks or active (instant) dried yeast. Baking applications of yeasts vary enormously. Bread-making procedures vary such that, at one extreme, yeasts need to be able to ferment and leaven doughs within a few minutes, whereas at the other extreme the baker requires the yeast to ferment more slowly but consistently for many hours (Stear 1990). Moreover, some doughs are made by mixing all ingredients at once, whereas others such as sponge and doughs, are manufactured in staged processes whereby yeast is exposed to variations in flour and water concentrations. An increasingly important bread making process involves frozen doughs, which are mixed, frozen and stored for perhaps months prior to thawing and baking. Bread recipes are also extremely varied. For example, some bread doughs contain no added sugars, requiring the yeast to adapt to maltose utilisation, whereas sweetened doughs have up to 30% sucrose added per wt. of flour, which represents a severe osmotic stress on yeasts. Antifungal preservatives may or may not be present. Sourdoughs require yeast to ferment at relatively low pH (Stear 1990).

There are a range of qualities that must be exhibited in order for any yeast strain to be economically useful. In industrial terms qualities needed include: efficient yield of a consistently good quality biomass, efficient dough leavening activities and production of good bread flavour characteristics in various bread-making conditions, robustness to withstand stresses encountered during production, transport and application, and keeping quality or shelf life, which is the ability of yeast biomass to maintain its dough-leavening activity in storage. Desirable characteristics of baker's yeast strains are listed in Table 1. See Evans (1990), and Reed and Nagodawithana (1991) for further information on the qualities required of baking strains.

Table 1. Some desirable characteristics of industrial baker's yeasts.

Biomass yield	High respiratory capacity and growth rate at temperatures in low- to mid-30s°C. Efficient utilisation of sucrose, glucose, fructose, raffinose and melibiose. Efficient nitrogen assimilation using urea, ammonia or ammonium phosphate. Low requirement for vitamins and metal ions. Resistance to inhibitors in molasses. Oxidative stress tolerance.
Down-stream processing	Ability to withstand nutrient limitation and prolonged starvation, temperature fluctuations, dewatering, compression and dehydration stresses.
Dough leavening	Fast fermentation of hexoses and of maltose in rapid or no-time plain doughs. Resistance to salt. Tolerance to high osmotic pressures, and rapid fermentation in sweetened doughs. Resistance to organic acid preservatives. Ability to maintain steady fermentation for several hours in traditional bread-making. Resistance to freeze-thaw stresses in frozen doughs. Tolerance to rehydration at various temperatures.
Keeping quality (shelf life)	Ability to maintain fermentative capacity under refrigerated and non-refrigerated conditions. Tolerance to starvation. Resistance to alcohol accumulated in compressed yeast blocks. Resistance to oxidative damage accumulated in stored dried yeast.

The more adaptable a given strain is to the requirements of the industry, the more useful it is to a producer. In biological terms the desirable characteristics are complex since they are affected by factors such as cell architecture, cell cycle, growth rate, glycolytic and respiratory fluxes, gluconeogenesis, storage carbohydrate metabolism, central nitrogen metabolism, stress gene response, levels of stress protectants, membrane status etc. Of course, these factors involve highly complex interactions between many genes and their proteins.

It could be argued that we have been making perfectly edible, nutritional and enjoyable breads for centuries and so why should we need to modify or manipulate the properties of baker's yeasts? The answer is that new strains are needed to meet varied and growing demands around the globe, especially as trends shift with regards to bread making processes. For example markets for frozen doughs are expanding and so demand for strains with improved freeze-thaw tolerance has increased. Applications of instant dried yeasts are also increasing, especially in Asian markets. At this time, the commercially available dry yeast strains tend to lose significant fermentative activity after drying and rehydration (especially at temperatures <18°C) leading to loss of leavening performance and problems of gas retention in doughs due to leakage of glutathione from yeast cells. Thus, strains with improved drying and rehydration tolerance are required. Strains with improved intrinsic tolerance of preservatives are also needed especially where sponge and dough processes are used (Stear 1990; Reed and Nagodawithana 1991). In this case the preservative resistance that may have been induced by baker's yeast producers during manufacture can be lost in the sponge phase where preservatives are absent, and consequently leavening activity is compromised when inhibitory agents are added at the dough stage. There are economies to be obtained by baker's yeast manufacturers if they can produce yeasts with significantly greater leavening capacity, since cream yeast products are sold on units of gassing activity rather than weight of yeast. Similarly, strains with broad dough applications are needed so that manufacturers have fewer production strains to manage in factories – currently strains tend to be useful in either plain (no added sugar) doughs or high sugar doughs, but not both. Although there has been experimental progress towards strains with broader dough (e.g. plain and high sugar) applications, these strains have not yet arisen in the market place. Thus, there is still the challenge of producing broad dough range yeasts that are commercially applicable.

Historically, improvement of baker's yeast strains has been empirical or limited to a few targeted phenotypes (Johnston and Oberman 1979; Evans 1990; Randez-Gil *et al.*,1999; Dequin 2001). Examples include introduction of melibiose utilisation to fully consume raffinose that is present in beet molasses, improvement of maltose utilisation to enhance fermentation in plain doughs, heightened glycolytic flux to increase fermentation rate, modulated balance of glycolysis and gluconeogenesis to increase the flow of sugar through to alcohol and CO_2, improved osmo-adaptation in sugar doughs and frozen doughs, improved freeze-thaw tolerance for frozen doughs, decreased low temperature fermentation to improve shelf life, improved trehalose storage to protect cells, resistance to organic acid preservatives, and improved flavour characteristics. Attempts to improve yeasts in these targeted areas have yielded limited success because although we can define broad targets for improvement there is still a long way to go before we can say we understand the molecular processes that determine these traits. Thus, the strategic question for today's baker's yeast researcher remains: which key genes and proteins determine the industrial usefulness of yeast strains? The answer(s) to this question will provide the rationales for highly targeted improvements of baker's yeasts. The fields of genomics and molecular genetics as applied to functional genetic analyses, will play a key role in providing the answers we need to develop strategies for deriving novel strains of yeasts. However, classical genetics will still play a major role in constructions of commercial strains, given the need to avoid the "Genetically Modified" tag in many food market applications.

2. YEAST GENOMICS AND FUNCTIONAL GENETICS

2.1 Genomics of *S. cerevisiae* in relation to industrial yeasts

Mapping of the yeast genome was carried out in the decades prior to genome sequencing by applying methods such as classical tetrad analysis, random spore analyses, trisomic analysis, mitotic crossing-over, and chromosome loss and transfer (reviewed by Mortimer and Schild 1981). The accumulated data provided information on chromosomal locations and orders of many of the genes with respect to centromeres. The derived maps were published in a series of regular updates by R.K. Mortimer and colleagues (all cited in Cherry *et al.* 1997). The cloning of chromosomal fragments and sequencing of the yeast genome provided for a physical map of the genome (Cherry *et al.* 1997), which is available from the Stanford University's *Saccharomyces* Genome Database (SGD; Web address given in Table 2). *S. cerevisiae* has 16 chromosomes making up its nuclear genome, circular mitochondrial

DNA and, in some strains, plasmids such as 2-μm. Based on the sequenced strain S288c, chromosome lengths vary from 230 kb (chromosome I) to 1532 kb (chromosome IV). The nuclear genome totals 12069 kb and the mitochondrial genome is 86 kb in length. There are many duplicated chromosomal regions in *S. cerevisiae* and this has led to the proposal that there was a whole genome duplication approximately one hundred million years ago (Wolfe and Shields 1997; Seoighe and Wolfe 1998). Initially, gene cluster analysis suggested there were about 55 major regions of clustered gene duplications not including telomeric and subtelomeric regions, which are well known for high level similarity (Mewes *et al.* 1997; Wolfe and Shields 1997). However, the number of smaller duplicated chromosomal regions may be considerably higher (Seoighe and Wolfe 1998). Ninety percent of the major duplications are in the same orientation relative to the centromere in the two copies, suggesting that reciprocal translocation has been the main form of large scale rearrangement within the yeast genome (Seoighe and Wolfe 1998).

Most strains of yeasts used in laboratories are pure haploids or diploids and have defined chromosome lengths and mutations. Industrial *S. cerevisiae* are characterised by high levels of genomic diversity. They are usually aneuploid, diploid or polyploid (Windisch 1962; Bakalinsky and Snow 1990; Codón *et al.* 1995) and carry chromosomal length polymorphisms (Rank *et al.* 1991; Bidenne *et al.* 1992; Rachidi *et al.* 1999). Baker's yeast strains show the highest level of chromosomal polymorphisms (Benítez *et al.* 1996; Codón *et al.* 1997). Considerable genomic diversity arises through the presence of transposable Ty elements in yeast genomes. These elements move from one genome location to another via an RNA intermediate. Strain S288c carries 33 copies of Ty1, 12 of Ty2, 2 of Ty3, 3 of Ty4, and none of Ty5 (details from the MIPS Comprehensive Yeast Genome Database - CYGD; Web address given in Table 2). Examples of similarities and differences in Ty patterns of different strains of baker's yeasts are shown in Fig. 1. Ty element transpositions are undoubtedly an important driver of chromosomal polymorphisms in industrial yeasts. Similarly, mitotic cross-overs, or mitotic gene conversions lead to genomic changes (discussed in reviews by Pretorius 2000; Pérez-Ortín et al. 2002). These genome-changing events, coupled with spontaneous mutations, are a major cause of genetic drift in industrial yeasts. However, the practical relevance of genetic drift in terms of performance parameters of industrial baker's yeasts is questionable. Rates of genome changes in polyploid yeasts are such that cells of baking strains do not go through sufficient generations during their lifetime in production and applications to exhibit relevant phenotypic changes.

Mating or hybridisation in yeasts provides a major source of genomic change. The diploid laboratory *S. cerevisiae* strains carry one copy of the mating-type locus MAT**a**, and one copy of MATα. These diploid cells undergo meiosis to form four-spored asci (tetrads) with two haploid spores of MAT**a** and two of MATα. Industrial yeasts are more complex, displaying

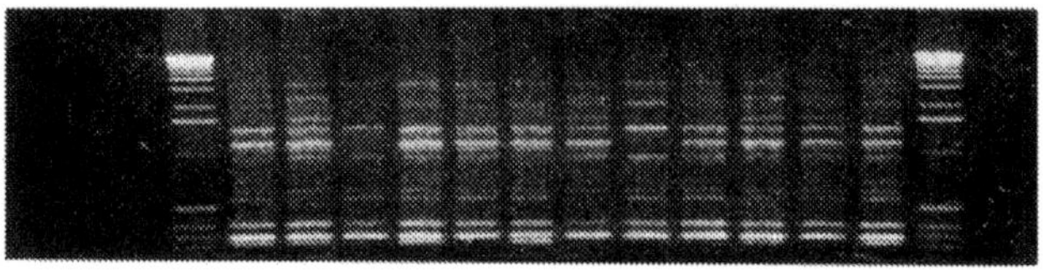

Fig. 1. RAPD analysis of Ty patterns in baker's yeast strains. Genomic DNAs isolated from 12 different baker's yeasts were subjected to PCR using oligonucleotide primers homologous to different Ty elements. The two outer lanes are DNA size markers, the remaining lanes show random patterns for the different strains of baker's yeasts.

genetic abnormalities and non-Mendelian segregation ratios (Lindgren 1949). Sporulation frequencies vary considerably between bakers' yeast strains, ranging from zero to >50% sporulation. Wine yeasts tend to give sporulation frequencies between 0 and 80% (Johnston *et al.* 2000; Pérez-Ortín *et al.* 2002). Strains of brewer's yeast often fail to sporulate or they sporulate at very low frequencies. Asci of sporulated industrial strains contain variable numbers of spores, indicative of complex ploidies. For example, asci of wine yeasts may contain two, three or four spores (Johnston *et al.* 2000). We have observed similar variability in baker's yeast, including rare eight-spored asci from some strains. Moreover, whereas spores of laboratory strains have a high level of viability, spores of industrial strains tend to give variable viability between zero and 95% (Johnston *et al.* 2000). Once germinated, only a proportion of the spores isolated from baker's yeast strains display a mating type, and this mating type is not always stable, nor exclusive to one or other type. In particular, as expected from a tetraploid parental strain, a high proportion of germinated spores will not have a mating type because they have a balanced **a**/α genetic make up. Sometimes the non-mating type spores can be further sporulated to generate true haploids with mating types. An added complexity is the homothallic or heterothallic nature of yeasts. In homothallic strains a single spore can switch mating type whilst forming a single colony, and mate with cells in the same colony that have not switched. This yields a diploid self-mated **a**/α strain. Heterothallic cells do not switch frequently if at all, and therefore can only mate with cells derived from a different spore of opposite mating type (reviewed by Herskowitz and Oshima 1981). Whereas wine yeasts are generally homothallic (Mortimer *et al.* 1994; Pretorius 2000), baker's yeasts tend to be heterothallic. Because the process of meiosis and homologous recombination introduces considerable genomic differences between spores of the same ascus leading to segregation of alleles, it is possible to eliminate spores that carry deleterious copies of genes (Fig. 2). However, it should be noted that the number of combinations of assorted genes possible through meiosis increases exponentially with the number of genes involved in the desired trait. Thus, 10 genes can be randomly assorted approximately one thousand ways (2^{10}), 20 genes can be randomly assorted in a million ways, and 30 genes can be assorted in a billion ways. Thus, if more than 10 genes affect a particular characteristic, then the number of germinated spores that need to be examined increases at a dramatic rate. In light of these facts it is salient to remember that an industrial yeast strain requires the combination of many multi-gene-determined characteristics. Nevertheless, we have found that provided suitable screening procedures can be put in place, it is possible to derive yeasts with improved industrial traits via classical means.

2.2 Functional Genetic Analyses of *Saccharomyces cerevisiae*

The sequencing of the genome of *S. cerevisiae* strain S288c revealed about 6000 genes or open reading frames (ORFs) (Dujon 1996; Goffeau *et al.* 1996; Oliver 1996). The exact number of coding ORFs remains uncertain (Basrai *et al.* 1997; Kowalczuk *et al.* 1999; Goffeau 2000; Ross-Macdonald 2000). Strain S288c is the most commonly used genetic background in research laboratories and was derived from a natural diploid isolate, EM93 isolated by Emil Mrak from a rotting fig in California in 1938 (Mortimer and Johnston 1986). The relationship between the sequenced *S. cerevisiae* strain and industrial strains is unclear, although it has been estimated that 85% of the S288c genome comes from EM93, which was probably a wine yeast (Mortimer 2000). Significantly, with regards to baking or brewing applications, strain S288c is incapable of utilising maltose, the major sugar available to yeasts in plain doughs or brewer's wort, even though the yeast genome sequencing project has identified two copies of the *MAL* loci (Volckaert *et al.* 1997; Feuermann *et al.* 1995). One likely reason why S288c is incapable of fermenting maltose appears to be that it lacks a functional *MAL* regulatory protein (V. Higgins, pers. comm.). Nevertheless, information obtained from studies of strain S288c and other laboratory strains does provide a starting position for understanding industrial yeasts.

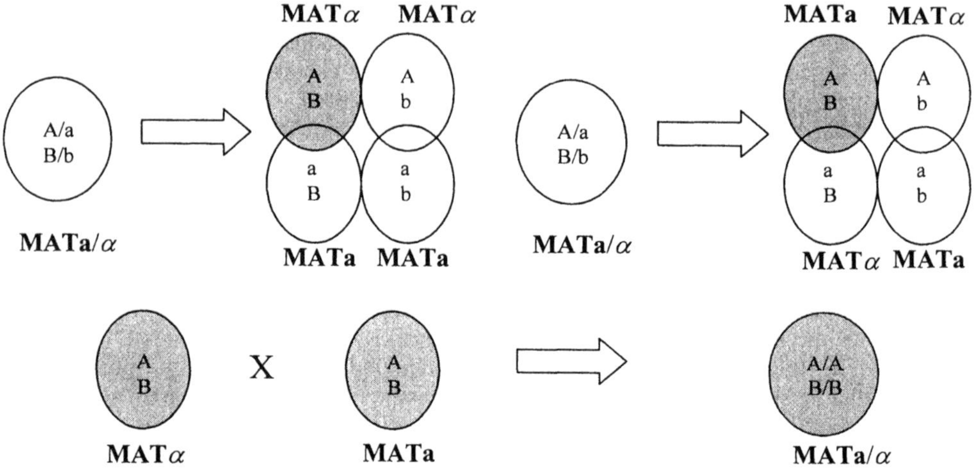

Fig. 2. Derivation of yeast strains homozygous for beneficial alleles. Two heterothallic cells heterozygous for desirable gene alleles undergo meiosis to yield haploid spores. Mating types **a** and α segregate randomly as do the strong (A or B) and weak (a or b) alleles. Random mating, or haploid pre-screening for traits and subsequent directed mating between spore types, leads to generation of 16 possible diploid varieties. In this example, only one of the possible 16 diploids would carry the homozygous A/A B/B genotype. For simplification, the progenitor cells are shown as diploids yielding four haploids each. In reality, the occurrence of true diploid and haploid states in industrial baker's yeast strains is rare. Nevertheless it is possible to obtain industrial baker's yeast segregants with optimised genetics by processes of classical mating if sufficient spores and progeny are analysed using proficient screens for desirable traits.

The fundamental challenge for yeast biologists has been to determine what the functions of the 6000 or so yeast genes are. At the time of the publication of the genome sequence, only about 30% of the ORFs corresponded to genes whose product functions were characterised by conventional means, including functional assay. About one half of the remaining ORFs showed some degree of relatedness (homologies or common motifs) to genes of yeast or other organisms whose products are functionally characterized, and as such a role in yeast can be inferred. The remaining 2000 or so ORFs had no known function. The complete data set of gene sequences for S288c has greatly enhanced the field of functional genetics allowing for a fully systematic approach to elucidating the functions of the novel genes (Winzeler and Davis 1997; Oliver *et al.* 1998; Goffeau 2000; Ross-Macdonald 2000; Oliver 2002). Approaches to gene functional analysis have included *in silico* comparisons between existing and developing databases, deletion-mutant libraries, transposon insertion libraries, DNA arrays, two-hybrid analysis and proteomics. *In silico* functional analysis can be carried out efficiently due to the availability of extensive databases of known gene and protein functions from a variety of organisms including yeast (Table 2). These databases enable the researcher to compare unknown ORF sequences with known gene sequences of yeast and other organisms, or, for example with gene sequences of various protein families (discussed by Ross-Macdonald 2000).

Functional analysis of the yeast genes has been greatly facilitated by the construction of a complete set of yeast mutants, each one carrying a deletion of one of the 6000 or so yeast genes (Winzeler *et al.* 1999). PCR was used to generate a DNA fragment containing a yeast selectable marker flanked at each end by a short (30-50 bp) sequence of homology to the target gene site. The flanking sequences allow the fragment to undergo homologous recombination at the target site, thus replacing the gene with a selectable marker. The construction of the deletants therefore relied upon the efficiency and accuracy of mitotic recombination in yeast. The selectable *kan*MX marker used for these constructions is without phenotypic effect in yeast. In addition, each of the deletants carries two unique 20 bp molecular barcode sequences (Shoemaker *et al.* 1996) that are not present in the yeast genome. These 20 bp sequences can be detected by PCR reaction on genomic DNA. This enables studies to be carried out on pools of strains as well as individuals (Baganz *et al.* 1998). Where the gene deletions give rise to viable strains, these are available as haploids of either mating type, homozygous diploid, and often as heterozygous diploid. Where disruptions cause inability to grow on rich medium, strains are available as heterozygous diploids. Almost 900 of the genes have been shown to be essential for viability and 3158 are non-essential. Yeast strains, plasmids and disruptant cassettes can be obtained through EUROSCARF, Research Genetics (Table 2), or the American Type Culture Collection. A concerted and systematic effort by more than 100 laboratories involves analysis of the deletant strains for phenotypes associated with varied growth conditions, temperature shifts, resistance to metal ions, to high osmotic pressure, to ethanol, to antibiotics and other drugs, meiosis, recombination repair, telomeric structure, transport, organelles, lipids, secretion, trafficking, cytoskeleton, cell wall, morphogenesis, sporulations, stationary phase recovery, mating, and genetic redundancy (Giaever *et al.* 2002; Goffeau 2000). The MIPS CYGD reported (6th May 2002) that 3400 of the sequenced yeast genes encoded known proteins, 230 encoded proteins with strong similarities to known proteins, 825 similar or weak similarity to known proteins, 1007 similar to unknown proteins, 516 with no similarity to other proteins, and 472 ORFs were questionable. Bianchi *et al.* (2001) carried out analysis of strains deleted for 564 of the poorly defined genes. About 30% of the deletants exhibited phenotypes associated with response to inhibitors/ stresses and many were pleiotropic. The data suggested identical contributions to cell functions of known and some unknown genes.

Yet another approach to gene function analysis has been the use of random transposon insertion (Burns *et al.* 1994; Ross-Macdonald 2000), involving the use of a mini-transposon designated mTn-3xHA/*LacZ* (modified bacterial Tn3) that contains the bacterial β-galactosidase-encoding *lacZ* gene without the initiating methionine codon and promoter elements. Insertion of the element into a transcribed and translated region of the yeast genome results in production of β-galactosidase. In-frame fusions between yeast genes and mTn-3xHA/*LacZ* also result in tagging of the fusion protein with a haemagglutinin (3xHA) epitope tag. This approach permits the study of gene disruption phenotypes, expression and protein subcellular localisation in concert (Ross-Macdonald 2000). Analysis of over 6000 random insertions provided novel data on previously unknown sporulation-induced genes and other previously uncharacterised genes, as well as subcellular localisation of in-frame 3xHA-tag insertions for over 200 constructs (Kumar *et al.* 2000). Phenotypic analysis has also been carried out by using transposons to generate random insertions in the yeast genome in a procedure termed genetic footprinting (Smith *et al.* 1995, 1996). In this procedure, the transposon mutagenised population is subjected to selection pressures and genomic DNA extracted. Transposon locations within the population's DNA are then mapped using PCR. Absence or loss of products in selected populations relative to control populations without selection pressure, is taken as indicative that those transposition events resulted in deleterious mutagenesis.

Table 2. Databases for yeast genome and functional analysis.

Database	Web address	Information
SGD	http://genome-www.stanford.edu/Saccharomyces/	1
MIPS CYGD	http://www.mips.biochem.mpg.de/proj/yeast/	1
YPD	http://www.proteome.com/databases/YPD	1
MITOP (mitochondrial)	http://www.mips.biochem.mpg.de/proj/	1
XREFdb	http://www.ncbi.nlm.nih.gov/XREFdb	2
Yeast protein function assignment	http://www.doe-mbi.ucla.edu/people/marcotte/ yeast.html	2
DIP	http://dip.doe-mbi.ucla.edu	2
Transposon insertion	http://ycmi.med.yale.edu/YGA/home.html	3
Yeast deletion project	http://sequence-www.stanford.edu/_group/ yeast_deletion_project/deletions3.html	3
Genetic footprinting	http://www.genome-www.stanford.edu/~rjk/ chrV/prechrV.html	3
Yeast SAGE	http://www.sagenet.org/	4
Yeast cell cycle project	http://cellcycle-www.stanford.edu/	4
Genome wide expression	http://web.wi.mit.edu/young/expression	4
Genome wide expression	http://www.biologie.ens.fr/yeast-publi.html	4
P.Brown laboratory	http://cmgm.stanford.edu/pbrown/	4
SWISS-2D PAGE	http://expasy.cbr.nrc.ca/ch2d/	5
Yeast protein map	http://www.ibgc.u-bordeaux2.fr/YPM	5
Two-hybrid interaction	http://portal.curagen.com	6
Two-hybrid interaction	http://depts.washington.edu/sfields/projects/YPLMhttp://w	6
EUROSCARF	ww.uni-frankfurt.de/FB/mikro/euroscarf	7
Research Genetics	http://www.resgen.com	7
Mini-transposon insertions	http://fondue.med.yale.edu/ygac/triples.htm	7

Adapted from Goffeau (2000) and Ross-Macdonald (2000). Information key: 1, central data bases for yeast genomics; 2, *in silico* analysis; 3, gene and protein functions; 4, expression; 5, proteomics; 6, protein interactions; 7, collections of yeast deletants, plasmids carrying cloned genes, gene disruptant cassettes, transposon-insertions library.

Perhaps the most productive approach to understanding gene functions on the genome-wide scale, has been the use of serial or mass gene expression analyses. Global gene expression studies reveal the mRNA transcripts that are produced under given conditions. This snap-shot of the genome-wide transcription is referred to as the transcriptome. Serial analysis of gene expression (SAGE) (Velculescu *et al.* 1997, 2000) involves the isolation of unique 15 base sequence tags from individual transcripts and concatenation of tags serially into long DNA molecules. Rapid sequencing of concatamer clones reveals individual tags and allows identification of gene transcripts. The observed frequency of a specific tag allows the researcher to infer the level of expression of a given gene under a given condition. The SAGE study by Velculescu *et al.* (1997) detected expression of 4665 genes in yeast cultures growing logarithmically or under cell cycle arrest. Kal *et al.* (1999) used SAGE to compare gene expression in yeast cells growing on either glucose or oleate. They identified 100 differentially regulated genes, many of which were associated with peroxisomal function, including 15 novel genes. As an alternative to SAGE, microarray analysis (reviewed by Bowtell 1999; Eisen and Brown 1999) offers an arguably easier and more rapid means of studying global gene expression. The microarray approach employs high-density DNA probes for virtually every ORF on glass or membrane supports (microarrays) to monitor gene expression (Shalon *et al.* 1996; De Risi *et al.* 1997; Wodicka *et al.* 1997). The probes are carefully selected for each ORF with criteria including sequence uniqueness relative to the rest of the genome, and absence of self-complementarity or clusters of single nucleotides.

The basic principle involves synthesis of fluorescently labeled cDNA from mRNA populations, and hybridisation to ORF probes on the microarray. The intensity of the fluorescent signal gives a quantitative measure of transcription from a particular gene. Microarray analysis offers the ability to analyse mass-gene function with respect to physiological and developmental conditions, to cluster genes on the basis of co-responsiveness, and to analyse global response to mutations. However, a major limitation of this technology is that small changes in transcript levels (below two-fold) are not held as experimentally significant. Nevertheless microarray analyses have already been successfully applied to the study of changes in global gene expression in relation to growth in rich versus minimal medium, diuaxic shift, meiosis, sporulation, ploidy, cell cycle, gene mutation, drug resistance, salt stress, osmotic stress, heat stress, oxidative stress, ethanol stress, overexpression or activation of transcription factors, and extended selection for mutations (see citations in Goffeau 2000; Ross-Macdonald 2000).

Key questions that relate to gene function are: what protein does an unknown ORF encode, how efficiently is the transcribed message translated, does the encoded protein interact with other proteins of known function and/ or does it undergo post-translational modification in response to changing physiological status or environmental conditions, how quickly is the protein degraded, and what is the steady state level of the protein in a cell under given conditions? Therefore other analytical methods are necessary in order to comprehend fully the role a gene and its encoded protein play in cellular biology. Here, the techniques for analysing the protein complement of cells under given circumstances (proteomics), and protein-protein interaction (two-hybrid analyses) are important complements to *in silico*, mutational and gene expression studies. Proteomics involves separation of proteins on the bases of their isoelectric points and molecular masses via two-dimensional (2-D) gel electrophoresis. This provides gels with proteins in discrete spots that can be removed, digested and analysed by mass-spectrometry (Joubert *et al.* 2001). Assayed peptide-fragment masses can be referred to protein databases, or the predicted protein masses derived from the yeast genome sequence data to provide the identity of a protein relative to its ORF. Sequencing of proteins from 2-D gels can also be used for confirmation of identity based on predicted sequences from sequenced ORF codons. Whilst there are still many technical limitations surrounding proteomics, this technology is developing rapidly and has become an important adjunct to genomic studies. Relative to gene expression analyses, protein studies can reveal post-translational alterations such as phosphorylation states of some proteins, which are important for activation/deactivation of activities. Moreover transcript levels do not always reflect the final level of activity provided by an encoded protein, such that proteomics may be more relevant to the cell's biological status than transcriptomics. Proteomics has been applied to study changes in protein complements through various growth conditions, diauxic shift, exposure of cells to heat shock (Garrels *et al.* 1997), hydrogen peroxide (Godon *et al.* 1998), and osmotic stress (Blomberg 1997).

Yet another important area of gene functional discovery is the determination of whether proteins interact with one-another. If the product of an unknown ORF is identified as interacting with a known protein this gives a major clue as to its biological function. The favoured method for analysing potential protein-protein interactions in yeast is the two-hybrid assay (Fields and Song 1989). This technique involves the fusing of proteins under study to functionally distinct domains of the *GAL4* transcription factor. If the proteins interact physically *in vivo* they bring the DNA binding domain and transcription activation domains of *GAL4* into proximity leading to expression of reporters or growth under whatever selection criteria have been designed (Fields and Song 1989; Lecrenier *et al.* 1998; Cagney *et al.* 2000). To conduct a full two-hybrid analysis on yeast would require about 36 million assays. Progress is being made towards this goal. Uetz *et al.* (2000) published the first array-based two-hybrid screen of a whole proteome describing the testing of over 1 x 10^6 possible interactions resulting in 281 potential protein-protein combinations. Ito *et al.* (2000, 2001) attempted to test all protein-protein pairs and identified about 4500 putative interactions with 841 being highly reliable. Two-hybrid analyses have also been used to determine protein interactions in mRNA metabolism pathways (Fromont-Racine *et al.* 2000). In May 2002, the MIPS CYGD indicated there were at least 9750 potential protein-protein interactions in yeast.

The major challenge associated with the various functional analyses is how to deal with the massive amount of data that they generate. Well over 1 billion data points have already been generated in microarray expression studies (Goffeau 2000). A major developing area of relevance to understanding yeast cell biology is bioinformatics, which involves computational methods for array design, image analyses, storage and organisation of experimental data, comparison of sets of data derived from different experiments, and functional interpretations, i.e. deriving biological meaning from the complex data sets (Tamames *et al.* 2002). Determining regulation and interaction of all genes and their products is the ultimate goal of functional analysis and this will not be achieved without the development of highly sophisticated informatic networks that allow modeling of gene and protein interactions under many different physiological and environmental conditions. Gene cluster analysis is a currently favoured approach used to imply functionality (Banerjee and Zhang 2002; Tamames *et al.* 2002). Array analyses can define sets or clusters of genes with similar expression patterns. The assumption is then made that these clusters are related by their involvement in particular biological processes (Lockhart *et al.* 1996). The process of defining the cellular role of the clusters typically involves *in silico* analysis by reference to different databases such as those listed in Table 2. Wu *et al.* (2002) have demonstrated the use of cluster analyses to identify new gene members of existing functional categories including 285 candidate proteins involved in transcription, processing and transport of non-coding RNA molecules. Several of these candidates appear to be involved in ribosomal RNA processing. Computational analysis of over 2700 putative protein-protein interactions derived from two-hybrid databases revealed a large network of 2358 interactions among 1548 proteins (Schwikowski *et al.* 2000). Proteins of known function and cellular locations were clustered together with 63% of interactions occurring between proteins with a common functional assignment and 76% between proteins in the same sub-cellular compartment.

The vast majority of gene deletions cause highly significant alterations in the expression of at least one other gene as revealed by whole-genome arrays (Hughes *et al.* 2000). This means that mutations can have many possible secondary effects beyond deleting the primary function of a gene's product, which adds another level of complexity to functional analysis. Featherstone and Broadie (2002) have used computational analysis to try and overcome this problem, and argue that gene expression forms a 'scale-free' network similar to artificial networks such as power grids and the internet. Their model predicts that the gene network structure is organised for robustness and helps make organisms resistant to deleterious effects of mutations, thus conferring evolutionary advantages.

The progression from genome-to-transcriptome-to-proteome status of yeast cells will only take us part of the way to understanding the biology of yeast. A further question is how the genetic information and protein complement of a cell affects its phenotype. Therefore, so-called metabolomics (profiling metabolites) becomes an important step in the quest to understand the interaction between genomics and environmental adaptation by cells. Metabolic analysis is useful for understanding the route and function of metabolic pathways in cells under various conditions. Measurement and comparison of concentrations and fluxes of metabolites also provides insight to the regulatory and interactive networks and the role of unknown genes (Phelps *et al.* 2002). Techniques employed include nuclear magnetic resonance spectroscopy, mass spectrometry, chromatographic analyses and metabolite network analysis models to estimate fluxes (Oliver 2002; Phelps *et al.* 2002). Biochemical systems theory (BST) offers a combination of computational and mathematical modeling that appears to be reliable for combining DNA microarray data with enzymatic processes to yield insights into metabolic pathway regulation. Voit and Radivoyevitch (2000) have used BST to rationalise variations in glycolytic gene expression patterns of heat shocked yeast. Their analysis provided evidence of benefits to be gained from observed changes (or not) in the activity of each enzyme and indicated that yeast cells have a highly effective, low-cost coordination of production of ATP, NADH and trehalose as well as control of intermediates under changing conditions.

To date all of the analytical techniques that are employed for functional analysis rely on sampling millions of cells *en masse*. Thus, the data we rely on is based on the mean response or status of cells within populations. The assumption is made that all cells within a pure clonal population will be similar. However, for the most part experimental samples are prepared by batch culture, which offers an extremely dynamic environment. We have shown by using flow cytometric analysis of individual cells within a batch cultured clonal population, that not all cells respond identically to mild and severe heat stresses (Attfield *et al.* 2001). Individual cells of yeast in pure culture gave differences in stress gene response of up to almost 40-fold. Similarly individual cells showed marked differences in sensitivity to heat stress. The cause of variations in cell-to-cell responses are not yet understood - they do not appear to correlate with the position that individuals are in with respect to cell cycle, and may even be chaotic. These analyses indicate however, that another level of complexity may need to be taken into account when drawing conclusions about functional analyses based on data derived in the form of means of cell populations. Heterogeneity among cells becomes especially important when considering production and application of yeast biomass in industrial situations, where consistency of performance at a defined standard is needed.

Clearly the emerging data on genomics and functional genetics of yeasts indicates that the scale of changes that occur in yeast cells and populations under different conditions is extremely complex. The level of complexity makes the application of the information to generate industrially relevant changes in yeasts via directed molecular strategies difficult.

2.3 Functional Genetic Studies in Relation to Baker's Yeast

Key aspects of modern baker's yeast performance include efficient fermentation of sugars to leaven bread doughs, and ability to cope with diverse environmental parameters that are encountered during production and application.

Maltose utilisation is a key determinant of leavening activity of baker's yeasts in plain (unsugared doughs). Genetic studies reveal that there are five unlinked maltose (*MAL*) loci in yeast and that at least one of these needs to be functional for maltose utilisation to occur (Needleman and Michels 1983; Charron *et al.* 1989). Each locus consists of a *MALx1* gene, encoding maltose permease, a *MALx2* gene coding for α-glucosidase (maltase), and a *MALx3* gene that encodes a positive regulatory protein. In laboratory strains of *Saccharomyces*, the *MALx1* and *2* genes are divergently transcribed from a bi-directional promoter. The *MALx3* product interacts with upstream activation sequences in the *MALx1/2* promoter region inducing transcription in the presence of maltose (Levine *et al.* 1992). Expression from the native *MAL* loci is induced by maltose, repressed by glucose and is basal (non-induced or -repressed) in galactose or ethanol (Needleman *et al.* 1984). Some industrial baker's yeast strains show a reduction in rate of CO_2 production in plain doughs after they have exhausted the relatively low concentrations of freely available glucose and fructose from the flour. These strains are termed "maltose lagging" (Hazell and Attfield 1999). Analysis of *MAL* loci in industrial baker's yeasts indicates they have different copy numbers of loci and probably different loci types. For example, restriction enzyme fragment data indicated that a lagging strain of baker's yeast carried *MAL2* and *MAL6*, whereas a non-lagging strain carried *MAL1*, *MAL2* and *MAL4* (unpublished data of this laboratory). Moreover, expression of *MALx1* and *Malx2* in some strains is attenuated not only via regulation by the glucose repressor encoded by *MIG1* (Hu *et al.* 1995), and induction by the *MALx3* encoded protein, but also through tandemly repeated 147b bp elements that cause structural and functional variation in the promoter region (Bell *et al.* 1997). Sequence comparisons of the *MAL* loci of industrial baker's yeasts indicate that considerable variations have arisen in the *MALx3* gene - most probably as a result of domestication of these yeasts and selection for rapid maltose fermentation over many generations. The sequence variations lead to glucose insensitivity and ability to express maltose permease and maltase at significantly higher levels in the absence of maltose that is seen in lagging strains (Higgins *et al.* 1999a and b; Danzi *et al.* 2000).

Central to the industrial usefulness of yeasts, are their abilities to adapt to the environmental challenges faced during production and applications (Attfield 1997). Thus, during production yeasts will be exposed variously, and sometimes in concert, to sugar and nutrient limitation, prolonged starvation, oxidative stress, heat and cold stress, low pH, dehydration/ desiccation, ionic stresses, osmotic stresses, exposure to organic acids, alcohols and other volatiles and heavy metals. During applications in baking yeasts will see osmotic stresses, salt stress, low pH, heat and cold stress, rehydration stresses at various temperatures, freezing and thawing, and exposure to organic acid preservatives. Yeasts, like all cellular organisms, need to maintain internal conditions within a defined range in order to grow and function optimally. The ability of a yeast strain to adapt to various challenges, which fall outside of the optimal environmental conditions for growth, will govern its usefulness and applicability across the spectrum of baking applications. In fact yeasts have numerous strategies designed to maintain internal conditions in the face of diverse external environments. Analyses of yeasts for their responsiveness to environmental challenges reveals a complex network of genetic and physiological responses involving processes of growth control, cell sensing, signal transduction, transcriptional control, post-translational modification of proteins, and synthesis of various protectants and repair factors (e.g. Piper 1993; Werner-Washburne *et al.* 1993; Thevelein 1994; Mager and De Kruijff 1995; Ruis and Schüller 1995; Moradas-Ferreira *et al.* 1996; Varela and Mager 1996; de Winde *et al.* 1997; Siderius and Mager, 1997; Attfield 1997, 1998, Attfield and Kletsas 2000; Estruch 2000; Hohmann 2002a and b; Mager and Siderius 2002). The Ras-adenylate cyclase pathway is of central importance to the regulation of cell responses to environmental conditions (Fig. 3).
Hyperosmotic stress response is particularly important for baker's yeasts because the aqueous phases of bread doughs contain high concentrations of salt and/ or sugars. In highly sweetened doughs, osmotic pressures and water activities can reach levels close to the limit of growth for *S. cerevisiae* (Myers *et al.* 1997; Attfield 1998). Baker's yeast strains vary markedly in their abilities to withstand the hyperosmotic pressures of sweetened doughs: whereas some strains are greatly inhibited, others are able to give commercially viable rates of gas production in sugar dough containing 25% or more sucrose (per wt flour). Hyperosmotic stress response is a well understood process in yeasts (reviewed by Estruch 2000; Hohmann 2002a and b; Mager and Siderius 2002). Upon hyperosmotic stress (e.g. immediately after inoculation into doughs) cells shrink and lose turgor pressure due to rapid efflux of water. Water is transferred from the vacuole into the cytoplasm as an immediate intracellular response. Several molecular events occur including growth arrest and closure

Optimal (unstressed) physiological conditions

Plentiful supply of glucose, fructose, mannose, or sucrose + nutrients, ~pH5.0, and 20-25°C.

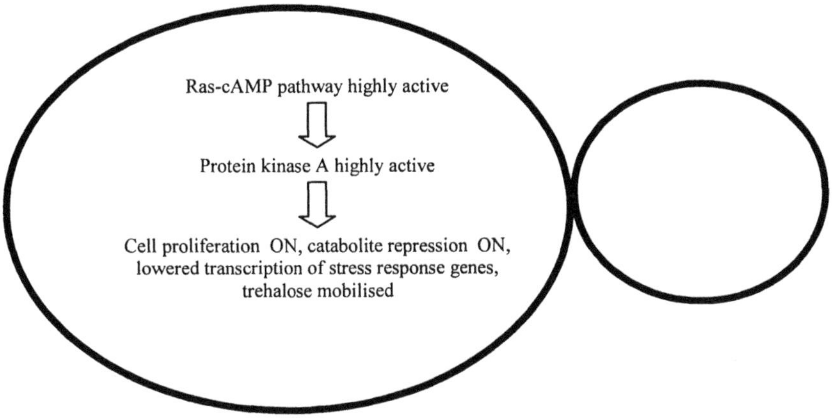

Stressed conditions

Respiratory or non-rapidly fermentable carbon sources, nutrient starvation, heat, oxidation, high osmolarity, freezing and thawing, drying and rehydration, organic acids, heavy metals.

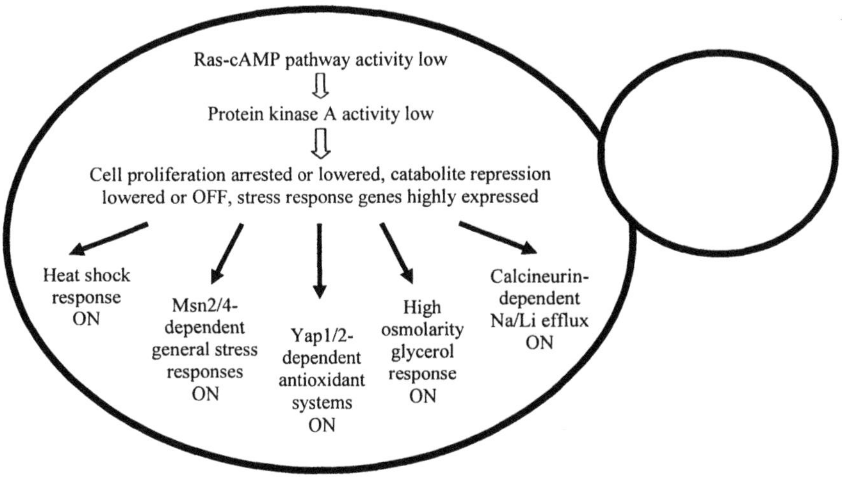

Fig. 3. The Ras-cAMP protein kinase A pathway is central to control of cell proliferation and adaptive or stress responses in yeast. In stressed conditions cell sensing and signalling factors effect activation of transcriptional regulators that induce expression of genes in the heat shock, general stress response, antioxidant and osmoresponsive pathways resulting in production of defense and repair proteins, modification of metabolism, accumulation of protectants etc.

of the Fps1p glycerol channel. Central to the ability of a strain to cope with high osmotic pressure is the generation of glycerol, which is the major osmoregulatory compound in *S. cerevisiae* (Brown 1978; Reed *et al.* 1987; Albertyn *et al.* 1994). Glycerol response is controlled by the high-osmolarity glycerol response pathway (HOG), a MAP kinase pathway that affects the expression of about 150 genes (Posas *et al.* 2000; Rep *et al.* 2000; Hohmann 2002a and b). Exposure of *S. cerevisiae* to Na^{+} also activates the Ca^{++}/calmodulin-dependent protein phosphatase, calcineurin, that appears to affect expression of over 160 genes. These calcineurin-dependent genes function in signaling pathways, ion/small molecule transport, cell wall maintenance, and vesicular transport (Yoshimoto *et al.* 2002). Yeast strains better suited to high sugar concentration doughs appear to be able to synthesise and retain glycerol, balance redox and adapt their glycolytic flux more efficiently than their plain dough counterparts in high sugar concentrations (Myers *et al.* 1997; Attfield 1998; Attfield and Kletsas 2000).

Reserve carbohydrate metabolism impacts on the quality of baker's yeast. Cellular concentrations of glycogen and trehalose influence the storage stability (shelf life) and drying and freezing tolerance of yeasts (Gadd *et al.* 1987; Gelinas *et al.* 1989; Hino *et al.* 1990; Wiemken 1990; Van Dijck *et al.* 1995). Cellular concentrations of glycogen and trehalose are responsive to a variety of factors including physico-chemical stresses, nutrient levels and carbon sources. Genetic studies reveal that the levels of reserve carbohydrates are controlled by complex, multi-gene regulatory systems involving transcriptional and post-translational mechanisms. The genes involved encode environmental sensing and signalling proteins, and enzymes for synthesis and degradation that determine patterns of accumulation or diminution of the reserve carbohydrates. Some of the genes are controlled by stress response elements. The reader is referred to François and Parrou (2001) for a comprehensive analysis of the physiology and genetics of storage carbohydrate metabolism in yeasts.

Transcriptome and proteome studies have served to underline the complexity of responses of *S. cerevisiae* cells to environmental changes. DNA microarray studies indicate massive and rapid genome-wide changes in gene expression when yeasts are exposed to various external stresses including temperature shifts, heat stress, osmotic stress, exposure to hydrogen peroxide, menadione, diamide, dithiothreitol, amino acid starvation, nitrogen depletion, stationary phase and alternative carbon sources (Gasch *et al.* 2000). In this study, cells responded with transient changes in transcript levels of hundreds of genes immediately after most environmental shifts. A significant fraction of the genome responded stereotypically to each of the stresses. Two clusters of genes, one set induced and the other repressed, showed nearly identical temporal responses. These clusters represented about 900 genes. Genes that were repressed by environmental stresses included those involved in growth, ribosomal and protein synthesis, RNA metabolism, nucleotide biosynthesis, secretion, and other metabolic processes. Over 300 genes that were induced by stresses included those known to be involved in intracellular signalling, carbohydrate metabolism, defense against oxidative radicals, cellular redox balance, cell wall modification, protein folding and degradation, vacuolar function, DNA repair, fatty acid metabolism, metabolite transport, and mitochondrial function. As well as commonly responding genes, there were specific responses to specific environmental challenges. Upon adaptation, cells appeared to return to a new steady-state level of transcription relative to unstressed cells. The level and duration of transient changes in the transcriptome varies with the degree of environmental change. Although the gene response to environmental stresses is stereotypical, regulation of response is gene- and condition-specific. Thus, expression of genes involves different transcription factors depending on the environmental conditions, and is governed by several different signalling pathways (Gasch *et al.* 2000). The implication is that yeast cells have a highly flexible genomic response to environmental change that is made up of many independent and simultaneous factors, which can be fine tuned to particular conditions. Microarray analysis of ethanol stressed cells reveals over 3% of the genes encoded in the yeast genome were upregulated, and similarly 3% down-regulated after 30 min exposure to 7% v/v alcohol (Alexandre *et al.* 2001). Ethanol-repressed genes included those with functions in protein biosynthesis, cell growth, RNA metabolism and cell biogenesis. Induced genes were those involved in energy metabolism, protein targeting, ionic balance and stress responses. Microarray analysis of yeast strains that show freeze-thaw resistance relevant to frozen dough applications indicated differences in expression of several ORFs involved in cellular organisation, metabolism, intracellular transport, transport facilitation, cell growth and division, rescue and defense, transcription, energy metabolism and ionic homoeostasis (Tanghe *et al.* 2000). Studies of gene expression in conditions relevant to wine fermentations (see Pérez-Ortín *et al.* 2002) revealed that genes involved in amino acid biosynthesis and purine biosynthesis generally gave high expression levels.

Hayes *et al.* (2002) have used hybridisation array technology to examine gene expression in chemostat-grown yeast under conditions of limited carbon or nitrogen supply and at different specific growth rates. Data derived from chemostat cultures was shown to be more reliable than batch culture for detecting gene expression changes due to the parameter being tested rather than any secondary effects arising from the uncontrolled dynamic nature of batch systems. Comparison of transcriptomes from chemostat cultures showed overexpression of several genes in carbon-limited cells relative to nitrogen-limited cells. Other genes were however, overexpressed in nitrogen-limited cells relative to carbon-limited cells. Several of the genes identified as being differentially expressed represented those with previously unassigned functions (Hayes *et al.* 2002).

The relative power of transcriptome and proteome techniques to reveal responses of yeasts to environmental challenges was exhibited by de Nobel *et al.* (2001) who compared data from DNA hybridisation analyses with 2-D protein gels of *S. cerevisiae* following exposure to the preservative sorbic acid. Proteomics revealed 10 proteins that were upregulated, and three that were absent in the presence of sorbic acid when compared with control cells. By contrast, transcriptomics revealed that 94 of the 6144 ORFs were induced 1.4-fold or more, and 72 had a reduced transcript level in cells exposed to sorbic acid relative to controls. Functional categories of genes that were induced by sorbic acid stress included stress genes (especially oxidative stress), transposon function, mating response and energy generation. Proteomic data yielded distinct information from transcript data. Only induction of Hsp26 was observed by both techniques (de Nobel *et al.* 2001). This suggests that concerted application of different analytical methods is desirable to obtain a full picture of cellular responses to environmental challenges.

Industrial applications of yeasts usually involve inoculation of stationary or anabiotic (e.g. dried) cells into the medium to be fermented - bread dough, grape juice or brewer's wort. Therefore the cellular events that occur upon inoculation and through the lag phase become important to the initiation and performance of fermentation. Brejning and Jespersen (2002) studied the protein expression of yeast during lag phase. Protein synthesis increased strongly during lag phase and the number of detectable protein spots on 2-D gels rose from about 500 at inoculation to over 1500 at the end of lag phase. Mass spectrometry-identified proteins that appeared in lag phase included those involved in carbohydrate metabolism, a ribosomal protein, translation, and biosynthetic reactions. This study represents early developments in lag phase analysis, but it underlines the importance of this approach to understanding the global changes in yeast proteome.

Joubert *et al.* (2000) reported on proteomic analysis of lager strains. The 2-D protein maps of industrial yeasts were compared with those of different *Saccharomyces* species. The proteomes of the brewing strains appeared to be a superimposition of two patterns. One pattern was derived from a *S. cerevisiae*-like genome and the other from a *S. pastorianus*.

Most of the available global data on functional genetics refers to laboratory strains of yeast grown in batch culture. It remains to be seen how transcriptomes and proteomes change when different industrial strains are exposed to conditions relevant to commercial applications. Therefore we are left with the conclusion that considerably more global expression, proteomic and metabolomic analyses need to be undertaken using industrially relevant strains and conditions, in order to define key molecular targets for industrial strain improvement. Clearly, the industrially important traits that require optimisation in yeasts will be influenced by multiple interactive factors. The effects of complex ploidies of industrial strains on gene interactions and regulatory networks remains to be elucidated.

3. MODIFYING THE GENETICS OF BAKER'S YEASTS

Genetic modifications of yeasts can be carried out using either classical genetic approaches or recombinant DNA (rDNA) techniques. Combination of classical and rDNA approaches is also feasible. From the point of view of the yeast researcher aiming to understand the biology of yeasts in order to make genetic improvements, a fundamental question is: what is the relationship between industrially relevant phenotypes and changes that need to be made in the genome of a strain? If, as in most cases, the relevant phenotype is governed by multiple genes then rDNA strategies become less feasible and classical genetics with carefully designed selection pressure(s) for the needed phenotype(s) is a more productive route to developing improved strains.

3.1 Issues Governing Choice of Classical or Recombinant Routes to Strain Improvements

Public perception and legislative matters surrounding 'Genetically Modified Organisms' (GMO's) lead to reticence in applying rDNA strategies, and preference for classical genetic approaches to improving industrial yeasts. The fact is that baker's yeast enjoys the important GRAS (generally recognised as safe) status and anything that detracts from this position could be counterproductive for manufacturers. Thus, even if an rDNA manipulations rendered a strain with outstanding properties under controlled testing conditions, a manufacturer would be reluctant to employ that strain in commercial situations until or unless legislative and public relations requirements were satisfied. Certainly the use of rDNA methods would need to result in strain properties that provide clear benefits to consumers as well as producers, for to date, regulatory authorities have tended to give approval for GMOs whilst the public have not. A major problem lies with our scientific community's inability to overcome the irrational and emotional arguments associated with "Frankenfood" imagery. The anti-GM lobby is far better prepared in terms of political lobbying and public relations, and even though many of its arguments are based on alarmist myths, the perception created is difficult to break down by using unemotional scientific facts. The fact is of course, that manipulation of native yeast genes by direct targeted mutagenesis through rDNA techniques leads to less genetic change or rearrangement than does classical mating or general mutagenesis. For example, it has proven possible to introduce *MEL* genes into baker's yeast strains via both recombinant technologies and classical breeding strategies (Liljestrom *et al.* 1991; Vincent *et al.* 1999). Direct comparison of the derived strains revealed similar fermentative capacity and similar expression levels of melibiase, yet the recombinant strain would be subject to many more legislative restrictions than the non- recombinant strain, despite the multiple "natural" genetic changes introduced into the non-recombinant strain through the process of meiosis and breeding. Table 3 lists some of the pros and cons of using classical or rDNA strain improvement strategies.

3.2 Technical considerations in yeast strain manipulation

In principle, the classical genetic manipulation of *S. cerevisiae* is easy and has been carried out for many decades. It involves the stages of sporulation to yield matable haploid forms, mating, counterselection of diploid hybrids from mixtures of haploid cell types and subsequent screening of purified hybrids for desirable traits. Although this seems simple, and indeed it is when using laboratory strains of yeasts, the technical difficulties that arise with industrial strains make classical genetics more challenging. A crucial stage in classical strain improvement is the obtention of haploids that carry useful industrial traits. This basically relies on having suitable screens that mimic industrial situations. Once suitable haploids or 'mating-type' strains have been identified that combine useful industrial characteristics, they can be mated with each other to generate novel hybrids. However, unlike laboratory yeast strains, industrial yeast strains do not generally have convenient auxotrophic markers, and parental strains, haploids and hybrids are all able to grow on the same media making it impossible to use standard complementary genetic markers to identify novel hybrids. This problem is exacerbated by the fact that some spore progeny, whilst showing a specific mating type, do not mate at high frequency and spontaneously lose their mating ability. Several alternative methods can be used to specifically identify hybrids. Most simply two strains of opposite mating type can be mixed together, incubated for several days, then streaked out to single colonies. Provided the mating reaction was efficient, and the mating types of the parental haploids were stable, hybrid strains can be readily identified from the single colonies by screening for mating type. If RAPD primers are available that can uniquely identify both parental spore progeny and offspring, the nature of the putative hybrids can be confirmed by PCR. However, problems occur if the mating type was not stable, or if mating occurred at very low frequencies.

Another method used to overcome the lack of auxotrophic markers is to introduce different antibiotic resistance markers into each of the parental yeast strains, and use dual antibiotic resistance to identify hybrid strains (Putrament *et al.* 1973). The major limitation to this method is the small range of suitable markers and the labor intensive nature of the procedures. Making one mating partner petite by deleting its mitochondrial DNA will also provide a means of selection. A novel flow cytometric based method has been developed that allows hybrids to be identified by using two-color flow cytometric cell sorting. In this procedure, one parental strain is labeled with a green fluorescent cell tracking dye, and the other is labeled with an orange fluorescent cell tracking dye. When mixed together under conditions that allow mating, hybrids can be identified and sorted on the basis of their dual orange and green fluorescence (Bell *et al.* 1998). The ability of flow cytometry to examine large numbers of cells allows this method to overcome problems associated with low mating efficiencies. Once hybrids have been constructed, a series of screening protocols are required to identify potentially improved yeast strains which can then enter complex highly controlled fermenter trials to determine the suitability of the strains for use in the industrial baking process.

The principle of genetic engineering of yeast is the same as that in any other organism. It involves isolation (cloning) of gene(s), manipulation of DNA *in vitro*, and introduction of the modified or cloned DNA into yeast in such a way that it can be expressed, replicated and transmitted to daughter cells at division. The stable inheritance of introduced DNA relies upon is introduction into the yeast genome, or linkage to a vector that is capable of autonomous replication and segregation to daughter cells. The methodologies used for rDNA manipulations of baker's yeast are essentially the same as those used for laboratory strains of yeasts (e.g. Beggs 1978; Hinnen *et al.* 1978; Botstein and Davis 1982; Ito *et al.* 1983; Struhl

Table 3. Approaches to genetic modification and their relative advantages and disadvantages.

Genetic approach	Advantages	Disadvantages
Classical	Public and political acceptance. Empirical nature means that there is low dependency on background knowledge of genetic cause and effect of industrially relevant traits	Poorly controlled outcomes - relies on statistical probability to achieve correct mutations or gene combinations. Extensive screening of very high numbers of mutants or hybrids required.
Mating/ breeding	Can be semi-directed if using known matable forms with desired phenotypes. Gives genetically stable progeny. Provides a means of combining optimal alleles, and polygenic traits	Requires sexually competent forms - yeast used in the baking industry are asexual. No convenient phenotypic markers for counter selection in industrial strains. Limited compatibility - intraspecific only. Problems occur with genetic dominance and recessiveness.
Mutagenesis (chemical or physical)	High level diversity.	Often leads to loss of desired phenotypes. Complex ploidies may require more than one gene copy to be mutagenised
Protoplast/ nuclear fusion	Intergeneric/interspecific.	Progeny can be genetically unstable for several generations. Problems with dominance and recessiveness.
Molecular	Highly directed outcomes. Reduced dependency on screening large numbers.	Currently unacceptable to public and politicians/ legislators.
rDNA techniques	Precise genetic changes to specifically targeted genes. Ability to alter control and strength of gene expression. Ability to eliminate unwanted genes. Constructs genetically stable.	Gene targets relating to industrially relevant phenotypes are poorly understood. Complex ploidies require more than one gene copy to be targeted if deleted functions are required.

1983; Webster and Dickson 1983; Parent *et al.* 1985; Williamson 1985; Evans and Attfield 1989; Pretorius 2000). There are numerous types of DNA vectors for use in yeasts. These include plasmids that integrate into the yeast genome in single- or low copy numbers, plasmids that replicate autonomously at moderate or high copy numbers, plasmids that carry centromeric sequences providing low copy number but high mitotic stability, specially designed plasmids for expressing novel or native genes under desired conditions, linear plasmids and artificial chromosomes. The choice of vector used depends upon the desired outcome of rDNA manipulation. Yeast cells can be rendered competent for introduction of exogenous DNA by treatment with chemicals such as Li salts, or via protoplasting or spheroplasting (enzymatic removal of cell wall components). Alternatively, physical methods of electroporation and biolistics are used to introduce DNA. Frequencies of transformation of industrial strains tend to be much lower than with laboratory strains. Selectable markers represent a problem because, unlike laboratory strains, the industrial yeasts are prototrophic and not haploid. Therefore, the standard auxotrophic markers such as *HIS3*, *LEU2*, *TRP1* and *URA3* cannot be used for complementation in industrial strains and there is usually a need to include a resistance marker of some sort in order to detect transformants. These markers, which include geneticin (G418), hygromycin, canavanine and heavy metal resistances can then out-recombined to leave only the desired, constructed DNA sequence. The faithful homologous recombination process in yeasts provides an excellent means of introducing novel genes, modified native genes or gene-deletions (Fig. 4).

3.3 Examples of Strain Improvements via Classical Techniques and Recombinant Techniques

Despite the technical challenges, classical genetic strategies have been used to produce yeasts that exhibit improved bread dough applications (Jacobson and Trivedi 1990; Oda and Ouchi 1990; Ejiofor *et al.* 1994). To improve the rate of adaptation to maltose in bread doughs without added sugars, laboratory yeast strains were mutagenised to generate constitutive *MAL* mutant strains that were then backcrossed to industrial yeast to ultimately construct an improved yeast strain capable of more rapid adaptation to maltose as a substrate (Johnston and Oberman 1979). However, laboratory strains in general, are unsuitable for use in baking applications due to relatively slow fermentation rates and poor abilities to utilise maltose (Bell *et al.* 2001). Angelov *et al.* (1996) used chemical mutagenesis to generate improved maltose-fermenting yeasts. Mating and screening strategies were used to generate new strains that combine high maltose-utilising activity required for plain dough leavening, with low invertase activity associated with efficient high sugar dough fermentation (Loiez *et al.* 1992). In an extension of this approach classical breeding and selection pressures were used to isolate pools of haploid yeast strains with strong plain dough or high sugar dough fermenting capacities. These were then mated *en masse* to generate a broad dough range strain that ferments efficiently in plain and high sugar doughs inspite of the fact that it has a high invertase activity (Higgins *et al.* 2001). The novel strain was efficient in high sugar doughs because its glycerol response was heightened.

Mutant selection programmes have also led to development of industrial baker's yeast strains that are substantially inactive at refrigeration temperatures, but which recover fermentative activity to normal levels at temperatures above 14°C (Hottinger *et al.* 1998; Gysler and Niederberger 2002). Such strains are potentially useful for prolonged storage as a chilled suspension or as refrigerated compressed blocks. A further example of mutant isolation is the obtention of baker's yeasts that show enhanced stress resistance during initiation of fermentation (Van Dijck *et al.* 2000). In normal strains of yeasts the onset of fermentation leads to a shut down of stress resistance factors as cells gear up for growth-associated activities. However, the so-called *fil* mutants, deficient in *f*ermentation *i*nduced *l*oss of stress resistance, offer potential advantages in frozen dough applications where preliminary fermentation occurs and cells can become stress-sensitive prior to doughs reaching freezing point. Others have also described various classical genetic strategies to obtain yeasts with improved potential for frozen dough applications (Hino *et al* 1987; Hahn and Kawai 1990; Nakagawa and Ouchi 1994).

To our knowledge there are no recombinant yeast strains being sold or used for baking. Even though the primary advantage of genetic engineering is that it allows single genetic changes to be made with high precision to yeast strains that already combine the majority of requisite industrial characteristics, targeted rDNA gene changes will not always achieve the proposed aims. The genetic redundancy that is prevalent in the *Saccharomyces* genome, coupled with ploidy issues of industrial strains, creates a considerable challenge for rDNA strategists. Large-scale functional analysis of yeast phenotypes reveals identical contributions to cell functions of some known and unknown genes in *S. cerevisiae* (Bianchi *et al.* 2001). It appears that a significant number of these genes may have involvement in cytoskeletal functions, functional networks or regulatory cascades that control different cell activities (Bianchi *et al.* 2001). Thus, trying to manipulate yeasts by specific gene knockout mutations may not be successful if an "unknown" gene can replace the function of deleted

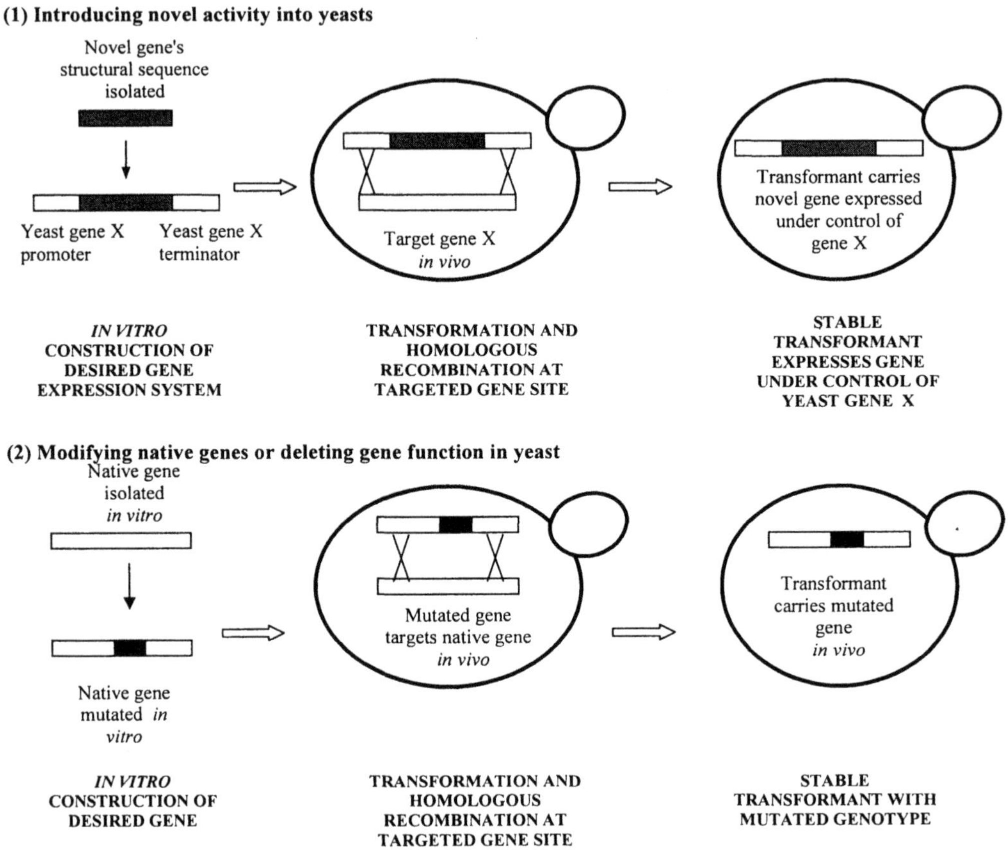

Fig. 4. Homologous recombination enables faithful targeting of genetic modifications in yeasts. 1) Novel activities (e.g. enzymes that broaden the carbohydrate utilisation abilities of yeasts) encoded by genes from other yeasts or other organisms are cloned and fused in-frame to yeast expression systems. The expression systems include promoters that are activated under desired conditions and transcriptional terminators. The gene constructs are then introduced into yeast cells and stable transformants selected. If the novel gene encodes a selectable trait such as new carbohydrate utilisation phenotype, transformants can be selected directly on that sole carbon source. 2) Native yeast genes are cloned and manipulated *in vitro*. Manipulation may involve altering the promoter or structural coding regions to modify expression or activity of a wanted gene function. Alternatively, if the gene function is unwanted, regions of the gene can be excised and replaced with other DNA (e.g. a nutritional marker such as *LEU2*, or some other marker such as an antibiotic or heavy metal resistance). This creates a stable deletion of the unwanted gene sequence. The replacement DNA is fused in-frame so that its encoded activity can be expressed faithfully. The construct is then introduced into yeast cells and transformants selected. Selection can be via the introduced deletion phenotype or the marker phenotype, or via screening for desired modified expression of the target gene.

gene copies. Alternatively, unless a desired gene manipulation leads to dominance, all copies of the target gene would need to be changed *in vivo*. Moreover, even before the vast complexities of gene regulation and networking were beginning to be unraveled, it became apparent that simplistic overexpression of target genes would not necessarily yield the desired improvements in a phenotype. For example, increasing fermentative activity by yeasts is a common goal for baking, brewing and wine-making industries and it was reasoned that overexpressing the glycolytic genes would achieve this. However, overexpression of the genes that encode the glycolytic enzymes failed to increase glycolytic flux in yeast (Schaaff *et al.* 1989). Recombinant DNA approaches have led to increased glycolytic flux, but only under conditions of increased ATP demand (Smits *et al.* 2000).

There are examples where recombinant techniques have been demonstrated to provide potential benefits to baker's yeast producers and bakers. For example, introduction of the *MEL* gene into industrial yeast strains has the potential to increase yields by up to 8% since melibiose is present (within raffinose) in relatively large quantities in beet molasses (Evans 1990). It is possible to construct a new industrial yeast strain using classical genetics since *MEL* genes are present in the species *S. cerevisiae*, although not in industrial baker's yeast strains (Vincent *et al.*, 1999). However, the classical process is laborious. By contrast, the *MEL* gene can be directly introduced into a current baker's yeast strain by transformation, leaving the many key industrial characteristics of the industrial strain unchanged (Liljestrom *et al.* 1991).

Two rDNA strategies have been used to overcome maltose lag in baker's yeasts. In the first strategy, expression of the maltose permease and maltase genes were put under control of heterologous promoters that were not subject to glucose repression (Osinga *et al.* 1989a and b). The promoters allowed the expression of the maltose permease and maltase genes under the conditions experienced in the fermenters. Consequently, the modified yeasts were able to rapidly use maltose immediately upon mixing into the dough. In the alternative strategy, investigations of the *MAL* regulatory gene indicated that non-lagging industrial yeast strains possessed a modified *MAL* regulatory gene that caused the yeast to express relatively high levels of maltose permease and maltase under non-repressing non-inducing conditions (Higgins *et al.* 1999a and b). As a result, strains possessing this regulatory mutation are preadapted to maltose utilisation prior to mixing into the dough. By cloning the *MAL* regulatory gene that conferred this phenotype on the host strains, it was possible to introduce this gene into strains that were maltose lagging. This overcame the maltose lag phenotype of the transformed strains (Higgins *et al.* 1999b).

In other developments various workers have attempted to improve the keeping and drying qualities of baker's yeast by manipulating trehalose content using rDNA techniques (Hohmann and Thevelein 1994; Londesborough and Vuorio 1995; Klionsky *et al.* 1997).

4. CONCLUSIONS

The rapidly growing knowledge of yeast genomics and functional genetics provides an excellent platform for understanding the cell biology relevant to yeast performance in industrial processes. However, it will always remain necessary to determine how relevant the findings with laboratory strains in experimental conditions are to industrial strains in conditions of their production and applications. While the unraveling of the biological processes that relate to key industrial traits is ongoing, classical genetics remains the authors' favoured approach for developing new strains. We are able to overcome problems associated with low sporulation and rare mating frequencies and this enables us to tap into a diverse gene pool from a wide variety of yeast strains that might normally be ignored for breeding programmes. Traditionally, breeding programmes have had limited success due of the difficulty of maintaining required characteristics through meiosis. However, by using high throughput screening and carefully designed enrichment/ selection techniques we can breed yeast strains with improvements in particular characteristics whilst maintaining other desirable traits.

The ability to modify the physiological characteristics of yeast strains by manipulationg environmental parameters during biomass production cannot be overlooked as an important strategy for improving yeast performance in bakery applications, i.e. by optimising the performance potential offered by a strain's genetic background. Global gene transcript and proteome analyses should help in this regard. For example, we can use these technologies to discover the gene expression and protein profiles of yeast biomass that delivers best industrial performance. Subsequent manipulation of cultures to achieve these profiles in a controlled way will provide protocols for optimising performance of yeast strains.

It is our opinion that the use of rDNA strategies for "commercial" strain improvement will only be relevant when very major benefits to bakers, consumers and yeast manufacturers can be proven, and legislation and public relations permit the release of such organisms. Moreover, the level of complexity of gene product interactions in yeasts also presents a challenge in designing highly specific genetic modifications that will provide improvement of an industrial baker's yeast strain's performance. For now, the important role of rDNA is in diagnosis of cause and effect between genes and physiology relevant to industrial situations. The knowledge gained from such studies is currently useful in designing screening, enrichment and selection protocols and is therefore an important adjunct to classical genetic strategies for obtaining novel strains.

REFERENCES

Albertyn J, Hohmann S, Thevelein JM, and Prior BA (1994). *GPD1*, which encodes glycerol-3-phosphate dehydrogenase, is essential for growth under osmotic stress in *Saccharomyces cerevisiae*, and its expression is regulated by the high-osmolarity glycerol response pathway. Mol Cell Biol 14:4135-4144.

Alexandre H, Ansanay-Galeote V, Dequin S, and Blondin B (2001). Global gene expression during short-term ethanol stress in *Saccharomyces cerevisiae*. FEBS Letts 498:98-103.

Angelov AI, Karadjov GI, and Roshkova ZG (1996). Strains selection of baker's yeast with improved technological properties. Food Res Int 29:235-239.

Attfield PV (1997). Stress tolerance: The key to effective strains of industrial baker's yeast. Nature Biotechnol 15:1351-1357.

Attfield PV (1998). Physiological and molecular aspects of hyperosmotic stress tolerance in yeasts. In: SG Pandalai, ed. Recent Developments in Microbiology. Trivandrum: Research Signpost, Vol 2, part 2, pp 427-442.

Attfield PV, Choi HY, Veal DA, and Bell PJL (2001). Heterogeneity of stress gene expression and stress resistance among individual cells of Saccharomyces cerevisiae. Mol Microbiol 40:1000-1008.

Attfield PV, and Kletsas S (2000). Hyperosmotic stress response by strains of baker's yeasts in high sugar concentration medium. Letts Appl Microbiol 31:323-327.

Baganz F, Hayes A, Farquhar R, Butler PR, Gardner DCJ, and Oliver SG (1998). Quantitative analysis of yeast gene function using competition experiments in continuous culture. Yeast 14:1417-1427.

Bakalinsky AT, and Snow R (1990). The chromosomal constitution of wine strains of *Saccharomyces cerevisiae*. Yeast 6:367-382.

Banerjee N, and Zhang MQ (2002). Functional genomics as applied to mapping transcription regulatory networks. Curr Opin Microbiol 5:313-317.

Basrai MA, Hieter P, and Boeke JD (1997). Small open reading frames: beautiful needles in the haystack. Genome Res 7:768-771.

Beggs JD (1978). Transformation of yeast by a replicating hybrid plasmid. Nature 275:104-109.

Bell PJL, Higgins VJ, Dawes IW, and Bissinger PH (1997). Tandemly repeated 147b bp elements cause structural and functional variation in divergent *MAL* promoters of *Saccharomyces cerevisiae*. Yeast 13:1135-1144.

Bell PJL, Deere D, Shen J, Chapman B, Bissinger P,H, Attfield PV, and Veal DA (1998). A flow cytometric method for rapid selection of novel industrial yeast hybrids. Appl Environ Microbiol 64:1669-1672.

Bell PJL, Higgins VJ, and Attfield PV (2001). Comparison of fermentative capacities of industrial baking and wild-type yeasts of the species Saccharomyces cerevisiae in different sugar media. Letts Appl Microbiol 32:224-229.

Benítez T, Martínez P, and Codón AC (1996). Genetic constitution of industrial yeast. Microbiologia 12:371-384.

Beudeker RF, Van Dam HW, Van Der Plaat JB, and Vellenga K (1990). Developments in baker's yeast production. In: H Verachtert, and R De Mot, eds. Yeast Biotechnology and Biocatalysis. New York: Marcel Dekker Inc., pp 103-146.
Bianchi MM, Ngo S, Vandenbol M, Sarton G, Morlupi A, Ricci C, Stafani S, Morlino GB, Hilger F, Carignani G, Slonimski PP, and Frontali L (2001). Large-scale phenotypic analysis reveals identical contributions to cell functions of known and unknown yeast genes. Yeast 18:1397-1412.
Bidenne C, Blondin B, Dequin S, and Vézinhet F (1992). Analysis of the chromosomal DNA polymorphism of wine strains of *Saccharomyces cerevisiae*. Curr Genet 22:1-7.
Blomberg A (1997). Osmoresponsive proteins and functional assessment strategies in Saccharomyces cerevisiae. Electrophoresis 18:1429-1440.
Botstein D, and Davis RW (1982). Principles and practice of recombinant DNA research with yeast. In: JN Strathern, EW Jones, and JR Broach, eds. Molecular Biology of the Yeast *Saccharomyces*: Metabolism and Gene Expression. New York: Cold Spring Harbor Laboratory Press, pp 607-636.
Bowtell DD (1999). Options available - from start to finish - for obtaining expression data by microarray. Nature Genet 21:25-32.
Brejning J, and Jespersen L (2002). Protein expression during lag phase and growth initiation in Saccharomyces cerevisiae. Int J Food Microbiol 75:27-38.
Brown AD (1978). Compatible solutes and extreme water stress in eukaryotic microorganisms. Adv Microb Physiol 17:181-242.
Burns N, Grimwade B, Ross-Macdonald PB, Choi EY, Finberg K, Roeder GS, and Snyder M (1994). Large-scale analysis of gene expression, protein localisation, and gene disruption in *Saccharomyces cerevisiae*. Genes Dev 8:1087-1105.
Burrows S (1970). Baker's yeast. In: AH Rose and JS Harrison, eds. The Yeasts, Vol 3. New York: Academic Press, pp 349-419.
Cagney G, Uetz P, and Fields S (2000). High-throughput screening for protein-protein interactions using two-hybrid assay. Meth Enzymol 328:3-14.
Charron MJ, Read E, Haut SR, and Michels CA (1989). Molecular evolution of the telomere-associated *MAL* loci of *Saccharomyces*. Genetics 122:307-316.
Chen SL, and Chiger M (1985). Production of baker's yeast. In: HW Blanch, S Drew and DIC Wang, eds. Comprehensive Biotechnology. New York: Pergamon Press, pp 429-462.
Cherry JM, Ball C, Weng S, Juvik G, Schmidt R, Adler C, Dunn B, Dwight S, Riles L, Mortimer RK, and Botstein D (1997). Genetic and physical maps of Saccharomyces cerevisiae. Nature 387 suppl: 67-73.
Cod6n AC, Benítez T, and Korhola M (1997). Chromosomal reorganisation during meiosis of Saccharomyces cerevisiae baker's yeasts. Curr Genet 32:247-259.
Cod6n AC, Gasent-Ramírez JM, and Benítez T (1995). Factors which affect the frequency of sporulation and tetrad formation in *Saccharomyces cerevisiae* baker's yeasts. Appl Environ Microbiol 61:630-638.
Danzi SE, Zhang B, and Michels CA (2000). Alterations in the *Saccharomyces MAL*-activator cause constitutivity but can be suppressed by intragenic mutations. Curr Genet 38:233-240.
De Nobel H, Lawrie L, Brul S, Klis F, Davis M, Alloush H, and Coote P (2001). Parallel and comparative analysis of the proteome and transcriptome of sorbic acid-stressed *Saccharomyces cerevisiae*. Yeast 18:1413-1428.
Dequin S (2001). The potential of genetic engineering for improving brewing, wine-making and baking yeasts. Appl Microbiol Biotechnol 56:577-588.
DeRisi JL, Iyer VR, Brown PO (1997). Exploring the metabolic and genetic control of gene expression on a genomic scale. Science 278:680-686.
De Winde JH, Thevelein JM, and Winderickx J (1997). From feast to famine: adaptation to nutrient depletion in yeast. In: S Hohmann, and WH Mager, eds. Yeast Stress Response. New York: Springer, RG Landes Co., pp 7-52.
Dujon B (1996). The yeast genome project: What did we learn? Trend Genet 12:263-270.
Ejiofor AO, Okafor N, and Ugwueze EN (1994). Development of baking yeast from Nigerian palm-wine yeast. World J Microbiol Biotechnol 10:199-202.
Eisen MB, and Brown PO (1999). DNA arrays for analysis of gene expression. Methods Enzymol 303:179-205.
Estruch F (2000). Stress-controlled transcription factors, stress-induced genes and stress tolerance in budding yeast. FEMS Microbiol Revs 24:469-486.
Evans IH (1990). Yeast strains for baking: Recent developments. In: JFT Spencer and DM Spencer, eds. Yeast Technology. Berlin:Springer-Verlag, pp 13-54.
Evans RJ, and Attfield PV (1989). Genetic engineering of yeasts: principles and applications. In. PL Rogers, and GH Fleet, eds. Biotechnology and the Food Industry. Melbourne: Gordon and Breach Science Publishers, pp 33-60.

Featherstone DE, and Broadie K (2002). Wrestling with pleiotropy: genomic and topological analysis of the yeast gene expression network. BioEssays 24:267-274.

Feuermann M, Charbonnel L, de Montigny J, Bloch JC, Potier S, and Souciet JL (1995). Sequence of a 9.8kb segment of yeast chromosome II including three genes of the *MAL3* locus and three unidentified open reading frames. Yeast 11:667-672.

Fields S, and Song O (1989). A novel genetic system to detect protein-protein interactions. Nature 340:245-246.

François J, and Parrou JL (2001). Reserve carbohydrates metabolism in the yeast *Saccharomyces cerevisiae.* FEMS Microbiol Rev 25:125-145.

Fromont-Racine M, Mayes A, Brunet-Simon A, Rain J-C, Colley A, Dix I, Decourty L, Joly N, Ricard F, Beggs JD, and Legrain P (2000). Genome-wide protein interaction screens reveal functional networks involving Sm-like proteins. Yeast 17:95-110.

Gadd GM, Chalmers K, and Reed RH (1987). The role of trehalose in dehydration resistance. FEMS Microbiol Letts 48:249-254.

Garrels JI, McLaughlin CS, Warner JR, Futcher B, Latter GI, Kobayashi R, Schwender B, Volpe T, Anderson DS, Mesquita-Fuentes R, and Payne WE (1997). Proteome studies of Saccharomyces cerevisiae: identification of abundant proteins. Electrophoresis 18:1347-1360.

Gasch AP, Spellman PT, Kao, CM, Carmel-Harel O, Eisen MB, Storz, G, Botstein D, and Brown PO (2000). Genomic expression programs in the response of yeast cells to environmental changes. Mol Biol Cell 11:4241-4257.

Gelinas P, Fiset G, LeDuy A, and Goulet J (1989). Effect of growth conditions and trehalose content on cryotolerance of baker's yeast in frozen doughs. Appl Environ Microbiol 55:2453-2459.

Giaever G (and 71 others) (2002). Functional profiling of the Saccharomyces cerevisiae genome. Nature 418:387-391.

Godon C, Lagniel G, Lee J, Buhler JM, Kieffer S, Perrot M, Boucherie H, Toledano MB, and Labarre J (1998). The H2O2 stimulon in *Saccharomyces cerevisiae.* J Biol Chem 273:22480-22489.

Goffeau A (2000). Four years of post-genomic life with 6000 yeast genes. FEBS Letts 480:37-41.

Goffeau A, Barrell BG, Bussey H, Davis RW, Dujon B, Feldmann H, Galibert F, Hoheisel JD, Jacq C, Johnston M, Louis EJ, Mewes HW, Murakami Y, Philippsen P, Tettelin H, and Oliver SG. (1996). Life with 6000 genes. Science 274:546, 563-567.

Gysler C, and Niederberger P. (2002). The development of low temperature inactive (Lti) baker's yeast. Appl Microbiol Biotechnol 58:210-216.

Hahn YS, and Kawai H (1990). Isolation and characterisation of freeze-tolerant yeast from nature available for the frozen-dough method. Agric Biol Chem 54:829-831.

Hayes A, Zhang N, Wu J, Butler PR, Hauser NC, Hoheisel JD, Lim FL, Sharrocks AD, and Oliver SG (2002). Hybridisation array technology coupled with chemostat culture: tools to interrogate gene expression in Saccharomyces cerevisiae. Methods 26:281-290.

Hazell BW, and Attfield PV (1999). Enhancement of maltose utilisation by *Saccharomyces cerevisiae* in medium containing fermentable hexoses. J Ind Microbiol Biotechnol 22:627-632.

Herskowitz I, and Oshima Y (1981). Control of cell type in *Saccharomyces cerevisiae*: Mating type and mating-type interconversion. In: Srathern JN, Jones EW, and Broach JR, eds. The Molecular Biology of the Yeast *Saccharomyces*: Life Cycle and Inheritance. New York: Cold Spring Harbor Laboratory, pp 181-209.

Higgins VJ, Braidwood M, Bell P, Bissinger P, Dawes IW, and Attfield PV (1999a). Genetic evidence that high noninduced maltase and maltose permease activities, governed by *MALx3*-encoded transcriptional regulators, determine efficiency of gas production by baker's yeast in unsugared dough. Appl Environ Microbiol 65:680-685.

Higgins VJ, Braidwood M, Bissinger P, Dawes IW, and Attfield PV (1999b). Leu343Phe substitution in the Malx3 protein of *Saccharomyces cerevisiae* increases the constitutivity and glucose insensitivity of *MAL* gene expression. Curr Genet 35:491-498.

Higgins VJ, Bell PJL, Dawes IW, and Attfield PV (2001). Generation of a novel *Saccharomyces cerevisiae* strain that exhibits strong maltose utilisation and hyperosmotic resistance using nonrecombinant techniques. Appl Environ Microbiol 67:4346-4348.

Hinnen, A, Hicks JB, and Fink GR (1978). Transformation of yeast. Proc Natl Acad Sci USA 75:1929-1933.

Hino A, Mihara K, Nakashima K, and Takano H (1990). Trehalose levels and survival ratio of freeze-tolerant versus freeze-sensitive yeasts. Appl Environ Microbiol 56:1386-1391.

Hino A, Takano H, and Tanaka Y (1987). New freeze-tolerant yeast for frozen dough preparations. Cereal Chem 64:269-275.

Hohmann S (2002a). Osmotic adaptation in yeast -control of the yeast osmolyte system. Int Rev Cytol 215:149-187.

Hohmann S (2002b). Osmotic stress signaling osmoadaptation in yeasts. Microbiol Molec Biol Revs 66:300-372.

Hohmann S, and Thevelein JM (1994). Souches de levures transformées de manière à posséder une résistance au stress et/ou un pouvoir fermentatif améloiré. European patent EPO 0577915A1.
Hottinger H, Gysler C, Niederberger P (1998). Baker's yeast having a low temperature inactivation property. US patent no. 5,827,724.
Hu Z, Nehlin JO, Ronne H, and Michels CA (1995). *MIG1*-dependent and *MIG1*-independent glucose regulation of *MAL* gene expression in *Saccharomyces cerevisiae*. Curr Genet 28:258-266.
HughesTR, Marton MJ, Jones AR, Roberts CJ, Stoughton R, Armour CD, Bennett HA, Coffey E, Dai H, He YD, Kidd MJ, King AM, Meyer MR, Slade D, Lum PY, Stepaniants SB, Shoemaker DD, Gachotte D, Chakraburtly K, Simon J, Bard M, and Friend SH (2000). Functional discovery via a compendium of expression profiles. Cell 102:109-126.
Ito H, Fukuda Y, Murata K, and Kimura A (1983). Transformation of intact yeast *Saccharomyces cerevisiae* cell treated with alkali cations. J Bacteriol 153:63-68.
Ito T, Chiba T, Ozawa R, Yoshida, Hattori M, and Sakaki Y (2001). A comprehensive two-hybrid analysis to explore the yeast protein interactome. Proc Natl Acad Sci USA 98:4569-4574.
Ito T, Tashiro K, Muta S, Ozawa R, Chiba T, Nishizawa M, Yamamoto K, Kuhara S, and Sakaki Y (2000). Toward a protein-protein interaction map of the budding yeast: a comprehensive system to examine two-hybrid interactions in all possible combinations between the yeast proteins. Proc Natl Acad Sci USA 97:1143-1147.
Jacobson GK, and Trivedi NB (1990). Yeast strains, method of production and use in baking. US patent 4 973 560.
Johnston JR, Baccari C, and Mortimer RK (2000). Genotypic characterisation of strains of commercial wine yeasts by tetrad analysis. Res Microbiol 151:583-590.
Johnston JR, and Oberman H (1979). Yeast genetics in industry. In: MJ Bull, ed. Progress in Industrial Microbiology. Amsterdam: Elsevier, Vol 15, pp 151-205.
Joubert R, Brignon P, Lehmann C, Monribot C, Gendre F, and Boucherie H (2000). Two-dimensional gel analysis of lager brewing yeasts. Yeast 16:511-522.
Joubert R, Strub J-M, Zugmeyer S, Kobi D, Carte N, van Dorsselaer A, Boucherie H, and Jaquet-Gutfreund L (2001). Identification by mass-spectrometry of two-dimensional electrophoresis-separated proteins extracted from lager brewing yeast. Electrophoresis 22:2969-2982.
Kal AJ, van Zonneveld AJ, Benes V, van den Berg M, Koerkamp MG, Albermann K, Strack N, Ruijter JM, Richter JM, Richter A, Dujon B, Ansorge W, and Tabak HF (1999). Dynamics of gene expression revealed by comparison of serial analysis of gene expression transcript profiles from yeast grown on two different carbon sources. Mol Biol Cell 10:1859-1872.
Klionsky D, Holzer H, and Struelle M (1997). Stress tolerant yeast mutants. International patent no. WO97/01626.
Kowalczuk M, Mackiewicz P, Gierlik A, Dudek MR, and Cebrat S (1999). Total number of coding open reading frames in the yeast genome. Yeast 15:1031-1034.
Kumar A, Cheung KH, Ross-Macdonald P, Coelho PS, Miller P, and Snyder M (2000). TRIPLES: a database of gene function in *Saccharomyces cerevisiae*. Nucl Acids Res 28:81-84.
Lecrenier N, Foury F, and Goffeau A (1998). Two-hybrid systematic screening of the yeast proteome. BioEssays 20:1-6.
Levine J, Tanouye L, and Michels CA (1992). The $UAS_{(MAL)}$ is a bidirectional promoter element required for the expression of both the *MAL61* and *MAL62* genes of the *Saccharomyces MAL6* locus. Curr Genet 22:181-189.
Liljeström PL, Tubb RS, and Korhola MP (1991). Construction of new alpha-galactosidase producing yeast strains and the industrial application of these strains. US patent no. 5,055,401.
Lindgren CC (1949). The yeast cell: its genetics and cytology. St. Louis USA: Education Publ. Inc.
Lockhart DJ, Dong H, Byrne MC, Follettie MT, Gallo MV, Chee MS, Mittmann M, Wang C, Kobayashi M, Horton H, and Brown PO (1996). Expression monitoring by hybridisation to high density oligonucleotide arrays. Nature Biotechnol 14:1675-1680.
Loiez A, Clement P, and Colavizza D (1992). Baker's yeast strains, their process of obtention, corresponding fresh and dry yeasts. European Patent EPO 0511108.
Londesborough J, and Vuorio O (1995). Method to increase the trehalose content of organisms by transforming them with the structural genes for the short and long chains of yeast trehalose synthase. US patent no. 5,422,254.
Mager WH, and De Kruijff AJ (1995). Stress-induced transcriptional activation. Microbiol Rev 59:506-531.
Mager WH, and Siderius M (2002). Novel insights into the osmotic stress response of yeast. FEMS Yeast Res 2:251-257.
Mewes HW, Albermann K, Bähr M, Frishman D, Gleissner A, Hani, J, Heumann K, Kleine K, Maierl A, Oliver SG, Pfeiffer F, and Zollner A (1997). Overview of the yeast genome. Nature 387 suppl: 7-8.

Moradas-Ferreira P, Costa V, Piper P, and Mager WH (1996). The molecular defense against reactive oxygen species in yeast. Mol Microbiol 19:651-658.
Mortimer RK (2000). Evolution and variation of the yeast (*Saccharomyces*) genome. Genome Res 10:403-409.
Mortimer RK, and Johnston JR (1986). Genealogy of principle strains of the Yeast Genetic Stock Center. Genetics 113:35-43.
Mortimer RK, Romano P, Suzzi G, and Polsinelli M (1994). Genome renewal: A new phenomenon revealed from a genetic study of 43 strains of *Saccharomyces cerevisiae* derived from natural fermentation of grape musts. Yeast 10:1543-1552.
Mortimer RK, and Schild D (1981). Genetic mapping in *Saccharomyces cerevisiae*. In: Srathern JN, Jones EW, and Broach JR, eds. The Molecular Biology of the Yeast *Saccharomyces*: Life Cycle and Inheritance. New York: Cold Spring Harbor Laboratory, pp 11-26
Myers DK, Lawlor DTM, and Attfield PV (1997). Influence of invertase activity and glycerol synthesis and retention on fermentation of media with high sugar concentration by *Saccharomyces cerevisiae*. Appl Environ Microbiol 63:145-150.
Nagodawithana TW, and Trivedi N (1990). Yeast selection for baking. In: CJ Panchal, ed. Yeast Strain Selection. New York: Marcel Dekker, pp 139-184.
Nakagawa S, and Ouchi K (1994). Construction from a single parent of baker's yeast strains with high freeze tolerance and fermentative activity in both lean and sweet doughs. Appl Environ Microbiol 60:3499-3502.
Needleman RB, and Michels CA (1983). Repeated family of genes controlling maltose fermentation in *Saccharomyces carlsbergensis*. Mol Cell Biol 3:796-802.
Needleman RB, Kaback DB, Dubin RA, Perkins EL, Rosenberg NG, Sutherland KA, Forrest DB, and Michels CA (1984). *MAL6* of *Saccharomyces*: a complex genetic locus containing three genes required for maltose fermentation. Proc Natl Acad Sci USA 81:2811-2815.
Oda Y, and Ouchi K (1990). Hybridisation of baker's yeast by the rare-mating method to improve leavening ability in dough. Enzyme Microbiol Technol 12:989-993.
Oliver SG (1996). From DNA sequencing to biological function. Nature 379:597-600.
Oliver SG (2002). Functional genomics: lessons from yeast. Phil Trans R Soc Lond B 357:17-23.
Oliver SG, Winson MK, Kell DB, and Baganz F (1998). Systematic functional analysis of the yeast genome. Trend Biotechnol 16:373-378.
Osinga KA, Beudeker RF, van der Plaat JB, and de Hollander JA (1989a). New yeast strains providing for an enhanced rate of the fermentation of sugars, a process to obtain such yeast and the use of these yeasts. European patent EPO 03060107.
Osinga KA, Renniers ACHM, Welbergen JW, Roobol RH, and van der Wilden W (1989b). Maltose fermentation in Saccharomyces cerevisiae. Yeast 5:S207-S212.
Oura E, Suomalainen H, and Viskari AK (1983). Breadmaking. In: AH Rose, ed. Economic Microbiology, Vol 7. London: Academic press, pp 84-146.
Parent SA, Fenimore CM, and Bostian KA (1985). Vector systems for the expression, analysis and cloning of DNA sequences in *S cerevisiae*. Yeast 1:83-138.
Pérez-Ortín JE, García-Martinez J, and Alberola TM (2002). DNA chips for yeast biotechnology: The case of wine yeasts. J Biotechnol. 98:227-241.
Phelps TJ, Palumbo AV, and Beliaev AS (2002). Metabolomics and microarrays for improved understanding of phenotypic characteristics controlled by both genomics and environmental constraints. Curr Opin Biotechnol 13:20-24.
Piper PW (1993). Molecular events associated with acquisition of heat tolerance by the yeast *Saccharomyces cerevisiae*. FEMS Microbiol Revs 11:339-356.
Posas F, Chambers JR, Heyman JA, Hoeffler JP, de Nadal E, and Arino J (2000). The transcriptional response of yeast to saline stress. J Biol Chem 275:17249-17255.
Pretorius IS (2000) Tailoring wine yeast for the new millennium: novel approaches to the ancient art of winemaking. Yeast 16:675-729.
Putrament A, Baranowska H, and Prazmo W (1973). Induction by manganese of mitochondrial antibiotic resistance mutation in yeast. Mol Gen Genet 126:357-366.
Rachidi N, Barre P, and Blondin B (1999). Multiple Ty-mediated chromosomal translocations lead to karyotype changes in a wine strain of *Saccharomyces cerevisiae*. Mol Gen Genet 261:841-850.
Randez-Gil F, Sanz P, and Prieto JA (1999). Engineering baker's yeast: room for improvement. Trends Biotechnol 17:237-244.
Rank GH, Casey, GP, Xiao W, and Pringle AT (1991). Polymorphism within the nuclear and 2-μm genomes of *Saccharomyces cerevisiae*. Curr Genet 20:189-194.
Reed G, and Nagodawithana TW (1991). Yeast Technology. 2nd ed. New York: Van Nostrand Reinhold, pp 261-368.

Reed RH, Chudek JA, Foster R, and Gadd GM (1987). Osmotic significance of glycerol accumulation in exponentially growing yeasts. Appl Environ Microbiol 53:2119-2123.

Rep M, Krantz M, Thevelein JM, and Hohmann (2000). The transcriptional response of *Saccharomyces cerevisiae* to osmotic shock. Hot1p and Msn2p/Msn4p are required for the induction of subsets of high osmolarity glycerol pathway-dependent genes. J Biol Chem 275:8290-8300.

Ross-Macdonald P (2000). Functional analysis of the yeast genome. Funct Integr Genomics 1:99-113.

Ruis H and Schüller C (1995). Stress signalling in yeast. Bioessays 17:959-965.

Schaaff I, Heinisch J, and Zimmerman FK (1989). Overproduction of glycolytic enzymes in yeast. Yeast 5:285-290.

Seoighe C, and Wolfe KH (1998). Extent of genomic rearrangement after genome duplication in yeast. Proc Natl Acad Sci USA 95:4447-4452.

Schwikowski B, Uetz P, and Fields S (2000). A network of protein-protein interactions in yeast. Nature Biotechnol. 18:1257-1261.

Shalon D, Smith SJ, and Brown PO (1996). A DNA microarray system for analysing complex DNA samples using two-colour fluorescent probe hybridisation. Genome Res 6:639-645.

Shoemaker DD, Lashkari DA, Morris D, Mittmann M, and Davis RW (1996). Quantitative phenotypic analysis of yeast deletion mutants using a highly parallel molecular bar-coding strategy. Nature Genet 4:450-456.

Siderius M, and Mager WH (1997). General stress response: in search of a common denominator. In: S Hohmann, and WH Mager, eds. Yeast Stress Response. New York: Springer, RG Landes Co., pp 213-230.

Smith V, Botstein D, and Brown PO (1995). Genetic footprinting: a genomic strategy for determining a gene's function given its sequence. Proc Natl Acad Sci USA 92:6479-6483.

Smith V, Chou KN, Lashkari D, Botstein D, and Brown PO (1996). Functional analysis of the genes of yeast chromosome V by genetic footprinting. Science 274:2069-2074.

Smits HP, Hauf J, Müller S, Hobley TJ, Zimmermann FK, Hahn-Hägerdal B, Nielsen J, and Olsson L (2000). Simultaneous overexpression of enzymes in the lower part of glycolysis can enhance the fermentative capacity of *Saccharomyces cerevisiae*. Yeast 16:1325-1334.

Spencer JFT, and Spencer DM (1983). Genetic improvement of industrial yeasts. Ann Rev Microbiol 37:121-142.

Stear CA (1990). Handbook of Breadmaking Technology. New York: Elsevier Applied Sciences.

Struhl K (1983). The new yeast genetics. Nature 305:391-397.

Tanghe A, Teunissen A, Van Dijck P, and Thevelein JM (2000). Identification of genes responsible for improved cryoresistance in fermenting yeast cells. Int J Food Microbiol 55:259-262.

Tamames J, Clark D, Herrero J, Dopazo J, Blaschke C, Fernández JM, Oliveros JC, and Valencia A (2002). Bioinformatics methods for the analysis of expression arrays: data clustering and information extraction. J Biotechnol 98:269-283.

Thevelein JM (1994). Signal transduction in yeast. Yeast 10:1753-1790.

Trivedi NB, Jacobson GK, and Tesch W (1986). Baker's Yeast. CRC Rev Biotechnol 4:75-109.

Uetz P, Giot L, Cagney G, Mansfield TA, Judson RS, Knight JR, Lockshon D, Narayan V, Srinivasan M, Pochart P, Qureshi-Emili A, Li Y, Godwin B, Conover D, Kalbfleisch T, Vijayadamodar G, Yang M, Johnston M, Fields S, and Rothberg JM (2000). A comprehensive analysis of protein-protein interactions in *Saccharomyces cerevisiae*. Nature. 403:623-7.

Van Dijck P, Colavizza D, Smet P, and Thevelein JM (1995). Differential importance of trehalose in stress resistance in fermenting and nonfermenting *Saccharomyces cerevisiae*. Appl Environ Microbiol 61:109-115.

Van Dijck P, Gorwa M-F, Lemaire K, Teunissen A, Versele M, Colombo S, Dumortier F, Ma P, Tanghe A, Loiez A, and Thevelein JM (2000). Characterisation of a new set of mutants deficient in fermentation-induced loss of stress resistance for use in frozen dough applications. Int J Food Microbiol 55:187-192.

Varela JCS, and Mager WH (1996). Response of *Saccharomyces cerevisiae* to changes in external osmolarity. Microbiology 142:721-731.

Velculescu VE, Vogelstein B, and Kinzler KW (2000). Analysing unchartered transcriptomes with SAGE. Trend Genet 16:423-425.

Velculescu VE, Zhang L, Zhou W, Vogelstein J, Basrai MA, Bassett DE Jr, Hieter P, Vogelstein B, and Kinzler KW (1997). Characterisation of the yeast transcriptome. Cell 88:243-251.

Vincent SF, Bell PJL, Bissinger P, and Nevalainen KMH (1999). Comparison of melibiose utilising baker's yeast strains produced by genetic engineering and classical breeding. Letts Appl Microbiol 28:148-152.

Volckaert G, Voet M, and Robben J (1997). Sequence analysis of a near-subtelomeric 35.4 kb DNA segment on the right arm of chromosome VII from *Saccharomyces cerevisiae* carrying the *MAL1* locus reveals 15 complete open reading frames, including *ZUO1*, *BGL2* and *BIO2* genes and an ABC transporter gene. Yeast 13:251-259.

Voit EO, and Radivoyevitch T (2000). Biochemical systems analysis of genome-wide expression data. Bioinformatics 16:1023-1037.

Webster TD, and Dickson RC (1983). Direct selection of *Saccharomyces cerevisiae* resistant to the antibiotic G418 following transformation with a DNA vector carrying the kanamycin resistance gene of transposon Tn903. Gene 26:243-252.
Werner-Washburne M. Braun E, Johnston GC, and Singer RA (1993). Stationary phase in the yeast *Saccharomyces cerevisiae*. Microbiol Rev. 57:383-401.
Wiemken A (1990). Trehalose in yeast, stress protectant rather than reserve carbohydrate. Antonie van Leeuwenhoek 58:209-2217.
Williamson DH (1985). Cloning in brewer's yeast, *Saccharomyces cerevisiae*. In: JM Walker, and EB Gingold, eds. Molecular Biology and Biotechnology. London: The Royal Society, pp.102-118.
Windisch S (1962). Genetic yeast research: methods and some new results. Wallerstein Comm 24:316-323.
Winzeler EA, and Davis RW (1997). Functional analysis of the yeast genome. Curr Opin Genet Dev 7:771-776.
Winzeler EA (and 51 others) (1999). Functional characterisation of the *S. cerevisiae* genome by gene deletion and parallel analysis. Science 285:901-906.
Wodicka L, Dong, H, Mittmann M, Ho M-H, and Lockhart DJ (1997). Genome-wide expression monitoring in *Saccharomyces cerevisiae*. Nature Biotechnol 15:1359-1367.
Wolfe KH, and Shields DC (1997). Molecular evidence for an ancient duplication of the entire yeast genome. Nature 387:708-713.
Wu LF, Hughes TR, Davierwala AP, Robinson MD, Stoughton R, and Altschuler SJ (2002). Large-scale prediction of *Saccharomyces cerevisiae* gene function using overlapping transcriptional clusters. Nature Genet 31:255-265.
Yoshimoto H, Saltsman K, Gasch AP, Li HX, Ogawa N, Botstein D, Brown PO, and Cyert MS (2002). Genome-wide analysis of gene expression regulated by the calcineurin/Crz1p signaling pathway in Saccharomyces cerevisiae. J Biol Chem 277:31079-31088.

Applied Mycology & Biotechnology
An International Series. Volume 3. Fungal Genomics

11

Enzyme Production in Industrial Fungi- Molecular Genetic Strategies for Integrated Strain Improvement

K.M. Helena Nevalainen and Valentino S. Jnr. Te'o
Department of Biological Sciences, Macquarie University, Sydney, NSW 2109, Australia.

Filamentous fungi have an established and central role in the industrial production of enzymes for various applications ranging from animal feed manufacture to pulp bleaching. Filamentous fungi are also of high interest as efficient expression hosts for a wide range of valuable gene products originating from other organisms. Progress aiming at reaching the full potential of filamentous fungi as production hosts depends on better understanding of gene regulation, protein modification, function of the secretory pathway, genetic and physiological aspects related to product fermentation and studies into functional genomics.

1. INTRODUCTION

The global market for industrial enzymes in 2000 was approximately $2 billion with an annual growth rate of 5-10 %. The US demand alone for enzymes is projected to exceed $2.6 billion in 2004, spurred by the advent of novel biocatalysts targeted at a multitude of end uses (e.g. textiles, cosmetics). Medical and diagnostic enzymes will remain on the top by value, while the enzyme industry will continue to benefit from biotechnology-based approaches to production. The industrial enzyme market comprises enzymes targeted for (i) technical applications such as pulp and paper, textile and laundry detergent industries, (ii) enzymes for food processes such as starch processing, brewing and cheese making and (iii) enzymes for animal feed production. The largest of the three key segments is technical enzymes, about $1 billion, of which two thirds are dominated by detergent enzymes. A considerable amount of these enzymes is produced by filamentous fungi using genetically modified strains.

Filamentous fungi, for example *Trichoderma reesei* and *Aspergillus niger* var. *awamori*, are among the most powerful secretors of extracellular protein in nature and have been extensively used in industry to produce various biocatalysts for over several decades. In addition to their secretion ability, these fungi grow on cheap undefined industrial media and provide a eukaryotic cell machinery for protein processing. Fungal enzymes are typically produced by submerged fermentation in volumes reaching 500,000 cubic meters.

The development of molecular biology tools for industrial fungi towards the end of 1980's revolutionized the development of new enzyme producers and products. Novozymes A/S (Denmark) was the first on the market in 1988 with a lipase product, Lipolase^{TM}, for detergents. The first act on genetic modification was implemented in 1986 by the Danish

Parliament. Today, regulations concerning genetically modified microorganisms are documented in international agreements, directives and recommendations, EC directives and decrees and, at the national level, by legislation and guidelines concerning varying aspects such as the manufacturing practises, quality control, safety issues and labelling of recombinant products.

Filamentous fungi currently used for large scale enzyme manufacture feature strains of *Aspergillus niger* var. *awamori*, *A. oryzae*, *Trichoderma reesei*, *Rhizomucor miehei and Humicola lanuginosa*. Recently, industrial expression systems have been developed to *Fusarium venenatum* (Blinkovsky *et al*., 1999) and *Chrysosporium lucknowense* (van Zeijl *et al.* 2001). Industrially–exploited production hosts are required to have GRAS status (Generally Regarded As Safe) or otherwise a long history of safe use in industry. Today, most industrial strains, both fungal and bacterial, are genetically modified organisms (GMO) tailored for overproduction of a particular enzyme with decreased or no production of undesired side activities. Contemporary fungal strain improvement draws from cell biology, molecular and genetic knowledge of a particular organism or group of organisms, fermentation physiology and functional genomic approaches to design strains for industrial enzyme production worldwide. In this review we will discuss both genetic and some fermentation strategies related to development of industrial fungal strains and improvement of product yields.

2. GENERAL AIMS IN FUNGAL STRAIN IMPROVEMENT

Relatively few fungal species and strains still dominate the enzyme production industry. However, novel production systems are currently being developed for the obvious reason to circumvent the tight patent protection around e.g. *Aspergillus* and *Trichoderma* production strains. The general aim using fungal producers is to achieve high level expression and effective secretion of a particular gene product, either homologous or heterologous in order to lower production costs and obtain a better product (Table 1).

Table 1. Goals of and technologies for fungal strain improvement.

Better yield	Methods
Homologous gene products	Sexual crossing (when applicable)
	Somatic crossing
	Random mutagenesis
	Genetic engineering
Heterologous gene products	Genetic engineering
	Random mutagenesis of transformants
	Crossing of transformants (when applicable)
Better product	Genetic engineering (enzyme profile modification)
	Protein engineering (improved protein)
Metabolic modification	Genetic engineering to introduce novel pathways or modify existing ones

A larger production volume of particular engineered-to-application enzymes such as proteases and lipases would be required, for example, for their application in enzymatic cleaning of industrial, medical, dental and veterinary premises and private households. These include enzymes homologous or heterologous to the production host. Fungal production of pharmaceutical and therapeutic compounds such as monoclonal antibodies, insulin and growth hormones would engage fungi as heterologous production hosts to benefit from the eukaryotic cell machinery. Here, production economy can be reached with lower product volumes, depending on the value of the product. Enzymatic hydrolysis of lignocellulosic biomass to fermentable sugars represents a case where all goals and tools displayed in

Table1 will become handy. In addition, innovative engineering, especially for simultaneous hydrolysis and fermentation is essential for the overall success. Special considerations apply for non-aqueous enzymology. Lipases that exhibit a high enantioselectivity, broad substrate specificity, do not require cofactors and are stable in organic solvents present the most widely used group of biocatalysts in organic chemistry.

An integrated part of industrial exploitation of fungi is the large-scale cultivation of fungal strains. The aspect of fermentation should be addressed fairly early on in the strain development programs in order to build towards an optimal process.

3. MOLECULAR APPROACHES TO STRAIN AND YIELD IMPROVEMENT

Molecular modification of a given fungal strain to perform better as an enzyme producer involves the use of strong promoters to drive gene expression, increasing and adjusting the copy number of the gene encoding the desired product, deletion of gene (s) encoding unwanted side activities, introduction of novel properties into the production strain, and improving the performance of the enzyme proteins by protein engineering and evolution. Genetic transformation systems have been developed for both ascomycetous and basidiomycetous fungi therefore facilitating the genetic modification of a wide variety of industrial enzyme producers (reviewed in Finkelstein, 1992; Nevalainen *et al.* 2002). The approach of cloning genes involved in the secretion with the longer term goal of improving secretion is still in its early stages (reviewed by Conesa *et al.* 2001 and Nevalainen *et al.* 2002).

3.1 The Importance of Gene Promoters

Among the strongest inducible promoters routinely used in industrial enzyme production are the main cellobiohydrolase 1 (*cbhl*) promoter from *Trichoderma reesei* (Harkki *et al.* 1991) and the glucoamylase A (*glaA*) promoter from *Aspergillus niger* var. *awamori* (Ward *et al.* 1990). Both promoters can drive expression of the gene product at several grams per liter level and are regulated by catabolite repression mediated by CRE, the repressor protein that binds to the sequence 5'-SYGGRG in the target promoters via the DNA binding domain consisting of zinc fingers (Cubero and Scazzocchio, 1994). The *cre* gene has been isolated and characterized from *Aspergillus* (Dowzer and Kelly, 1991; Drysdale *et al.* 1993), *Trichoderma* species (Ilmén *et al.* 1996; Strauss *et al.* 1995; Takashima *et al.* 1996) and *Humicola grisea* (Takashima et al. 1998) among industrially-exploited fungal genera. The hypercellulolytic mutant strain *T. reesei* RUT-C30 developed by traditional mutagenesis and screening (Montenecourt *et al.* 1981) has been found to be mutated in the *cre1* gene (Ilmén *et al.* 1996). The strain was originally described as "derepressed" in relation to cellulase production (Montenecourt *et al.* 1981) Therefore, current molecular technologies have also proven helpful in explaining particular characteristics of fungal strains produced by random mutagenesis.

Molecular attempts to release some of the strong promoters from catabolite repression while retaining their strength have not produced desired results to date. Conversely, natural constitutive or carbon catabolite repression insensitive promoters comparable in strength to the *cbhl* and *glaA* have not yet been described. An intriguing novel approach into isolation of condition-specific strong promoters has been described by Curach *et al.* (2002). The approach is based on the assumption that a gene expressed strongly under e.g. a given cultivation medium producing a dominant protein spot, must therefore have a strong promoter. The most prominent protein in a 2-D proteomic display of cell-envelope associated proteins from *T. reesei* mycelia grown on glucose medium (catabolite repression) was identified as HEX1, the major protein in the fungal Woronin body (Buller 1933). The *T. reesei hex 1* gene together with its promoter and terminator regions was subsequently isolated by chromosomal walking PCR using oligonucleotides designed based on the peptide sequence obtained from the spot. We are currently evaluating the application of the *hex1* promoter for fungal gene expression in a broader sense and comparing its performance with other known strong promoters.

In addition to effective gene expression, discovery of novel promoters with application potential contribute to formulation of fermentation practices to maximize enzyme production (product synthesis) throughout the whole fermentation process. For example, expression of a heterologous gene under the constitutive glycolytic promoter *pkiA* (pyruvate kinase A) allows synthesis of the gene product under conditions (medium containing high amounts of glucose and ammonium) where the production of most of the extracellular proteases is repressed (van den Hombergh *et al.* 1994). Combination of a growth-correlated promoter (*glaA*) together with a growth rate independent promoter (a trypsin-like protease) also exhibiting different pH optima allowed extended production of a recombinant glucoamylase in a fed batch culture of *Fusarium venenatum* (Gordon et al. 2001). The multiple promoter strategy appears attractive and can be applied to other fungi as well where suitable promoters are available.

Recent advances in computer based programs may allow future simulation of regulation of a given promoter *in officio* before embarking into experimental research. For example, Agger and Nielsen (1999) reported development of a genetically structured model for the expression of the inducible alcohol dehydrogenase I (*alcA*) promoter in *A. nidulans* that was successfully shown to simulate the experimental data.

3.2 Transcriptional Regulation

Most of the strong promoters used for protein production in filamentous fungi such as *glaA* and *cbh1* discussed above are controlled at transcriptional level by induction and (glucose) repression. The difference in expression levels between the repressed and induced conditions can be several thousand-fold as shown for the cellulase promoter *cbh1* of *T. reesei* (Ilmén *et al.* 1997). In addition to glucose repression mediated by CRE, a number of other regulatory proteins exists whose overexpression, mutation or deletion may contribute to increased enzyme yields. For example, overexpression of *alcR*, the regulator of genes involved in ethanol utilization in *A. nidulans* (Felenbok *et al.* 2001) improved expression of the *alcA* gene. This implies a possibility that positively acting regulatory factors may become limiting for gene expression by "dilution" of a regulatory protein especially in strains carrying multiple copies of the expression cassette (Mathieu and Felenbok 1994). Overexpression of the *xlnR* gene encoding XlnR, a potent activator of several cellulase and hemicellulase genes of *A. niger* and *A. oryzae* (van Peij *at al.* 1998b; Marvi *et al.* 2002) increased expression of xylanases and cellulases in *A. niger* (Gielkens *et al.* 1999b; van Peij *et al.* 1998a).

In *T. reesei*, deletion of the cellulase and xylanase regulator *ace1* increased gene expression in the presence of cellulose or the inducer sophorose (Saloheimo *et al.* 2000; Aro *et al.* 2002). The gene *ace2* is involved in activation of the genes on cellulose medium (Aro *et al.* 2001). *A. nidulans* proteases, xylanases (MacCabe *et al.* 1998) and arabinofuranosidase (Gielkens *et al.* 1999a) have been reported to be controlled by *pacC* which acts as an activator at alkaline pH and prevents expression of genes that are normally expressed in acidic conditions (Caddick *et al.* 1986, Tilburn *et al.* 1995). *AreA* from *Aspergillus* (Christensen *et al.* 1998) is involved in utilization of nitrogen sources and factors known to be involved in induction of amylase genes include AMYR (Petersen *et al.* 1999) and SREB (Tani *et al.* 2000). Gene regulation apparently involves a complex genetic network, therefore, attempts to increase gene expression by operating with a single gene at a time may not be enough to result in extensive improvements in product yield. Interestingly, there is at least one example from filamentous fungi where overexpression of a gene encoding a transcription factor, in this particular case *hacA* from *Aspergillus awamori* mediating the UPR (unfolded response) induction of chaperone and foldase genes, was found to help increase expression of a heterologous gene product (Valkonen *et al.* 2002). When the UPR-induced *hacA* cDNA was overexpressed in *A. niger* var. *awamori* producing laccase enzyme originating from *Trametes versicolor*, production levels for the foreign protein were increased up to 7.6 fold.

Emerging technologies such as proteomic displays and gene arrays addressing complex genetic networks relating to the expression will undoubtedly identify novel targets for the manipulation of gene regulation and thus provide tools to work around some present problems using fungal hosts.

3.3 Gene Fusions and Matching the Codon Usage in the Production of Heterologous Gene Products

In the expression of heterologous gene products in filamentous fungi, transcription as such has not been shown to propose a limiting factor for gene expression (Jeenes *et al.* 1991). However, good steady-state levels of messenger RNA, affected by transcription efficiency and mRNA stability, are elementary for obtaining good product yields. In eukaryotes, mRNA stability and translability are linked to the level of mRNA polyadenylation (Wickens *et al.* 1999). A widely used strategy that has resulted in a considerable yield increase of heterologous gene products produced in fungal hosts is in-frame fusion of a homologous carrier such as GLA in *A. niger* (Ward *et al.*1990; Jeenes *et al.* 1993; Gouka *et al.* 1997), and CBHI in *T. reesei,* to the 5' end of the heterologous gene (Harkki *et al.* 1989; Nyyssönen *et al.*1993; Nyyssönen and Keränen, 1995). In addition to stabilization of the recombinant mRNA, the N-terminal fungal fusion partner has been proposed to facilitate the translocation of foreign proteins in the secretory pathway and protect the heterologous part from degradation (Nyyssönen *et al.* 1993; Gouka *et al.* 1997; Penttilä 1998). Translational fusions have improved the synthesis of gene products originating from non-fungal organisms by 5-1000 fold, up to hundreds of milligrams per liter, but appear not necessary for efficient expression of fungal proteins in heterologous fungal hosts (Conesa *et al.* 2000; Faria *et al.* 2002) where gram (s) per liter levels have been obtained.

The importance of codon usage in heterologous gene expression has been addressed only recently. The wide variation of codon usage between different genes and organisms would warrant mapping out the codon usage patterns in the intended host and the gene to expressed and subsequent modification of codons according to the preference of the host right at the start. For example, Te'o *et al.* (2000) reported successful expression in *T. reesei* of an AT rich xylanase (*xynB*) gene from the thermophilic bacterium *Dictyoglomus thermophilum* only after changes were made to 115 nucleotides in the 630 bp *xynB* coding region. Importantly, Te'o *et al.* (2000) also showed that functionality of the synthetic gene could be first tested in *E. coli* before embarking to time consuming fungal transformation for testing purposes. The correct codon usage may facilitate translation initiation and efficiency at the ribosome and circumvent a potential problem of the non-availability of isoacceptor tRNAs in the production host suitable for codons of the native foreign gene.

3.4 Copy Numbers and Gene Targeting

Integration of the gene to be expressed in an endogenous locus known to promote efficient transcription has been reported to increase expression of the gene product. An example is the targeting of *T. reesei egl1* genes encoding the main endoglucanase into the endogenous *cbh1* locus (Harkki *et al.* 1991). On the other hand, there are also cases where the integration site has had no notable effect on product yields, as exemplified by a study addressing heterologous phytase expression in *T. reesei* (Nevalainen *et al.* 1994). Therefore, there seems to be no universal rule and the effect of integration at the gene locus remains to be tested at experimental level. Further to increasing product yields, targeted integration has been used to delete unwanted genes that may have adverse effects in production strains. For example, *Aspergillus* strains deficient in the main protease aspergillopepsin have been constructed to be used as production hosts for a range of heterologous gene products that may be sensitive to host proteases (Berka *et al.* 1990a; Moralejo *et al.* 1999).

3.5 Getting through the Secretory Pathway

Molecular tools in hand leading to visualisation of secretion include gene fusions to fluorescing proteins such as Green Fluorescent Protein (GFP; Lorang *et al.* 2001; Gordon *et al.* 2000) and the application of immunoelectron microscopy (Nykänen *et al.* 1997, 2002a,b) have contributed to mapping of the fungal secretion pathway. However, even after these studies, the details of production bottlenecks are not known. Methods for quantification of the amount of a heterologous protein in the secretory organelles have only recently been applied to filamentous fungi (Nykänen 2002b). Elementary quantitation studies are imperative in order to gain information on the production dynamics in the fungal hyphae secreting foreign proteins.

3.5.1 Protein folding

There is plenty of experimental evidence suggesting that several foreign proteins expressed in filamentous fungi are lost in the secretory pathway. This may be because of incorrect processing or misfolding that will result in their elimination by cellular quality control mechanisms (reviewed in Archer and Peberdy 1997; Gouka *et al.* 1997). These observations have led to research programs addressing cellular pathways for unfolded protein response (UPR; reviewed in Chapman *et al.* 1998; Welihinda *et al.* 1999) as well as cloning of genes encoding gene products assisting in protein folding and quality control. Overexpression of chaperones and foldases in fungal hosts has been trialled in order to increase heterologous protein production. In spite of high hopes originally put on these studies, successful results have come so far mainly from studies involving overexpression of PDI (protein disulfide isomerase) that catalyzes the formation and rearrangement of disulfide bridges in proteins during folding. For example, the expression of thaumatin, a plant protein containing eight disulfide bridges, about five-fold increase in thaumatin yield was obtained in an *A. awamori* transformant showing increased PDI expression reaching 150 mg per liter in a fermenter cultivation (Moralejo *et al.* 2001). It is highly likely that several chaperones and foldases are needed for a collective effort to aid foreign proteins progress through the secretory pathway, therefore, the approach involving one or few genes at a time will not produce the hoped result. Here again, the holistic proteomic and transcriptomic approaches are expected to provide further clues to working towards better understanding of the factors related to protein folding and to obtaining better yields.

3.5.2 Protein glycosylation

Considering the industrial importance of fungal hydrolases, surprisingly little information is available on the sites, type and composition of enzyme glycosylation that may affect secretion, structure and stability, immunological properties, intracellular processing and activity and proteolytic degradation of enzyme proteins (Lis and Sharon 1993). Recent research with fungi has addressed the form and content of glycans added to secreted fungal proteins (Takegawa *et al.* 1991; Chiba *et al.* 1993; Maras *et al.* 1997a,b and references therein; Harrison *et al.* 1998, 2002; Klarskov *et al.* 1997). Briefly, the fungal N-linked glycan core has shown to be identical to the mammalian N-linked core ($Man_3GlcNac_2$). Some fungal strains synthesize large amounts of high-mannose type glycans whereas others have only a single *N*-acetyl-glucosamine added on an effectively secreted enzyme such as the main cellobiohydrolase CBHI of *T. reesei* (Harrison *et al.* 1998; Klarskov *et al.* 1997). This observation suggests either that strains of *T. reesei* N-glycosylate CBHI differently or that glycan trimming enzymes are secreted to the culture medium. This in turn, opens up a possibility of choosing a suitable host strain or cultivation condition for the synthesis of a particular foreign gene product. There is also evidence that glycosylation in, for example, *Trichoderma* is different to *Aspergillus* (Maras *et al.* 1997b and references therein; Nevalainen *et al.* 1994), and therefore, a choice can be made between different fungal species depending on the glycosylation requirement of the gene product in hand. Importantly, fungi seem to produce core glycans suitable for extension to glycan structures of mammalian type and the *in vivo* synthesis of complex N-glycans with terminal N-acetylglucosamine residues has been demonstrated in *T. reesei* (Maras *et al.* 1999). For a more detailed discussion on implications of fungal glycosylation on enzyme properties, see Nevalainen *et al.* (2002).

3.5.3 Proteolytic processing in the secretory pathway

A number of eukaryotic proteins contain propeptides that have been suggested to have an important role in secretion, folding and organelle targeting (Baker et al. 1993; Chang et al. 1994; reviewed in Conesa *et al.* 2001 and references therein). Activation of several eukaryotic enzymes such as lipases and proteases, including those produced by fungi require cleavage of the propeptide. The relatively low level of production of proteins originating from mammals and plants in filamentous fungi may indicate a problem in their processing during secretion. In depth studies into the mechanism and cellular basis for protein processing are therefore elementary for understanding the bottlenecks in heterologous protein expression and secretion and to be able to devise improvement strategies.

Intracellular processing of the majority of propeptides by Kex2p-like proteases occurs at a dibasic cleavage site after Lys-Arg (KR) or Arg-Arg (RR). Dibasic sites that resemble the Kex2p target sites are frequently found in sequences of secretory proteins in filamentous fungi (Goller *et al.* 1998; Calmels *et al.* 1991). Studies with the endogenous *T. reesei* xylanases exhibiting proprotein processing sequences showed that secretion was inhibited by aminophenylmethylsulfonyl fluoride (pAPMSF), which inhibits dibasic endopeptidase activity (Goller *et al.* 1998). In addition to Kex2p like proteases, experimental data obtained from *T. reesei* studies point to existence of yet unidentified endoproteolytic enzymes in the fungal hyphae (Nykänen *et al.* 2002a, Nyyssönen *et al.* 1993).

A Kex2p cleavage site has been introduced into fusion proteins at the fusion junction to separate the foreign protein from the endogenous carrier in filamentous fungi (Contreras *et al.* 1991; reviewed in Gouka *et al.* 1997; Paloheimo *et al.* 1998; J. Te'o, unpublished). The *kexB* gene encoding a kexin-like maturase was isolated from *A. niger* (Jalving *et al.* 2000) and used to produce *A. niger* strains either overexpressing or lacking the *kexB* gene. The gene product clearly has a role in protein processing since expression of a glucoamylase-human interleukin-6 fusion protein with an engineered Kex2p in a *kexB* disruptant was affected by the inability of the hosts to process the fusion protein at the dibasic target site. Engineering of Kex2p cleavage sites into constructs expressing the catalytic subunit of bovine enterokinase and human mucus inhibitor protein fused to the *glaA* (glucoamylase) was shown to result in the correct processing of the fusion protein at the Kex2p site in *A. niger* (Krasevec *et al.* 2000; Mikosch *et al.* 1996). However, levels of heterologous proteins secreted in the culture medium remained at 3 and 5 mg per liter respectively implying that yet additional factors were involved. In a study using the *T. reesei* secretion machinery, Nykänen *et al.* (2002a) demonstrated that proteolytic processing of the barley cysteine endoproteinase (EPB) occured by Kex2p-like cleavage at three of the four potential dibasic sites in the enzyme sequence and that fungal glycosylation of EPB interfered with the final processing of the protein by an unknown peptidase resulting in a decreased recombinant enzyme activity. It has been shown that maturation of heterologous fungal proteins such as the *Hormoconis resinae* glucoamylase (Nykänen 2002b) and *Humicola grisea* xylanase (Faria *et al.* 2002) occurred by Kex2p-like processing in *T. reesei.* Considering the evidence available so far, it is apparent that Kex2p-like cleavage in the trans-Golgi has an important role in the protein processing in filamentous fungi which in turn affects the quality and characteristics of the heterologously produced protein.

3.5.4 Attempts to avoid proteolytic degradation in fungal cultivation media

Several studies have been carried out especially with *Aspergillus* to characterize the proteases produced by the fungus and thereby give insights into how to reduce proteolytic degradation of especially foreign gene products produced in the fungal host (reviewed in van den Hombergh *et al.* 1997a). Both the classical mutagenesis and screening technique and targeted gene inactivation have resulted in fungal strains that produce decreased amounts of protease activity in general or are deficient in the production of a particular protease such as the aspartyl protease which seems to represent the main protease activity in industrially important filamentous fungi (Mäntylä *et al.* 1984, 1998; van den Hombergh *et al.* 1997a; Berka *et al.* 1990a,b). However, the question of proteases and how to deal with them has not been addressed in a systematic way. A quick test to start with could include analyzing potential specific protease cleavage sites present in a foreign peptide sequence first and then knocking out the most harmful activity/activities from the fungal host. Alternatively, one would need a series of protease deficient host strains to choose from according to the characteristics and sensitivity of a particular foreign protein to be expressed (van den Hombergh *et al.* 1997b).

Successful attempts to suppress protease production by the means of bioprocess engineering include 25% reduction of extracellular protease secretion in *A. niger* by immobilization of the hyphae (Liu *et al.* 1998) and inhibition of their secretion by pelleted growth in liquid fermentation (Xu *et al.* 2000).

4. IMPROVEMENT OF THE PROPERTIES OF ENZYME PROTEINS BY PROTEIN ENGINEERING AND DIRECTED EVOLUTION

Industrial requirements for optimal performance of enzymes in dedicated applications may feature characteristics that are not selected for in nature. Some properties that are beneficial in an industrial setting may never become selected for in nature or may even prove harmful. Functional criteria for industrial enzymes include specificity, suitable pH and temperature characteristics for a particular application as well as stability and activity (reaction rate) under the required conditions such as presence of solvents, detergents and heavy metals. In order to improve these characteristics, two approaches have been used, protein engineering (rational design) and gene shuffling or directed evolution (irrational design).

4.1 Protein Engineering

Protein engineering involves premeditated change of amino acids and is usually based on the known 3-D structure of a given protein and its biochemically established catalytic mechanism. The preferred approach is site directed mutagenesis of the gene encoding the target enzyme. The properties of industrial microbial enzymes changed by site directed mutagenesis include substrate specificity, thermostability, laundry wash performance, protease stability, activity in alkaline and acid solutions and oxidative stability (reviewed in Leisola *et al.* 2000). An example of a fungal enzyme for which the properties have been modified by site directed mutagenesis based on the crystal structure includes a lipase from *Humicola lanuginosa* used in household detergents to improve lipid removal (Boel and Jensen 1989). The *H. lanuginosa* lipase (LipolaseTM) has been extensively mutated to improve the washing performance by replacing the negatively charged residues in the lipid contact zone by mainly hydrophobic or positively charged residues. The improved mutant enzyme (Lipolase UltraTM) shows better surface activity. A strategy to improve *H. lanuginosa* protease stability involved replacing the labile loops (cleavage sites) with non labile loops. Modified enzymes are produced in large scale by a surrogate high-secreting host *A. oryzae* after cloning of the enzyme-encoding genes into this organism.

An example of an industrial enzyme of which the thermostability has been improved by designed mutagenesis is presented by a *T. reesei* xylanase II. The increased thermostability (about 200 times at 70ºC) was achieved by stabilizing the alpha-helix region and the N-terminus of the enzyme protein. At the same time, the pH shifted towards the alkaline region by one pH-unit (Turunen *et al.* 2002).

4.2 Directed Evolution

While protein engineering relies on established knowledge from studied genes and proteins, directed evolution explores either natural or mutated gene pools in order to select for desired properties. Evolution is the culprit for sometimes not finding the "right" genes and gene products in nature. Enzymes are optimized and often highly specialized for specific biological functions within the context of a living organism. Biotechnology, in contrast, needs enzymes, which are stable over long periods of time, enzymes which are active in non-aqueous solvents, and enzymes which can accept different substrates that may not be found in nature.

Directed protein evolution works around these problems by creating gene libraries and applying mutagenesis/recombination and/or gene and domain shuffling techniques (Gibbs *et al.* 2001; Kolkman and Stemmer 2001) in order to isolate the right DNA encoding protein with the desired properties (Arnold 1996; Farinas *et al.* 2001; Joern *et al.* 2002). The transformation frequencies with filamentous fungi, regardless of the method used, are typically in the region of 10-100 transformants per microgram of DNA and even though those with yeast can reach 10^5 per microgram of transforming DNA, they are not advanced enough to allow for effective screening for enzyme evolution purposes. Therefore, existing molecular evolution programs are mainly carried out in *E. coli* with which transformation frequencies and robotized screening methods are highly developed to enable compilation of mutant libraries and effective high throughput screening of gene products. Directed evolution has been applied to a number of microbial enzyme proteins such as bacterial proteases (Zhao and Arnold 1999) and lipases (Liebeton *et al.* 2000), and commercial companies, e.g. Diversa Corporation (San Diego, CA, USA) are capitalizing on the technology.

5. PROSPECTING FOR NOVEL ENZYMES AND GENES

The global demand for enzymes has resulted in scouring the globe for biocatalysts with superior characteristics that can replace those currently used in a variety of commercial applications. There is ever increasing strive for obtaining better products and the need to keep up with technical developments in industry and product end use. Enzymes carry a bonus by being environmentally friendly compared with, for example, chemicals traditionally used in some large volume applications such as bleaching of wood pulp. Areas where enzymes will no doubt have a big impact in the near future are improvement/formation of flavours and aroma, production of bulk organic materials and production of fragrances and cosmetics. New applications in the medical and diagnostic arena such as enzyme replacement therapy, treatment of cancer and synthesis of antimicrobial compounds will also continue to attract enzymes.

5.1 Techniques and Sources for the Isolation of Novel Genes

Traditionally, novel genes have been isolated by constructing genomic DNA libraries in *Escherichia coli* first before transferring the DNA into the expression host for the screening of enzyme activity. Such a technique can be time-consuming and usually relies on the genomic DNA quality and availability. Some mircoorganisms cannot be cultured in the laboratory, therefore hindering production of microbe specific genomic libraries. However, isolation of novel genes from unculturable organisms is possible provided that genomic DNA can be extracted, for example, from their growth environment.

5.1.1 Expression Cloning

Genes from filamentous fungi encoding industrially relevant enzymes have been cloned by expressing cDNA libraries in the yeast *Saccharomyces cerevisisae* (Saloheimo *et al.* 1994, 1997; Dalboge 1997). The approach has resulted in the isolation of genes coding for endoglucanases, xylanases, pectinases, proteases, hemicellulases and rhamnogalacturonan-degrading enzymes reviewed in Dalboge (1997). The yeast system has also been applied for the cloning of novel genes encoding fungal transcription factors as ACEI and ACEII involved in the regulation of one of the most extensively utilized expression promoters, the main cellobiohydrolase *cbh1* promoter from *T. reesei* (Saloheimo *et al.* 2000; Aro *et al.* 2001). Isolation of genes coding for fungal transcription factors and for example, their subsequent overexpression in fungal hosts opens up a new route for industrial strain modification. Expression cloning is also making possible to discover enzyme genes originating from organisms which have not been established in pure culture in the laboratory. Examples of this type include genes encoding enzymes active in the gut of the termite larvae and in the cow rumen that represent complex ecosystems. These genes can then be inserted into suitable vectors for their expression in filamentous fungi.

5.1.2 Molecular screening

Molecular screening of enzyme encoding sequences is largely based on gene alignments of genes in order to find specific areas of conserved DNA sequence to be used as for the design of PCR primers for the desired type of gene. The PCR based strategy combined with chromosomal walking PCR has been successfully used for the cloning of a number of enzyme encoding genes from thermophilic microorganisms (Peek *et al.* 1992). The PCR based approach can also be used to discover variants of the same type of genes from other microorganisms. Novozymes used molecular screening to find as many as 48 new microorganisms that produced a cellulase of interest, belonging to Family 45 (Lange *et al.* 1999).

5.2 Looking into Extreme Environments for Fungal Enzyme Activities

Extreme environments, especially hot pools, have been an excellent source for a number of bacterial genes encoding economically relevant enzymes. However, the highest temperatures that fungi have been found to thrive at, are between 45 and 55°C (Maheswari *et al.* 2000) which makes them tolerate higher temperatures than most other eukaryotic organisms. Despite this feature and the fact that thermophilic fungi are a rich source for enzymes that degrade plant biomass, they have not provoked a great amount of research interest. Some thermophilic fungi could find uses as production hosts for thermophilic and thermolabile proteins.

5.2.1 Fungi from Antarctica as a Source for Cold-active Enzymes

Filamentous fungi from cold environments have been even less studied for their physiological basis for cold tolerance and as sources of novel enzymes than those thriving in hot environments. A variety of filamentous fungi have been isolated from Antarctica (reviewed by Vishniac 1996; Azmi and Seppelt 1998) among which are representatives of several industrially-exploited genera such as *Penicillium* and *Trichoderma*. Screening for hydrolase activities secreted at different temperatures by three isolates of *Penicillium*, *Phoma*, *Alternaria* and two isolates of *Trichoderma* sp. by Bradner *et al.* (1999a) indicated the presence of cold-adapted enzymes amongst these fungi. More detailed studies on hemicellulase activity showed clearly that the temperature optimum for hemicellulase activity in the Antarctic strains was, in general, between 10-30°C lower than that of the mesophilic reference strain (Bradner *et al.* 1999b). Other fungal isolates from the Antarctic include fungi collected from fuel-contaminated soils (J. Aislabie, personal communication) that can provide a source for bioremediation activities in a cold climate.

6. FILAMENTOUS FUNGI AS PRODUCTION HOSTS

The natural ability of filamentous fungi to secrete effectively enzymes into their environment, the availability of strong fungal promoters and the eukaryotic protein modification machinery makes them attractive as hosts for the expression of various gene products originating from bacteria, plants and animals. So far, the studies have concentrated on mesophilic fungi.

6.1 Tailoring Homologous Enzyme Profiles

Modification of endogenous enzyme profiles in long-standing industrial fungi such as *T. reesei* and *A. niger* var. *awamori* is a routine procedure today. Different combinations involving increasing the copy numbers of particular genes and eliminating others are mainly restricted by the availability of enzyme-encoding genes and transformation markers. Gene replacement technology (Ward 1989; Karhunen *et al.* 1993; Suominen *et al.* 1993) provides a tool for simultaneous multiplication of one gene while inactivating another (Figure 1).

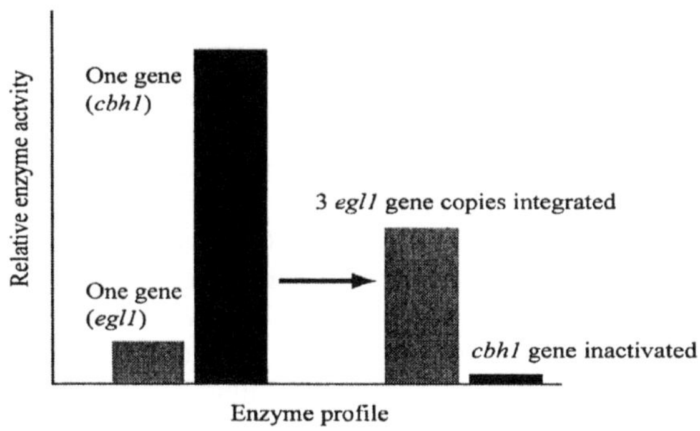

Fig. 2. Modification of the enzyme profile in *Trichoderma reesei*. Integration of three copies of the *egl1* gene into the endogenous *cbh1* locus results in a considerable change in the ratio EG1 and CBHI that are being produced.

Effective gene replacement has typically relied on the 5' and 3' homology of the incoming DNA to the chromosomal locus targeted for gene replacement. More recently, gene replacement strategies using fungal PCR based techniques have been introduced for *Aspergillus nidulans* (Chaveroche *et al.* 2000) and the plant pathogen *Ashbya gossypii* (Wenland *et al.* 2000). A considerable amount of microbial enzyme preparations currently on the market are produced by genetically modified microorganisms and tailored to a particular application (e.g. http://www.novo.dk/backgrou/position/list.htm).

6.2 Points to Consider in Heterologous Gene Expression

It is generally viewed that transcription is not the main factor restricting product yields and the bulk of a foreign product seems to be lost in the secretory pathway (revieved in Conesa *et al.* 2001). Some of the factors affecting product yields and quality have been discussed above. At this point of time, there is no overall solution to guarantee a good yield of a heterologous gene product - the best results can be achieved by addressing the obvious restricting factors such as codon usage and proteolytic degradation. A checklist to work around some limiting factors is presented in Table 2.

Table 2. A checklist to address factors limiting heterologous protein yields.

Question	Procedure
Does the codon usage of the incoming gene match that of the host?	Change codons by PCR
Does the protein require post-translational modifications for activity?	Analyze biochemical data and/or amino acid sequence
	Choose a suitable strain e.g. for low/high glycosylation
Are there subcellular postal addresses?	Make changes /eliminate if required
	Choose a suitable expression vector
Is the foreign protein to be expressed sensitive to host proteases?	Incubate the gene product with a series of host culture supernatants
	Express as a fusion protein to an endogenous carrier
Does the protein require extensive folding?	Express as a fusion protein to an endogenous carrier
	Co-express with suitable foldases

As long as the detailed molecular basis for yield improvement and production bottlenecks are not known, random mutagenesis and screening of transformant strains for improved yields of a particular gene product such as a hydrolytic enzyme, will provide an option for further improvement. For example, *T. reesei* transformants producing a thermophilic proteinase originating from *Thermus* sp. (Saul *et al.* 1996), treated with UV and screened on skim-milk containing plates (protease substrate) at +85°C shoved improved secretion of the thermophilic enzyme (Nevalainen, unpublished). Automated screening technology, essential to be able to reach the critical numbers of colonies screened, is also available to filamentous fungi. Most likely, such programs have been carried out for filamentous fungi expressing heterologous gene products, however, this work has remained largely unpublished.

There are also a number of technologies successfully used with unicellular organisms and now revisited with a view of applying them to filamentous fungi. For example, a window of opportunity has been identified that is suitable to assess metabolic activity in filamentous fungi using fluorescent stains and flow cytometry (Bradner and Nevalainen, in press). This will pave the way to mass screening of, for example, fungal transformants and mutant strains.

7. TOWARDS A BIGGER PICTURE

Gene expression is influenced, for example, by transcriptional activators and repressors of which the activity is influenced by yet other gene products. Therefore, expression of a particular gene in a fungal host involves a complex genetic network. This genetic network is in turn connected to, if not dictated by, the physiological status of the cell, and the physiological and stress responses that, for example, production of a foreign protein causes for the organism. Such networks can be studied, for example, by transcriptional profiling (gene expression), proteomic analysis (protein profiling) and using computational modeling. Most genes and gene products have multiple functions and can occur in multiple forms due to posttranslational processing. The combination of gene expression microarrays with a proteomic approach resulting in a 2 dimensional map of proteins provides a powerful tool for understanding gene functions and networking under different circumstances. Cultures carried out in fermenters allow careful control of the physiological state of the fungus which can be complemented by metabolite and metabolic flux analyses.

On the side of linking protein structure to function, data is being collected concerning a wide range of proteins. One example of classification of enzymes based on their structural features and thereby looking into evolutionary relationships and functional characteristics is the classification of glycoside hydrolases (http://afmb.cnrs-mrs.fr//CAZY/). Members of the same hydrolase family have the same stereoselectivity indicating that they share a common general fold, active site topology and catalytic mechanism (Gebler *et al.* 1992).

The broad goal of linking genes, genomes, expression, structure and function represents a huge computational challenge that is being addressed by tackling smaller tasks related to the analysis of genomic and proteomic data.

7.1 Genomics approaches

Examples of filamentous fungi for which genomic sequencing programs are underway include genetically well known fungi such as *Neurospora crassa* (e.g http.//www.mips.biochem.mpg.de/proj/neurospora/) and *Aspergillus nidulans* (Roe *et al.* 1999), the ligninolytic *Phanerochaete chrysosporium* (http://www.jgi.doe.gov/programs/whiterot/whiterot_mainpage.html) for which the task has been completed, plant pathogens such as *Magnaporthe grisea* (Martin *et al.* 2002), opportunistic human pathogens such as *Aspergillus fumigatus* (Brookman and Denning 2000) and industrially relevant fungi *A. niger* and *T. reesei* (Chambergo *et al.* 2002).

The work by Chambergo *et al.* (2002) used EST analysis and cDNA microarrays of *T. reesei* to find answers to the central question why glucose is habitually metabolized by respiration rather than fermentation in multicellular organisms. *T. reesei* proposes an especially interesting organism for these studies since it is highly cellulolytic being able to hydrolyze cellulose effectively to glucose that can be fermented to ethanol. Therefore, the findings would provide good pointers towards metabolical engineering of cellulolytic microorganisms for the production of bioethanol. In *T. reesei*, the metabolism was directed towards oxidation of pyruvate via the TCA cycle instead of reduction of pyruvate to ethanol by fermentation. Also, instead of being channeled to ethanol, acetaldehyde may be metabolized to acetate which would prevent regeneration of NAD^+ required for anaerobic metabolism. According to the authors, regulation of glucose metabolism has been a likely target for evolution directing the flow either towards respiration or fermentation.

The white rot fungus *Phanerochaete chrysosporium* of which the genome is approximately 30 Mb and organized in 10 chromosomes is the first basidiomycetous filamentous fungus of which the whole genome has been sequenced. *P. chrysosgenum* has been widely studied for its potential uses in pulp and paper industry for biopulping and as a source for extracellular lignin peroxidases (reviewed in Eriksson 1997). *P. chrysogenum* is also capable of degrading a wide variety of toxic waste compounds such as pentachlorophenol, TNT, nitroglycerin, DDT, naphthalene, Arcolor 1242 (polychlorinated biphenyl, 42%) and Arcolor 1254, polychlorinated biphenyl, 54%) which make the fungus a potential organism for bioremediation and related environmental applications.

7.2 Proteomics

The application of proteomics in fungal biotechnology is in its early childhood. Even though a considerable amount of work has been carried out with yeast, filamentous fungi have gained fairly little attention. This may be direct reflection of the prementioned fact that not enough genome sequencing data are available for reliable protein identification.

However, the situation is likely to change in the near future and not all applications actually require protein identification.

Table 3. Production proteomics

Area of study	**Task**
Post-translational modification of gene products	Linking form and function
Establishing markers for high producing strains	Strain diagnostics
Identification of gene products	Product secretion
- specific for particular functions	Heterologous production
- responding to a particular environmental condition	Disease
	Pollution
Isolation of condition	Protein expression on chosen medium, pH, etc.

The first ever reported proteome for a filamentous fungus was that for cell envelope-associated proteins in *T. reesei* (Lim *et al.* 2001). Since then, proteomic approaches have been used to map out proteins associated with the response of *A. nidulans* to the antibiotic concanamycin A (Melin *et al.* 2002), and facilitate whole proteome analysis of *T. reesei* by deglycosylation of proteins en mass to aid in their identification by mass spectrometry (Fryksdale *et al.* 2002). Glycosylation of the acetylxylan esterase (AXE) in *T. reesei* was analysed in detail using 2-D gel electrophoresis by Harrison *et al.* (2002). It was observed that that two protein spots for each of the linker-substrate binding domain and core peptides were consistent with the identification of partial sulfation of the linker and phosphorylation of the *N*-linked glycan on the core peptide. These examples highlight the uses of proteome analysis related to biotechnological goals, called here production proteomics (Table 3).

7. CONCLUSIONS

Inherent characteristics of filamentous fungi such as excellent protein secretion and the ability to grow on cheap cultivation media makes them an economical choice for bulk protein production when compared to other eukaryotic systems available. After the initial success of traditional strain development, further improvement is very much dependent on better basic knowledge of the fungal systems, especially gene regulation and the secretory pathway. It is evident that the next big leap forward will draw from these studies including aspects of functional genomics. Filamentous fungi will undoubtedly retain their position as the high-yielding hosts or industrial enzyme production and strengthen their role as a surrogate host for efficient expression and secretion of valuable heterologous gene products originating from bacteria, plants and animals.

REFERENCES

Agger T, and Nielsen J (1999). Genetically structured modeling of protein production in filamentous fungi. Biotechnol Bioeng 66:164-170.

Archer DB, and Peberdy JF (1997). The molecular biology of secreted enzyme production by filamentous fungi. Crit Rev Biotechnol 17:273-306.

Arnold FH (1996). Directed Evolution: Creating Biocatalysts For The Future. Chem Eng Sci 51:5091-5102.

Aro N, Saloheimo A, Ilmèn M, and Penttilä M (2001). ACEII, a novel transcriptional activator involved in the regulation of cellulase and xylanase genes of *Trichoderma reesei*. J. Biol. Chem. 276:24309-24314.

Aro N, Ilmén M, Saloheimo A and Penttilä M (2002). ACEI of *Trichoderma reesei* is a repressor of cellulase and xylanase expression. Appl Environ Microbiol in press.

Azmi OR, and Seppelt RD (1998). The broad distribution of microfungi in the Windmill Islands region, continental Antarctica. Polar Biol 19:92-100.

Baker DA, Shiau AK, and Agard DA (1993). The role of pro regions in protein folding. Curr Opin Cell Biol 5:966-970.

Berka R, Hayenga K, Lawlis VB, and Ward M (1990a). Aspartic proteinase deficient filamentous fungi. WO 90/00192.
Berka RM, Ward M, Wilson LJ, Hayenga KJ, Kodama KH, Carlomagno LP and Thompson SA (1990b). Molecular cloning and deletion of the gene encoding aspergillopepsin A from *Aspergillus awamori*. Gene. 86:153-162.
Blinkovsky AM, Buyn T, Brown KM, and Golitghly E (1999). Purification, characterization and heterologous expression in *Fusarium venenatum* of a novel serine carboxypeptidase from *Aspergillus oryzae*. Appl Environ Microbiol 65:3298-3303.
Boel, E and Huge Jensen, I.B. (1989). Recombinant *Humicola* lipase and process for the production of recombinant *Humicola* lipases. European patent application EP 0305216.
Bradner JR, Gillings M, and Nevalainen H (1999a). Qualitative assessment of hydrolytic activities in antarctic fungi at different temperatures on solid media. World. J Microbiol Biotechnol 15:143-145.
Bradner JR, Sidhu RK, Gillings M, and Nevalainen H (1999b). Hemicellulase activity of antarctic microfungi. J Appl Microbiol 87:366-370.
Bradner JR, and Nevalainen H (2002). Metabolic activity in filamentous fungi can be analysed by flow cytometry. J Microbiol Meth in press.
Brookman JL and Denning DW (2000). Molecular genetics of *Aspergillus fumigatus*. Curr Opin Biotechnol 3:468-474.
Buller AHR (1933). Researches in Fungi. New York: Hafner.
Caddick MX, Brownlee AG, and Arst HN Jr (1986). Regulation of gene expression by pH of the growth medium in *Aspergillus nidulans*. Mol Gen Genet 203:346-353.
Calmels TPG, Martin F, Durand H, and Tiraby G (1991). Proteolytic events in the processing of secreted proteins in fungi. J Biotechnol 17:51-66.
Chambergo FS, Bonaccorsi, ED, Ferreira AJS, Ramos ASP, Ribamar Ferreira Jr J, Abrahao-Neto, J, Simon Farah JP, and El-Dorry H (2002). Elucidation of the metabolic fate of glucose in the filamentous fungus *Trichoderma reesei* using EST analyisi and cDNA microarrays. J Biol Chem 277: 13983-1388.
Chang SC, Chang PC, and Lee YH (1994). The role of propeptide in maturation and secretion of Npr protease from *Streptomyces*. J Biol Chem 269:3548-3554.
Chapman R, Sidrauski C, and Walter P (1998). Intracellular signalling from the endoplasmic reticulum to the nucleus. Annu Rev Cell Dev Biol 14:459-485.
Chaveroche MK, Ghigo JM, and d'Enfert C (2000). A rapid method for efficient gene replacenment in the filamentous fungus *Aspergillus nidulans*. Nucleic Acids Res 28:E97.
Chiba Y, Yamagata Y, Iijima S, Nakajima T, and Ichishima E. (1993). The carbohydrate moiety of the acid carboxypeptidase from *Aspergillus saitoi*. Curr Microbiol 27:281-288.
Christensen T, Hynes MJ, and Davis MA (1998). Role of the regulatory gene *areA* of *Aspergillus oryzae* in nitrogen metabolism. Appl Environ Microbiol 64:3232-3237.
Conesa A, van den Hondel CAMJJ, and Punt P (2000). Studies on the production of fungal peroxidases in *Aspergillus niger*. Appl Environ Microbiol 66:3016-3023.
Conesa A, Punt PJ, van Luijk N, and van den Hondel CAMJJ (2001). The secretion pathway in filamentous fungi: a biotechnological view. Fung Genet Biol 33:155-171.
Contreras R, Carrez D, Kinghorn JR, van den Hondel CAMJJ, and Fiers W (1991). Efficient KEX2-like processing of a glucoamylase-interleukin-6 fusion protein by *Aspergillus nidulans* and secretion of mature interleukin-6. Bio/Technology 9:378-381.
Cubero B and Scazzocchio C (1994). Two different, adjacent and divergent zinc finger binding sites are necessary for CREA-mediated carbon catabolite repression in the proline gene cluster of *Aspergillus nidulans*. Embo J 13:407-15.
Curach N, Te'o VJS, Bergquist PL, and Nevalainen KMH (2002). *Hex1*, a new promoter for gene expression in *Trichoderma reesei*. Abstracts of the 6th European Conference on Fungal Genetics. Abstract IIo5.
Dalboge H (1997). Expression cloning of fungal enzyme genes; a novel approach for efficient isolation of enzyme genes of industrial relevance. FEMS Microbiol Rev 1:29-42.
Dowzer CE and Kelly JM (1991). Analysis of the *creA* gene, a regulator of carbon catabolite repression in *Aspergillus nidulans*. Mol Cell Biol 11:5701-5709.
Drysdale MR, Kolze SE and Kelly JM (1993). The *Aspergillus niger* carbon catabolite repressor encoding gene, *creA*. Gene 130:241-245.
Eriksson K-E (1997). Biotechnology in the pulp and paper industry: An overview. ACS Symp Ser 687:2-14.
Felenbok B, Flipphi M, and Nikolaev I (2001). Ethanol catabolism in *Aspergillus nidulans*:a model for studying gene regulation. In: Progress in Nucleic Acid Research and Molecular Biology, Vol 69. Academic Press, pp 149-204.

Faria FP, Te'o VJS, Bergquist PL, Azevedo MO, and Nevalainen KMH (2002). Expression and processing of a major xylanase (XYN2) from the thermophilic fungus *Humicola grisea* var. *thermoidea* in *Trichoderma reesei.* Lett Appl.Microbiol 34:119-123.

Farinas ET, Bulter T, and Arnold FH (2001). Directed enzyme evolution. Curr Opin Biotechnol 12:545-551.

Finkelstein DB (1992). Transformation. In: DB Finkelstein and C Ball, eds. Biotechnology of Filamentous Fungi, Technology and Products. MA: Butterworth-Heinemann, pp 113-156.

Fryksdale BG, Jedrzejewski PT, Wong DL, Gaertner AL, and Miller BS (2002). Impact of deglycosylation methods on two-dimensional gel electrophoresis and matrix assisted laser desorption/ionization-time of flight-mass spectrometry for proteomic analysis. Electrophoresis 23:2184-2193.

Gebler J, Gilkes NR, Claeyssens M, Wilson DB, Beguin P, Wakarchuk WW, Kilburn DG, Miller RC Jr, Warren RA, and Withers SG (1992). Stereoselective hydrolysis catalyzed by related beta-1,4-glucanases and beta-1,4-xylanases. J Biol Chem 267:12559-12561.

Gibbs MJ, Nevalainen KMH, and Bergquist P L (2001). Degenerate oligonucleotide gene shuffling (DOGS): A method for enhancing the frequency of recombination with family shuffling. Gene 271:13-20.

Gielkens M, Gonzalez-Candelas L, Sanchez-Torres P, van de Vondervoort P, de Graaff L, Visser J, and Ramon D (1999a). The *abfB* gene encoding the major alpha-L-arabinofuranosidase of *Aspergillus nidulans*: nucleotide sequence, regulation and construction of a disrupted strain. Microbiology 145: 735-741.

Gielkens MM, Dekkers E, Visser J, and de Graaff H (1999b). Two cellobiohydrolase-encoding genes from *Aspergillus niger* require D- xylose and the xylanolytic transcriptional activator XlnR for their expression. Appl Environ Microbiol 65:4340-4345.

Goller SP, Schoisswohl D, Baron M, Parriche M, and Kubicek CP (1998). Role of endoproteolytic dibasic proprotein processing in maturation of secretory proteins in *Trichoderma reesei.* Appl Environ Microbiol 64:3202-3208.

Gordon CL, Archer DB, Jeenes DJ, Doonan JH, Wells B, Trinci APJ, and Robson GD (2000). A glucoamylase:GFP gene fusion to study protein secretion by individual hyphae of *Aspergillus niger.* J Microbiol Methods 42:39-48.

Gordon C, Thomas S, Griffen A, Robinson GD, Trinci PJ, and Wiebe MG (2001). Combined use of growth rate correlated and growth rate independent promoters for recombinant glucoamylase production in *Fusarium venenatum.* FEMS Microbiol Lett 194:229-234.

Gouka RJ, Punt PJ, and van den Hondel CA. (1997). Efficient production of secreted proteins by *Aspergillus*: progress, limitations and prospects. Appl Microbiol Biotechnol 47:1-11.

Harkki A, Uusitalo J, Bailey M, Penttilä M, and Knowles J (1989). A novel fungal expression system: secretion of active calf chymosin from the filamentous fungus *Trichoderma reesei.* Bio/Technol 7:596-603.

Harkki A, Mäntylä A, Muttilainen S, Bühler R, Suominen P, Knowles J, and Nevalainen H (1991). Genetic engineering of *Trichoderma* to produce strains with novel cellulase profiles. Enzyme Microb Technol 13: 227-233.

Harrison M.J, Nouwens AS, Jardine DR, Zachara NE, Gooley AA, Nevalainen H, and Packer NH (1998). Glycosylation of cellobiohydrolase I from *Trichoderma reesei.* Eur J Biochem 256:119-127.

Harrison MJ, Wathugala IM, Tenkanen M, Packer NH, and Nevalainen KMH (2002). Glycosylation of acetylxylan esterase from *Trichoderma reesei.* Glycobiology 12:291-298.

Ilmén M, Thrane C, and Penttilä M (1996). The glucose repressor gene *cre1* of *Trichoderma*: isolation and expression of a full-length and a truncated mutant form. Mol Gen Genet 251:451-460.

Ilmén M (1997). Molecular mechanisms of glucose repression in the filamentous fungus *Trichoderma* reesei. VTT Publications 315. Espoo, Finland.

Jalving R, van den Vondervoort PJI, Visser J, and Schaap PJ (2000). Characterization of the kexin-like maturase of *Aspergillus niger.* Appl Environ Microbiol 66:363-368.

Joern JM, Meinhold P, and Arnold FH (2002). Analysis of shuffled gene libraries. J Mol Biol. 316:643-656.

Jeenes DJ, Mackenzie DA, Roberts IN, and Archer DB (1991). Heterologous protein production by filamentous fungi. Biotechnol Genet Eng Rev 9:327-367.

Jeenes DJ, MacKenzie DA and Archer DB (1993). A truncated glucoamylase gene fusion for heterologous protein secretion from *Aspergillus niger.* FEMS Microbiol Lett 107:267-271.

Karhunen T, Mäntylä A, Nevalainen H, and Suominen P (1993). High frequency one-step gene replacement in *Trichoderma reesei.* I. Endoglucanase I overproduction. Mol Gen Genet 241: 515-522.

Klarskov K, Piens K, Ståhlberg J, Høi PB, van Beeumen J, and Claeyssens M (1997). Cellobiohydrolase I from *Trichoderma reesei*: identification of an active-site nucleophile and additional information on sequence including glycosylation pattern of the core protein. Carbohydr Res 304:143-154.

Kolkman JA, and Stemmer WP (2001). Directed evolution of proteins by exon shuffling. Nat Biotechnol 19:423 -428.

Krasevec N, van den Hondel CA, and Komel R (2000). Can hTNF-alpha be successfully produced and secreted in filamentous fungus *Aspergillus niger*? Pflugers Arch 439 (3 Suppl):R84-86.

Lange L., Skjøt, M., Schülein, Kattila P., Kauppinen, S. (1999). Cellulase discovery and 18sDNA studies of five cythrids. Fungal Genetics Newsletter 46 Suppl, p 56.
Leisola M, Jokela J, Pastinen O, Turunen O and Shoemaker H (2002). Industrial use of enzymes. http://www.hut.fi/ Units/Biotechnology/Kem-70.415/ INDUSTRIAL_USE_OF_ENZYMES.DOC
Liebeton K, Zonta A, Schimossek K, Nardini M, Lanf D, Dijkstra BW, Reetz M, and Jaeger K-E (2000) directed evolution of an enantioselective lipase. Chem Biol 7:709-718.
Lim D, Hains P, Walsh B, Bergquist P, and Nevalainen H (2001). Proteins associated with the cell envelope of *Trichoderma reesei*: A proteomic approach. Proteomics 1:899-910.
Lis H and Sharon N (1993). Protein glycosylation. Structural and functional aspects. Eur J Biochem 218:1-27.
Liu F, Li W, Ridgway D, Gu T, and Moo-Young M (1998). Inhibition of extracellular protease secretion by *Aspergillus niger* using immobilization. Biotechnol Lett 20:539-542.
Lorang JM, Tuori RP, Martinez JP, Sawye, TL, Redman RS, Rollins JA, Wolpert TJ, Johnson KB, Rodriguez RJ, Dickman MB, and Ciuffetti LM (2001). Green fluorescent protein is lighting up fungal biology. Minireview. Appl Environ Microbiol 67:1987-1994.
MacCabe AP, Orejas M, Perez-Gonzalez JA, and Ramon D (1998). Opposite patterns of expression of two *Aspergillus nidulans* xylanase genes with respect to ambient pH. J Bacteriol 180:1331-1333.
Maheswari R, Bharadwaj G, and Bhat MK (2000). Thermophilic fungi: their physiology and enzymes. Microbiol Mol Biol Revs 64:461-488
Mathieu M, and Felenbok B (1994). The *Aspergillus nidulans* CREA protein mediates glucose repression of the ethanol regulon at various levels through competition with the AlcR-specific transactivator. EMBO J 13:4022-4027.
Maras M, de Bruyn A, Schraml J, Herdewijn P, Claeyssens M, Fiers W, and Conteras W (1997a). Structural characterization of N-linked oligosaccharides from cellobiohydrolase I secreted by the filamentous fungus *Trichoderma reesei* RUTC 30. Eur J Biochem 245:617-625.
Maras M, De Bruyn A, Schraml J, Herdewijn P, Piens K, Claeyssens M, Uusitalo J, Penttilä M, Fiers W, and Contreras R (1997b). Engineering of the carbohydrate moiety of fungal proteins to a mammalian type. In: M Claeyssens, W Nerinckx, and K Piens, eds. Carbohydrates from *Trichoderma reesei* and other microorganisms. Structures, biochemistry, genetics and applications. Cambridge UK: The Royal Society of Chemistry, pp 323-326.
Maras M. De Bruyn A, Vervecken W, Uusitalo J, Penttilä M, Busson R, Herdewijn P and Contreras R (1999). *In vivo* synthesis of complex N-glycans by expression of human N-acetylglucosaminyltransferase I in the filamentous fungus *Trichoderma reesei*. FEBS Lett 452:365-370.
Martin SL, Blackmon BP, Rajagopalan R, Houftek TD, Sceeles RG, Denn SO, Mitchell TK, Brown DE, Wing RA, and Dean RA (2002). *Magnaporthe*DB: a federated solution for integrating physical and genetic map data with BAC and derived sequences for the rice blast fungus *Magnaporthe grisea*. Nucleic Acids Res 30:121-124.
Marvi J, Tanaka A, Mimura S, de Graaff L, Visser J, Kitamoto N, Kato M, Kobayashi T, and Tsukagoshi N (2002). A transcriptional activator, AoXlnR, controls the expression of genes encoding xylanolytic enzymes in *Aspergillus oryzae*. Fung Genet Biol 35:157-169.
Melin P, Schnürer J, and Wagner EGH (2002). Proteome analysis of *Aspergillus nidulans* reveals proteins associated with the response to the antibiotic concanamycin A, produced by *Streptomyces* species. Mol Gen Genom 267:965-702.
Mikosch M, Klemm P, Gassen HG, van den Hondel CAMJJ, and Kemme M (1996). Secretion of active human mucus propteinase inhibitor by *Aspergillus niger* after KEX2-like processing of a glucoamylase inhibitor fusion protein. J Biotechnol 52:97-106.
Moralejo FJ, Cardoza RE, Gutierrez S, and Martin JF (1999). Thaumatin production in *Aspergillus awamori* by use of expression cassettes with strong fungal promoters and high gene dosage. Appl Environ Microbiol 65:1168-1174.
Moralejo FJ, Watson AJ, Jeenes DJ, Archer DB, and Martin JF (2001). A defined level of protein disulfide isomerase expression is required for optimal secretion of thaumatin by *Aspergillus awamori*. Mol Genet Genomics 266:246-253.
Montenecourt BS, Nhlapo SD, Trimino-Vazquez H, Cuskey H, Schamhart DHJ and Eveleigh DE (1981). Regulatory controls in relation to overproduction of fungal cellulases. In: A Hollaender, R Rabson, P Rogers, A san Pietro, R, Valentine, and R Wolfe, eds. Trends in the Biotechnology of Fermentations for Fuels and Chemicals. New York: Plenum Publishing, pp 33-53.
Morris DD, Gibbs MD, Chin CW, Koh MH, Wong KK, Allison RW, Nelson PJ and Bergquist PL (1998). Cloning of the *xynB* gene from *Dictyoglomus thermophilum* Rt46B.1 and action of the gene product on kraft pulp. Appl Environ Microbiol 64:1759-1765.
Mäntylä A, Saarelainen R, Fagerström R, Suominen P, and Nevalainen H (1994). Cloning of the aspartic protease gene from *Trichoderma reesei*. 2nd European Conference on Fungal Genetics, Lunteren.

Mäntylä A, Paloheimo M, and Suominen P (1998). Industrial mutants and recombinant strains of *Trichoderma reesei*. In: GR Harman, and Kubicek CP, eds. *Trichoderma & Gliocladium*. Vol 2, Enzymes, biological control and commercial applications. London: Taylor and Francis, pp 291-309.

Nevalainen H, Paloheimo M, Miettinen-Oinonen A, Torkkeli T, Turunen M, Fagerström R, Cantrell M, Piddington C, and Rambosek J (1994). Production of phytate degrading enzymes in *Trichoderma*. WO94/03612.

Nevalainen H, Te'o V, and Penttilä M (2002) Application of genetic engineering for strain improvement in filamentous fungi. New York: Marcel Dekker, in press.

Nykänen M, Saarelainen R, Raudaskoski M, Nevalainen H, and Mikkonen A (1997). Expression and secretion of barley cysteine endopeptidase B and cellobiohydrolase I in *Trichoderma reesei*. Appl Environ Microbiol 63:4929-4937.

Nykänen M, Raudaskoski M, Nevalainen H, and Mikkonen A (2002a). Maturation of barley cysteine endopeptidase expressed in *Trichoderma reesei* is distorted by incomplete processing. Can J Microbiol 48:138-50.

Nykänen M (2002b). Protein secretion in *Trichoderma reesei*. Expression, secretion and maturation of cellobiohydrolase I, barley cysteine endoproteinase and calf chymosin in Rut-C30. PhD Dissertation, University of Jyväskylä, Finland.

Nyyssönen E, Penttilä M, Harkki A, Saloheimo A, Knowles JKC, and Keränen S (1993). Efficient production of antibody fragments by the filamentous fungus *Trichoderma reesei*. Bio/Technology 11:591-595.

Nyyssönen E and Keränen S (1995). Multiple roles of the cellulase CBHI in enhancing production of fusion antibodies by the filamentous fungus *Trichoderma reesei*. Curr Genet 28:71-79.

Paloheimo M, Mäntylä A, Vehmaanperä J Hakola S, Lantto R, Lahtinen T, Parkkinen E, Fagerström R, and Suominen P (1998). Thermostable xylanases produced by recombinant *Trichoderma reesei* for pulp bleaching. In: M Claeyssens, W Nerinckx and K Piens, eds. Carbohydrates from *Trichoderma reesei* and other microorganisms. Structures, biochemistry, genetics and applications. Cambridge UK: The Royal Society of Chemistry pp 255-264.

Peek K, Ruttersmith LD, Daniel RM, Morgan HW, and Bergquist PL (1992). Thermophilic enzymes as industrial catalysts. Int J Biotechnol 9: 466-470.

Penttilä M (1998). Heterologous protein production in *Trichoderma*. In: GR Harman and CP Kubicek, eds. *Trichoderma & Gliocladium*. Vol 2, Enzymes, biological control and commercial applications. London: Taylor and Francis, pp 365-382.

Petersen KL, Lehmbeck J, and Christensen, T (1999). A new transcriptional activator for amylase genes in *Aspergillus*. Mol Gen Genet 262:668-676.

Roe BA, Kupfer D, Zhu H, Gray J, Clifton S, Prade R, Loros J, and Dunlap J (1999). *Aspergillus nidulans* sequencing project. http://www.genome.ou.edu/fungal.html.

Saloheimo A, Henrissat B, Hoffrén A-M, Teleman O, and Penttilä M (1994). A novel, small endoglucanase gene, *egl5*, from *Trichoderma reesei* isolated by expression in yeast. Mol. Microbiol 13:219-228.

Saloheimo M, Nakari-Setälä T, Tenkanen M, and Penttilä M (1997). cDNA cloning of a *Trichoderma reesei* cellulase and demonstration of endoglucanase activity by expression in yeast. Eur J Biochem 429:584-591.

Saloheimo A, Aro N, Ilmèn M, and Penttilä M (2000). Isolation of the *ace1* gene encoding a Cys_2-His_2 transcription factor involved in regulation of activity of the cellulase promoter *cbh1* of *Trichoderma reesei*. J Biol Chem 275:5817-5825.

Saul DJ, Williams LC, Toogood HS, Daniel R, and Bergquist PL (1996). Sequence of the gene encoding a highly thermostable neutral proteinase from *Bacillus* sp. Strain EA1: expression in *Eschericia coli* and characterization. Biochim Biophys Acta 1308:74-80.

Strauss J, Mach RL, Zeilinger S, Hartler G, Stoffler G, Wolschek M, and Kubicek CP (1995). Cre1, the carbon catabolite repressor protein from *Trichoderma reesei*. FEBS Lett 376:103-107.

Suominen P, Mäntylä A, Karhunen T, and Nevalainen H (1993). High frequency one-step gene replacement in *Trichoderma reesei*. II. Effects of deletion of individual cellulase genes. Mol Gen Genet 241:522-530.

Takashima S, Nakamura A, Iikura H, Masaki H, and Uozumi T (1996). Cloning of a gene encoding a putative carbon catabolite repressor from *Trichoderma reesei*. Biosci Biotechnol Biochem 60:173-176.

Takashima S, Nakamura A, Hidaka M, Masaki H, and Uozumi T (1998). Isolation of the *creA* gene from the cellulolytic fungus *Humicola grisea* and analysis of CreA binding sites upstream from the cellulase genes. Biosci Biotechnol Biochem 62:2364-2370.

Takegawa, K, Kondo A, Iwamoto H, Fujiwara K, Hosokawa Y, Kato I, Hiromi K, and Iwahara S (1991). Novel oligomannose-type sugar chains derived from glucose oxidase of *Aspergillus niger*. Biochem Int 25:181-190.

Tani S, Kawaguchi T, Kato M, Kobayashi T, and Tsukagoshi N (2000). A novel nuclear factor, SREB, binds to a cis-acting element, SRE, required for inducible expression of the *Aspergillus oryzae* Taka-amylase A gene in *A. nidulans*. Mol Gen Genet 263:232-238.

Te'o VJS, Cziferszky AE, Bergquist PL and Nevalainen KMH (2000). Codon optimization of xylanase gene *xynB* from the thermophilic bacterium *Dictyoglomus thermophilum* for expression in the filamentous fungus *Trichoderma reesei*. FEMS Microbiol Lett 190:13-19.

Tilburn J, Sarkar,S, Widdick DA, Espeso EA, Orejas M, Mungroo J, Penalva MA, and Arst HN Jr (1995). The *Aspergillus* PacC zinc finger transcription factor mediates regulation of both acid- and alkaline-expressed genes by ambient pH. Embo J 14:779-790.

Turunen O, Vuorio M, Fenel F and Leisola M. (2002). Engineering of multiple arginines into the Ser/Thr surface of *Trichoderma reesei* endo-1,4-beta-xylanase II increases the thermotolerance and shifts the pH optimum towards alkaline pH. Protein Eng 15:141-145.

Valkonen M, Penttilä M, and Saloheimo M (2002). The effect of inactivation and consitutive expression of the unfolded protein response pathway on protein production in *Saccharomyces cerevisiae*. Submitted.

van den Hombergh JPTW, Jarai G, Buxton FP, and Visser J (1994). Cloning, characterization and expression of *pepF*, a gene encoding a serine carboxypeptidase from *Aspergillus niger*. Gene 151:73-79.

van den Hombergh JPTW, van den Vondervoort PJI, Fraissinet-Tachet L and Visser J (1997a). *Aspergillus* as a host for heterologous protein production: the problem of proteases. TIBTECH 15:256-263.

van den Hombergh JPTW, Fraissinet-Tachet L, van de Vondervoot PJI, and Visser J (1997b). Production of the homologous pectin lyase B protein in six genetically defined protease-deficient *Aspergillus niger* mutant strains. Curr Genet 32:73-81.

van Peij NN, Gielkens MM, de Vries RP, Visser J, and de Graaff LH (1998a). The transcriptional activator XlnR regulates both xylanolytic and endoglucanase gene expression in *Aspergillus niger*. Appl Environ Microbiol 64:3615-3619.

van Peij, NN, Visser J, and de Graaff LH (1998b). Isolation and analysis of *xlnR*, encoding a transcriptional activator co- ordinating xylanolytic expression in *Aspergillus niger*. Mol Microbiol 27:131-142.

van Zeijl C, Punt P, Emalfarb M, Burlinghame R, Sinitsyn A, Parriche M, Bousson JC, and van den Hondel CAMJJ (2001). *Chrysosporium lucknowense*, a new fungal host for protein production. Fungal Genetics Newsletter 48 Suppl, p 89.

Vishniac HS (1996) Biodiversity of yeasts and filamentous fungi in terrestrial Antarctic ecosystems. Biodiversity and Conservation 5:1365-1378.

Ward M (1989) Heterologous gene expression in *Aspergillus*. In: H Nevalainen and M Penttilä, eds. Molecular Biology of Filamentous Fungi, Proceedings of the EMBO-Alko Workshop, Espoo Finland. Foundation for Biotechnical and Industrial Fermentation Research, Vol. 6, pp 119-128.

Ward M, Wilson LL, Kodama KH, Rey MW, and Berka RM (1990). Improved production of chymosin in *Aspergillus* by expression as a glucoamylase-chymosin fusion. Bio/Technology 8:435-440.

Welihinda AA, Tirasophon W, and Kaufman RJ (1999). The cellular response to prtein misfolding in the endoplasmic reticulum. Gene Express 7:293-300.

Wendland J, Ayad-Durieux Y, Knechtle P, Rebischung C, and Philippsen P (2000). PCR-based gene targeting in the filamentous fungus *Ashbya gossypii*. Gene 242:381-391.

Wickens M, Anderson P, and Jackson RJ (1999). Life and death in the cytoplasm: messages from the 3' end. Curr Opin Genet Dev 7:220-232.

Xu J, Wang L, Ridgway D, Gu T, and Murray M-Y (2000). Increased heterologous protein production in *Aspergillus niger* fermentation through extracellular proteases inhibition by pelleted growth. Biotechnol Prog 16:222-227.

Zhao H and Arnold FH (1999). Directed evolution converts subtilisin E into a functional equivalent of a thermitase. Protein Eng 12: 47-53.

Applied Mycology & Biotechnology
An International Series. Volume 3. Fungal Genomics

12

Global Expression Profiling of the Lignin Degrading Fungus *Ceriporiopsis subvermispora* for the Discovery of Novel Enzymes

Debbie Sue Yaver, Barbara Weber and Jeff Murrell
Novozymes Biotech Inc., 1445 Drew Avenue, Davis, California 95616-4880 USA

The unique ability of white rot fungi to degrade all components of wood, including lignin, has attracted considerable biotech interest for several decades. Many studies have focused on the isolation, characterization, production and commercial use of enzymes secreted by white rot basidiomycetes (Kirk and Jeffries, 1996; Tuor *et al.*, 1995) as well as the use of white rot fungi for pretreatment of wood chips (biopulping) (Blanchette *et al.*, 1988; Messner and Srebotnik, 1994; Scott and Swaney, 1998; Breen and Singleton, 1999; Scott *et al.*, 2000) *Ceriporiopsis subvermispora* is among the most selective lignin-degraders (Otjen *et al.*, 1987), and pretreatment of wood chips with *C. subvermispora* prior to mechanical pulping has been shown to reduce energy consumption by 30-45%. Using DNA microarray technology, global gene expression profiling of *C. subvermispora* was performed to discover novel enzymes whose expression is induced during growth on thermomechanical pulp. An array containing shotgun genomic clones of *C. subvermispora* was prepared and interrogated with probes from RNA which was isolated from cultures grown on either minimal medium or hardwood mechanical pulp. Hybridization to the first 20,000 clones identified 20 clones whose expression is induced due to growth on pulp. Sequence analysis has shown that these 20 clones actually represent 11 unique clones. One of these clones has significant homology to manganese peroxidases from white rot fungi. This chapter reviews this shotgun approach to identifying novel enzymes induced on complex substrates as well as the characterization of the novel peroxidase.

1. INTRODUCTION

The unique ability of white rot fungi to degrade all components of wood, including lignin, has attracted considerable biotech interest for several decades. Studies of lignin-degrading enzymes have focused on oxidoreductase enzymes including lignin and manganese peroxidases, laccases and enzymes involved in the generation of hydrogen peroxide. These classes of enzymes have been identified, purified and characterized from many white-rot fungi. *C. subvermispora,* a selective lignin degrader shows great potential for industrial applications where removal of lignin components is desirable while cellulose fiber strength is maintained (Sethuraman *et al.*, 1998). Several enzymes capable of degrading plant cell walls are produced. *C. subvermispora* is known to produce both laccase and manganese peroxidase activity (Rüttimann *et al.*, 1992; Rüttimann-Johnson *et al.*, 1993; Lobos *et al.*, 1994; Salas *et al.*, 1995; Fukushima and Kirk, 1995) but no lignin peroxidase activity has been detected although lip-like genes have been identified (Rajakumar *et al.*, 1996). Up to five isoenzymes of laccase are produced when *C. subvermispora* is grown in defined medium or on wood chips (Salas *et al.*, 1995; Fukushima and Kirk, 1995; Lobos *et al.*, 1994), and a single gene coding for a laccase has been cloned (Karahanian *et al.*, 1998). When grown in defined medium seven isoenzymes of of manganese peroxidase can be detected, while only four are produced when *C. subvermispora* is grown on wood (Lobos *et al*, 1994). The amino terminal sequences of the peroxidase isoenzymes produced in liquid or on wood are clearly distinct (Lobos *et al*, 1994). Three manganese peroxidase genes (*mnp1, mnp2* and *mnp3*) have been cloned to date and two alleles of *mnp2* have been identified (Lobos *et al.*, 1998; Tello *et al.*, 2000). MnP1 and MnP3 appear to be similar to isoenzymes produced on wood based on predicted amino terminal amino acids. In addition, oxalate oxidase has been detected in mycelia of *C. subvermispora* (Aguilar *et al.*, 1999); oxalic acid is produced by several white-rot fungi and is thought to facilitate the release of Mn^{3+} from the active site of manganese peroxidase as well as stabilization of the metal ion. Urzúza *et al.* (1998) proposed a model in which Mn^{3+} reacts with oxalic acid to produce carbon dioxide and a formate radical. This radical is postulated to react with oxygen, generating superoxide that is subsequently reduced by Mn^{2+} yielding hydrogen peroxide and Mn^{3+}. The Mn^{3+} and hydrogen peroxide are utilized to further accelerate manganese peroxidase reactions.

Pretreatment of wood chips with *C. subvermispora* prior to mechanical pulping has been shown to reduce energy consumption by 30-45% (Messner and Srebotnik 1994; Scott *et al.* 1998, Scott *et al.* 2000). This 'biomechanical pulping' process is under commercial development, although strain and process improvements are continuously sought (Scott *et al.*, 2000). The mechanism(s) of biopulping are unknown (Blanchette, 1994). However, it's firmly established that the major benefit of fungal colonization, i.e. energy savings, is realized within one week of chip inoculation, well before bulk lignin degradation has occurred. Summarizing many separate studies, *C. subvermispora* colonization of wood involves at least two stages; an initial rapid colonization which somehow 'softens' wood without affecting fiber strength or lignin content, and later an efficient depolymerization and mineralization of lignin. The enzyme systems involved in both stages may have considerable value in developing new energy efficient pulping processes and in delignification/bleaching technologies. Efforts to elucidate the mechanism have been stymied by difficulties investigating enzyme activities in wood, and several recent studies clearly show that studies of enzyme expression in defined laboratory media have little or no relevance towards expression on wood substrates (Janse *et al.*, 1998).

DNA microarray technologies allow the simultaneous monitoring of changes in gene expression of thousands of genes under different physiological conditions (DeRisi *et al.* 1997; DeRisi and Iyer, 1999). The completion of genomic sequences for many microorganisms as well as large EST sequencing programs has provided the opportunity to use global transcription profiling to study response to stress, development programs, pathogenesis and many other physiological conditions. Because the *C. subvermispora* genome is not completed we created a shotgun genomic array as previously described for the malaria parasite *Plasmodium falciparum* by Hayward *et al.* (2000). Clones from a random genomic library of *C. subvermispora* were used to generate the shotgun array. The arrays were probed with differentially labeled cDNAs prepared from polyA RNA isolated from cells grown on a defined laboratory medium or on thermomechanical hardwood pulp to identify key genes involved in colonization and growth on pulp. These genes might include novel secreted enzymes that may have commercial applications in either augmenting biopulping or other stages in the pulp and papermaking process (Kirk and Jeffries, 1996). Our initial studies have identified several genes whose expression is induced during growth on hardwood thermal mechanical pulp.

2. SHOTGUN GENOMIC ARRAY OF *CERIPORIOPSIS* FOR ENZYME DISCOVERY

A random genomic library containing 2 to 3 kb inserts of *C. subvermispora* (FPL104807SS-5 obtained from Forest Products Laboratory, Madison, WI) was created by partial digestion with *Tsp*509I and ligation into pUC19. Fifty thousand independent clones were picked, which assuming an average insert size of 2.5 kb would cover 125 Mb. Plasmid DNA from each clone was spotted onto poly-L-lysine coated glass microscope slides using the equipment and methods that are described on the web site of Professor P.O. Brown of Stanford University (http://cmgm.stanford.edu/pbrown/protocols). The density of spots was 10,000 per slide.

RNA was isolated from cultures of *C. subvermispora* grown on either basal minimal medium or on 20 grams of hardwood thermomechanical pulp. The inoculum was generated by growth in basal minimal medium as described by Rüttimann *et al.* (1992) at 28°C without shaking for 15 days. Mycelia mats were harvested from the flasks and homogenized in sterile water. Homogenized mycelia were added to a 1 liter flask containing 20 gm of hardwood thermomechanical pulp that had previously been processed in a Waring blender, and the mixture was stirred. For the minimal medium cultures, homogenized mycelia were added to 15 ml of basal minimal medium in 1 liter flasks. The cultures were incubated at 30°C for 30 days without shaking and the entire pulp culture containing pulp plus mycelia was frozen quickly in liquid nitrogen. After 30 days of incubation, the pulp was visibly bleached compared to an uninoculated control. Mycelia from the minimal medium culture were harvested by filtration and quickly frozen in liquid nitrogen. Total RNA was isolated as described by Timberlake and Barnard (1981) and poly-A RNA was isolated using a mRNA Separator kit (Clonetech, Palo Alto, CA).

Fluorescent probes for hybridization to the arrays were prepared by reverse transcription of 1 µg of polyA RNA from *Ceriporiopsis subvermispora* to incorporate aminoallyl-dUTP into first strand cDNA. The amino- cDNA products were subsequently labeled by direct coupling to either Cy3 or Cy5 monofunctional reactive dyes (Amersham Pharmacia Biotech, Arlington Heights, IL). The details of this protocol are described at http://cmgm.stanford.edu/ pbrown/protocols. The probes were hybridized to the microarray at 65°C overnight. Before scanning, the arrays were washed consecutively in 1X SSC with 0.03% SDS, 0.2X SSC, and 0.05X SSC, and centrifuged for 2 minutes at 500 rpm to remove excess liquid. The slides were imaged using a custom-built confocal laser microscope (Eisen and Brown 1999). Hybridizations were repeated at least four times to assure reproducibility of results.

Hybridizations to the first 20,000 clones identified 20 clones that were induced due to growth on pulp. Nucleotide sequences of the 20 clones were determined, and the results demonstrated that the 20 clones represented 11 unique clones. The insert size of the clones ranged from 1595 to 3000 bp. Inserts of the 11 unique clones were translated in all six reading frames and queried against the public databases. One of the clones, pCsubHP1 contained a genomic fragment, which shared considerable identity to manganese peroxidases from white rot fungi. Five of the clones found no hits. Clones pCsubHP2, 3 and 4 shared homology to a hydrophobin, oxidoreductases and isocitrate lyase, respectively.

2.1 GENOMIC AND cDNA PEROXIDASE CLONES

The putative peroxidase clone, pCsubHP1, was a partial genomic clone. A full-length cDNA clone was constructed using 5' and 3' RACE (Rapid Amplification of cDNA Ends) products amplified from pulp-specific total RNA using a GeneRacer™ kit (Invitrogen, Carlsbad, CA). A full-length genomic clone was constructed from the pCsubHP1 clone and a genomic fragment containing the 5' end of the gene obtained by PCR.

The peroxidase cDNA clone (Csubmnp4) encodes an open reading frame of 1131 bp, which is 60% GC, and a predicted polypeptide of 377 amino acids. Using the SignalP program (Nielsen *et al.*, 1997, *Protein Engineering* 10: 1-6), a signal peptide of 18 residues is predicted, therefore, the predicted mature peroxidase is composed of 359 amino acids.

A comparative alignment of peroxidase sequences was performed using the Clustal method using the LASERGENE™ MEGALIGN™ software (DNASTAR, Inc., Madison, WI) with the identity table (Figure 1). *Ceriporiopsis subvermispora* peroxidase (MnP4, GenBank Accession Number AY217670) shares 81% identity with the manganese peroxidase 2 from *Phanerochaete chrysosporium* (EMBL Accession number L29039). Comparative alignment also showed that MnP4 shares approximately 80% identity with the manganese peroxidases 1, 3 and 4 from *Phanerochaete chrysosporium* (EMBL Accession Numbers M60672, U70998, and J04980) and manganese peroxidases 1 and 2 from *P. sordida* (EMBL accession numbers AB078604 and AB078605). In addition it shares 78, 72 and 69% identities with *C. subvermispora* manganese peroxidase 1, 2 and 3, respectively.

The mature protein has predicted MW of 37.5 kDa and pI of 4.191. Like the *C. subvermispora* MnP1, the MnP4 protein has a large number of aspartic and glutamic acid residues relative to lysine and arginine residues consistent with the acidic pI determined for the manganese peroxidases secreted by the fungus when grown on pine wood chips (Lobos *et al.*, 1994: Lobos *et al.*, 1998). In contrast to the other *C. subvermispora* manganese peroxidases, MnP4 does not contain the four amino acid insert after glycine 266 (Tello *et al.*, 2000; Figure 1). The amino terminal sequence does not match the consensus for manganese peroxidases secreted by *C. subvermispora* during cultivation in either defined medium or on wood chips (Lobos *et al.*, 1994); however, it appears closest to the manganese peroxidases found on wood. MnP4 contains the two glutamic acid residues for manganese binding (Figure 1). The distal arginine and both proximal and distal histidine residues involved in the peroxidase catalytic core are also conserved (Figure 1). The consensus aromatic binding site of L/V/I-P-Xaa-P is present in MnP4 and identical to the sequence (IPEPQD, Figure 1) found in MnP2 and MnP3 from *C. subvermispora* as well as *P. chrysosporium* Mnp1 (Lobos *et al.*, 1998; Tello *et al.*, 2000). MnP4 does not contain a four or five C-terminal extension as found in the other *C. subvermispora* MnPs which may influence the catalytic activities of these enzymes due to its close proximity to the manganese binding site (Tello *et al.*, 2000). There are 4 potential N-linked glycosylation sites (Asn-X-Ser/Thr) within the *Ceriporiopsis subvermispora* peroxidase.

The open reading frame of Mnp4 is interrupted by six introns, ranging in size from 55 to 68 bp, in contrast to *mnp1, mnp2* and *mnp3* which all have seven introns (Tello *et al.*, 2000). Intron three which splits the phenylalanine and histidine codons (70 and 71) in *mnp1, mnp2* and *mnp3* is missing. All the splice sites conform to the GT-AG rule and most of the putative internal lariat sites conform to the consensus CTRAY.

3. HETEROLOGOUS EXPRESSION OF PUTATIVE SECRETED PEROXIDASE

3.1 *Aspergillus oryzae* Transformants Expressing the Peroxidase

In order to characterize the enzyme activity of MnP4, the protein was expressed in *Aspergillus oryzae* as previously described (Yaver *et al.*, 1999). An expression vector containing the *A. oryzae* α-amylase promoter, the MnP4 open reading frame including its

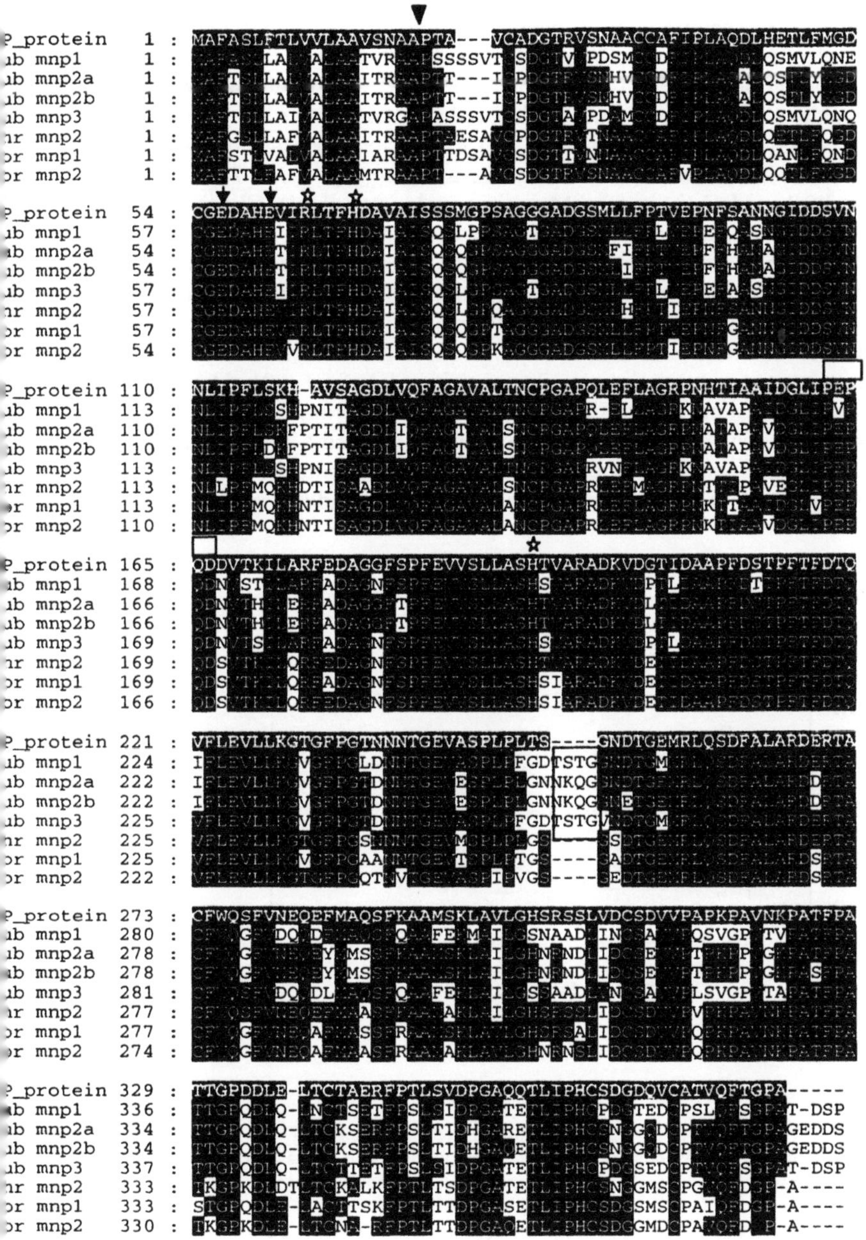

ure 1. Alignment of the primary structures of MnPs from *C. subvermispora* and *P. chrysosporium* MnP2, *P.* *dida* MnP1 and *P. sordida* MnP2. Numbers correspond to residues in the precursor protein. MnP is MnP4, and ub mnp1, mnp2a, mnp2b and mnp3 are the previously identified MnPs from *C. subvermispora*. Pchr and Psor nd for *P. chrysosporium* and *P. sordida*. The symbols denote the following: ▼, beginning of mature protein as dicted by SignalP; ↓, residues involved in manganese binding; ☆, residues involved in peroxidase catalytic cycle d □, aromatic binding site. The four amino acid insertion found in C. sub MnP1, MnP2a, MnP2b and MnP3 is ked. The residues identical to Csub MnP4 are shaded in black.

own signal sequence for secretion, the *A. niger* glucoamylase terminator and the *A. nidulans amdS* gene for selection was used to transfom *A. oryzae*. Twenty five transformants were pore purified and tested for secretion of peroxidase activity. The production media used was ASP03 supplemented with either hemin or hemoglobin (Andersen *et al.*, 1992). Cultures were grown for 7 days with sampling daily and expression levels were measured by the oxidation of 2,6-dimethoxyphenol (2,6-DMP). The assay consisted of 0.2 mM 2,6-DMP, 0.5 mM $MnSO_4$, and 0.1 mM H_2O_2 in 50 mM sodium malonate (pH 4). Change in absorbance was measured at 468 nm, and the untransformed expression host was used as a negative control to determine background peroxidase activity. Eight of the 25 cultures produced peroxidase levels 5-10 times greater than the negative controls. Figure 2 shows peroxidase activity in culture supernatants for three transformants (#22, 27 and 34) as well as the untransformed host strain at different timepoints during shake flask cultivations. The untransformed strain gave secreted peroxidase activity of around 0.01 dA/min. The three transformants showed distinct kinetics of production. Transformant Csub22 had a lag with peroxidase activity peaking at 132 hours while Csub 27 showed a peak of production at 84 hours. Csub34 showed peak production at 60 hours. Cultures grew and expressed peroxidase at levels 10 to 30 fold higher when hemoglobin was added compared to hemin.

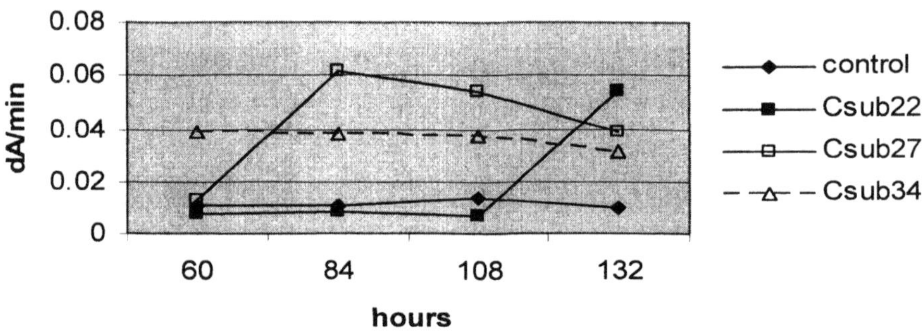

Fig. 2. Peroxidase activity secreted by *A. oryzae* transformants expressing MnP4 grown in shake flasks. Culture supernatant samples were taken at the times indicated on the x-axis and assayed for peroxidase activity using 2,6-DMP as substrate. The untransformed strain was used as a negative control.

3.2 CHARACTERIZATION OF THE RECOMBINANT PEROXIDASE

To begin characterization of the peroxidase, supernatants of transformants producing peroxidase activity were pooled for purification, concentrated and desalted with 20 mM sodium succinate (pH 4.5) by ultrafiltration with a 10 kDa cutoff filter. Initial protein purification was performed on a MonoQ column in 20 mM sodium succinate (pH 4.5) and elution with a linear gradient of 0.3M NaCl in the running buffer was performed. Fractions were assayed for peroxidase activity using 2,6-DMP as a substrate, and the fraction enriched for activity contained upwards of 60% pure protein as observed by electrophoresis (Figure 3).

The partially purified preparation has an apparent MW of 62 kDa as determined by SDS-PAGE. In addition, the protein was biochemically characterized and does not have an absolute requirement for manganese, showing only a slight increase in activity with the addition of manganese. The enzyme has an acidic pH optimum (< pH3) and can oxidize various dyes such as phenol red and 2,2'-azinobis-(3-ethylbenzthiazoline-6-sulfonic acid), albeit at a slower rate than 2,6-DMP.

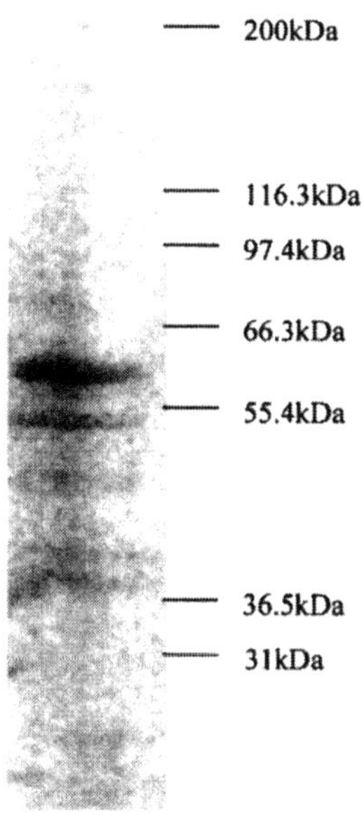

Fig. 3. SDS-PAGE of recombinant MnP4. A sample of the fraction from the MonoQ column enriched for peroxidase activity activity was treated with β-mercaptoethanol, boiled, and applied to an 8-20% Tris-glycine gel.

4. CONCLUSIONS

Shotgun genomic DNA microarrays appear to be a viable approach to identify novel enzymes involved in the degradation of complex substrates. Using this method a novel peroxidase from *C. subvermispora* whose expression is induced during growth on hardwood thermomechanical pulp was cloned. The peroxidase shares significant homology to manganese peroxidases from other white rot fungi, but from initial characterization of the recombinant protein produced in *A. oryzae* does not appear to have an absolute requirement for manganese. Additional purification is required for unequivocal demonstration of manganese independent peroxidase activity. MnP4 was identified as being induced in *C. subvermispora* when grown on hardwood thermomechanical pulp for 30 days; it will be interesting to observe expression of the gene over a timecourse after inoculation to see when *mnp4* is induced as well as when the protein/enzyme activity can be detected. In addition, only 20,000 of the 50,000 shotgun genomic clones were interrogated to date; further hybridizations to the other genomic clones should be completed to determine if there are other pulp-induced clones. Once the promoter of *mnp4* has been cloned it will be worthwhile to compare it to the promoters of the other pulp-induced clones to identify motifs as well as to compare it to the promoters of the *C. subvermispora mnp1, mnp2a, mnp2b* and *mnp3*.

Future studies will also include the characterization of the other ten clones identified as being induced in a pulp-grown culture in addition to the optimization of fermentation conditions for the recombinant strains, and testing of the peroxidase in commercial applications.

Acknowledgements: We would like to thank Shari Brody Karpin for the large number of robotic-assisted plasmid preparations and Beth Nelson for performing the automated DNA sequencing. Also thank you to Carrie Vierra for her assistance in rearraying the clones for array printing.

REFERENCES

Aguilar C, Urzúa U, Koenig C, and Vicuña R (1999). Oxalate oxidase from *Ceriporiopsis subvermispora*: Biochemical and cytochemical studies. Arch Biochem Biophys 366:275-282.

Alic M, Akileswaran L, and Gold MH (1997). Characterization of gene encoding manganese peroxidase isozyme 3 from *Phanerochaete chrysosporium*. Biochimica et Biophysica Acta 1338:1-7.

Andersen HD, Jensen EB, Welinder KG, Dalboege H and Dalboge H (1992). Production of haem proteins, esp. *Coprinus* peroxidase using a DNA expression vector in filamentous fungi for high level expression. Patent Number WO9216634.

Blanchette RA (1991) Delignification by wood-decaying fungi. Annu Rev Phytopathol 29:381-398.

Blanchette RA (1994) Degradation of the lignocellulose complex in wood. Can J Bot 73:S999-S1010.

Blanchette RA, Burnes TA, Leatham GF and Effland MJ (1988). Selection of white-rot fungi for biopulping. Biomass 15:93-101.

Blanchette RA, Krueger EW, Haight JE, Akhtar M, and Akin DE (1997). Cell wall alterations in loblolly pine wood decayed by the white-rot fungus, Ceriporiopsis subvermispora. J Biotechnol 53:203-213.

Breen A and Singleton FL (1999). Fungi in lignocellulose breakdown and biopulping. Curr Opinion Biotech 10:252-258.

Cullen D. (1997). Recent advances on the molecular genetics of ligninolytic fungi. J Biotechnol 53:273-289.

DeRisi JL and Iyer VR (1999). Genomics and array technology. Curr Opin Oncol 11:76-79.

DeRisi JL, Iyer VR and Brown PO (1997). Exploring the metabolic and genetic control of gene expression on a genomic scale. Sciene 278:680-686.

Eisen, MB and Brown PO (1999). DNA arrays for analysis of gene expression. Methods Enzymol. 303: 179-205.

Fukushima Y and Kirk TK (1995). Laccase component of the *Ceriporiopsis subvermispora* lignin-degrading system. Appl Environ Microbiol 61:872-876.

Hayward RE, DeRisi JL, Alfadhli S, Kaslow DC, Brown PO and Rathod K (2000). Shotgun DNA microarrays and stage-specific expression in *Plasmodium falciparum* malaria. Mol Microbiol 35:6-14.

Janse BJH, Gaskell J, Akhtar M and Cullen D (1998). Expression of *Phanerochaete chrysosporium* genes encoding lignin peroxidases, manganese peroxidases, and glyoxal oxidase in wood. Appl Environ Microbiol 64:3536-3538.

Karahanian E, Corsini G, Lobos S and Vicuña R (1998). Structure and expression of a laccase gene from the ligninolytic basidiomycete *Ceriporiopsis subvermispora*. Biochimica et Biophysica Acta 1443:65-74.

Kirk TK and Jeffries TW (1996). Roles for Microbial Enzymes in Pulp and Paper Processing. In: TW Jeffries and L Viikari, eds. Enzymes for Pulp and Paper Processing. Washington D.C.: American Chemical Society, pp2-14.

Larrondo LF, Lobos S, Stewart P, Cullen D, and Vicuña R (2001). Isoenzyme multiplicity and characterization of recombinant manganese peroxidase from *Ceriporiopsis subvermispora* and *Phanerochaete chrysosporium*. Appl Environ Microbiol 67:2070-2075.

Lobos S, Larraín J, Salas L, Cullen D and Vicuña R (1994). Isoenzymes of manganese-dependent peroxidase and laccase produced by the lignin-degrading basidiomycete *Ceriporiopsis subvermispora*. Microbiol 140:2691-2698.

Lobos S, Larrondo L, Salas L, Karahanian E and Vicuña R (1998). Cloning and molecular analysis of a cDNA and the Cs-mnp1 gene encoding a manganese peroxidase isoenzyme from the lignin-degrading basidiomycete *Ceriporiopsis subvermispora*. Gene 206:185-193.

Messner K and Srebotnik E (1994). Bioplulping: An overview of developments in an environmentally safe paper-making technology. FEMS Microbiol Rev 13:351-364.

Nielsen H, Engelbrecht J, Brunak S and von Heijne G (1997). Identification of prokaryotic and eukaryotic signal peptides and prediction of their cleavage sites. Protein Eng 10:1-6.

Otjen L, Blanchette R, Effland M and Leatham G (1987). Assessment of 30 white-rot basidiomycetes for selective lignin degradation. Holzforschung 41:343-349.

Rajakumar S, Gaskell J, Cullen D, Lobos S, Karahanian E and Vicuña R (1996). Lip-like genes in Phanerochaete sordida and *Ceriporiopsis subvermispora*, white rot fungi with no detectable lignin peroxidase activity. Appl Environ Microbiol 62:2660-2663.

Rüttimann-Johnson C, Cullen D and Lamar RT (1994). Manganese peroxidases of the white rot fungus *Phanerochaete sordida*. Appl Environ Microbiol 60:599-605.

Rüttimann-Johnson C, Salas L, Vicuña R and Kirk TK (1993). Extracellular enzyme production and synthetic lignin mineralization by *Ceriporiopsis subvermispora*. Appl Environ Microbiol 59:1792-1797.

Rüttimann C, Schwember E, Salas L, Cullen D and Vicuña R (1992). Ligninolytic enzymes of the white rot basidiomycetes *Phlebia brevispora* and *Ceriporiopsis subvermispora*. Biotechnol Appl Biochem 16:64-76.

Sala C, Lobos S, Larraín J, Salas L, Cullen D and Vicuña R (1995). Properties of laccase isoenzymes produced by the basidiomycete *Ceriporiopsis subvermispora*. Biotechnol Appl Biochem 21:323-333.

Scott GM and Swaney R (1998). New technology for papermaking: biopulping economics. TAPPI J 81:153-175.

Scott GM, Akhtar M, Swaney RE and Houtman CJ (2000). Recent Developments in Biopulping Technology at Madison, WI. In: L Viikari and R Lantto, ed. Progress in Biotechnology 21: Biotechnology in the Pulp and Paper Industry: 8th ICBPPI Meeting. Amsterdam: Elsevier Science B.V. pp 61-71.

Sethuraman A, Akin DE and Eriksson KL (1998). Plant-cell-wall-degrading enzymes produced by the white-rot fungus *Ceriporiopsis subvermispora*. Biotechnol Appl Biochem 27:37-47.

Tapia J and Vicuña R (1995). Synthetic lignin mineralization by *Ceriporiopsis subvermispora* is inhibited by an increase in the pH of the cultures resulting from fungal growth. Appl. Environ Microbiol 61:2476-2481.

Tello M, Corsinin G, Larrondo LF, Salas L, Lobos S and Vicuña R (2000). Characterization of three new manganese peroxidase genes from the ligninolytic basidiomycete *Ceriporiopsis subvermispora*. Biochimica et Biophysica Acta 1490:137-144.

Timberlake WE and Barnard EC (1981). Organization of a gene cluster expressed specifically in the asexual spores of *Aspergillus nidulans*. Cell 26:29-37.

Tuor U, Winterhalter K and Fiechter A (1995). Enzymes of white-rot fungi involved in lignin degradation and ecological determinants for wood decay. J Biotechnol 41:1-17.

Urzúa U, Larrondo LF, Lobos S, Larraín J and Vicuña R (1995). Oxidation reactions catalyzed by manganese peroxidase isoenzymes from Ceriporiopsis subvermispora. FEBS Letters 371:132-136.

Urzúa U, Kersten PJ and Vicuña R (1998). Manganese peroxidase-dependent oxidation of glyoxylic and oxalic acids synthesized by *Ceriporiopsis subvermispora* produces extracellular hydrogen peroxide. Appl Environ Microbiol 64:68-73.

Yaver DS, Overjero MDC, Xu F, Nelson BA, Brown SH, and Kauppinen S (1999). Molecular Characterization of Laccase Genes from the basidiomycete *Coprinus cinereus* and heterologous expression of the laccase Lcc1. Appl Environ Microbiol 65:4943-4948.

Applied Mycology & Biotechnology
An International Series. Volume 3. Fungal Genomics

13

Microarrays: Technologies and Applications

Leming Shi[1], Weiming Hu[1], Zhenqiang Su[1], Xianping Lu[1], and Weida Tong[2]
[1]Chipscreen Biosciences, Ltd., Research Institute of Tsinghua University, Suite C301, Shenzhen, Guangdong 518057, China (lmshi@chipscreen.com); [2]National Center for Toxicological Research, Food and Drug Administration, 3900 NCTR Road, Jefferson, Arkansas 72079, USA (wtong@nctr.fda.gov).

With the completion of genome sequencing of more and more organisms, research focus has now been shifted from sequencing to delineating the biological functions of all genes coded within the genome of a particular organism. Methodologies of biological research are evolving from "one gene in one experiment" to "multiple genes in one experiment" paradigm. Microarrays, including DNA microarrays, protein microarrays, cell microarrays, and tissue microarrays, have proven to be extremely powerful tools for analyzing thousands of unique molecules in a biological system in a highly parallel and high throughput fashion, making it possible to gain a global picture of the system under study. In this chapter, we first discuss different formats of the microarray technologies, and then the steps involved in a microarray experiment, such as selection of probes, array-making, target preparation, hybridization, signal readout, image processing, and informatics. Finally, the applications of microarray technologies in biological research, medical diagnostics, drug discovery and development, and toxicology will be discussed.

1. INTRODUCTION

It is understood that thousands of genes and their products (i.e., RNAs and proteins) in a given living organism function in a complicated and orchestrated way that creates the mystery of life. However, traditional methods in molecular biology generally work on a "one gene in one experiment" basis, indicating that the throughput is very limited and the "global picture" of gene functions is hard to obtain. In the past several years, a new technology, called microarray, has attracted tremendous interests among scientists in biological research and other scientific fields. This technology promises to monitor the whole genome on a single chip so that researchers can have a much broader and better view of the interactions among thousands of genes simultaneously.

The fundamental concept of microarray technology is to miniaturize traditional bioanalytical detection system so that hundreds or even thousands of biomolecules with unique identity can be detected simultaneously in one single experiment by using a tiny amount of test sample. Therefore, it is essential to achieve high sensitivity for a tiny amount of analyte in test samples. However, the common wisdom in the early 1990s was that, it was necessary "to bind the majority of the analyte present in a (test) sample" in order to achieve high sensitivity (Hay *et al.*, 1991). Today few microarray technologies conform to this concept (Ekins and Chu, 1999).

The key concept underlying microarray technologies' emergence is that high sensitivities are achievable using far smaller amounts of "binding agent" (located at a high surface density on a solid support) than have, for decades, been regarded as obligatory. To the best of our knowledge, Roger Ekins is the first to have laid down the theoretical foundations of microarray-based analysis (Ekins, 1987, 1989; Ekins and Chu, 1991, 1999; Ekins *et al.*, 1990). Ekins recognized that, using high-specific-activity (e.g. fluorescent) labels, sufficient "capture" (probe) agent could be accommodated on a "microspot" a few microns in diameter to achieve ultrasensitive detection of a target analyte. This allows the construction of microarrays, each microspot therein recognizing a different analyte. By using simple microspotting and confocal scanning techniques, Ekins and colleagues had demonstrated, described, and patented the construction and use of sensitive microarray-based assays before 1989 (Ekins, 1987, 1989; Ekins and Chu, 1991, 1999; Ekins *et al.*, 1990). Although the main focus of Ekins *et al.* was immunological assays, the underlying principles apply to other assay formats. The seminal paper by Schena *et al.* (1995) on "Quantitative monitoring of gene expression patterns with a complementary DNA microarray" leads to the popularity of using DNA microarray as a powerful research tool (Lockhart, 2000; Schena, 1999, 2002).

2. DNA MICROARRAYS

2.1. Definition

An array is an orderly arrangement of features. DNA microarray is one form of the generic "Ekins binding assay". It provides a medium for hybridization of known with unknown DNA samples based on base-pairing rules and automating the process of identifying the unknowns. An array experiment can make use of common assay systems such as microplates or standard blotting membranes, and can be created by hand or by making use of robotics to deposit the samples. In general, arrays are described as macroarrays or microarrays, the difference being the size of the deposited sample spots. Macroarrays contain sample spot sizes of about 300 microns in diameter or larger and can be easily imaged by existing gel and blot scanners. The membrane-based arrays or filter arrays are fallen into this category. The sample spot sizes in microarray are typically less than 200 microns in diameter and these arrays usually contain thousands of spots. Microarrays require specialized robotics and imaging equipment that generally are not commercially available as a complete system.

Terminologies that have been used in the literature to describe this technology include, but are not limited to: biochip, DNA chip, DNA microarray, and gene array. Affymetrix, Inc. owns a registered trademark, GeneChip®, which refers to its high density, oligonucleotide-based DNA arrays. However, in some articles appeared in professional journals, popular magazines, and the WWW, the term "gene chip(s)" has been used as a general terminology that refers to the microarray technology. Affymetrix strongly opposes such usage of the term "gene chip(s)". A few years ago we used the term "genome chip", indicating that this technology is meant to monitor the whole genome on a single chip (http://www.gene-chips.com).

DNA microarrays, or DNA chips are fabricated by high-speed robotics, generally on glass but sometimes on nylon substrates, for which probes with known identity are used to determine complementary binding, thus allowing massively parallel gene expression and gene discovery studies. An experiment with a single DNA chip can provide researchers information on thousands of genes simultaneously - a dramatic increase in experiment throughput.

In the literature there exist at least two confusing nomenclature systems for referring to hybridization partners. Both systems use common terms: "probes" and "targets". According to the nomenclature recommended by B. Phimister (1999) of *Nature Genetics* and generally accepted by the DNA microarray community, a "probe" is the tethered nucleic acid with known sequence or identity, whereas a "target" is the free nucleic acid sample whose identity and/or abundance are being detected. More specifically or commonly, the probes are the DNAs that are immobilized on a substrate, whereas the targets are the mRNAs extracted from a sample (see the following text). We follow this recommendation throughout the discussions in this chapter.

There are two major application forms of the DNA microarray technology: identification of sequences (genes or gene mutations) and determination of expression levels or abundance of genes. A variety of microarray technologies or formats have been developed, depending on the specific combinations of the parameters listed in Table 1.

Table 1. Parameters determining the nature of a microarray technology.

No.	Parameter	Option
1	Probes: features arrayed on the microarray substrate that have known identity or sequence	cDNA, oligonucleotides, proteins, peptide nucleic acids, small molecules, cells, tissues, and organisms
2	Fabrication: techniques to array probes on the microarray substrate	*In situ* synthesis, robotic deposition
3	Targets: samples to be analyzed against the probes	DNA, mRNA, proteins, enzymes, small molecules
4	Assays: principles based on which the targets are being analyzed	Hybridization, electrophoresis, flow cytometry, ELISA, TaqMan
5	Signal readout: principles based on which the assay results can be detected	Fluorescence (confocal microscope scanner and CCD camera), chemiluminescence, mass spectrometry, radioactivity, electrochemistry
6	Image processing: signal intensities of hybridized array spots are quantified from the scanning image	Software for image processing
7	Informatics: computational tools with which the huge amount of data generated from a microarray experiment can be effectively stored and interpreted	Database management system, data mining and visualization, interpretation of biological meaning

A DNA microarray experiment is conducted according to the steps shown in Fig. 1. Details of the steps are explained in the following sections. Part of a pseudo-color image from a DNA microarray experiment is shown in Fig. 2.

2.2. Probes

Probes are features arrayed on the microarray substrate with known identity or sequence. They are used to capture the targets (defined in the following text) of complementary nature. There are two major types of probes used for DNA microarray analysis. The first is cDNA. Each clone is generally of 500~5,000 base pairs in length. The second is oligonucleotides, generally of 25~80-mer. The advantages and disadvantages of cDNA- and oligonucleotide-based microarray technologies are compared in Table 2. cDNA probes, due to its readily availability, was once the mainstream probes used in microarray technologies in the late nineties. In addition to cDNAs that have been characterized by sequences, anonymous sequences obtained from cDNA libraries, are often arrayed on the DNA microarray, allowing for discovering

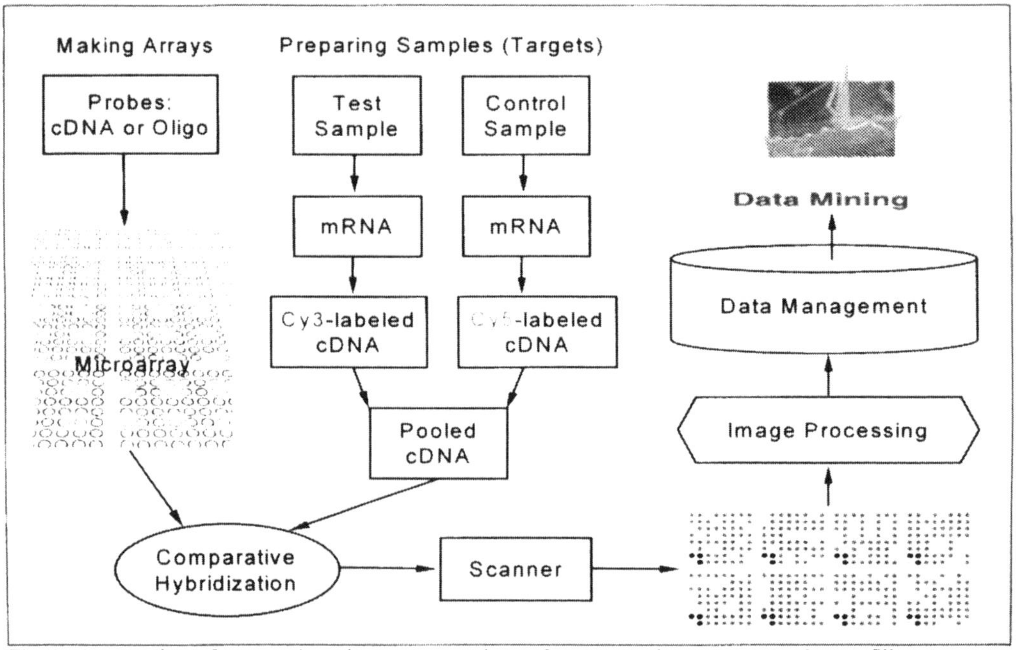

Fig. 1. An overview of a two-color microarray experiment for comparative gene expression profiling.

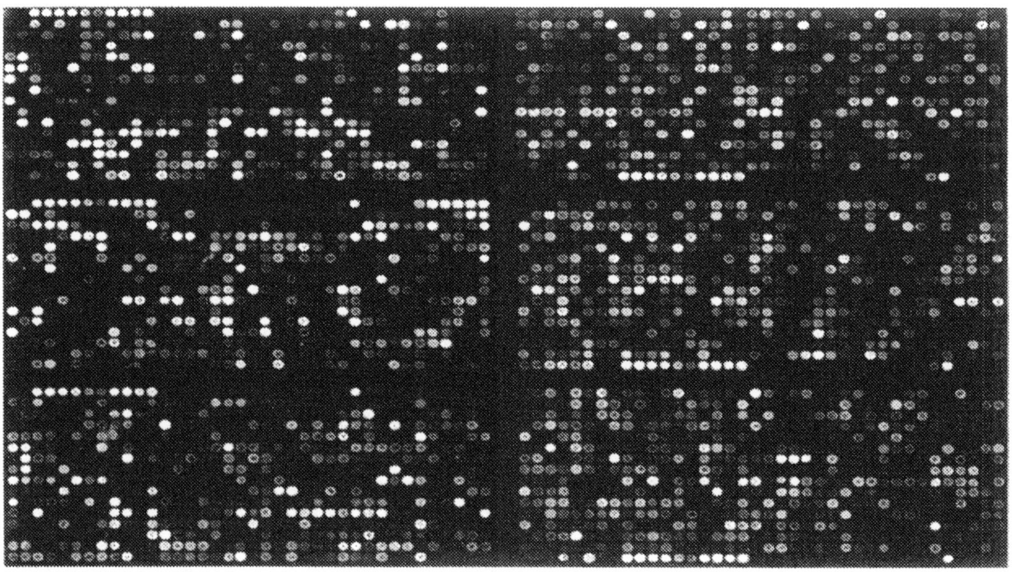

Fig. 2. A portion of a two-color DNA microarray image. The intensity and color of each spot represent the relative expression level of that gene expressed in the test sample compared to the control sample.

Table 2. Pros and cons of cDNA- and oligonucleotide-based microarray technology.

	cDNA microarray	**Oligonucleotide microarray**
Advantages	Flexible array design and construction Good for genes of uncharacterized sequences Clones available for signal validation Low cost per array Simultaneous two-color hybridization Good for gene expression profiling	High density High specificity Less cross-hybridization Less RNA sample required Less reagent / sample cost Good for gene expression profiling and sequence detection No PCR amplification Optimized probe set Good reproducibility
Disadvantages	EST clone library contamination Cross-hybridization (homologs and alternatively spliced transcripts) PCR amplification – costly and time-consuming Low specificity Large amount of RNA required High reagent costs (Taq and fluors) Expensive lab setup	Good for genes or non-coding regions of known sequences only Less flexibility for array design and fabrication Higher cost per array Lower signal intensity

functional roles for as yet uncharacterized genes. In fact, most people in the microarray field consider the work published by the Stanford University groups on cDNA microarrays (Schena *et al.*, 1995; Schena *et al.*, 1996) in the middle nineties as a breakthrough of the microarray technology. However, the intrinsic problems of cross-hybridization of cDNA probes to homologs in the target sample make the oligonucleotide-based microarrays an ideal choice to study organisms that have been sequenced. Advanced bioinformatic tools are available for designing and selecting an optimal set of oligonucleotides with minimal chance of cross-hybridization.

2.3. Fabrication

Probes (cDNAs or oligonucleotides) can be arrayed on a microarray substrate, generally a chemically modified microscope glass slide, silicon wafer, or other material, in two different ways. The first method, called "printing", utilizes a gridding robot equipped with a series of pins, made of stainless steel, glass, or other materials, to transfer tiny amounts (a few nanoliters or less) of a concentrated DNA solution from the wells of microtiter plates to a very small spot (100~200 microns) on the microarray surface. Both cDNAs and oligonucleotides can be deposited by using the "printing" approach, the only difference being surface chemistry. Ink-jetting printing technology has also been used to accurately deposit oligonucleotides on the microarray substrate. The second method, called "*in situ* synthesis", is only applicable in arraying oligonucleotide probes on the microarray surface. The most well-known *in situ* synthesis approach was initially developed for combinatorial chemistry by Affymax, Inc. (Fodor *et al.*, 1991), from which the much more famous Affymetrix, Inc. was spun off. In the Affymax *in situ* synthesis approach, photolithographic technology widely used in the semiconductor industry was applied to chemically synthesizing and addressing oligonucleotides on a silicon wafer. In this case, each array requires a set of photolithographic masks. Precursors are photosensitive hydroxyl-protected deoxynucleotides, which are tethered at the 5'-end and are reactive at the 3'-OH end after light activation (Fodor *et al.*, 1991). Synthesis of oligonucleotides is taken place in a light-induced deprotecting and masking areas.

There are several other techniques available for *in situ* synthesis of oligonucleotides. One example is the Maskless Array Synthesizer (MAS) technology (Singh-Gasson *et al.*, 1999) adopted by NimbleGen Systems, Inc. (http://www.nimblegen.com). Ink-jetting printing technology has also been used to synthesize oligonucleotide probes on the microarray by Agilent (http://www.agilent.com). Synthetic yield in each step of the *in situ* oligonucleotide synthesis is of critical importance. The oligonucleotide probes deposited or synthesized on the microarray constitute the immobile phase DNA.

2.4. Targets

A "target" is the free nucleic acid sample whose identity (i.e. sequence) and/or abundance (i.e. mRNA expression level) is being detected. Standard protocols are available for extracting and labeling RNA from cells or tissue samples. Cell, tissue, or organ is sampled and immediately broken into small fragments in phenol like Trizol or Qiagen buffer. Total RNA or mRNA is extracted with chloroform and isopropanol precipitation or following the standard Qiagen RNA purification protocol. Target molecules (total RNA or mRNA) should be isolated as rapidly as possible to avoid any potential changes in transcript profiles during the procedure. For the Affymetrix chip, poly(A)+ mRNA is enriched from total RNA in a single round using the Qiagen Qligotex kit. Double-stranded cDNA synthesis is carried out incorporating the T7 RNA-polymerase promoter in the first round. This cDNA is then used as template for *in vitro* transcription, which amplifies the RNA pool and incorporates biotinylated ribonucleotides required for the staining procedures after hybridization.
For the cDNA microarray slides, there are two labeling methods. One is direct labeling of cDNAs, which incorporates either Cy3- or Cy5-labeled nucleotides into the first strand cDNA transcribed from total RNA, mRNA, or *in vitro* amplification. The other is indirect labeling of cDNAs, which couples either Cy3 or Cy5 monoreactive fluors to the aminoallyl linker incorporated in the first strand cDNA reverse-transcribed from total RNA or mRNA. The target samples to be hybridized with probe array constitute the mobile phase DNA.

2.5. Assay/Hybridization

For DNA analysis, the most frequently used assay approach is hybridization. In DNA microarray analysis, the analysis is usually performed in a comparative way. That is, the RNA is extracted from each cell or tissue sample independently and labeled with one specific fluorescent dye. Two different targets are equally mixed (pooled) together and hybridized onto the same microarray slide arrayed with probe DNAs. After competitive hybridization with the mixed target, the microarray slide is washed to remove unbound and unspecific bound targets from the microarray surface. Only gene-specific targets remain on the microarray spots and are detected.

2.6. Signal Readout and Scanner

The amount of targets bound to the microarray spots correlates the expression level of genes under investigation. The signal intensity, which represents the amount of target bound to the microarray spots, is measured through fluorescent emission from the hybridized targets under laser excitation. The most widely used imaging method in microarray analysis is typically a dedicated confocal microscope scanner. In fact, as early as the eighties Ekins began to use confocal microscope scanner for highly sensitive microarray signal detection (Ekins, 1987, 1989, 1999). Because of the high cost of confocal microscope scanner, CCD cameras have recently become an alternative option for microarray imaging. It should be kept in mind that microarray analysis is a very dedicated trace analysis technique. It is important for the signal detection system to have a high sensitivity and reliability with a wide dynamic range so that relatively rarely expressed genes may be detected. Since the microarray spot is usually less than 200 microns in diameter, the resolution of the scanner should be good enough to capture the signal quantitatively. The resolution of most scanners is 5~10 microns.

This resolution allows the signal intensity for each spot to be represented by hundreds of pixels in the 2-D image file, which is generated by the scanner.

2.7. Image Processing

In the image processing step, the fluorescent intensity of each spot is quantified by using specially designed software packages. Each image file is composed of many pixels. The value of each pixel represents the signal intensity of that particular area on the microarray. The depth of the image can be 8-, 12-, or 16-bits, which means that the maximum value for a particular pixel is $2^8 - 1 = 255$, $2^{12} - 1 = 4095$, or $2^{16} - 1 = 65535$, respectively, which is related to the dynamic range of a scanner.

There are three important functions for an image analysis software: to distinguish a spot from its background; to quantify the overall intensity for each spot and its background intensity by summing the intensity values of all the pixels falling into that spot; to calculate a statistical confidence value that identifies the quality of the spot. Data analysis should focus only on spots with high quality. Most scanners come with an image analysis software. However, the ImaGene software from BioDiscovery (http://www.biodiscovery.com) has emerged as one of the industrial standards in automated image quantification.

Another function of image processing software is to generate a pseudo-color image from the two white-and-black images generated from the Cy3 and Cy5 channels.

2.8. Informatics

2.8.1. Databases

In a single microarray experiment a large number of data points are generated. Therefore, informatics tools become essential and critical for handling huge amount of data and for extracting meaningful information from them. All raw data (images and intensity values) and information pertinent to the experiment need to be stored in a database system for subsequent data analysis. The output from an image quantification software is usually a large 2D spreadsheet. Each row is associated with a particular gene on the chip and each column is a particular sample. The expression level of a gene for a sample is displayed in the corresponding cell. For comparative gene expression analysis, the results are usually represented as the ratios of the intensities of two targets labeled with different fluorescent labels representing two separate samples to be compared.

2.8.2. Normalization

Because of the difference in dye behavior and labeling efficiency, it is necessary to normalize the relative fluorescence intensities in each of two channels. The normalization process eliminates systematic difference in the intensities from the Cy3 and Cy5 scanning channels (Quackenbush, 2001, 2002). The simplest and most widely used normalization method is "total intensity normalization", which assumes that the quantity of initial mRNA is the same for both labeled samples. Compared to the control sample, some genes in the test sample are upregulated and some are downregulated, but the net result for thousands of genes on the array should be balanced, leading to the assumption that each of the Cy3 and Cy5 channel should show the same overall fluorescent signal intensity. Therefore, a normalization factor, defined as the ratio of overall Cy3 channel intensity over overall Cy5 channel intensity, can be calculated to re-scale the signal intensity for each gene on the microarray.

Another normalization method uses regression techniques, both linear and nonlinear regression (Quackenbush, 2001, 2002). For mRNA derived from closely related samples, most of the genes are expected to express at similar levels. Therefore, the correlation between Cy3 and Cy5 intensities across these genes should be along a straight line of slope 1.

Any systematic deviation from such a straight line can be corrected by using a linear regression method. In reality, the correlation between Cy3 and Cy5 intensities are nonlinear. In such cases, LOWESS (Locally Weight Scatterplot Smoothing) regression is more suitable for such corrections. Fig. 3 shows the original and LOWESS normalized intensity data from the Cy3 and Cy5 channels.

2.8.3. Ratio Calculation

Considering the variability in each of the experimental procedures, microarray gene expression analysis has been performed largely on a comparative basis, i.e., the ratio of two signals from the Cy3 and Cy5 channels provides a relative measure of the difference between the control and test samples. Most of the statistical analysis of microarray data is based on the ratio data rather than on the absolute intensities.

2.8.3. Data mining and visualization

Data mining and knowledge discovery in databases (KDD) are new terms that have been used to describe the research efforts of turning raw data into useful knowledge for decision-making in scientific research (Hu and Kamber, 2001). Data mining or KDD is defined as "the nontrivial extraction of implicit, previously unknown, and potentially useful information from data" (Frawley *et al.*, 1992). "Nontrivial" means that data mining is not a simple task. Usually, data mining and KDD are used interchangeably, although, generally speaking, data mining focuses on the algorithms and KDD deals with the whole process that includes data storage, retrieval, pre-processing, and analysis. To get the hidden, previously unknown information from data requires special expertise. Visualization is an integrated component of the data mining process. Data mining results are often communicated to researchers via a convenient, easy-to-perceive visual interface. Mining of microarray experimental data starts with a data table illustrated in Fig. 4. The table usually contains thousands of rows, which represent the genes being monitored on the microarray. Each column of the table represents a particular experiment or sample under which the expression levels of thousands of genes are monitored simultaneously. Each data entry in the table represents the mRNA expression ratio for a particular gene in a particular experiment.

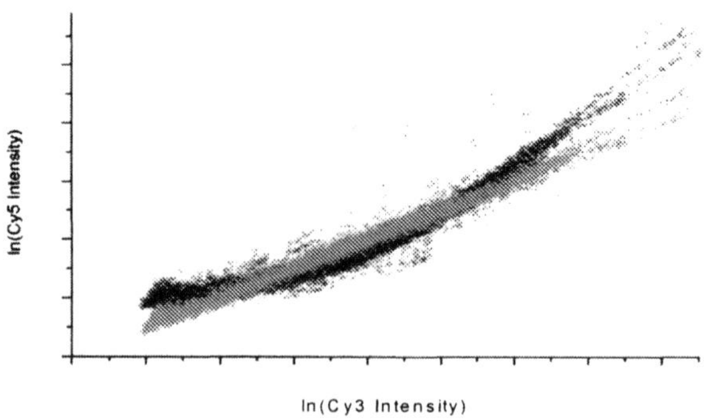

Fig. 3. Normalization of the Cy3 and Cy5 channel intensities using LOWESS. Data points shown in dark represent original intensity values, whereas data points in gray represent LOWESS normalized intensity values.

There are three typical questions that researchers may ask about the microarray data. The first question is which genes are differentially expressed between test and control samples. To answer this question, we need to examine the expression data and sort out those genes that are either over-expressed or under-expressed. Because of the inherent variation in the microarray experiment, it is difficult to make a judgment solely based on the results of one experiment. In other words, how much fold change for a gene could be considered biologically relevant? Statistically, it is necessary to perform replicates to assess the variation that comes from the microarray technology itself so that one will not take experimental fluctuation as the real difference in gene expression. In practice, some researchers use a rational fold-difference cutoff value, such as 2.0, to determine which genes are differentially expressed.

The second question is which genes are co-expressed. Here, two genes are compared in terms of the expression profiles across the P experiments. In other words, two rows are compared. The Euclidean distance and/or the Pearson correlation coefficient are two widely used metrics for gene comparison. The assumption is that co-expressed genes should have similar expression profiles across experiments under different conditions.

The third question is which genes form the same gene clusters. The fundamental principle to identify gene clusters is to assume that genes in a cluster share similar function. At this level, data for multiple genes across multiple experimental conditions are considered simultaneously. In other word, the data matrix shown in Fig. 4 is taken into consideration altogether.

A bottleneck problem in a microarray experiment is how to make biological sense of the

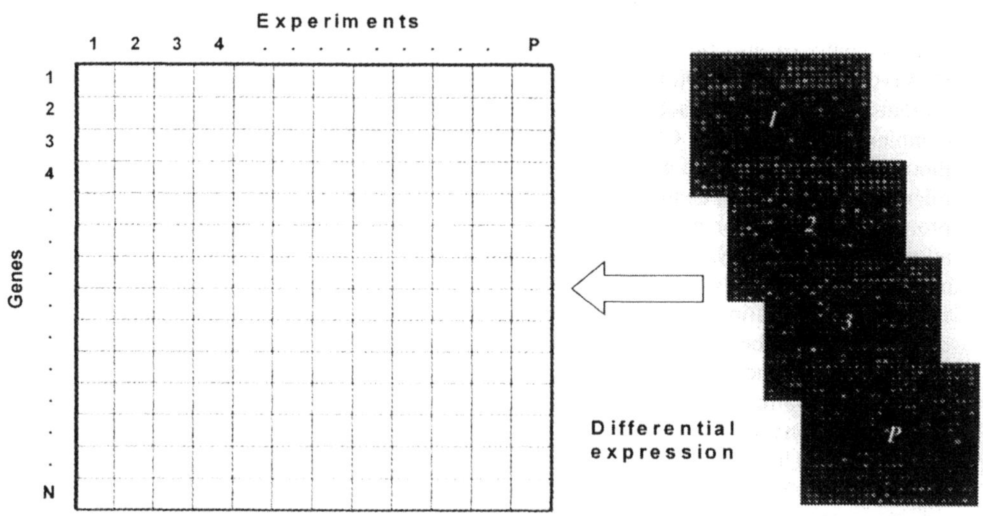

Fig. 4. Data format for DNA microarray data mining. Each gene is characterized by its expression profile across the P experiments (samples), whereas each experiment (sample) is characterized by the relative mRNA expression levels of the N genes monitored by the microarray. Each data entry in the table represents the mRNA expression ratio for a particular gene in a particular experiment.

massive expression data. Many statistical, machine learning, and visualization techniques that have been used in data mining for drug discovery (Shi, 2000, 2002a) have been applied successfully for the analysis of microarray data. Table 3 lists some of the most widely used data mining and visualization methods for this purpose. Our experience shows that there is no method that is suitable for all problems. Instead, each method has its own advantages and limitations. For a particular problem one method may be better than the other. It is up to the researcher to identify the most appropriate method(s) for her/his particular problem, usually by exploring many different methods for a data set. In fact, a very important task for a bioinformatics researcher is to identify the best method(s) for analyzing the available data set. GeneSpring was developed by Silicon Genetics, Inc. (http://www.silicongenetics.com) specifically for the analysis of gene expression data. Spotfire is another widely used data mining and visualization package (http://www.spotfire.com).

For a comprehensive review of microarray data analysis, the readers are encouraged to refer to a few recent review articles by Quackenbush (2001) and Zhang (2002) and the book edited by Lin and Johnson (2002).

3. RELATED MICRARRAY TECHNOLOGIES

The power of DNA microarray technology lies in its capability of analyzing thousands of genes in a sample in one single experiment. In terms of the number of genes that can be monitored simultaneously, DNA microarray is truly a very high throughput technology for analyzing DNA and RNA samples. It seems straightforward to apply the same microarraying format to analyze other types of molecules of biological significance in a parallel and high throughput fashion. This leads to several other powerful microarray formats to be discussed in the following sections.

3.1. Protein Microarrays

The mystery of life is coded in the DNA sequences of an organism's genome. Studying mRNA levels of an organism does provide a spectrum of the biology at functional level. However, most of the biological functions of a living organism are performed by many proteins in a highly interacted way. Profiling mRNA alone is insufficient to understand the complexity of biology. Clearly, it would be advantageous if the biological functions of thousands of proteins in a cell can be studied simultaneously in a single experiment. Protein microarrays have been designed exactly for such purposes (Templin *et al.*, 2002). The idea of protein microarrays or protein chips is not new. In fact, well before DNA microarray was adopted for DNA/RNA analysis, Ekins has laid down the theoretical foundations for the analysis of multi-analytes on microspots (Ekins, 1987, 1989). Ekins and colleagues utilized robotic and ink-jetting systems for microspotting and laser confocal microscope scanner for detecting fluor-labeled proteins in immunoassays (Ekins, 1987, 1989). These antibody microarrays are one form of the protein microarrays widely used today, as shown in Table 4.

Proteins immobilized on a microarray can be used to study protein-protein interactions and protein functions, including protein-nucleic acids and receptor-ligand interactions. Ge (2000) described the Universal Protein Array (UPA) for studying protein functions. MacBeath and Schreiber (2002) spotted more than 10,000 protein spots on a microarray glass slide using a robotic spotter as used for DNA microarray production. The protein array was used to identify protein-protein and protein-drug interactions. The difficulty is how to obtain thousands of purified proteins and maintain them in their natural conformation. Zhu *et al.* (2001) cloned 5,800 yeast open reading frames (ORFs) and overexpressed and purified their corresponding proteins. The proteins were printed onto slides at high spatial density to form a yeast proteome microarray and screened for their ability to interact with proteins and phospholipids.

Table 3. Data mining and visualization techniques applied for microarray data..

No.	Method	Application*
1	Principal Component Analysis (PCA)	DR, Viz
2	Multidimensional Scaling (MDS)	DR, Viz
3	Singular value decomposition (SVD)	DR, Viz
4	Pattern Recognition (PR)	Class, Cluster, DR, Viz
5	Hierarchical Cluster Analysis (HCA)	Cluster, Viz
6	Non-Hierarchical Cluster Analysis (*K*-Means, Jarvis-Patrick)	Cluster, Viz
7	Correlation Analysis (Pearson, Spearman)	Class
8	Multiple Linear Regression (MLR)	Class
9	Partial Least-Squares Regression (PLS)	Class, Cluster, DR
10	Soft Independent Modeling of Class Analogy (SIMCA)	Class, Cluster, DR
11	*K*-Nearest Neighbors (KNN)	Class, Cluster
12	Artificial Neural Networks (ANN)- Back Propagation (BP) and Self-Organizing Maps (SOM)	Class; Cluster, Viz
13	Classification and Regression Trees (CART)	Class, Viz
14	Multivariate Adaptive Regression Splines (MARS)	Class, Viz
15	Genetic Algorithms (GA)	Class, DR
16	Cross-Validation (CV) and Bootstrapping	Class
17	Support Vector Machines (SVM)	Class, Cluster, Viz
18	Clustered Image Maps (CIM)	Viz

*DR: dimension reduction; Class: classification; Cluster: clustering; Viz: visualization

Table 4. Comparison of various protein array systems.

	Antibody Array	Peptide Array	Tissue Array	UPA
Arrayed Features	Specific antibodies	Synthetic peptides	Crude tissue extracts	Purified active proteins
Applications	Protein expression	Protein-protein interaction	Protein expression	Protein functions
Analysis	Tissue extracts; Body fluids	Proteins; nucleic acids; small molecules	Antibodies	Proteins; nucleic acids; small molecules
Focus	Disease-related proteins	Peptide-targeted molecules	Disease-related proteins	Protein-targeted molecules

3.2. Cell Microarrays

Cell microarrays are a new tool for functional genomic studies (Wu *et al.*, 2002). On a cell microarray slide each of the thousands of cell clusters is transfected with a defined DNA, resulting in either the overproduction or the inhibition of a particular gene product. By using a variety of appropriate detection assays, the phenotypic consequences of perturbing each gene in mammalian cells can be probed in a systematic and high-throughput fashion. Combining well-established methods for cellular investigation with the miniaturization and multiplexing capabilities of microarrays, cell microarrays are a versatile tool that can be useful in many cell-biological applications. With cell microarrays single cells can be deposited on a microarray substrate or manipulated in an arrayed structure so that physiological behavior of single cells can be studied. The interaction between single cells can also studied with cell microarrays (Offenhausser, 2002).

3.3. Tissue Microarrays

Tissue microarrays (Kononen *et al.*, 1998) were first reported by a research group led by Kallioniemi at the U.S. National Human Genome Research Institute (NHGRI), NIH. In contrast to DNA microarrays where thousands of genes are profiled on a single slide for a single sample, tissue microarrays have hundreds of tissue samples arrayed on a microscope slide by a robotic system and are analyzed against a single molecular target, usually through parallel *in situ* detection of DNA, RNA, and protein targets in each specimen on the array. Tissue microarrays are high throughput screening tool in terms of the number of tissue samples that can be analyzed in a single experiment. The combination of cDNA microarrays and tissue microarrays proved to be very powerful. cDNA microarrays can be used to identify a small subset of "interesting" genes from thousands of genes on the microarray. But this tool can only comprehensively study one or two samples at a time. However, analysis of hundreds of specimens from patients at different stages of a disease is often needed to establish the diagnostic, prognostic, and therapeutic importance of each of the emerging markers. Tissue microarray technology could be a powerful tool for validating the findings from DNA microarrays (Kallioniemi *et al.*, 2001).

3.4. Chemical Microarrays

The immobilized features on chemical microarrays are small molecules that are screened against proteins for discovering drug candidates. MacBeath *et al.* (1999) used a robotic spotter to print small molecules on a microarray slides in order to detect protein-ligand interactions on a large scale. These small molecules can selectively capture proteins to which the small molecules can bind with high affinity. The difference compared to protein microarrays for protein-ligand study is which counterpart of the reaction pair is immobilized on the microarray surface. Graffinity Pharmaceutical Design GmbH uses chemical microarrays as screening tools to enhance the understanding of protein binding specificity, based on diversity label-free detection (http://www.graffinity.com). It is important to keep the small molecules in the right orientation that is reachable by the proteins.

3.5. Lab-on-a-chip Systems

An attractive goal for researchers in applying microtechnology for the analysis of samples of general interests is to integrate all the components of an analytical lab into one single system. "Lab-on-a-chip" or Micro Total Analysis Systems (uTAS) (Auroux *et al.*, 2002; Reyes *et al.*, 2002) integrate microfluidic channels with pumps, valves, and detectors. Microfluidic systems are composed of fluid channels and chambers with critical dimensions of tens to hundreds of microns. Microfluidic systems are used for sample preparation for mass spectrometric analysis, fluid and particle routing, detection and control of chemical reactions, mixing of solutions, and separations. These operations are used in a number of different techniques, such as process analysis, environmental monitoring, clinical diagnostics, drug discovery, cell culture, cell manipulations, protein analysis, polymerase chain reaction (PCR), DNA sizing, and sequencing (Cheng *et al.*, 1998; Meldrum and Holl, 2002).

4. APPLICATIONS

The applications of microarray technology have flourished dramatically during the past several years, demonstrating the superior power of this technology for large-scale parallel analysis of biological samples or the identification of genes and their functions or their mutations. The applications of DNA microarrays can be classified into two categories: gene expression profiling and sequence identification. Gene expression profiling can be performed with either cDNA microarrays or oligonucleotide microarrays. However, sequence identification or mutation detection can only be done with the use of oligonucleotide microarrays.

4.1. Biological Research

Microarray-based analytical tools allow researchers to perform biological experiments in a much faster and larger scale fashion. Combined with information on a large number of

genes, it is possible to delineate the complex relationships in an organism at DNA, RNA, and protein levels. One important trend in microarray applications is that the data-driven approach is helping pursue hypothesis-driven inquiries through broad genomic surveys. It is hard to image how functional genomics would look like without microarray-based tools. In fact, the very first applications of DNA microarrays for gene expression profiling of yeast cell cycles helped to identify cell cycle specific genes and the dynamic changes of gene expression (Spellman *et al.*, 1998). Using periodicity and correlation algorithms, Spellman *et al.* identified 800 genes that met an objective minimum criterion for cell cycle regulation.
The response of human fibroblasts to serum at the gene expression level was explored with a cDNA microarray containing 8,600 different human genes (Iyer *et al.*, 1999). By the withdrawal and addition of the serum supply that nourished the cultures of human fibroblasts, Iyer *et al.* monitored which DNAs in the microarray bound to the targets and were able to identify which genes were active and when they are active. With the aid of a computer program that examined the 500 most active genes, they grouped those with similar activity patterns and drew a conclusion that the fibroblasts essentially reacted to exposure to serum in culture much as they would in the body if blood had seeped into a fresh skin wound. Another interesting application is to reveal DNA copy number alteration in human breast tumors (Pollack *et al.*, 2002). Genomic DNA copy number alterations are key genetic events in the development and progression of human cancers. Pollack *et al.* employed the DNA microarray technology to study genome-wide copy number variation in a series of primary human breast tumors. Parallel microarray analysis revealed the remarkable degree to which variation in gene copy number contributed to variation in gene expression in tumor cells, which contributed to the development or progression of cancer.

4.2. Medical Diagnostics and Personalized Medicine

Synthetic oligonucleotide microarrays have been used to identify large numbers of specific DNA markers by molecular hybridization. Specifically, genotyping for point mutations, single nucleotide polymorphisms (SNPs), and short tandem repeats (STRs) are being detected by using oligonucleotide microarrays. Array-based assays are rapidly becoming routine tools in modern molecular biology laboratories. It is evident that these tools will also be used in molecular diagnostics for the routine screening and identification of diseases of genetic origins. Both the academic and commercial communities have been putting a lot of efforts and resources in producing a high-quality map of the genetic markers, known as SNPs. This will provide the scientific and medical community worldwide with a powerful new tool to enhance the understanding of disease processes and to facilitate discovery and development of safer and more effective therapies. The SNP map can be used to identify specific genes and mutations involved in both common and rare diseases. In some cases a single base pair mutation in the sequence of a gene can have profound impact on the normal function of the gene product.

Genes in the cytochrome P450 family have been under intense studies (Rushmore *et al.*, 2002). Mutation in these genes can dramatically affect drug metabolism. The understanding of the genetic variations that predict response to therapy or drug metabolism serves as the basis for developing novel diagnostic tests, many of which are array-based. Many companies are developing more rapid and less expensive SNP scanning technologies for potential medical diagnosis.

Until recently, diagnostic and prognostic assessment of diseased tissues and tumors had relied heavily on indirect indicators that permitted only general classifications into broad histologic or morphologic subtypes and did not take into account the alterations in individual gene expression. Global expression analysis using microarrays now allows for simultaneous interrogation of the expression of thousands of genes in a high-throughput fashion and offers unprecedented opportunities to obtain molecular signatures of the state of activity of diseased cells and patient samples (Macgregor and Squire, 2002). Microarray analysis may provide invaluable information on disease pathology, progression, resistance to treatment, and response to cellular microenvironments and ultimately may lead to improved early diagnosis and innovative, personalized therapeutic approaches for cancer (Shi, 2001).

Cancer is a family of genetic diseases and has been the most intensively studied field with DNA microarray technology. Alon *et al.* (1999) used Affymetrix Hum6000 chip with >6,500 genes to profile gene expression of 40 tumor and 22 normal colon tissues. Data on 2,000 most varied genes across the 62 samples were used in the analysis. An efficient, divisive, two-way HCA algorithm, based on the deterministic-annealing algorithm, was implemented in MatLab and had computation time of $O\{N\log(N)\}$. The algorithm was applied to both the genes and the tissues, revealing broad coherent patterns that suggested a high degree of organization underlying gene expression in these tissues. Coregulated families of genes clustered together, as demonstrated for the ribosomal proteins. Clustering also separated cancerous from noncancerous tissues and cell lines from *in vivo* tissues on the basis of subtle distributed patterns of genes even when expression of individual genes varied only slightly between the tissues. Two-way clustering is useful both in classifying genes into functional groups and in classifying tissues based on gene expression.

Based on gene expression profiling using Affymetrix chip of 6,817 genes, Golub *et al.* (1999) were able to classify acute myeloid leukemia (AML) and acute lymphoblastic leukemia (ALL) samples. In this study, 38 samples (27 ALL and 11 AML) were used as the training set and 34 samples (20 ALL and 14 AML) as the prediction set. "Neighborhood (correlation) analysis" method identified 50 “informative genes” for self-organizing maps. 36 of the 38 training samples and 29 of the 34 prediction samples were correctly classified.

Alizadeh *et al.* (2000) used a cNDA microarray of 17,856 clones to classify B-cell lymphoma. 96 normal and malignant lymphocyte samples were hybridized to 128 Lymphochip microarrays. 4,026 array elements were used in data analysis and Pearson correlation was used as the distance metric for cluster analysis. There was diversity in gene expression among the tumors of diffuse large B-cell lymphoma (DLBCL) patients. The authors identified two molecularly distinct forms of DLBCL that had gene expression patterns indicative of different stages of B-cell differentiation. One type expressed genes characteristic of germinal center B cells (germinal center B-like DLBCL); the second type expressed genes normally induced during *in vitro* activation of peripheral blood B cells (activated B-like DLBCL). Patients with germinal center B-like DLBCL had a significantly better overall survival than those with activated B-like DLBCL. This correlation validated the significance of clustering based on expression profiles.

Bittner *et al.* (2000) used a cDNA microarray of 8,150 features (representing 6,971 unique genes) to study malignant melanomas. Thirty-eight samples (31 melanomas and 7 controls) were used in the study. Global transcript analysis could identify unrecognized subtypes of cutaneous melanoma and predicted experimentally verifiable phenotypic characteristics that might be of importance to disease progression. Using a cDNA microarray with 8,102 human genes, Perou *et al.* (2000) performed 84 gene expression experiments with 65 surgical specimens of human breast tumors from 42 different individuals. 84 experiments and 1,753 genes were analyzed. Twenty of the tumors were sampled twice, before and after a 16-week course of doxorubicin chemotherapy, and two tumors were paired with a lymph node metastasis from the same patient. Gene expression patterns in two tumor samples from the same individual were almost always more similar to each other than either was to any other sample. Sets of coexpressed genes were identified for which alteration in mRNA levels could be related to specific features of physiological difference. The tumors could be classified into subtypes distinguished by pervasive differences in their gene expression patterns. Ramaswamy *et al.* (2001) performed a very comprehensive study on multiclass cancer diagnosis using tumor gene expression signatures. To determine whether the diagnosis of multiple common adult malignancies could be achieved purely by molecular classification, Ramaswamy *et al.* used oligonucleotide microarrays to perform gene expression analysis of 218 tumor samples, spanning 14 common tumor types, and 90 normal tissue samples. The expression levels of 16,063 genes and expressed sequence tags were used to evaluate the accuracy of a multiclass classifier based on a support vector machine algorithm. Overall classification accuracy was 78%, far exceeding the accuracy of random classification (9%). Poorly differentiated cancers resulted in low-confidence predictions and could not be accurately classified according to the origin of tissues, indicating that they are molecularly distinct entities with dramatically different gene expression patterns compared with their well-differentiated counterparts. These results demonstrate the feasibility of accurate, multiclass molecular cancer classification and suggest a strategy for future clinical implementation of molecular cancer diagnostics.

This technology was recently employed for the delineation of prognostic biomarkers in prostate cancer, wherein the clustering of tumors according to their mRNA expression profiles was found to correspond to the four different clinical stages of the prostate specimens (Dhanasekaran SM *et al.*, 2001). Traditionally, the serum level of prostate-specific antigen (PSA) is used as a biomarker for the diagnosis and monitoring of prostate cancer. However, a false-positive diagnosis can result because elevated PSA levels can also exist in certain nonmalignant conditions of the prostate, such as benign prostatic hyperlasia. The authors identified many statistical associations between genes and prostate cancer. Associations between two genes (hepsin, a transmembrane serine protease; and pim-1, a serine/threonine kinase) and prostate cancer were further investigated using a tissue microarray consisting of over 700 prostate cancer specimens. The tissue microarray was stained with either antihepsin antibody or anti-pim-1 antibody. The expression of hepsin and pim-1 proteins could effectively predict the clinical outcome of patients. This study clearly demonstrated the power of expression DNA microarray in identifying tumor biomarkers for diagnostic purpose followed by the validation of the biomarkers using tissue microarray. The ability to predict clinical outcome by biomarkers identified by DNA microarrays sets the foundation for accurate and fast molecular diagnostics and personalized medicine.

Why some drugs work better in some patients than in others? And why some drugs may even be highly toxic to certain patients? Pharmacogenomics or pharmacogenetics (Chicurel and Dalma-Weiszhausz, 2002) is the hybrid of functional genomics and molecular pharmacology. One main goal of pharmacogenomics is to find the correlations between therapeutic responses to drugs and the genetic profiles (expression profile, SNPs, etc.) of patients. The promise of personalized medicine, "finding the right drug for the right patient at the right time", relies on the availability of reliable diagnostic tools based on microarrays.

4.3. Drug Discovery and Development

The pharmaceutical industry has been a major driving force for the acceptance and widespread utility of microarray technology. The process of discovering and developing a new drug is time-consuming and expensive (Drews, 2000; Smith, 1992). According to information provided at the Web site (http://www.phrma.org) of the Pharmaceutical Research and Manufacturers Association (PhRMA), on average it costs a company \$500~800 million dollars and 10~15 years to get a new drug from the laboratory to patients. Among the 5,000~10,000 compounds that are synthesized and screened, only about 250 enter into preclinical animal testing and five of them are advanced to human testing, i.e., Phase I - Phase III clinical trials. Only one of those five compounds is approved by the Food and Drug

Administration (FDA) for marketing. Furthermore, only three of ten approved drugs can generate profit that is enough to cover the research and development costs.
On average, the pharmaceutical industry is investing 15~20% of its annual revenue in R&D. Not surprisingly, every pharmaceutical company is trying hard to increase its success rate and shorten the time required for FDA approval in order to remain competitive on the market. Competition in the pharmaceutical industry forces each company to adopt innovative technologies, e.g. computer-aided drug design, combinatorial chemistry, high throughput screening, and microarrays, which are promising to bring lead molecules to the marketplace in a shortened period of time. The parallel nature of DNA microarray technology enables researchers to investigate the biological effects of drug candidates at a genome-wide scale. Making sense of the huge amount of gene expression data from DNA microarray technology has become one of the most challenging tasks of bioinformaticians. Identifying novel drug targets can give pharmaceutical companies many advantages in developing new therapeutics. For example, drugs against a novel target may solve the problem of drug resistance encountered with previous drugs. It has been seen in the development of new AIDS drugs against different targets. There are only several hundreds of targets for currently available drugs, but the number of druggable targets is estimated to be at least in thousands. Understandably, the potential of identifying novel targets for drug discovery with microarray technologies has been extremely attractive to the pharmaceutical industry. For example, Millennium Pharmaceuticals, Inc. has invested tremendously in cDNA microarrays and Affymetrix GeneChip® technology to identify and validate novel drug targets both for its internal drug discovery programs and for strategic alliance with big pharmaceutical companies like Bayer, Roche, and AHP.

One important change in the application of microarrays is the shift from drug discovery to drug development. Although gene expression analysis for drug discovery is still a very important area for microarray applications and will remain so for many more years to come, researchers are combining their microarray studies with other data, then trying to map all those data to biological pathways and systems. Meanwhile, an increasing number of microarray studies are being done at the early development stage, e.g. in lead optimization and preclinic evaluation, specifically, in toxicological studies.

At Chipscreen Biosciences Ltd. (http://www.chipscreen.com) we have developed an integrated drug discovery platform based on chemical genomics principles. Central to Chipscreen's drug discovery platform is its capability of integrating computer-aided drug design, medicinal chemistry, parallel multi-target high throughput screening, global gene expression profiling, and informatics to rapidly and effectively advance the drug discovery process. A very important aspect of applying microarray technology in our drug discovery and development process is gene expression profiling for candidate evaluation. Gene expression profiles of our own lead compounds and controls, i.e. drugs on the market, candidates from our competitors, and similar drugs with adverse effects, are compared. The rationale is that for a candidate to move forward in the R&D, it should not show any gene expression profile similar to drugs causing adverse effects. In addition, specific genes related to the mechanism of toxicological effects are being investigated in great detail. To fulfill Chipscreen's drug discovery needs, we have developed an integrated biochemoinformatics software system to effectively store and analyze various types of experimental data including chemical structures, biological activity fingerprints, and gene expression profiling (Shi, 2002b; Shi *et al.*, 2003). Early evaluation of toxicological profile of drug candidates has been the major application of DNA microarray technology in Chipscreen's internal drug discovery and development projects.

4.4. Toxicogenomics

An emerging field called toxicogenomics (Nuwaysir *et al.*, 1999; Lakkis *et al.*, 2002) has attracted the attentions of academic institutions, the pharmaceutical industry, and regulatory agencies. Toxicogenomics is the hybrid of functional genomics and molecular toxicology. The goal of toxicogenomics is to find correlations between toxic responses to toxicants and changes in the genetic profiles of the objects exposed to such toxicants. By using large-scale gene expression profiling with Affymetrix GeneChip® microarrays on the analysis of clinical samples and drug-treated samples, Gene Logic, Inc. (http://www.genelogic.com) has built up a large database that is available upon subscription. It is up to the subscribers to mine the database and interpret the biological meaning of their findings. Iconix Pharmaceuticals, Inc., in collaboration with Incyte Genomics, Inc., has built the DrugMatrix database based on a similar concept (http://www.iconixpharm.com). In September 2000, the US National Institute of Environmental Health Sciences (NIEHS) of the NIH established the National Center for Toxicogenomics (NCT, http://www.niehs.nih.gov/nct), whose mission is to coordinate a nationwide research effort for the development of a toxicogenomics knowledge base. There are five goals of the NCT:

1. To facilitate the application of gene and protein expression technology;
2. To understand the relationship between environmental exposures and human disease susceptibility;
3. To identify useful biomarkers of disease and exposure to toxic substances;
4. To improve computational methods for understanding the biological consequences of exposure and responses to exposure; and
5. To create a public database of environmental effects of toxic substances in biological systems.

A compendium of gene expression data enhanced by complete proteomic analysis will enable investigators to probe the complexities of the mechanisms of normal genetic and metabolic pathways, and subsequently, to learn how disease occurs when these pathways malfunction. When combined with information on gene and protein groups, functional pathways and networks, and human genetic polymorphisms, these data will confer new knowledge of gene-environment interactions and human health risks (Hamadeh *et al.*, 2002).

The US Food and Drug Administration (FDA)'s National Center for Toxicological Research (NCTR) has recently created the Center for Toxicoinformatics for handling the huge amount of data being generated from toxicogenomic studies as well as other "omic" researches at the NCTR/FDA. They are developing a Toxicoinformatics Integrated System (TIS) that integrates diverse data types from toxicogenomics studies, conventional toxicological endpoints, and public data. A prototype of the system, called ArrayTrack for DNA microarray data management and analysis has been recently released for testing within the FDA centers, and will be available to the public in the near future. The first FDA/NCTR toxicoinformatics workshop on Toxicogenomics Database, Study Design, and Data Analysis was held at the NCTR on December 4, 2002 (http://www.fda.gov/nctr/nctr_eventinfo.html).

4.5. Other Applications

The detection of viral pathogens is of critical importance in biology, medicine, and agriculture. To comprehensively detect viral prevalence, Wang *et al.* (2002) developed a genomic strategy for highly parallel viral screening that was based on the use of a long oligonucleotide (70-mer) DNA microarray capable of simultaneously detecting hundreds of viruses. Using virally infected cell cultures, they were able to efficiently detect and identify many diverse viruses. Related viral serotypes could be distinguished by the unique hybridization pattern generated by each virus. Furthermore, by using microarray elements derived from highly conserved regions within viral families, individual viruses that were not explicitly represented on the microarray were still detected, making it possible for virus discovery. This method greatly expanded the spectrum of detectable viruses in a single assay while simultaneously providing the capability to discriminate among viral subtypes.

Microarray technology has been used to elucidate the mechanistic roles of genes in the pathogenesis of infectious disease, to identify the genes involved in pathogenicity by studying host-pathogen interaction, to figure out the evolutionary relationship between species and to integrate the clinical and genomic data. Smoot *et al.* (2002) arrayed DNA from 36 serotypes M18 *A Streptococcus* strains and genome-wide analyzed the relation between the variation regions and the disease Acute Reheumatic Fever (ARF). Discovery of new putative virulence-related variations and analysis of their distribution among strains provided a more complete view of the molecular armaments. This work provides a critical foundation for accelerated research into ARF pathogenesis and the relevance to studies of host-pathogen interactions.

Understanding the host and pathogen interaction is important in medical research. The pathogen *Bordetella pertussis* is known as the causative agent of whooping cough. Belcher *et al.* (2000) studied the pathology caused by *B. pertussis* at the molecular level and revealed pathogenic mechanisms by gene expression profiling of the early transcriptional responses. In this study, they examined the interaction of *B. pertussis* with a human bronchial epithelial cell line and measured host transcriptional profiles by using high-density DNA microarrays. Host genomic transcriptional profiling provides insight into the complex interaction of host and pathogen.

5. Challenges and Opportunities

Profiling gene expression in biological systems by the use of microarrays plays a crucial role in our understanding of the gene networks that control developmental, physiological, and pathological processes. However, the full promise of microarray technology has yet to be realized, as the superficial simplicity of the concept belies considerable problems.

Data quality and reproducibility are fundamental requirements for any scientific measurement system. An experiment with microarray technology involves many steps and each step introduces variability that propagates to the final experimental measurement, e.g. fluorescent intensity from a laser confocal microscope scanner. Biological explanation of experimental data is based on and will be affected by the quality and reproducibility of microarray data. Quality control and reproducibility issues have been brought to the attentions of researchers recently (Piper *et al.*, 2002; Pritchard *et al.*, 2001; Yang *et al.*, 2002; Tu *et al.*, 2002). There is much room for improvement both in experimental design and variability control. Only reproducible and reliable data can lead to advancement in science. We expect that these issues will continue to be a subject of many investigations. In fact, researchers should not try to explain the biological significance from their unreliable, irreproducible microarray data.

It is interesting to point out that except for Affymetrix chips, most of the gene expression analysis system uses relative expression ratio to compare the difference between test and control samples. The main reason for doing so is to "cancel out" the systematic variability in the two samples and the detection system. However, such relative values cause some problems in practice. For example, it may be difficult to compare experiments from lab to lab, and from experiment to experiment because the difference in reference or control sample could make all the measurement incomparable. It would be desirable to have some sort of "absolute" measurement (e.g. mRNA copy number per cell). Dudley *et al.* (2002) used a new hybridization procedure to overcome this difficulty. First, instead of cohybridizing labeled test and control samples, they hybridize each test sample against labeled oligonucleotides complementary to every microarray feature. Ratios between test sample intensities and intensities of the oligonucleotide reference measure sample RNA levels that relates to their absolute abundance. Previously, DNA microarray protocols use a cDNA reference that has variable and unknown abundances. They demonstrated that results from this type of hybridization were accurate and retained absolute abundance information far better than conventional microarray ratios.

Full Moon BioSystems, Inc. (http://www.fullmoonbio.com) developed a microarray scanner calibration slide for users to perform quantitative evaluation of their microarray scanners. Specifically, it can be used for determining dynamic range, detection limit, and uniformity of microarray scanners. It can also be used for detecting laser channel cross-talk and laser stability. With this calibration slide, it is possible to set up a common standard for measuring absolute gene expression.

The disadvantage of cross-hybridization of cDNA microarray has been generally recognized and oligonucleotide arrays seem to be able to solve this problem quite well. However, there are still different opinions on the best length of oligonucleotide probes. Different vendors offer oligonucleotide arrays with different lengths. For example, Affymetrix uses 25-mer, MWG uses 50-mer, Agilent uses 60-mer, Operon uses 70-mer, and Clontech uses 80-mer for their oligonucleotide microarray fabrication. Peck (2002) showed that 150-mer oligonucleotide arrays achieved the best balance between sensitivity and specificity. More data are needed to find a probe length that best balances specificity and sensitivity. Other factors such as surface chemistry must be considered at the same time.

Microarray is still an evolving technology and common standards have yet to be established, e.g. nomenclature, experimental protocols, data exchange, and bioinformatic tools. There are some activities in the scientific community to address these issues. For example, the International Union of Pure and Applied Chemistry (IUPAC) has set up a Task Force on the nomenclatures related to nanotechnology and biochips. The Microarray Gene Expression Data (MGED) Society (http://www.mged.org) is an international organization for facilitating the sharing of microarray data from functional genomics and proteomics experiments. The MGED society has set up the Minimum Information About a Microarray Experiment (MIAME) guideline for authors, editors, and reviewers of microarray gene expression publications. These guidelines are based on the MIAME document developed by the MGED society (Brazma *et al.*, 2001). The MIAME standard contains all the information necessary for interpreting microarray results and verifying a microarray experiment. Public repository for microarray data has been available, e.g. the Gene Expression Omnibus (GEO) of the NIH (http://www.ncbi.nlm.nih.gov/geo/). Other types of hybridization arrays and microfluidic devices, e.g. microelectronic arrays, bead-based arrays, are being developed. There is a convergence of lab-on-a-chip and microarray systems.

Bioinformatics will continue to play a critical role in microarray experiments from which regulatory networks of genes may be deduced. However, novel algorithms and tools are needed to make maximal uses of huge amount of expression data.

Some companies, e.g. Incyte and Motorola, have chosen to get out of the microarray business. However, new companies with novel microarray technologies continue to emerge. These new companies, such as NimbleGen, are competing with big players like Affymetrix.

The number of genes being monitored on the current available microarrays is limited and incomplete. However, with the advancement of manufacturing technology, it will become possible to put all genes in a genome onto one microarray slide. Human Genome U133 Set (HG-U133) from Affymetrix, consisting of two GeneChip® arrays, contains almost 45,000 probe sets representing more than 39,000 transcripts derived from approximately 33,000 well-substantiated human genes. Before the whole genome chip is available, selection of genes to be put on a microarray is biased. Reducing manufacturing costs will be important for microarrays to gain wider acceptance.

For more information on the microarray technologies and recent developments in this fast-evolving field, the readers are strongly encouraged to consult the following web sites: http://www.gene-chips.com, http://www.biochipnet.de, and http://www.lab-on-a-chip.com.

6. CONCLUSIONS

Microarray technology, in which microspots of probe molecules are immobilized in an array format on a solid support and exposed to samples containing the corresponding target molecules, is revolutionizing the way biological research is performed. It allows the simultaneous analysis of thousands of molecules of unique identity within a single experiment. Different formats of microarrays, including DNA microarrays, protein microarrays, cell microarrays, and tissue microarrays, have been applied successfully alone or in conjunction with each other in biological research, medical diagnostics, drug discovery and development, and toxicogenomics. Although there is still much room for technological improvement of the microarrays, results and conclusions already coming out from published studies clearly demonstrated the utilities and power of these miniaturized tools. Reliable and inexpensive microarray-based clinical diagnostic tools will become a reality and serve as the basis of personalized medicine.We firmly believe that the microarray technology will gain more and more interesting applications in life sciences. The power of the microarray technology should only be limited by a researcher's imagination.

Acknowledgements: This work was supported in part by the National Hi-Tech ("863") Project of China. We are grateful to Drs. Hong Fang, Zhibin Li, Zhiqiang Ning, Desi Pan, Lingwen Zeng, and Qiang Zheng for helpful discussions during the preparation of this chapter. Megan Cao, Wei Qiao, and Dajie Zhang are acknowledged for their contributions to the TIS and biochemoinformatics projects.

REFERENCES

Alizadeh AA, Eisen MB, Davis RE, Ma C, Lossos IS, Rosenwald A, Boldrick JC, Sabet H, Tran T, Yu X, Powell JI, Yang L, Marti GE, Moore T, Hudson J Jr, Lu L, Lewis DB, Tibshirani R, Sherlock G, Chan WC, Greiner TC, Weisenburger DD, Armitage JO, Warnke R, Staudt LM, *et al.* (2000). Distinct types of diffuse large B-cell lymphoma identified by gene expression profiling. Nature 403:503-511.

Alon U, Barkai N, Notterman DA, Gish K, Ybarra S, Mack D, and Levine AJ (1999). Broad patterns of gene expression revealed by clustering analysis of tumor and normal colon tissues probed by oligonucleotide arrays. Proc Natl Acad Sci USA 96:6745-6750.

Auroux PA, Iossifidis D, Reyes DR, and Manz A (2002). Micro total analysis systems. 2. Analytical standard operations and applications. Anal Chem 74:2637-2652.

Belcher CE, Drenkow J, Kehoe B, Gingeras TR, McNamara N, Lemjabbar H, Basbaum C, and Relman DA. The transcriptional responses of respiratory epithelial cells to *Bordetella pertussis* reveal host defensive and pathogen counter-defensive strategies. Proc Natl Acad Sci USA 97:13847-13852.

Bittner M, Meltzer P, Chen Y, Jiang Y, Seftor E, Hendrix M, Radmacher M, Simon R, Yakhini Z, Ben-Dor A, Sampas N, Dougherty E, Wang E, Marincola F, Gooden C, Lueders J, Glatfelter A, Pollock P, Carpten J, Gillanders E, Leja D, Dietrich K, Beaudry C, Berens M, Alberts D, and Sondak V (2000). Molecular classification of cutaneous malignant melanoma by gene expression profiling. Nature 406:536-540.

Brazma A, Hingamp P, Quackenbush J, Sherlock G, Spellman P, Stoeckert C, Aach J, Ansorge W, Ball CA, Causton HC, Gaasterland T, Glenisson P, Holstege FC, Kim IF, Markowitz V, Matese JC, Parkinson H, Robinson A, Sarkans U, Schulze-Kremer S, Stewart J, Taylor R, Vilo J, and Vingron M (2001). Minimum information about a microarray experiment (MIAME) - toward standards for microarray data. Nat Genet 29:365-371.

Cheng J, Sheldon EL, Wu L, Uribe A, Gerrue LO, Carrino J, Heller MJ, and O'Connell JP (1998). Preparation and hybridization analysis of DNA/RNA from *E. coli* on microfabricated bioelectronic chips. Nat Biotechnol 16:541-546.

Chicurel ME and Dalma-Weiszhausz DD (2002). Microarrays in pharmacogenomics - advances and future promise. Pharmacogenomics 3:589-601.

Dhanasekaran SM, Barrette TR, Ghosh D, Shah R, Varambally S, Kurachi K, Pienta KJ, Rubin MA, and Chinnaiyan AM (2001). Delineation of prognostic biomarkers in prostate cancer. Nature 412:822-826.

Drews J (2000). Drug discovery: a historical perspective. Science 287:1960-1964.

Dudley AM, Aach J, Steffen MA, and Church GM (2002). Measuring absolute expression with microarrays with a calibrated reference sample and an extended signal intensity range. Proc Natl Acad Sci USA 99:7554-7559.

Ekins RP (1987). Determination of ambient concentrations of several analytes. UK patent application GB8803000.

Ekins RP (1989). Multi-analyte immunoassay. J Pharm Biomed Anal 7:155-168.

Ekins R, Chu F, and Biggart E (1990). Fluorescence spectroscopy and its application to a new generation of high sensitivity, multi-microspot, multianalyte, immunoassay. Clin Chim Acta 194:91-114.

Ekins RP and Chu FW (1991). Multianalyte microspot immunoassay - microanalytical "compact disk" of the future. Clin Chem 37:1955-1967.

Ekins R and Chu FW (1999). Microarrays: their origins and applications. Trends Biotechnol 17:217-218.

Fodor SP, Read JL, Pirrung MC, Stryer L, Lu AT, ans Solas D (1991). Light-directed, spatially addressable parallel chemical synthesis. Science 251:767-773.

Frawley WJ, Piatetsky-Shapiro G, and Matheus CJ (1992). Knowledge discovery in databases: an overview. AI Magazine 13:57-70.

Ge H (2000). UPA, a universal protein array system for quantitative detection of protein-protein, protein-DNA, protein-RNA and protein-ligand interactions. Nucleic Acids Res 28:e3.

Golub TR, Slonim DK, Tamayo P, Huard C, Gaasenbeek M, Mesirov JP, Coller H, Loh ML, Downing JR, Caligiuri MA, Bloomfield CD, and Lander ES (1999). Molecular classification of cancer: class discovery and class prediction by gene expression monitoring. Science 286:531-537.

Hamadeh HK, Bushel PR, Jayadev S, DiSorbo O, Bennett L, Li L, Tennant R, Stoll R, Barrett JC, Paules RS, Blanchard K, and Afshari CA (2002). Prediction of compound signature using high density gene expression profiling. Toxicol Sci 67:232-240.

Han J and Kamber M (2001). Data mining: concepts and techniques. San Francisco: Morgan Kaufmann Publishers, pp 1-550.

Hergenrother PJ, Depew KM, and Schreiber SL (2000). Small-molecule microarrays: covalent attachment and screening of alcohol-containing small molecules on glass slides, J Am Chem Soc 122:7849-7850.

Iyer VR, Eisen MB, Ross DT, Schuler G, Moore T, Lee JC, Trent JM, Staudt LM, Hudson J Jr, Boguski MS, Lashkari D, Shalon D, Botstein D, and Brown PO (1999). The transcriptional program in the response of human fibroblasts to serum. Science 283:83-87.

Kallioniemi OP, Wagner U, Kononen J, and Sauter G (2001). Tissue microarray technology for high-throughput molecular profiling of cancer. Hum Mol Genet 10:657-662.

Kononen J, Bubendorf L, Kallioniemi A, Barlund M, Schraml P, Leighton S, Torhorst J, Mihatsch MJ, Sauter G, and Kallioniemi OP (1998). Tissue microarrays for high-throughput molecular profiling of tumor specimens. Nat Med 4:844-847.

Korbel GA, Lalic G, and Shair (2001). Reaction microarrays: a method for rapidly determining the enantiomeric excess of thousands of samples, J Am Chem Soc 123(2); 361-362.

Lakkis MM, DeCristofaro MF, Ahr HJ, and Mansfield TA (2002). Application of toxicogenomics to drug development. Expert Rev Mol Diagn 2:337-345.

Lin SM and Johnson KF (2002). Methods of microarray data analysis. Boston: Kluwer Academic Publishers, pp 1-189.

Lockhart DJ and Winzeler EA (2000). Genomics, gene expression and DNA arrays. Nature 405:827-836.

MacBeath G, Koehler AN, and Schreiber SL (1999). Printing small molecules as microarrays and detecting protein-ligand interactions en masse, J Am Chem Soc 121:7967-7968.

MacBeath G and Schreiber SL (2000). Printing proteins as microarrays for high-throughput function determination. Science 289:1760-1763.

Macgregor PF and Squire JA (2002). Application of microarrays to the analysis of gene expression in cancer. Clin Chem 48:1170-1177.

Meldrum DR and Holl MR (2002). Microscale bioanalytical systems. Science 297:1197-1198.

Murphy D (2002). Gene expression studies using microarrays: principles, problems, and prospects. Adv Physiol Educ 26:256-270.

Nuwaysir EF, Bittner M, Trent J, Barrett, JC, and Afshari, CA (1999). Microarray and toxicology: the advent of toxicogenomics. Mol Carcinog 24:153-159.

Offenhausser A (2002). Cells on silicon – functional coupling of biology with microelectronics. Technical Digest of International Forum on Biochip Technologies, Beijing, November 9-13, 2002, pp 84-92.

Peck K (2002). Use of DNA arrays for genetic analysis and clinical diagnosis. Technical Digest of International Forum on Biochip Technologies, Beijing, November 9-13, 2002, pp 208-213.

Perou CM, Jeffrey SS, van de Rijn M, Rees CA, Eisen MB, Ross DT, Pergamenschikov A, Williams CF, Zhu SX, Lee JC, Lashkari D, Shalon D, Brown PO, and Botstein D (1999). Distinctive gene expression patterns in human mammary epithelial cells and breast cancers. Proc Natl Acad Sci USA 96:9212-9217.

Phimister B (1999). The chipping forecast. Nat Genet 21(Suppl):1-60.
Piper MD, Daran-Lapujade P, Bro C, Regenberg B, Knudsen S, Nielsen J, and Pronk JT (2002). Reproducibility of oligonucleotide microarray transcriptome analyses. An interlaboratory comparison using chemostat cultures of *Saccharomyces cerevisiae*. J Biol Chem 277:37001-37008.
Pritchard CC, Hsu L, Delrow J, and Nelson PS (2001). Project normal: defining normal variance in mouse gene expression. Proc Natl Acad Sci USA 98:13266-13271.
Pollack JR, Sorlie T, Perou CM, Rees CA, Jeffrey SS, Lonning PE, Tibshirani R, Botstein D, Borresen-Dale AL, and Brown PO (2002). Microarray analysis reveals a major direct role of DNA copy number alteration in the transcriptional program of human breast tumors. Proc Natl Acad Sci USA 99:12963-12968.
Quackenbush J (2001). Computational analysis of microarray data. Nat Rev Genet 2:418-427.
Quackenbush J (2002). Microarray data normalization and transformation. Nat Genet (Suppl) 2:496-501.
Ramaswamy S, Tamayo P, Rifkin R, Mukherjee S, Yeang CH, Angelo M, Ladd C, Reich M, Latulippe E, Mesirov JP, Poggio T, Gerald W, Loda M, Lander ES, and Golub TR (2001). Multiclass cancer diagnosis using tumor gene expression signatures. Proc Natl Acad Sci USA 98:15149-15154.
Reyes DR, Iossifidis D, Auroux PA, and Manz A (2002). Micro total analysis systems. 1. Introduction, theory, and technology. Anal Chem 74:2623-2636.
Rushmore TH and Kong AN (2002). Pharmacogenomics, regulation and signaling pathways of phase I and II drug metabolizing enzymes. Curr Drug Metab3:481-490.
Schena M, Shalon D, Davis RW, and Brown PO (1995). Quantitative monitoring of gene expression patterns with a complementary DNA microarray. Science 270:467-470.
Schena M, Shalon D, Heller R, Chai A, Brown PO, and Davis RW (1996). Parallel human genome analysis: microarray-based expression monitoring of 1000 genes. Proc Natl Acad Sci USA 93:10614-10619.
Schena M (1999). DNA microarrays: a practical approach. New York: Oxford University Press, pp 1-210.
Schena M (2002). Microarray analysis. New York: Wiley.
Shi LM, Fan Y, Lee JK, Waltham M, Andrews DT, Scherf U, Paull KD, and Weinstein JN (2000). Mining and visualizing large anticancer drug discovery databases. J Chem Inf Comput Sci 40:367-379.
Shi L (2001). Arrays, molecular diagnostics, personalized therapy and informatics. Expert Rev Mol Diagn 1:363-365.
Shi L (2002a). Data mining: an integrated approach for drug discovery. In: W.L Xing and J. Cheng, ed. Biochip Technology, : Springer-Verlag Press, in press.
Shi L (2002b). An integrated biochemoinformatics system for drug discovery. Technical Digest of International Forum on Biochip Technologies, Beijing, November 9-13, 2002, pp 143-148.
Shi L, Su Z, Xie A, Liao C, Qiao W, Zhang D, Li Z, Ning Z, Hu W, and Lu X (2003). Integrating chemical structures, biological activity fingerprints, and gene expression profiling for drug discovery. Abstract for 225th ACS National Meeting, Session on "Informatics challenges in pharmacogenomics". New Orleans, LA, USA, March 23-27, 2003.
Singh-Gasson S, Green RD, Yue Y, Nelson C, Blattner F, Sussman MR, and Cerrina F (1999). Maskless fabrication of light-directed oligonucleotide microarrays using a digital micromirror array. Nat Biotechnol 17:974-978.
Smith CG (1992). The process of new drug discovery and development. Boca Raton: CRC Press.
Smoot JC, Barbian KD, Van Gompel JJ, Smoot LM, Chaussee MS, Sylva GL, Sturdevant DE, Ricklefs SM, Porcella SF, Parkins LD, Beres SB, Campbell DS, Smith TM, Zhang Q, Kapur V, Daly JA, Veasy LG, and Musser JM (2020). Genome sequence and comparative microarray analysis of serotype M18 group *A Streptococcus* strains associated with acute rheumatic fever outbreaks. Proc Natl Acad Sci USA 99:4668-4673.
Spellman PT, Sherlock G, Zhang MQ, Iyer VR, Anders K, Eisen MB, Brown PO, Botstein D, and Futcher B (1998). Comprehensive identification of cell cycle-regulated genes of the yeast *Saccharomyces cerevisiae* by microarray hybridization. Mol Biol Cell 9:3273-3297.
Templin MF, Stoll D, Schrenk M, Traub PC, Vohringer CF, and Joos TO (2002). Protein microarray technology. Trends Biotechnol 20:160-166.
Tu Y, Stolovitzky G, and Klein U (2002). Quantitative noise analysis for gene expression microarray experiments. Proc Natl Acad Sci USA 99:14031-14036.
Wang D, Coscoy L, Zylberberg M, Avila PC, Boushey HA, Ganem D, and DeRisi JL (2002). Microarray-based detection and genotyping of viral pathogens. Proc Natl Acad Sci USA 99:15687-15692.
Wu RZ, Bailey SN, and Sabatini DM (2002). Cell-biological applications of transfected-cell microarrays. Trends Cell Biol 12:485-488.
Yang IV, Chen E, Hasseman JP, Liang W, Frank BC, Wang S, Sharov V, Saeed AI, White J, Li J, Lee NH, Yeatman TJ, and Quackenbush J (2002). Within the fold: assessing differential expression measures and reproducibility in microarray assays. Genome Biol 3(11):research0062.

Zhang M (2002). Extracting functional information from microarrays: a challenge for functional genomics. Proc Natl Acad Sci USA 99:12509-12511.

Zhu H, Bilgin M, Bangham R, Hall D, Casamayor A, Bertone P, Lan N, Jansen R, Bidlingmaier S, Houfek T, Mitchell T, Miller P, Dean RA, Gerstein M, and Snyder M (2001). Global analysis of protein activities using proteome chips. Science 293:2101-2105.

Applied Mycology & Biotechnology
An International Series. Volume 3. Fungal Genomics

14

Fungal Germplasm and Databases

Kevin McCluskey
Fungal Genetics Stock Center, Research Assistant Professor, Department of Microbiology, University of Kansas Medical Center, Kansas City, Kansas, USA (kmcclusk@kumc.edu).

Fungal culture collections exist in a wide variety of scales and with a variety of purposes. From the smallest collections of individual researchers to the largest international Biological Resource Centers, fungal collections provide materials and services to researchers promoting research around the world. The nature of fungal culture collections has changed over the last twenty years as the ability to publish an electronic database of cultures presented itself. In 2002, there are over seventy large fungal culture collections worldwide and many smaller collections, usually in individual laboratories. There are over 385,000 living strains of filamentous fungi and yeasts in these collections. The ability to access an electronic database over the internet makes the smaller collections more relevant. There have been a number of efforts to publish the holdings of smaller collections, and these have taken different approaches and met with differing levels of success. As more and more information becomes available online, the ability to use the exact strain of interest should make research more fruitful and reproducible.

1. INTRODUCTION

Culture collections have long served as foci for biological science research (Samson *et al.*, 1966). They serve a number of functions to support the advancement of research including the provision of uniform biological materials, timely sharing of new materials, and maintenance of key materials long past the career of any individual scientist. Many collections have provided these functions admirably for nearly a century but with the advent of molecular genetics and the internet, novel functions have evolved for culture collections. Among these novel functions are the distribution of gene libraries, cloned genes, cloning vectors, chemicals, ESTs, and clones from sequencing projects. Many collections originated as the working collection of an individual researcher and developed into the working assets that they are today. For the present consideration, we will look at the numbers of collections around the world and describe the larger collections. The large number of collections makes a case-by-case description unwieldy and while there are many smaller collections around the world, the reader is referred to the global databases and collection organizations for access to the smaller collections.

2. CULTURE COLLECTION OVERVIEW

According to the most recent data released by the World Data Centre for Micro-organisms (WDCM), there are nearly 470 culture collections in sixty two countries at present. This does not include collections in independent researcher's laboratories. There are over 2,300 people working in culture collections worldwide and these people maintain over one million cultures. While the average number of cultures per worker is around 430, some collections have 5,000 stocks per worker and some have closer to twenty or thirty strains per worker. This is, in part, related to the amount of work required to maintain individual stocks, as well as the nature of other responsibilities each individual has. Many smaller collections are a small part of a larger research effort and there is no way of identifying such collections other than by looking at the number of individual publications from a researcher, laboratory or collection.

Of the nearly 470 culture collections, 175 are supported by government and 149 are supported by universities. Others are considered semi-governmental, private or industrial. While bacteria make up the majority of cultures held in these collections, fungi are a close second place. There are over 350,000 fungal strains in collections around the world, as of October 2002 (WDCM Data).

Materials are often deposited in several collections. There are several reasons for this with the most common being convenience for the researchers involved. Cross depositing to collections in different countries has allowed researchers to avoid issues such as paying import fees or clearing customs repeatedly . There is also an issue of security. By depositing strains in multiple collections, researchers can insure that the strain would survive a loss if one particular collection suffers a catastrophic failure. Some collections, such as the FGSC, avoid this type of loss by keeping a copy of each strain at a second site. Collections also make possible certain types of research that would not be possible because they last longer than the typical research career. In 1999 the author was able to open vials of Neurospora that had been sealed in 1946 describing a record for lyophilized fungal spore viability (McCluskey, 2000). This emphasizes the long-term nature of collections. Some collections are considered to be 'at risk' and there are places for such collections to look for help. Among these are the national and international federations for culture collection.

2.1 Categories of fungal culture collections

There are several categories of fungal culture collections. The first, and most common, is a type collection. This sort of collection emphasizes holding large numbers of species while having only a limited number of strains from an individual species. Collections of this type include the American Type Culture Collection and the Centraalbureau voor Schimmelcultures in The Netherlands.

Another type of collection is a mission based collection such as the International Culture Collection of Arbuscular and Vesicular Arbuscular Mycorrhizal Fungi (INVAM) in West Virginia, or the Laboratory of Molecular Genetics and Breeding of Edible Mushrooms in Bordeaux, France. Collections such as this typically have a variety of strains whose biological characteristics have been cataloged without carrying out traditional genetics on the strains. Strains in collections such as this may have tremendous economic value because of their individual traits.

Genetic culture collections, such as the Yeast Genetic Stock Center or the Fungal Genetics Stock Center (FGSC), typically emphasize a tremendous depth in one or a few organisms. The FGSC, by way of example, holds over 5,000 mutant strains, mainly of Neurospora. Among fungi, there are relatively few genetic collections, but there are many others for research organisms, such as *Escherichia coli*, *Caenorhabditis elegans* or *Zea mays*. While these collections are focused in their emphasis, they serve a broad constituency. They are also likely to have expanded to include molecular resources and to publish materials such as genetic maps or newsletters.

The final type of collection is the Patent Depository (also called International Depository Authorities). These are collections who agree to hold materials according the Budapest treaty on the international recognition of the deposit of microorganisms for the purposes of patent procedure which was effective April 28, 1977. Briefly, this treaty was enacted to allow those seeking a patent that included a micro-organism to deposit the strain or strains in an internationally recognized repository and to hold certain rights to the strain while still allowing for the full disclosure required by the patent process. The collections that are Patent Depositories have to meet certain requirements, according to Article 7 of the Budapest Treaty. The ATCC was the first international depository, having gained approval on January 31, 1981 although the USDA Repository and ATCC have been accepting patented strains for deposit according to the rules of the United States Patent and Trademark Office since 1949. Most patent depositories are already recognized international repositories and there are 64 recognized patent depositories in operation. Many collections are actually hold several different types of collections under one roof and maintain differing degrees of separation among the subsets. For example, most Patent Depositories are established collections that hold patent strains along with their main holdings.

3. BIOLOGICAL RESOURCE CENTERS

As part of the Organization for Economic Cooperation and Development (OECD) theme in Biotechnology and the environment, a panel of experts was convened to explore the development of biological resource centers. The panel envisions the development of new biological materials and is proposing that a global network of centers will make these materials available to the furtherance of research and development around the world. Moreover, there is the goal that biological diversity will be effectively exploited and due credit will be given to the originator of the material. The Convention on Biological Diversity (adopted in 1992) described biological resource centers as an ex situ collection of diversity and emphasizes that the tremendous value of diverse biological material is only fully appreciated when it can be described and reproduced. Biological Resource Centers are of special utility in preserving diversity of human genetic material and this type of diversity is a focus of their establishment.

4. MAJOR CULTURE COLLECTIONS

While there are many fungal culture collections around the world (Table 1), the largest are

Table 1. Independent Fungal Culture Collections

NAME	LOCATION	HOLDINGS
American Type Culture Collection (ATCC)	Manassas, Virgina USA	27,000 filamentous fungal strains and yeast strains
USDA ARS Culture Collection (NRRL)	Peoria, Illinois, USA	55,000 filamentous fungal strains and 10,000 yeast strains
Fusarium Research Center	Department of Plant Pathology Penn State University	16,000 Fusarium stocks
ARS Collection of Entomopathogenic Fungi	USDA - ARS Plant Soil Nutrition Laboratory Ithaca, New York, USA	6,400 fungal cultures
Fungal Genetics Stock Center (FGSC)	Department of Microbiology, University of Kansas Medical Center	16,000 fungal cultures plus cloned genes and gene libraries
Yeast Genetic Stock Center	Now at ATCC	1,200 Saccharomyces stocks
International Culture Collection of Arbuscular & Vesicular Arbuscular Mycorrhizal Fungi (INVAM)	West Virginia University Morgantown, USA	1,061 VA Mycorrhizal Fungal strains
Scherer Candida Collection	University of Minnesota	1,899 Candida stocks

Canadian Collection of Fungal Cultures (CCFC)	Eastern Cereal & Oilseed Research Centre, Ottawa	11,000 Fungal strains
University of Alberta Microfungus Collection and Herbarium (UAMH)	Edmonton, Alberta, Canada	>9,900 living strains
Coleccion de Microhongos (INIF)	Centro de Investigaciones Forestales y Agropecuarias del Distrito Federal Ciudad de Mexico, Mexico	>1,000 fungal strains (mostly Aspergillus, Fusarium and Alternaria)
Collection of *Histoplasma capsulatum* Strains	Department of Microbiology and Parasitology National Autonomous University of Mexico, Mexico	>180 *Histoplasma capsulatum* strains from patients and the environment
Forest Pathology Culture Collection (DFF)	Pacific Forest Research Centre, Canadian Forestry Service Victoria, British Columbia, Canada	500 fungal cultures
Labatt Culture Collection Labatt Brewing Company	London, Ontario, Canada	2,000 yeast cultures
IIB-INTECH Collection of Fungal Cultures	IIB-INTECH Instituto de Investigaciones Biotecnologicas sede Chascomus alia, Argentina	>450 fungal cultures
Embrapa Genetic Resources and Biotechnology Collection of Fungi of Interest to Biological Control	Parque Estacao Biologica, Brasilia, DF, Brazil	>850 fungal cultures
Colecao de Culturas de Basidiomicetos (CCB)	Instituto de Botanica Sao Paulo, Brazil	600 fungal cultures
The Belgian Co-ordinated Collections of Micro-organisms (BCCMTM)	BCCMTM/MUCL - Agro/industrial Fungi & Yeast Collection Mycoth□que de l'Universite catholique de Louvain Belgium	>25,000 strains of filamentous and yeast-like fungi
The Belgian Co-ordinated Collections of Micro-organisms (BCCMTM)	BCCMTM/IHEM - Biomedical Fungi and Yeast Collection Scientific Institute of Public Health - Louis Pasteur Belgium	>6,500 strains of filamentous and yeast-like fungi
Culture Collection of Basidiomycetes (CCBAS)	Institute of Microbiology Prague, Czech Republic	>700 fungal strains
Culture Collection of Fungi	Institution Department of Botany Charles University Praha, Czech Republic	1,800 fungal strains
IBT Culture Collection of Fungi	BioCentrum, Technical University of Denmark Lyngby, Denmark	22,000 fungal strains
VTT Biotechnology, Culture Collection	VTT Biotechnology and Food Research Finland	800 filamentous fungal strains and 800 yeast strains
Collection of Yeasts of Biotechnological Interest	INRA-Institut National de la Recherche Agronomique Thiverval-Grignon, France	500 yeast strains
Laboratory of Molecular Genetics and Breeding of Edible Mushrooms	University of Bordeaux 2, INRA d'Ornon, France	3,500 filamentous fungal strains
Fungal Strain Collection	Museum National d'Histoire Naturelle Paris, France	4,000 Filamentous fungal strains
Deutsche Sammlung von Mikroorganismen und Zellkulturen (DSMZ)	Braunschweig, Germany	2,400 filamentous fungal strains and 500 yeast strains
Athens University Mycology	National and Kapodistrian	500 fungal strains

	University of Athens Greece	
National Collection of Agricultural and Industrial Microorganisms	Szent István University Budapest, Hungary	300 fungal strains and 1,100 yeast strains
Fungal Collection of the Department of Plant Biology	University of Turin, Italy	2,500 strains representing nearly all classes of fungi
Industrial Yeasts Collection	Dipartimento di Biologia Vegetale e Biotecnologia Agroambientale Perugia, Italy	4,500 yeast strains
Centraalbureau voor Schimmelcultures (CBS)	Utrecht, The Netherlands	28,000 filamentous fungal strains and 4,500 yeast strains
Culture Collection of Industrial Microorganisms	Institution National Institute of Industrial Technology (INETI) Lisbon, Portugal	>450 fungal strains and 200 yeast strains
Portuguese Yeast Culture Collection	Center of Microbiological Resources, Faculty of Sciences and Technology/New University of Lisbon Lisbon, Portugal	2,000 yeast strains
All-Russian Collection of Microorganisms (VKM)	Moscow, Russsia	>3,300 filamentous fungal strains and 2,300 yeast strains
Yeast Collection of the Department of Soil Sciences, (YBP)	Moscow State University Moscow, Russia	>2,000 yeast strains
Collection of Marine Microorganisms (KMM)	Vladivostok, Russia	500 marine yeast strains
Culture Collection of Basidiomycetes of the Komarov Botanical Institute (LE(BIN))	St. Petersburg, Russia	>1,000 Basidiomycete
Peterhoff Genetic Collection of Yeasts (PGCY)	St. Petersburg, Russia	600 genetically marked yeast strains
Culture Collection of the Institute of Plant Protection (VIZR)	St. Petersburg, Russia	>1,000 Plant pathogenic fungal strains
Institute of Genetics and Selection of Industrial Microorganisms (VKPM)	Moscow, Russia	750 fungal strains
Culture Collection of Yeasts Institution Institute of Chemistry	Slovak Academy of Sciences	3,800 yeast strains
Microbial Culture Collection of National Institute of Chemistry (MZKI)	National Institute of Chemistry, Hajdrihova, Slovenia	2,000 filamentous fungal strains and 300 yeast strains
ZIM Culture Collection of Industrial Microorganisms	University of Ljubljana Ljubljana, Slovenia	>1,700 yeast strains
The Spanish Type Culture Collection (CECT)	University of Valencia Valencia, Spain	1,500 fungal strains and 2,500 yeast strains
Uppsala University Culture Collection of Fungi	Botanical Museum University of Uppsala Sweden	3,000 fungi and 200 lichen
Fungal Cultures University of Goteborg (FCUG)	Botanical Institute Goteborg, Sweden	9,000 fungal strains
CABI Bioscience	Egham, Surrey, UK	22,000 living fungi and 300,000 preserved specimens
The International Bank for the Glomales Formerly known as La Banque Européenne des Glomales (BEG)	International Institute of Biotechnology University of Kent Campus, Kent, UK	500 fungal strains

National Collection of Pathogenic Fungi	PHLS Mycological Reference Laboratory, London, UK	1,100 fungal strains and 200 yeast strains
National Collection of Yeast Cultures	Institute of Food Research Norwich, UK	3,000 Yeast strains
Glasgow Aspergillus Collection	Division of Molecular Genetics Anderson College University of Glasgow, UK	575 Aspergillus strains
Agricultural Culture Collections of China (ACCC)	Bejing, China	>2,000 strains
China Center for Industrial Culture Collection	China National Research Institute of Food and Fermentation Industries Beijing, China	>1,700 filamentous fungal strains and yeast cultures
The University of Hong Kong Culture Collection	Dept. of Ecology and Biodiversity Hong Kong, China	5,000 fungal strains
Chiba University Research Center for Pathogenic Fungi and Microbial Toxicoses	Chiba, Japan	>10,000 fungal strains
IAM Culture Collection	Institute of Molecular and Cellular Biosciences, The University of Tokyo, Japan	>1,300 fungal and yeast strains
Japan Collection of Microorganisms	RIKEN (The Institute of Physical and Chemical Research), Saitama, Japan	> 1,200 filamantous fungal strains and >2,100 yeast strains
National Research Institute of Brewing	Hiroshima, Japan	>1,000 filamentous fungal and yeast strains
Institute for Fermentation (IFO)	Osaka, Japan	All biological materials transferred to NBRC, June 2002
AHU Culture Collection	Graduate School of Agriculture, Hokkaido University Japan	>1,300 filamentous fungal strains and >800 yeast strains
MAFF Genebank, National Institute of Agrobiological Sciences	Ibaraki, Japan	10,000 fungal strains and >500 yeast strains
NITE Biological Resource Center (NBRC)	National Institute of Technology and Evaluation, Kisarazu, Chiba, Japan	>8000 fungal strains and >3150 yeast strains (including IFO strains)
Korean Agricultural Culture Collection (KACC)	National Institute of Agricultural Science and Technology Korea	>1,500 fungal strains
Bioresource Collection and Research Center (BCRC, formerly CCRC)	Hsinchu, Taiwan	>3,000 filamentous fungal strains and >1,500 yeast strains
BIOTEC Culture Collection	National Center for Genetic Engineering and Biotechnology Bangkok, Thailand	3,000 fungal strains
Australian National Reference Laboratory in Medical Mycology (AMMRL)	The Royal North Shore Hospital of Sydney Sydney, Australia	>1,100 strains of filamentous fungi and 175 strains of yeast
Mycology Culture Collection	Women's and Children's Hospital North Adelaide, Australia	2,000 strains of filamentous fungi and yeast
Flinders University Smut Collection	School of Biological Sciences Bedford Park, Australia	1,500 smut fungi
Defence Materials & Stores Research & Development Establishment Culture Collection	Defence R&D Organization, New Delhi, India	>1,100 fungal strains
Division of Mycology and Plant	Indian Agricultural Research	2,500 fungal strains

Pathology	Institute New Delhi, India	
National Collection of Industrial Microorganisms	National Chemical Laboratory Pune, India	950 fungal strains and 600 yeast strains
University of Indonesia Culture Collection	Department of Biology, University of Indonesia Depok, Indonesia	300 fungal strains
Forest Research Culture Collection	New Zealand Forest Research Institute, Rotorua, New Zealand	1,500 filamentous fungal strains
International Collection of Microorganisms from Plants	Plant Diseases Division DSIR Auckland, New Zealand	4,700 fungal strains
National Collection of Fungi: Culture Collection	ARC-Plant Protection Research Institute, Pretoria, South Africa	4,500 fungal strains

Data from WFCC-MIRCEN World Data Centre for Microorganisms and from other sources.

in United States, The Netherlands, Germany, Russia and China. Other notable collections exist in, Australia, Japan, the UK, and other countries in Europe. Most national scale collections are supported by their governments, at least to the extent required above user fees. The largest collections are required to have a variety of departments to handle issues such as shipping, billing, regulatory compliance, and operations. They have special requirements that dictate to some extent that they be located in a major metropolitan area. Among these requirements are access to a scientific infrastructure (to provide reagents such as liquid nitrogen) and a pool of potential trained employees.

4.1 Collections in the Americas

In the United states, the ATCC is the largest independent culture collection. It houses many collections beyond the fungal collection and is both a Patent Depository and a Biological Resource Center. The ATCC holds 27,000 filamentous fungal and yeast strains in addition to its collection of 18,000 bacterial strains, 4,000 cell lines, 1,200 hybridoma lines, as well as viruses, protozoa, algae and plant lines. The ATCC holds the Yeast Stock Center collection of 1,200 yeast mutants which moved there following the retirement, from the University of California, of Dr. R. Mortimer.

The ATCC also acts as a clearinghouse for collections from the Johns Hopkins University and the Wistar Institute. The ATCC offers many services beyond standard culture deposition and distribution. They offer genomic DNA for most cultures as well as gene libraries for select organisms. The ATCC has established offices in a number of countries to facilitate distribution, payment, and compliance with local regulations. Research is a part of the ATCC mission and they have a variety of areas in which they publish including the development of the human cell collection (Hay, 1996), molecular identification of fungi (Molina, 1994), and more.

The United States Department of Agriculture maintains the largest American fungal culture collection in Peoria Illinois. This collection is called the USDA Agricultural Research Service Culture Collection or the NRRL, an acronym for the original name of the Peoria laboratory, the Northern Region Research Laboratory. The collection began as a working collection when Drs. Charles Thom and Margaret B. Church began cataloging strains associated with cheese production in 1904. These strains formed the foundation of the USDA collection when it opened under the direction of Dr. Kenneth B. Raper.
Since then, individual strains have been deposited as well as entire collections, such as the Blakslee collection of Mucorales and the US Army Quartermaster Collection of filamentous fungi. The NRRL was the first Patent Depository in the USA, having accepted strains for this purpose as early as 1949.

The NRRL presently holds 15,000 yeast strains, 55,000 filamentous fungal strains as well as 10,000 actinomycete and 10,000 bacterial strains. The NRRL has an additional 6,000 strains in its patent collection. The curators of the NRRL maintain active research programs and publish regularly on topics such as taxonomy (Logrecio *et al*, 1995 and Kurtzman, 2000), strain characteristics (Ito *et al*, 1998), collection maintenance (O'Donnell and Peterson, 1992).

The Fusarium Research Collection at the Department of Plant Pathology of Penn State University is the main repository devoted solely to Fusarium. This collection was established through the dedicated effort of Dr. Paul Nelson and holds 17,000 Fusarium stocks from around the world.

The Fungal Genetics Stock Center (FGSC) at the Department of Microbiology in the University of Kansas Medical Center is one of the truly genetic collections. Housing over 5,000 mutant strains, as well as 11,000 other strains, the FGSC has grown from a small collection of Neurospora mutants in 1960 to an internationally respected resource supporting genetic research with fungi. The FGSC collection has grown largely through the strains being deposited by researchers who wanted both the assure that they were available and who wanted to be relieved of the burden of distributing useful strains following every request. The FGSC holds mutants of *Neurospora crassa*, *N. intermedia* and *N. tetrasperma* as well as an extensive collection of *Aspergillus nidulans* and *A. niger* mutants. In recent years, the FGSC has acquired an extensive collection of wild Neurospora strains from around the world. The FGSC has also responded to the needs of its research community by holding and distributing genomic DNA libraries and cDNA libraries. These have a tremendous added value as researchers have published the location of particular genes on cosmids in the library allowing others to obtain the library and immediately have the location of many key genes. These libraries also formed the backbone of the physical map used in the Neurospora genome project at the Whitehead Institute Center for Genome Research. As such, a researcher can find a gene of interest at the Whitehead genome web-site and obtain a cosmid or BAC clone carrying the gene of interest from the FGSC in a matter of days. Another function provided by the FGSC is the publication of a peer-reviewed journal, the Fungal Genetics Newsletter (FGN) in both print and electronic formats. The FGSC also coordinates publication of abstracts from the biannual Fungal Genetics Conference and the Biannual Neurospora Conference as supplements to the FGN. The FGSC has recently developed and described a database that has allowed users to search or browse the collection online (McCluskey, 2000).

The USDA Agricultural Research Service collection of Entomopathogenic fungi at the USDA Plant Soil Nutrition Laboratory in Ithaca New York maintains a specialized collection of 5,500 strains that are pathogenic on insects. This collection was established in the early 1970s to provide characterized biological material for use in biological control of insects, other arthropods and nematodes. In recent years, the strains have become recognized as a source of secondary metabolites and compounds of interest to agriculture and medicine. This is a useful example of how the value of a collection may lie outside its original focus.

The International Culture Collection Arbuscular and Vesicular Arbuscular Mycorrhizal Fungi (INVAM), housed at the West Virginia University, holds over 1,500 stocks of fungi symbiotic with plant roots. Because of the nature of these fungi, most of the collections are maintained either as living cultures associated with plants or as collected macrospores at 4°C. INVAM distributes nearly 400 stocks each year.

The University of Alberta Microfungus Collection and Herbarium houses nearly 10,000 living strains. The UAMH specializes in ascomycetes and hyphomycetes but also has fungi associated with human disease and collections from specific habitats including endophytes and plant symbiots.

Canadian Collection of Fungal Cultures (CCFC) maintains over 11,000 strains from 2,500 species. These are part of a larger group of collections at the Eastern Cereal & Oilseed Research Centre (ECORC) in Ottawa, Canada. Other collections at ECORC include the Canadian National Collection of Insects, Arachnids and Nematodes (CNC), a mycology herbarium holding over 300,000 specimens, a vascular plant herbarium and the Canadian branch of the Glomales *in vitro* Collection.

The Forest Pathology Culture Collection at the Pacific Forest Research Centre holds 500 fungal cultures emphasizing wood-destroying hymenomycetes.

An example of a private collection is the Labatt culture collection of the Labatt brewing company. Their collection of 2,000 yeast cultures is held at their Ontario Canada facility. Few private culture collections are listed in the databases or documentation of organizations such as WFCC, although personnel from private collections have an active role in such organizations.

Brazil hosts two major culture collections. The Colecao de Culturas de Basidiomicetos in the Instituto de Botanica in Sao Paulo holds 600 fungal cultures. The Embrapa Genetic Resources and Biotechnology Collection of Fungi of Interest to Biological Control in Brasilia holds over 850 cultures.

The *Histoplasma capsulatum* collection in the Department of Microbiology and Parasitology of the National Autonomous University of Mexico is an important resource for the study of this important and emerging pathogen. While only made up of 180 strains, they vary in their origin, including strains from patients and from the environment. The Coleccion de Microohongos at the Centro de Investigaciones Forestales y Agropecuarias del Distro Federal is comprised of over 1,000 strains. These are mainly environmental microfungi such as Aspergillus, Fusarium and Alternaria.

4.2 European Culture Collections

The use of fungi has a long history in European culture and several different fungi were even found on the Tyrolean Iceman (Peintner et al, 1998) who was thought to be 5,000 years old. There are now several large fungal culture collections in Europe. The Centraalbureau voor schmieel cultures (CBS) in Utrech, The Netherlands holds 35,000 strains of fungi and, though their focus is narrow when compared to collections like the ATCC, the CBS is one of the largest and most diverse collections in the world. It employs nearly 50 scientists and support personnel and receives support both from user fees and from the Royal Netherlands Academy of Arts and Sciences. The CBS has been a patent depository since 1955. The CBS was established after the first world war. It moved to Utrecht in 2000. The CBS has a significant research component to its mission and CBS scientists and their collaborators published nearly 60 articles in 2000 and nearly 90 in 1999. In addition to their phylogenetic and taxonomic research they have developed databases of use to other researches (Wuyts, 2001 and Wuyts, 2002).

They also have an ambitious project to generate a DNA database of type strains. The IBT Culture Collection of Fungi, in Lyngby Denmark, holds 22,000 fungal strains. The collection is part of a large mycology group and includes over 1,000 Aspergillus and Penicillum strains identified to species level. They were in the process of putting their catalog online in late 2002.

A collection of 2,300 fungi is held in Estonia at the Tartu Fungal Culture Collection. This collection is in the Institute of Zoology and Botany which also holds a plant herbarium and an insect collection.

France hosts a number of collections, notably Fungal Strain Collection at the Museum National d'Histoire Naturelle (MNHN) in Paris. This collection is comprised of 4,000 strains of filamentous fungi in the Laboratory of Cryptogamy. Also at the MNHN are collections of

Fish, Nematodes, Vascular plants and Meteorites. Edible mushrooms are the focus of the 3,500 strains held at the Laboratory of Molecular Genetics and Breeding of Edible Mushrooms at the University of Bordeaux 2. This laboratory is a part of the French Institut National de la Recherche Agronomique. The Collection de Champignons at the Institut Pasteur is notable, as is the Hoechst Marion Roussel collection at Romainville which holds hundreds of fungi in addition to several thousand bacterial strains.

In Germany, the Deutsche Sammlung von Mikroorganismen und Zellkulturen (DSMZ) is the largest collection, including over 9,000 bacterial strains and nearly 2,500 fungal strains. It also holds plant and animal viruses, animal cell lines, plasmids, and hybridomas. As such, the DSMZ is a large scale Biological Resource Center that employs over 50 scientists and support personnel. The DSMZ is a patent depository and receives support from user fees and from the German Federal Ministry of Research and Technology. Other collections in Germany include the Institut fur Pflanzenschutz im Forst, Biologische Bundesans which holds 500 fungal strains, the Bayerische Landesanstalt fur Weinbau und Gartenbau which specializes in yeast, the Institute for Microbiology und Landscape Ecology, Justus-Liebig-Universitat which holds fungi and yeast, in addition to their bacterial collection. Other smaller university based collections exist and can be found through other online databases (see below).

In Italy, the Industrial Yeast Collection at the Dipartimento di Biologia Vegetale holds 4,500 yeast strains. Founded in the 1920's the collection holds yeasts isolated from a variety of substrates including fermenting grapes, flowers, fruits, soils, air, water, compost, dung, animal and human organs, and various foods. This collection is a patent depository and provides screening and identification services.

Portugal has a collection of yeast strains at the Center of Microbiological Resources in the New University of Lisbon as well as a smaller collection of industrial microorganisms at the Institute of Industrial Technology in Lisbon.

Russia hosts a number of fungal collections including the Collection of Marine Microorganisms in Vladivostok which holds 500 marine yeast strains among others. The All-Russian Collection of Microorganisms in Moscow (VKM) holds over 3,300 filamentous fungal strains and 2,300 yeast strains. These are in addition to over 3,500 bacterial strains. Like many culture collections, the research at the VKM emphasizes taxonomy. The VKM is an international patent depository as well as a private safe-keeping service. VKP provides consultation services as well as identification of strains for clients. The Culture Collection of Basidiomycetes of the Komarov Botanical Institute holds a collection of basidiomycetes numbering over 1,100. Among these strains are isolates belonging to 395 species, 142 genera and 31 families. Moscow State University hosts the yeast collection of the department of Soil Science. This collection is comprised of over 2,000 strains. Many Russian collections are linked by the Consolidated Catalogue of Microbial Cultures Held in Russian Non-medical Collections, hosted at the VKM web-site (www.vkm.ru). This site also provides descriptions of a variety of collections, with different interests, in Russia.

In Ljubljana, Slovenia, the ZIM collection of industrial microorganisms includes over 1,700 yeast strains. Hajdrihova is home to the National Institute of Chemistry and its Microbial Culture Collection (MZKI) which includes 2,300 fungal strains.

The home of the Spanish Type Culture Collection (La Colección Española de Cultivos Tipo, CECT) is the University of Valencia. This collection holds, among other things, 1,500 fungal strains and 2,500 yeast strains. This collection has moved three times since it's founding in 1960 in Madrid. It moved in 1968 to Salamanca, then again in 1974 to Bilbao where it stayed until 1980 when it moved to it's current home in Valencia. In 1992 the CECT became a patent depository.

Sweden is home to the Uppsala University Culture Collection of Fungi at the Botanical Museum of the University of Uppsala. This collection holds 3,000 fungal isolates and an additional 200 lichen isolates. A larger collection exists at the Botanical Institute in Goteburg. The Fungal Cultures University of Gogeburg holds 9,000 strains. This collection is mostly Basidiomycetes in the Corticiaceae and Polyporaceae. They have generated ribosomal RNA sequence for a number of strains and include this in their database, upon collaboration.

The United Kingdom is home to a number of culture collections, including the CABI collection in Surrey. CABI Bioscience is a large organization incorporating a number of agricultural service agencies including IMI. Their collection of fungi includes strains with unique characteristics including mating type testers, mutants, parasites, assay strains, strains that produce unique metabolites and more.

Also in the UK is the large collection of yeast cultures at the Institute of Food Research in Norwich. This collection is comprised of over 3,000 yeasts including *Saccharomyces cerevisiae* and *Schizosaccharomyces pombe*. The National Collection of Yeast Cultures is a patent depository and offers safe storage as well as a variety of services and consultancy related to yeast identification, culture, and storage. The National Collection of pathogenic Fungi at the PHLS Mycological Laboratory in London holds 1,100 fungal strains and 200 yeast strains. These are predominantly medically relevant fungi. The online database for this collection, as well as for the National Collection of Wood-rotting Macrofungi (in Garston) is via the United Kingdom United Culture Collection site which "co-ordinates the activities, marketing and research of the UK national service collections of microbial organisms."

The International Bank for the Glomales, housed in the International Institute of Biotechnology at the University of Kent, was formerly known as the Banque Europeenne des Glomales. In addition to their 171 registered isolates, they host a web-site with protocols and translations into several languages. They have an international oversight committee.

Overall, Europe is host to a variety of culture collections with broad international support for the effort. Many funding agencies support mycological research with different interests including medical, and industrial mycology as well as a variety of agricultural emphases including forestry, pathology, mushroom culture and the investigation of symbiotic fungi.

4.3 Asian Culture Collections

Study and cultivation of fungi in Asia goes back thousands of years and cultivation of mushrooms for food was apparently practice in China as early as 1,500 years ago (Chang, 1993). Red Yeast Rice is considered to be a traditional Chinese food and medicine and was recently shown to contain chemicals known to lower cholesterol in humans (Ma, 2000). In keeping with their long history of mycology, there are many culture collections in asia. The largest collection in China is the University of Hong Kong Culture Collection in the Department of Ecology and Biodiversity. Their holdings amount to 5,000 fungal strains

In China, the Agricultural Culture Collection of China (ACCC) holds over 2,000 strains of fungi. This collection operates under the auspices of the China Committee for Culture Collections of Microorganisms (CCCCM) and was established in 1980. The ACCC has 4 laboratories, two of which specialize in fungi. These are the fungi and the edible fungi laboratories. Other laboratories specialize in different aspects of prokaryotic biology. The China Center for Industrial Culture Collection in the National Research Institute of Food and Fermentation Industries at Beijing houses over 1,700 filamentous fungi and yeast strains. This collection, established in 1979, is associated with several other national centers including Food Quality Supervision & Testing, National Information Center for Food and Fermentation Industries, the National Center for Food and Fermented Products Standardization, and the Edible Fungus Research Center of China.

Hsinchu, Taiwan is home to the Bioresource Collection and Research Center (BCRC) which holds over 4,000 fungal and yeast stocks in addition to several thousand bacteria, plasmids, and plant and animal cell lines. The collection is part of the Food Industry Research and Development Institute which was established in 1965. The BCRC is a patent depository and provides a variety of services beyond its depository mission. There is a large research component to the activities of BCRC as well.

The Taiwan Agricultural Research Institute houses the Arbuscular Mycorrhizal Fungal Collection center in Taiwan, a collection which includes over 600 isolates from 20 species. The isolates are from a variety of sites including America, Bangladesh, Indonesia, Japan, and Nepal.

Japan has several different culture collections with the two biggest housing 10,000 stocks each. these are the Culture collection of the Chiba University Research Center for Pathogenic Fungi and Microbial Toxicoses and the National Institute of Agrobiological Sciences at Ibaraki. The former is part of Chiba University and is part of a large research organization including a variety of topics. The latter is part of the Ministry of Agriculture, Forestry and Fisheries and exists alongside large collections of plants, animals and a variety of micro-organisms. The next largest collection in Japan is the NITE Biological Resource Center which is part of the National Institute of Technology and Evaluation. Housed in Chiba, this collection holds over 8,000 stocks with their focus being on industrially important organisms. They incorporated into their collection the holdings of the Osaka Institute for Fermentation in 2000. Other smaller collections exist in Japan and most serve a specific niche, such as brewing or specific university departments.

In Korea, a major collection is the Korean Agricultural Culture Collection at the National Institute of Agricultural Science and Technology in Suwan. Founded in 1995, this is a patent depository for Korea and serves researchers in academic institutions as well as those in Korea's Rural Development Administration. Korea also has a type collection called the Korean Collection for type cultures which operates as part of the Korean Federation of Culture Collections. Also in Korea is the Korean Culture Center for Microorganisms. The two latter collections have their holdings described online only in Korean.

4.4 Other Culture Collections

India has a long history of studying fungi and research with fungi is carried out at a variety of institutes including the Centre for Cellular and Molecular Biology in Hyderabad and the Indian Institute of Technology. Collections in India include the National Collection of Industrial Microorganisms in Pune as well as the collection of the 2,500 fungal strains in the Division of Mycology and Plant Pathology at the Indian Agricultural Research Institute in New Delhi.

Several non fungal collections exist in India as well as mycological herbaria. In addition, the National Bureau of Agriculturally Important Microorganisms will establish a collection in the coming years.

In Indonesia, the most significant collection is the University of Indonesia Culture Collection in the Department of Biology. This is one of the smaller collections listed, holding only 300 cultures.

In the South Pacific, Australia and New Zealand house several important culture collections. The most exotic is the Australian Collection of Antarctic Microorganisms which emphasizes bacteria. Other collections include the Wine Research Institute which houses several hundred wine yeast strains. Given the high quality of wine being exported from Australia, this program is certainly seeing some success. At the other end of the spectrum is the several hundred clinical specimens at the Mycology Culture Collection, Women's and Children's Hospital in Adelaide and the Australian National Reference Laboratory in Medical

Mycology in The Royal North Shore Hospital of Sydney which houses over 1,000 fungi. The Commonwealth Scientific and Industrial Research Organization (CSIRO) Insect Pathogen Culture Collection holds over 1,000 fungi as a small part of the CSIRO mission. The Flinders University Smut collection is another unique but important collection in Australia. Overall there are 37 Australian collections listed with the World DataCentre for Microorganisms.

While New Zealand is a small country, their culture collections system is well supported. They have collections of microorganisms from plants, forest microorganisms and a variety of agriculturally relevant collections. Several operate under the umbrella of the New Zealand Reference Culture Collection which has different sections for different organisms. Many of these collections emphasize bacteria. The New Zealand Forest Service maintains two fungal collections. The first, the Forest Research Culture Collection, is comprised of over 3,000 specimens of pathogenic and saprophytic fungi from native temperate forests, plantations and urban gardens. The second, the New Zealand Fungal Herbarium, houses over 65,000 specimens for taxonomic reference.

5. UNIFIED CULTURE DATABASES

The WFCC-MIRCEN World Data Centre for Microorganisms (WDCM) is the most central and complete database for culture collections in the world. Their coverage, however, depends on the effort of individuals from each individual collection to enter data about holdings and services and addresses. As such, some of the data is out of date and some is overstated. For example, collections of every academic department in Thailand are listed while in the United States there are fewer collections listed. The strain database available at their web-site (http://www.wdcm.nig.ac.jp/) lists whether a particular organism is held by a particular collection, but offers relatively little information about cultures. This information would be of particular use to scientists looking for diverse holdings of a organism.

Several different countries have put together unified culture databases (Table 3) with differing degrees of success. One factor that seems to limit the success of this sort of effort is that it has relatively easy to get money to set up databases, but not to provide the longterm curation they require.

In the United States, the Microbial Germplasm Database was launched in the late 1980s and offers information about materials in large and small collections in the US. It has not been updated in recent years, but is still available online. Also in the US, the Germ Plasm Information Resource (GRIN, http://www.ars-grin.gov/) serves as a portal to the US Department of Agriculture collections.

The UK has seen more success with their United Kingdom National Culture Collection (UKNCC) database. Online at http://www.ukncc.co.uk/, the database lists over 70,000 stocks. While these are not limited to fungi, the database is easy to use and offers a convenient way to look for materials using a simple interface. This includes the IMI/CABI database among 10 different collections.

Also in the UK is the Microbial Strain Data Network (MSDN) which provides access to collections from around the world. While the goals are broad, the databases have not been recently available online. The MSDN is an initiative of the United Nations Environment Program and was integrated with CABRI (Common Access to Biological Resources and Information). The data management portion of this project has largely been assumed by the individual collections (http://www.cabri.org/collections.html).

In Canada, the Canadian Microbial Genetic Resources Information System provides access to a number of collections, including the Canadian Collection of Fungal Cultures. This database, however, has not been updated since 1996.

Table 3. Online Databases

Name	Location	Focus
World Data Centre for Microorganisms (WDCM)	http://www.wdcm.nig.ac.jp/	A comprehensive directory of culture collections and databases
Microbial Germplasm Database (MGD)	http://mgd.nacse.org/cgi-bin/mgd	Broad, includes small collections
The United Kingdom National Culture Collection (UKNCC)	http://www.ukncc.co.uk/	Lists several UK collections including databases of strains
USDA GRIN	http://www.ars-grin.gov/)	Links to USDA sites
Canadian Collection of Fungal Cultures	http://sis.agr.gc.ca/brd/ccc/	Listings of Canadian resources
Collnet	http://www.collnet.cnrb.it/	Provides search of Italian biological resource centers
All-russian Collection of Microorganisms - VKM	http://www.vkm.ru/	Consolidated Catalogue of Microbial Cultures Held in Russian Non-medical Collections

The All-Russian Collection of Microorganisms offers catalogs of holdings of a number of collections throughout Russia. This was supported by a biodiversity grant from the Russian government and its databases are more up to date, some having been updated as recently as 2002. In addition to providing catalogs of the holdings of individual collections, the All-Russian Collection of Microorganisms provides a web-interface for many of Russia's collections.

Other smaller regional databases exist, such as the Microbial Information Network of China, but the most global in scale is the WFCC-MIRCEN World Data Centre for Microorganisms (WDCM) and new collections that are seeking to develop their databases would be best served by trying to assure compatibility with the standards in place at WCDM.

6. NATIONAL, REGIONAL AND GLOBAL CULTURE COLLECTION ORGANIZATIONS

The World Federation for Culture Collections (WFCC, http://www.wfcc.info/) operates as an umbrella organization to promote the interests and services of culture collections. The WFCC is a branch of the International Union of Biological Sciences. The WFCC has been instrumental in the development of the WDCM and provides a forum for global discussion of issues relevant to the operation of collections. The WFCC has been proactive in addressing questions of security and works to assure that collections will be able to distribute cultures to scientists who need them.

The US Federation for Culture Collections (USFCC, http://www.usfcc.us/) exists in the United States to promote the interests of culture collections. It also publishes a newsletter in print and electronic formats and sponsors workshops and courses to promote the development and maintenance of culture collections.

7. CONCLUSIONS

While the number of culture collections specializing in fungi is impressive, many collections are small and local in emphasis. While some of these have been overlooked in the present treatment, the internet has allowed many of these smaller local collections to make their holdings readily available. The number of collections with online databases is growing. This also allows global access to previously local culture collections. There is also an effort to develop global databases which will list resources from a variety of collections.

The current era marks a paradigm shift in fungal genetics and mycology where the access to information about fungi becomes as important as access to cultures themselves. The proliferation of fungal genomes that are available over the internet has empowered researchers in every branch of mycology. In turn, the availability of genome data has made the cultures of increasing value. It is essential, however, that a reliance on genome data does not allow cultures collections to be marginalized. The ability to integrate genome data with observation of live biological materials demands that the strains whose genomes were sequenced are available. This is the sort of function that culture collections are able to provide. Beyond genome information, the tremendous value of biological materials offers resources including documented mutations, including many that would not be found in systematic gene-knockout efforts, as well as diversity of global scale. The future of culture collections likely is that the biggest ones will become Biological Resource Centers. This does not threaten the future of smaller collections, but rather should allow them to share their resources more widely by having both a center to look to for help in challenges such as publishing their catalogs online and following shipping regulations and more.

Acknowledgement: The author wishes to cite the NSF support of the FGSC by their grant #9726962.

REFERENCES

Chang, ST (1993). Mushrooms and Mushroom biology. Pp. 1-13. IN: Genetics and Breeding of Edible Mushrooms. Eds S.T. Chang, J. A. Buswell, and P.G. Miles. Gordon & Breach, Philadelphia

Hay RJ (1996) Human cells and cell cultures: availability, authentication and future prospects. Hum Cell 1996 Sep;9(3):143-52

Ito Y, Peterson SW, and Goto (1998). Properties of *Aspergillus tamarii, A. caelatus* and related species from acidic tea field soils in Japan.Mycopathologia 1998-99;144(3):169-75

Kurtzman CP. (2000) Systematics and taxonomy of yeasts. Contrib Microbiol. 2000;5:1-14.

Logrieco A, Peterson SW, and Bottalico A (1995) Phylogenetic relationship within *Fusarium sambucinum* Fuckel sensu lato, determined from ribosomal RNA sequences.Mycopathologia 1995;129(3):153-8

Ma J, Li Y, Ye Q, Li J, Hua Y, Ju D, Zhang D, Cooper R, and Chang M (2000) Constituents of red yeast rice, a traditional Chinese food and medicine. J Agric Food Chem 2000 Nov;48(11):5220-5

McCluskey, K (2000). Long term viability of *Neurospora crassa* at the FGSC. Fungal Genetics Newsletter 47:110

McCluskey, K (2000). A relational database for the FGSC. Fungal Genetics Newsletter 47:74-78

Molina FI, Geletka LM, Jong SC, and Zhang Y (1994) Use of a nested primer pair as control for PCR amplification
of ribosomal DNA internal transcribed spacers in fungi. Biotechniques 1994 Jun;16(6):998-1000

O'Donnell K, and Peterson SW (1992). Isolation, preservation, and taxonomy. Biotechnology 1992;21:7-39

ORGANISATION FOR ECONOMIC CO-OPERATION AND DEVELOPMENT (2001) Biological Resource Centres: Underpinning the Future of Life Sciences and Biotechnology. ISBN 9264186905

Peintner, U, Poder, R and Pümpel T (1998). The Iceman's fungi. Mycological Research 102: 1153-1162

Rollo, F Sassaroli,S and Ubaldi,M (1995). Molecular phylogeny of the fungi of the Iceman's grass clothing. Cur Genet, 28(3): p. 289-97.

Culture collections to improve the quality of life (eds. R.A. Samson et al.). Proceedings of the eighth International
Congress for Culture Collections. 520 pp., 1966.

Wuyts J, De Rijk P, Van de Peer Y, Winkelmans T, and De Wachter R (2001). The European Large Subunit Ribosomal RNA database. Nucleic Acids Res. 29(1):175-177

Wuyts J, Van de Peer Y, Winkelmans T, and De Wachter R (2002). The European database on small subunit ribosomal RNA.Nucleic Acids Res. 30, 183-185.

Keyword Index

www.ingramcontent.com/pod-product-compliance
Ingram Content Group UK Ltd.
Pitfield, Milton Keynes, MK11 3LW, UK
UKHW020238290926
15470UKWH00002B/610